New Directions For Sma

New Directions For Smallholder Agriculture

Edited by

Peter Hazell and Atiqur Rahman

Great Clarendon Street, Oxford, OX2 6DP,
United Kingdom

Oxford University Press is a department of the University of Oxford.
It furthers the University's objective of excellence in research, scholarship,
and education by publishing worldwide. Oxford is a registered trade mark of
Oxford University Press in the UK and in certain other countries

First Edition published in 2014

Impression: 1

Published in the United States of America by Oxford University Press
198 Madison Avenue, New York, NY 10016, United States of America

British Library Cataloguing in Publication Data

Data available

Library of Congress Control Number: 2013956958

ISBN 978–0–19–968934–7 (Hbk.)
ISBN 978–0–19–968935–4 (Pbk.)

Printed in Great Britain by
Clays Ltd, St Ives plc

FOREWORD

The environment in which smallholder agriculture operates across the developing world has seen significant changes in recent years. Food markets across the globe have become more complex and better integrated, driven by growing demand and by technological and institutional change. Environmental pressures and climatic shocks have created new challenges for food production systems in many regions. Both globally and at the country level, a number of factors have combined to produce higher and more volatile food prices. All these developments have focused global and national political debates on how to ensure food security. An important issue in these debates is how to raise agricultural productivity and improve the efficiency of global, regional, and national food markets. In this context, the role of smallholder agriculture in achieving national and global food security has moved towards the centre of policy discourse—including, most recently, in debates on a post-2015 global development agenda.

Meanwhile, increasing demand for food and rising food prices have drawn new domestic and international private investors to agriculture. As the policies in many developing countries have become more welcoming to corporate investors, we are witnessing the growth of large-scale investments in agriculture and the expansion of market and trading opportunities.

At the local level, non-farm activities have become an increasingly important part of rural economies. They provide services to the farm economy, an important complement to farm-based livelihoods, as well as new opportunities—particularly for rural youth and rural women. While large numbers of rural people remain underserved by formal finance institutions, in many places these are spreading into rural areas, offering products and services targeted to the investment needs of small rural entrepreneurs, whether in agriculture or other sectors.

These changes present both opportunities and challenges to smallholder agriculture, which today continues to support some 2 billion people globally. Whether or not smallholders can meet the challenges and seize the opportunities of a rapidly changing environment depends on whether appropriate measures and policies are in place to enable them to overcome their constraints. Today, the people who work the world's estimated 450 million small farms lack access to high-quality technology, secure access and control over productive land, access to water, quality education, markets, and financial resources. Often they live in marginal areas, and many of them endure absolute poverty and food insecurity. The policy and institutional environments at times work against rather than for them. And climate-induced changes are increasing the risks that they face, negatively impacting their investment capacity, their livelihoods, and their food security.

Despite these constraints, given appropriate technical, financial, and institutional support, smallholders—women and men alike—can raise their productivity and output and get their product to market efficiently. Furthermore, IFAD's work, as well as that of others, has demonstrated the contribution smallholder agriculture can make to environmental sustainability, as well as the feasibility of climate change adaptation in this sector.

A dynamic smallholder agricultural sector would ensure food security at the farm level and make a major, and in many contexts the principal, contribution to food security at the national, regional, and global levels. It would also be an important driver of inclusive and sustainable economic growth, and foster job-rich rural development as a complement to thriving urban economies. And as custodians of a large part of the world's natural resources, smallholders can and need to play a major role in the environmental sustainability and climate mitigation agendas that are now of increasing global concern. These perspectives must be part of the policy debate on food security, both globally and at country level, both today and looking ahead beyond 2015.

IFAD organized the 2011 Conference on 'New Directions for Smallholder Agriculture' to provide a forum for renowned specialists and IFAD's own experts to discuss the many challenges and opportunities that smallholder agriculture faces in a rapidly changing world. I am delighted that the conference was able to debate many critical issues, both old and new. These included topics such as: how efficient smallholders are vis-à-vis their larger counterparts; how to factor sustainability into smallholder development; how smallholders can thrive in competition with capital-intensive larger farmers; and what support smallholders need from development agencies and governments to embark on a path of sustainable development. In short, the conference addressed the question: what is the future of smallholders in agriculture?

While the conference could not be expected to provide ready-made answers to all these questions, it was nonetheless successful in expounding key findings and insights that can guide policymaking today and in the future. The conference noted the great diversity in the policy, institutional, economic, and social environments of smallholders across Africa, Asia, and Latin America. Yet, it also stressed commonalities in the opportunities and challenges that all smallholders face. I would like to highlight a few of these.

First, it is evident that the social profile of smallholder households is changing. Increasingly, around the globe, and particularly in Africa, smallholder farmers are women, who take on new responsibilities in farming as a consequence of male migration—although in many cases without having gender-equal access to land, inputs, finance, and other resources. In addition, high unemployment and population growth rates in rural areas have left large numbers of rural youths unemployed, many of whom cannot engage in farming on increasingly fragmented family plots. These youth represent a resource of immense value for a modernized, more climate-adapted, sustainable agriculture, which is also well connected to a thriving non-farm rural sector.

Second, smallholders continue to have difficulties securing access to essential resources such as land and water, due to pressures from increasing population, competition with larger investors, and environmental degradation. For smallholder farming to be sustainable and to drive inclusive economic growth in the coming years, institutions will need to address these problems—focusing on easier but also more secure access to land and on respect and recognition of rights and entitlements. It is also very important to improve access to financial services, technology, and other inputs. This agenda requires the collaboration of governments, development agencies, and the private sector to develop and scale-up products and processes that are well suited to the needs of smallholders—women and men alike—with particular attention to supporting smallholders in managing the risks they face.

Third, it is essential that smallholder farmers be seen for what they are—operators of small businesses, and thus a core part of the rural private sector. Other private-sector actors should be seen as potential partners and enablers of smallholder farmers, as they can facilitate access to inputs, services, capacity development opportunities, and markets. Experiences around the world, including IFAD's own experience, demonstrate that such private–private partnerships are possible, although the transaction costs may require third-party support and facilitation, especially to ensure the inclusion of poor farmers.

Fourth, it is quite evident today that many smallholders are not engaged solely in agricultural activities. An increasing number are diversifying into non-farm activities to supplement their farm incomes. Sustainable development of smallholders in agriculture has to take this into account, and promote the policy, financial, and infrastructural support needed for the growth of a thriving non-farm rural economy as a necessary complement of smallholder agriculture, and as a critical link between inclusive rural and urban economies.

Fifth, smallholders often lack strong institutional and social capital to influence policy- and decision-making processes or to negotiate favourable terms of market participation or access to key inputs and services. Development agencies, supported by an enabling policy framework, can help address this challenge, along with partners within civil society, starting from farmers' organizations themselves. IFAD's experience amply shows the transformative potential of such partnerships and the centrality of farmers' empowerment in smallholder development.

The authors who have contributed to this book address these and other issues in depth, drawing on the papers they presented at the 2011 IFAD Conference and on additional research. Our aim in publishing this book is to promote further discussion on the future of smallholder agriculture in a rapidly changing world. I am confident that this volume will greatly enrich the ongoing policy debate and dialogue, helping us to chart a path that will lead to the full integration of smallholder farmers in a global agenda for sustainable and inclusive development.

Kanayo F. Nwanze
President, International Fund for Agricultural Development (IFAD)

Rome, Italy
December 2013

ACKNOWLEDGEMENTS

This volume owes a lot to the vision of the President of IFAD, Kanayo Felix Nwanze, who believes that smallholder agriculture of developing countries can not only be self-sustaining, but can also contribute towards food self-sufficiency of poor countries and be the driving force for economic growth and overall development if market and entrepreneurial skills of smallholder farmers can be harnessed. With appropriate support, smallholdings can be developed as farm business enterprises.

He encouraged the Strategy and Knowledge Management Department of IFAD, under the leadership of Henock Kifle, to organize a conference on the future of smallholder agriculture in 2011. The conference, conceptualized and guided by Henock Kifle, and enthusiastically supported and contributed to by the members of the Department, attracted considerable attention among development experts, academicians, civil society representatives, and the private sector. The conference participants and the chairpersons/facilitators contributed greatly towards understanding various issues relating to smallholder development. The four young rapporteurs to the conference (Arindam Banerjee, Antonio Ferreira, Mateo Mier, and Jennifer Smolak) ably captured and shared with all the participants the essence of the two-day conference. All these helped to shape the focus of this edited volume. We would like to thank them all for their contributions.

The papers presented at the conference provided much of the material for the book; however, some key gaps in analysis emerging from the conference were filled up by a number of post-conference papers from experts. We thank all the authors for their contributions.

Carlos Sere who took charge of the Strategy and Knowledge Management Department from Henock Kifle kept up the momentum for the book, leading, guiding, and providing the right support as needed. He went through some of the chapters and provided valuable comments. Without his enthusiastic support, this volume would not have materialized.

The chapters were reviewed by a number of referees, whose comments and suggestions were very helpful in improving the quality of the chapters. We note the contributions made in this respect by Derek Byerlee, Alain de Janvry, John Farrington, Kjell Havenvik, Saleemul Huq, Nurul Islam, Stephen Klerkx, Jonathan Mitchell, S. R. Osmani, Frank Place, Collin Poulton, Agnes Quisumbing, Caludia Ringler, Ashwani Saith, Geoff Tyler, and Sajjad Zohir. We also thank the four anonymous referees of the Oxford University Press whose comments were very helpful in strengthening the focus of the book and tightening its structure.

Pierre-Justine Kouka, and earlier Thomas Elhaut, of SKM provided the much-needed managerial support towards the final, most critical part of the book, pushing it forward over some administrative hurdles. We acknowledge their contributions in this regard.

Alfredo Baldoni provided the administrative support, kept track of expenses, and made sure that the support services were in place. Birgit Plockinger of the Communication Division led the design team in preparing all the graphics in high resolution with her usual quiet efficiency. Anthony Lambert carefully went through all the chapters, editing them and ensuring that they meet the guidelines of the Oxford University Press. The legal matters in publishing the book and other publication-related issues were ably handled and supported by Daniela Ronchetti and Bruce Murphy respectively. We thank them all.

Beyond these named persons, many more provided support in the book's preparation. We express our gratitude to them all for their contributions at different stages.

The findings, interpretations, and conclusions expressed here are those of the authors and do not necessarily reflect the views of IFAD, as an institution, or its Governing Bodies.

Peter Hazell
Atiqur Rahman

TABLE OF CONTENTS

LIST OF FIGURES

LIST OF PLATES

LIST OF TABLES

LIST OF BOXES

LIST OF ABBREVIATIONS/ACRONYMS

ADB	Asian Development Bank
AFSI	L'Aquila Food Security Initiative
AGRA	Alliance for a Green Revolution for Africa
AICD	The Africa Infrastructure Country Diagnostic
AIS	Agricultural Innovation Systems
APEDA	Agricultural and Processed Food Products Export Development Authority
APR	Annual Performance Review
APSIM	Agricultural Production Systems Simulator Model
ARU	African Rural University
ASARECA	Association for Strengthening Agricultural Research in Eastern and Central Africa
ASIF	African Seed Investment Fund
ATMA	Agricultural Technology Management Agency
AZMJ	A microfinance institution based in the USA
AusAid	Australian Government Overseas Aid Program
BAAC	Bank of Agriculture and Agricultural Co-operatives
BCs	Bank Correspondents
BecA	Biosciences eastern and central Africa (BecA) Hub
BIH	Basic Irrigated Hectares
BMGF	Bill and Melinda Gates Foundation
BRAC	Bangladesh Rural Advancement Committee
BRC	British Retail Consortium
CAADP	The Comprehensive Africa Agricultural Development Programme
CBA	Community-Based Adaptation
CCs	Collection Centres
CDC	Commonwealth Development Corporation
CDF	Central Distribution Facility
CEPAL	Comisión Económica para América Latina y el Caribbean
CGIAR	Consultative Group on International Agricultural Research
CIAT	the International Centre for Tropical Agriculture
CIMMYT	International Maize and Wheat Improvement Center
CNFA	Citizens' Network for Foreign Affairs

COMESA	Common Market for Eastern and Southern Africa
CRO	Certificate of Right of Occupancy
CSPR	Centrale de Securitisation des Paiements et du Recouvrement
CSR	corporate social responsibility
CSV	Created Shared Value
DBI	Doing Business Index
DFBA	Dairy Farmers Business Associations
DLB	District Land Bureaux
DOAE	Department of Agricultural Extension
DSS	Decision-support systems
EAs	Environmental Assessments
EAC	East African Community
EADD	East Africa Dairy Development
ECLAC	Economic Commission for Latin America and the Caribbean
ECOWAS	Economic Community of West African States
ESC	Electronic Silo Certificates system
ECX	Ethiopian Commodity Exchange
F&V	Fruits and vegetables
FAO	the Food and Agricultural Organisation of the United Nations
FAPRO	Farmers' Produce Promotion Society
FARA	the Forum for Agricultural Research in Africa
FDI	Foreign Direct Investment
FIPS	Farm Inputs Promotion Service
FLO	Fairtrade Labelling Organisation
FOCAC	China–Africa Cooperation
FoSHoL	the Food Security at Household Level
FTA-K	Fair Trade Alliance-Kerala
FWWB	Friends of Women's World Banking
GASFP	Global Agriculture and Food Security Program
GCI	Global Competitiveness Index
GDI	Gender-related Development Index
GDP	Gross Domestic Product
GEM	Gender Empowerment Measure
GFTA	Grand Free Trade Area
GHG	Greenhouse gas
GIS	Geographic Information Systems
GLASOD	Global Assessment of Soil Degradation

GlobalGAP	Global Good Agricultural Practices
GM	Genetically modified
GOB	Government of Bangladesh
GPS	Global positioning system
GTZ	Deutsche Gesellschaft für Internationale Zusammenarbeit (German Agency for International Cooperation)
HACCP	Critical Control Practices
HAS	Hindu Succession Act
HVCs	High-Value Crops
HOPSCOM	Co-operative Marketing and Processing Society Limited
IAASTD	International Assessment of Agricultural Knowledge, Science, and Technology for Development
IAR4D	integrated agricultural research for development
IBGE	Brazilian Institute of Geography and Statistics
ICARDA	International Center for Agricultural Research in the Dry Areas
ICIPE	International Centre for Insect Physiology and Ecology
ICRISAT	International Crops Research Institute for the Semi-Arid Tropics
ICT	Information and Communication Technology
IDE	International Development Enterprises
IFAD	International Fund for Agricultural Development
IFAD-UPU	IFAD-Universal Postal Union
IFC	International Finance Corporation
IFPRI	International Food Policy Research Institute
IIASA	International Institute of Applied Systems Analysis
IIED	International Institute of Environment and Development
IIMA	Indian Institute of Management, Ahmedabad
ILRI	International Livestock Research Institute
ILTAB	The International Laboratory for Tropical Agricultural Biotechnology
IMF	International Monetary Fund
IPCC	The Intergovernmental Panel on Climate Change
IRRI	International Rice Research Institute
ISIC	International Standard Industrial Classification
ITC	Indian Tobacco Company
IWMI	International Water Management Institute
KACE	Kenya Agricultural Commodity Exchange
KARI	Kenya Agricultural Research Institute

KGPL	KNIDS Green Private Ltd
KTDA	Kenya Tea Development Agency (formerly Authority)
LAC	Latin America and the Caribbean region
LDCs	Least Developed Countries
LSC	Land Settlement Co-operative
LSMS	Living Standard Measurement Survey
M&E	Monitoring and Evaluation
MCA	Millennium Challenge Account
MCP	Millennium Challenge Corporation
MDFVL	Mother Dairy Fruit and Vegetables Limited
MENA	Middle East and North Africa region
MFIs	Microfinance Institutions
MIDA	Millennium Development Authority, Ghana
MNO	Multinational Organizations
MOCO	Mumias Outgrower Company
M-PESA	Mobile money transfer system
MSAMB	Maharashtra State Agricultural Marketing Board
NAADS	National Agricultural Advisory Development Service
NARI	National Agricultural Research Institutes
NARS	National Agricultural Research System
NBARD	National Bank for Agriculture and Rural Development
NBHC	National Bulk Handling Corporation
NCDC	National Co-operative Development Corporation
NCDP	The Northwest Crop Diversification Project
NCEUS	National Commission for Enterprises in the Unorganized Sector
NDC	National Development Council
NDDB	National Dairy Development Board, in north India.
NEPAD	New Partnership for Africa's Development
NF	Namdhari Fresh
NGOs	Non-Government Organisations
NHB	National Horticulture Board
NIB	Nib International Bank SC
NLC	National Land Centre
NOC	No-Objection Certificate
NRM	Natural Resources Management
NSS	National Sample Survey
NTEX	Non-Traditional Exports

OBCs	Other Backward Castes
ODA	Official Development Assistance
ODEPA	La Oficina de Estudios y Políticas Agrarias
OECD	Organisation for Economic Co-operation and Development
OLL	Organic Land Law
OSS	Office of Strategic Services
PA	Procuraduria Agraria
PABRA	Pan African Bean Research Alliance
PASS	Programme for Africa's Seed Systems
PGC	Potato Growers' Cooperative
PHL	Post Harvest Loss
PI	Participation Index
PIDA	Programme for Infrastructure Development in Africa
PKSF	Palli Karma Sahayak Foundation (Rural Employment Support Foundation)
PLA	Participatory Learning and Action
PMO	Primary Marketing Organisation
PO	Producer Organisations
PPP	Public–Private Partnerships
PRA	Participatory Rural Appraisal
R&D	Research and Development
RAN	Registro Agrario Nacional
REDD	Reducing Emissions from Deforestation and Forest Degradation
REDS	Rural Economic and Demographic Survey
RNFE	Rural Non-farm Economy
ROPPA	Network of Farmers' and Agricultural Producers' Organisations of West Africa
ROSCAs	Rotating Savings and Credit Associations
SACCO	Savings and Credit Co-operatives
SACRED	Sustainable Agricultural Centre for Research Extension and Development in Africa
SADC	Southern African Development Community
SAKSS	Strategic Analysis and Knowledge Support Systems
SBI	The State Bank of India
SCODP	Sustainable community-oriented Development Program
SCs	Scheduled Castes
SEA	South-East Asia

SEDC	Sarawak Economic Development Corporation
SHG	Self-Help Group
SME	Small and Medium Enterprise
SMS	Short Message Service
SOFESCA	Soil Fertility Consortium for Southern Africa
SSA	Sub-Saharan Africa
T&V	Training and Visit
TSBF	Tropical Soil Biology and Fertility Programmes
UNCTAD	United Nations Conference on Trade and Development
UNDP	United Nations Development Programme
UNECA	United Nations—Economic Commission for Africa
UNFCCC	United Nations Framework Convention on Climate Change
URDT	Uganda Rural Development and Training Programme
USAID	US Agency for International Development
VCF	Value-chain finance
VCR	value to cost ratio
WAAIF	West Africa Agricultural Investment Fund
WARDA	West Africa Rice Development Association
WEF	World Economic Forum
WEI	A Women's Empowerment Index
WDI	World Development Indicators
WDR	World Development Report
WII	Weather Indexed Insurance
WRMS	Weather Risk Management Services
WTO	World Trade Organization
XAAH	The Agricultural Bank of Mongolia

LIST OF CONTRIBUTORS BY CHAPTER

Chapter 1

Peter Hazell *(PhD, Cornell University)* has held various research positions at the World Bank and the International Food Policy Research Institute (IFPRI) in Washington DC. After retiring from IFPRI he became a Visiting Professor at Imperial College London and a Professorial Research Associate at the School of Oriental and African Studies (SOAS). Peter's extensive and widely cited publications include works on mathematical programming, risk management; insurance; the impact of technological change on growth and poverty; the rural non-farm economy; sustainable development strategies for marginal lands; the role of agriculture in economic development; and the future of small farms.

E-mail: p.hazell@cgiar.org

Atiqur Rahman *(PhD, Cantab)* is an ex-staff member of IFAD. He held various managerial, strategy and policy formulation, research, and teaching positions with IFAD and earlier with Bangladesh Institute of Development Studies. He was a visiting Professor at the University of California, Riverside and recently an Adjunct Professor at John Cabot University, Rome. He consulted for the Asian Development Bank, International Labour Organization, United Nations Secretariat, World Bank, FAO, IFAD, and others. He was the lead co-coordinator and researcher of the IFAD 2001 Rural Poverty Report. He has many publications to his credit, including an edited volume with Oxford University Press.

E-mail: atiqur1150@gmail.com

Chapter 2

Gordon Conway (*PhD, University of California, Davis*) is a Professor of International Development at Imperial College, London and Director of Agriculture for Impact. Earlier he was Chief Scientific Adviser to the Department for International Development. Previously he was President of The Rockefeller Foundation and Vice-Chancellor of the University of Sussex.

E-mail: g.conway@imperial.ac.uk

Chapter 3

Geoffrey Livingston is currently working as IFAD's Regional Economist for East and Southern Africa at IFAD. He earlier worked for Chemonics International, a Washington-based development consulting firm where he served as a long-term technical advisor and project director on agribusiness development projects in Rwanda, Mali, and Senegal. Mr Livingston's areas of research interest include competitiveness, value chain development, export marketing, transport policy, and project management.

E-mail: g.livingston@ifad.org

Steven Schonberger is currently Lead Operations Officer in the Africa Region of the World Bank. He was a core author of the Comprehensive Framework for Action to guide the response of the UN-Bretton Woods High Level Task Force in Response to the Food Price Crisis. Mr Schonberger has contributed to several books, and magazine and journal articles focused on issues of agriculture and rural development.

E-mail: sschonberger@worldbank.org

Sara Delaney is a Programme Officer and works on Agriculture and Food Security for Episcopal Relief & Development in New York. She previously worked with IFAD as a Knowledge Management consultant, and in London on the Gates Foundation and DFID-funded research and advocacy projects. Sara served as a Peace Corps volunteer in Mali.

E-mail: sara.delaney@gmail.com

Chapter 4

Ganesh Thapa *(PhD, Cornell University)* is the Regional Economist for Asia and the Pacific Division of the International Fund for Agricultural Development where his work focuses on policy analysis and strategy formulation for agricultural development and rural poverty reduction in Asia and the Pacific region. Earlier he worked as Country Director for Winrock International in Nepal, and led a project on policy analysis in agriculture and natural resource management.

E-mail: g.thapa@ifad.org

Raghav Gaiha is an ex-Professor of Public Policy at Faculty of Management Studies, University of Delhi. He was Visiting Fellow at MIT, Harvard and Stanford Universities. He has served as a consultant with the World Bank, ILO, FAO, IFAD, WIDER, and ADB. His research interests are in poverty, inequality, nutrition, rural institutions, agricultural research, and emerging Asian economies.

E-mail: raghavdasgaiha@gmail.com

Chapter 5

Julio A. Berdegué *(PhD, Wageningen University)* is the Principal Researcher at Rimisp-Latin American Center for Rural Development, Santiago, Chile. He is Chairman of the Board of Trustees of CIMMYT, the International Maize and Wheat Improvement Center, serves on the Board of Trustees of the International Institute for Environment and Development (IIED, London), and is a member of ICCO's International Advisory Council.

E-mail: jberdegue@rimisp.org

Ricardo Fuentealba is a sociologist, mastering in human geography at University of Bristol, UK. He has research experience in the development of local governments in Chile, urban–rural interactions, and Latin American development issues in general.

(Contact the lead author, Julio A. Berdegué)

Chapter 6

Maximo Torero is the Division Director of the Markets, Trade, and Institutions Division at the International Food Policy Research Institute, and leader of the theme on Linking Small Producers to Markets in the CGIAR research programme on Policies, Institutions, and Markets.

E-mail: m.torero@cgiar.org

Chapter 7

Sukhpal Singh is Associate Professor and Chairperson, Centre for Management in Agriculture (CMA), Indian Institute of Management (IIM), Ahmedabad. He was a visiting fellow at IDS, Sussex (UK), Chulalongkorn University, Bangkok, and University of Manchester; has been a member of various committees/working groups of the Planning Commission of India and a member of a globally networked research project on global production and trade networks called 'Capturing the Gains'; and is the founding editor of Millennial Asia—an international journal of Asian studies.

E-mail: sukhpal@iimahd.ernet.in

Chapter 8

Atiqur Rahman *(PhD, Cantab)* is an ex-staff member of IFAD. He held various managerial, strategy and policy formulation, research, and teaching positions with IFAD and earlier with Bangladesh Institute of Development Studies. He was a visiting Professor at the University of California, Riverside and recently an Adjunct Professor at John Cabot University, Rome. He consulted for the Asian Development Bank, International Labour Organization, United Nations Secretariat, World Bank, FAO, IFAD, and others. He was the lead co-coordinator and researcher of the IFAD 2001 Rural Poverty Report. He has many publications to his credit, including an edited volume with Oxford University Press.

E-mail: a.rahman@tin.it; at.rahman@ifad.org

Jennifer Smolak is an independent consultant specializing in international development, food security, and monitoring and evaluation. Jennifer's professional experience includes international consultancy and staff postings for the International Fund for Agricultural Development (IFAD), Oxfam Australia, the Foundation for International Community Assistance (FINCA), and Deloitte.

E-mail: jennifer.smolak@mail.mcgill.ca

Chapter 9

Akinwumi Adesina *(PhD, Purdue University)* is the Minister of Agriculture and Rural Development of Nigeria. He was previously the Vice President of the Alliance for a Green Revolution in Africa and Associate Director at the Rockefeller Foundation. He won the Yara Prize in 2007.

E-mail: adesina1234@gmail.com

Augustine Langyintuo is a Senior Operations Officer - Agribusiness of the International Finance Corporations (IFC) - the World Bank Group. He was previously a Senior Policy Officer with AGRA. He is an Associate Editor of the African Journal of Agricultural and Resource Economics and serves as an Advisory Board Member of the DTMA Project of

CIMMYT, where he previously worked as an Economist. Augustine holds a PhD in Agricultural Economics from Purdue University.

E-mail: Alangyintuo@ifc.org

George Bigirwa is a Senior Programme Officer, AGRA. He is in charge of Seed Production and Dissemination for Eastern and Southern Africa, and is based in Nairobi, Kenya.

(Contact Akinwumi Adesina and Augustine Langyintuo)

Nixon Bugo holds an MBA in Finance and is a Certified Public Accountant.

(Contact Akinwumi Adesina and Augustine Langyintuo)

Kehinde Makinde is a Programme Officer with AGRA based at Accra, Ghana.

(Contact Akinwumi Adesina and Augustine Langyintuo)

John Wakiumu is a Programme Officer whose main role at AGRA is to develop innovative ways of working with financial institutions to leverage and mobilize financing for the development of agricultural value chains.

(Contact Akinwumi Adesina and Augustine Langyintuo)

Chapter 10

Martin Evans *(PhD, Cantab)* studied natural sciences, agriculture, and agricultural economics. He taught at the University of Papua New Guinea before joining the Asian Development Bank in Manila. The greater part of Martin Evans' professional career, however, was spent with Booker Tate Ltd, an international agribusiness management company. Since retiring from Booker Tate as new business director nine years ago, Martin Evans has been working as a private consultant. He is a former Chair of Farm Africa.

E-mail: martin.evans@agriprojects.com

Chapter 11

John K. Lynam *(PhD, Stanford University)* has been an independent consultant since 2007. He has consulted for World Bank, FAO, ILRI, the World Fish Centre, the World Vegetable Centre, and the CGIAR. Mr Lynam currently is Chair of the Board of Trustees of the World Agroforestry Centre, and on advisory panels for the Collaborative Crop Research Program of the McKnight Foundation, and the Monitoring and Evaluation Division of the Alliance for a Green Revolution in Africa.

E-mail: johnklynam@gmail.com

Stephen Twomlow *(PhD, Birkbeck College, University of London)* is a Systems Agronomist with more than twenty-five years of agro-ecosystems research, development, and training for disadvantaged farming communities in the developing world, particularly in sub-Saharan Africa, Mexico, and Asia. He's currently the Regional Climate and Environmental Specialist, providing technical support to IFAD's East and Southern African portfolio.

E-mail: s.twomlow@ifad.org

Chapter 12

Calestous Juma *(DPhil in Science and Technology Studies from the University of Sussex)* is Professor of the Practice of International Development at Harvard Kennedy School and director of the Agricultural Innovation in Africa Project at the Kennedy School's Belfer Centre for Science and International affairs. He is founding executive director of the African Centre for Technology Studies in Nairobi and former executive secretary of the UN Convention on Biological Diversity. He is author of *The New Harvest: Agricultural Innovation in Africa* (Oxford University Press, 2011).

E-mail: calestous_juma@harvard.edu

David J. Spielman *(PhD in Economics, American University)* is currently a Senior Research fellow in IFPRI, Washington DC. His research agenda covers a range of topics, including agricultural science, technology, and innovation policy; seed systems and input markets; and community-driven rural development.

E-mail: d.spielman@cgiar.org

Chapter 13

Edward Heinemann currently works as a Senior Policy Advisor in IFAD's Policy and Technical Advisory Division. He was formerly in IFAD's Strategy and Knowledge Management Department, where he was the lead author of IFAD's 2011 Rural Poverty Report.

E-mail: e.heinemann@ifad.org

Chapter 14

Klaus Deininger is a Lead Economist at the World Bank. His research focuses on income and asset inequality and its relationship to poverty reduction and growth; access to land, land markets, and land reform, and their impact on household welfare and agricultural productivity; and land tenure and its impact on investment, including environmental sustainability for policy analysis and evaluation, mainly in the Africa, Central America, and East Asia regions.

E-mail: kdeininger@worldbank.org

Chapter 15

Mahabub Hossain *(PhD, Cantab)* is currently the Executive Director of BRAC. Before joining BRAC in 2007, he was Head of the Social Sciences Division at the International Rice Research Institute (IRRI), and Director General at the Bangladesh Institute of Development Studies. He has authored and co-authored eleven books, and over one hundred and forty papers in various journals and edited books.

E-mail: mahabub.hossain@brac.net

W. M. H. Jaim *(PhD, University of London)* is a Professor in the Department of Agricultural Economics, Bangladesh Agricultural University, Mymensingh. Earlier, he was the Dean of the Faculty of Agricultural Economics and Rural Sociology of Bangladesh Agricultural University at Mymensingh and the Director of Research and Evaluation Division of BRAC, Bangladesh.

E-mail: wmh-jaim@brac.net

Chapter 16

Camilla Toulmin *(DPhil, Oxon)* is Director of the International Institute for Environment and Development (IIED) based in London. An economist by training, she has worked mainly in Africa on agriculture, land, climate, and livelihoods.

E-mail: camilla.toulmin@iied.org

Chapter 17

Steve Wiggins (*PhD, Reading*) is a Research Fellow at the Overseas Development Institute (ODI) London. As an agricultural economist he has worked on rural livelihoods, poverty, food security, and nutrition in the developing world, above all in Africa and Latin America. He has more than thirty-five years' experience of working in the developing world, as well as teaching and researching agricultural and rural development at the University of Reading and the Overseas Development Institute. He has published on food prices, climate change, and commercialization of smallholders in Africa, among others.

E-mail: steve-wiggins@ntlworld.com; s.wiggins@odi.org.uk

Chapter 18

(See Chapter 1.)

1 Introduction

PETER HAZELL AND ATIQUR RAHMAN

Small family farms abound around the developing world and have done so since the beginning of settled agriculture. Yet they have long spawned debate about their efficiency and viability. Karl Marx viewed small farms as self-exploiting, and a naive belief that large-scale mechanized farming necessarily means greater efficiency and productivity led some policy-makers to seek to consolidate holdings, often through compulsory means or land seizures. These have ranged from large state farms in some post-independence African countries, large settler farms in colonies or new territories, cooperatives and state collectives in communist regimes, to contemporary land grabs by corporate interests and sovereign wealth funds.

Although few of these entities survive, the view that small farms are inefficient and an impediment to development continues to be widely held, and development experts like Paul Collier (2009) argue it is still wiser to promote large-scale operations in sub-Saharan Africa. This position ignores the enormous diversity of small-farm situations around the world today, and hence fails to take into account the substantial prospects for small-farm development in many developing countries. Nor does it suggest how a rapid exit of small farms could be managed without leading to a much larger number of people becoming trapped in rural poverty and urban ghettos.

1. What role for small farms?

Small farms matter because they exist in huge numbers, and exercise a strong influence over a whole range of development issues. There are nearly 450 million farmers today who farm less than 2 hectares of land, and many more family farms larger than 2 ha which struggle to make an adequate living from farming. Small farms are predominantly concentrated in Asia and Africa, and are home to some 2 billion people, including half the world's undernourished people and the majority of people living in absolute poverty (IFPRI, 2005). Despite facing difficult challenges, small farms are proving remarkably resilient and in many countries are becoming more numerous and smaller than ever (see Eastwood et al., 2010; and Chapters 3–5 in this book).

Clearly, policies and investments designed to promote economic growth, greater food security, the achievement of the Millennium Development Goals (MDGs), or to curtail global warming must all be cognizant of this vast army of small farmers. But there is continuing debate about the potential role that small farms can play in the development process. Advocates argue that small farms offer several critical advantages to poorer countries. They are more efficient than large farms, create large amounts of productive employment, reduce rural poverty and food insecurity, support a more vibrant rural non-farm economy (including rural towns), and help to contain rural–urban migration.

The efficiency of smaller farms has been demonstrated by an impressive body of empirical studies showing an inverse relationship between farm size and land productivity across Asia and Africa (Eastwood et al., 2010; Binswanger-Mkhize and McCalla, 2010; and Chapter 4 in this book). Moreover, small farms typically achieve their higher land productivity with lower capital intensities than large farms. They exploit labour using technologies that increase yields (hence land productivity) and they use labour-intensive methods rather than capital-intensive machines. These are important efficiency advantages in poor countries where land and capital are scarce relative to labour. The greater efficiency of small farms stems from the absence of economies of scale in most types of farming,[1] and their greater abundance of family labour per hectare farmed. Family workers are typically more motivated than hired workers, and provide higher-quality and self-supervising labour (Tomich et al., 1995).

In poor, labour-abundant economies, not only are small farms more efficient, but because they also account for large shares of the rural poor, small farm development can be a 'win-win' proposition for growth and poverty reduction. Asia's Green Revolution demonstrated how agricultural growth that reaches large numbers of small farms can transform rural economies and raise enormous numbers of people out of poverty (Rosegrant and Hazell, 2000). Recent studies also show that a more egalitarian distribution of land not only leads to higher economic growth, but also helps ensure that the growth that is achieved is more beneficial to the poor (World Bank, 2007).

Small farms also contribute to greater food security, particularly in areas with poor infrastructure where high transport costs make locally produced foods less costly and less risky than many purchased foods. Because they

[1] Some plantation crops are an exception (see Hayami, 2010).

produce more output per hectare than large farms, they also contribute to greater national food self-sufficiency in land-scarce countries.

Small farm households with cash incomes also have more favourable expenditure patterns than large farms for promoting growth of the local non-farm economy, including rural towns. They spend higher shares of their incremental income on locally produced goods and services, many of which are labour intensive (Mellor, 1976; Hazell and Roell, 1983). These demand patterns generate additional income and jobs in the local non-farm economy which can be beneficial to the poor.

Advocates recognize that the efficiency advantages of small farms slowly disappear as countries develop. As per capita income rises, economies diversify and workers leave agriculture, rural wages go up, and capital becomes cheaper relative to land and labour. It then becomes more efficient to have progressively larger farms. Economies of scale in mechanized farming eventually kick in, accelerating this trend. The result is a natural economic transition towards larger farms over the development process, but one that depends critically on the rate of rural–urban migration, and hence on the growth of the non-agricultural sector (Eastwood et al., 2010; Huang, 1973). Historically, this transition to larger farms does not normally begin until countries have grown out of low-income status. A common misdiagnosis about the role of small farms stems from overlooking this broader economic context for determining the economics of farm size.

Small farms survive longer into the transformation process if they can adapt to the changing economic environment. Key adjustments include buying or renting additional land, diversifying into higher-value production activities (e.g. fruits and vegetables, and niche markets such as organics), and expanding into non-farm sources of income or employment. Fortunately, opportunities to diversify into a broader range of farm and non-farm activities also grow as countries become richer. This is because the demand for more diverse and higher-value foods increases with per capita incomes and urbanization, and the non-farm economy grows more quickly than agriculture.

Small-farm sceptics argue that these historical patterns are less relevant in today's globalized economy (e.g. Maxwell et al., 2001; Collier, 2009). Agricultural marketing chains are changing and small farmers are increasingly being asked to compete in markets that are more demanding in terms of quality and food safety, more concentrated and integrated, and much more open to international competition. Supermarkets, for example, are playing an increasingly dominant role in controlling access to urban retail markets (Reardon et al., 2003), and direct links to private exporters are often essential for accessing high-value export markets. Large farms are often the preferred suppliers in these integrated market chains, and many small farms struggle to diversify into higher-value products because they have only small quantities to sell and cannot meet the food quality and safety requirements of such demanding markets, both at home and overseas. These changes offer new opportunities to small farmers who can successfully compete in the transformed markets, but the requirements are a direct threat to the many small farms that cannot.

Sceptics also note that large numbers of small farmers have become too small to make a viable living out of farming and have anyway diversified their livelihoods away from agriculture to the point where farming now accounts for only a small share of their total income. Moreover, research results about the stronger regional growth linkages arising from the consumption patterns of small farms were generated at a time when small farms were about twice as large as they are today, and the results may not hold for many of today's subsistence-oriented small farms. Sceptics also argue that development programmes to assist small farms are too difficult, as evidenced by the many failed attempts of the past (Maxwell et al., 2001; Collier and Dercon, 2009). As such, they argue it is better to encourage private investments in large-scale farm operations and to direct public assistance towards helping small farmers diversify out of agriculture, including helping more workers migrate and settle in urban areas, and providing additional support for safety-net programmes to assist in the transition (Maxwell et al., 2001; Ellis and Harris, 2004; Collier, 2009).

Counter arguments are based on the fact that in most poor countries (e.g. much of Africa), domestic markets lag behind the rest of the world in terms of their integration and spread of supermarkets, and that most small farmers still grow and sell traditional foods in local markets. Moreover, there are many successful examples of organizing small farmers into producer groups that can successfully link to modern high-value markets and input chains (Joshi et al., 2007; Swinnen and Maertens, 2007; Chapter 7 in this book). It is also argued that for many countries, the problem is not that their small farms are inherently unviable in today's marketplace, but that they face an increasingly tilted playing field that, if left unchecked, could lead to their premature demise. A major problem has been that structural adjustment and privatization programmes have left many small farmers without adequate access to key inputs and services, including farm credit. The removal of state agencies that provided many of these marketing and service functions to small farms has left a vacuum that the private sector has yet to fill adequately in many countries (Kherallah et al., 2002). The removal of subsidies has also made some key inputs, such as fertilizer, prohibitively expensive for many small farmers, and the removal of price stabilization programmes has exposed many farmers to greater downside price risks. These problems are especially difficult for small farmers living in more remote regions with poor infrastructure and market access, and for women farmers. If greater action were taken to restore many of these key functions, say through innovative public–private partnerships, then many small farms would again become more competitive.

As for rural income diversification into non-farm activities, this is not entirely a new phenomenon,[2] nor is it an unequivocally positive phenomenon. On the one hand, diversification may reflect a successful structural transformation in which rural workers are gradually absorbed into more lucrative non-farm jobs, such as teaching, milling, or welding. Entry into these formal jobs often requires some capital, qualifications, and/or possibly social contacts (Start, 2001). On the other hand, in Africa, diversification into the non-farm economy is often driven by growing land scarcity, declining wages, and poor agricultural growth (Bryceson and Jamal, 1997; Start, 2001). Migration driven by a stagnant agricultural and rural environment or due to growth in low-productivity, urban-sector activity, particularly informal services, is often a dead end, which Lipton characterizes as 'the migration of despair'. In this case, migration 'depresses wage rates, denudes rural areas of innovators, and hence, while it may briefly relieve extreme need, seldom cuts chronic poverty' (Lipton, 2004, p.7).

Finally, while the experience with small farm development programmes has been mixed, there have been sufficient successes and at scale to show what is possible and how, and at costs that can be justified by the benefits (e.g. Haggblade and Hazell, 2010; Spielman and Pandya-Lorch, 2010; Fan, 2008; and many chapters in this book).

Resolution of the small farm debate depends in part on recognizing that the role that small farms might play varies by country context. Two key roles can be identified (Hazell et al., 2007). One is a growth, or development role. This role arises when commercially oriented small farms are efficient and can compete in the market. This is most likely in countries where agriculture is the lead growth sector, and small-farm growth opportunities will arise primarily in the domestic market for food staples and in high-value export markets, at least during the early stages of development when the domestic market for high-value products is still nascent. Commercial opportunities for small farms in high-value markets arise in a broad range of successfully growing countries that are rapidly diversifying their diets.

A second role for small farms arises from their potential social contributions. Small farms can provide a way for governments to spread the benefits from a large mineral or urban-based manufacturing sector during the early stages of development when most people are still engaged in agriculture. As

[2] The first large-scale rural household survey in Africa conducted in 1974–5 in Kenya found that smallholders derived at least half of their incomes from sources other than from the farming of their own lands (Kenya Central Bureau of Statistics, 1977). A similar situation is also reported by Reardon et al. (1994) from a series of studies in eight West African countries. Even in many Asian countries, farmers were highly diversified before the Green Revolution (e.g. see evidence from India in Ravallion and Datt, 1996).

economies grow, small farms can also serve as a useful reserve employer until sufficient exit opportunities exist—a role that can be especially important in fast-growing countries such as India or China. Finally, small farms may provide a social safety net, or subsistence living, for many of the rural poor, even when they are too small to be commercially viable. These social roles are most important in countries with a poor agricultural productivity potential, an equitable distribution of land, or a large mineral or urban-based manufacturing sector. These social roles do not necessarily require that small farms be commercially viable, and in fact subsistence-oriented small farms may be the most appropriate ones to target. Direct support to subsistence-oriented small farms may be a more cost-effective alternative to other forms of income transfers and social safety nets. For example, food aid, donors' common response to distress, typically costs more than US$250 for each metric tonne of cereals delivered in rural areas, compared with typical smallholder production costs of US$100 or less (Hazell et al., 2007, p.23).

The role that small farms can play will also depend critically on government and donor policies, and whether they help level the playing field so that more small farms can compete in today's markets. Recent years have seen significant growth in the types of rural support that promote social protection, but a decline in support for developing the kinds of commercially oriented agriculture that can create new market opportunities for small farms. This can lead to self-fulfilling outcomes in terms of poorly performing small-scale farming, and a better balance of these two types of support is often needed.

2. **Structure of the book**

IFAD, conscious that the effectiveness of its own development assistance projects for small farmers depends on resolving many of these key issues surrounding the future of small farms, organized an international conference in Rome, 24–25 January 2010, on *New Directions for Smallholder Agriculture.* The conference addressed five key questions:

1. Why should we care about small farms?
2. How fast should the transition to large farms be?
3. What is a viable small farm?
4. What help do small farmers need?
5. Does it pay to invest in small farms?

This book brings together some of the key papers presented at the conference, as well as papers commissioned after the event to fill important gaps.

The book examines both the development and social roles of small farms identified, and addresses these issues at country and regional levels that capture wide variations in the agro-climatic, social, and economic conditions of small farms around the developing world. It also draws on recent and innovative experiences of key players (governments, producer organizations, the private sector, donors, and NGOs) in assisting small farms, showing what is possible as well as lessons about best practices.

Part I of the book reviews the state of smallholder agriculture around the developing world. In Chapter 2, Conway explores the characteristics of a typical small farm, and their aspirations and constraints in meeting family food and energy needs and generating cash income. This is followed by chapters examining trends in farm size distributions in Africa, Asia, and Latin America (Chapters 3–5), and the conditions, opportunities, and threats that small farms face in each of these regions. A common question that arises in these chapters is how to define a small farm in ways that relate to viable livelihoods, and which can be compared over time and between countries. There is no easy answer to this question and no single metric that works everywhere. The most common definitions of size are based on land holding, primarily because these kinds of data are widely available from agricultural censuses. But if size is to reflect the ability of a small farm household to create viable livelihoods, then this inevitably varies enormously with the type of farming that is possible at any location. A 'viable' small cereal farm, for example, might vary from just a couple of hectares in parts of Asia or Africa, to 100 or more hectares in parts of Latin America. Size is even more difficult to define in pastoral areas. Small is often equated with family farming, but the two are not always equivalent. Even in countries where land is highly concentrated (e.g. Brazil and the US), much of the land is still farmed by family farms when defined by their ownership, management, and primary source of labour; it is just that they have grown large with the aid of mechanization (Brookfield, 2008).

There is a continuum of small farms ranging from commercially oriented farm businesses that are market driven and provide the major if not sole source of livelihood for the family, through part-time farmers who combine farming with other sources of employment, to poor people who are trying to subsist on a farm base and who are often net buyers of food (Figure 1.1). The motives and contributions of each group differ. Emergent or commercially viable small farms are market driven, and in Asia and Africa they generate significant marketed surpluses and are a powerful engine of rural economic growth, creating jobs for others in both the farm and rural non-farm economy. Small farms do not need to be full time to provide a viable farm business opportunity, but they do need access to markets and entrepreneurial skills.

Judgments about who are viable farmers based on existing patterns of farming can be very misleading because they are circumscribed by existing

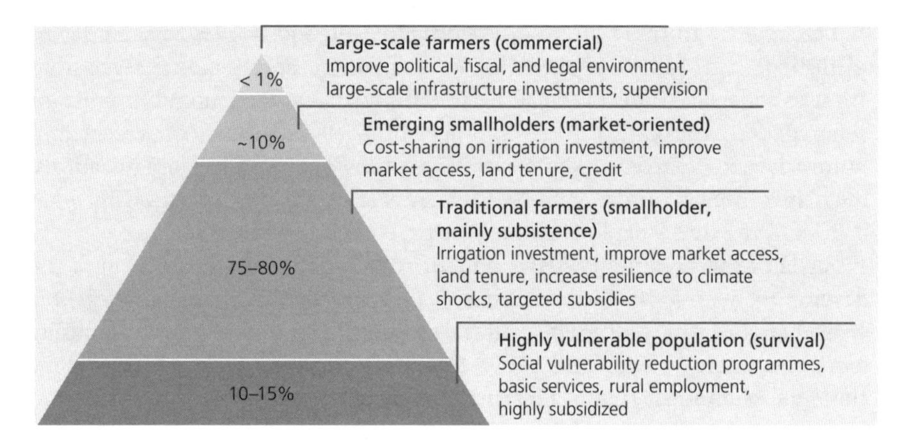

Figure 1.1 A functional classification of small farms based on market orientation

Source: Adapted from Water and the Rural Poor (FAO and IFAD, 2008).

opportunities. There are countless examples of subsistence-oriented small farms seizing new commercial opportunities when given the chance. Given the right kinds of support, many of the traditional small farms in Figure 1.1 have the possibility of becoming emerging farmers.

Subsistence-oriented farming plays important social roles in feeding and employing many poor people and providing them with a home base from which they can diversify their livelihoods. Many small farms in this category are very small or are located in marginal areas and are unlikely to become viable commercial businesses, yet if neglected, they can become a poverty trap and a cause of considerable environmental damage. Smallness in combination with poverty can, over time, cause downward spirals of worsening degradation and poverty (Cleaver and Schreiber, 1994).

Since not all small farms are equally viable as farm businesses, Torero (Chapter 6) explores methods of differentiating and mapping micro-regions of small farms so that appropriate types of assistance programmes can be targeted to each. His method takes into account agro-climatic conditions, market access, poverty incidence, and the management efficiencies of the farmers themselves, and provides a basis for spatially targeting different types of business or socially oriented interventions.

In Part II of the book, Chapters 7–10 explore the problems and options for helping small farms as businesses in linking to value chains for modern inputs, financial services, and selling products. Small farms may be the more efficient producers in many countries, but they face major disadvantages in accessing modern value chains. These include low volumes of business, variable quality and seasonality of production, limited storage facilities, high transaction costs, high risk, poor market information and contacts, and limited ability to meet the high credence requirements of many high-value outlets. Although local produce markets still exist, the best business opportunities often lie with farmers who can organize for urban and export markets. As explored in Chapters 7 and 10, promising alternatives include contract farming arrangements with large farms or marketing and processing agents, voluntary producer groups, and marketing cooperatives.

Chapters 8 and 9 explore the experience with providing small farmers with financial services and modern inputs, especially fertilizer and seeds, and discuss innovative new approaches like smart subsidies, loan guarantee systems, weather index insurance, and support programmes for networks of local seed and fertilizer dealers, that can help overcome some of the usual bottlenecks to small farm access.

Since private corporate-scale investment in agricultural value chains has increased significantly in recent years in many developing countries, involving foreign as well as domestic investors, Chapter 10 explores the types of policies that can help ensure greater compatibility of these investments with small farm development.

In addition to improved access to value chains, small farmers also need access to new and game-changing technologies that can substantially increase the productivity of their land and labour on a sustainable basis. So far, agricultural R&D systems have been much more successful in transforming small-farm productivity in irrigated areas than in rain-fed areas, and Chapter 11 explores the reasons for this and the kinds of R&D approaches and institutional reforms that are needed to resolve these problems in the future.

While access to value chains and improved technologies are crucial if small farms are to prosper as businesses, it is also important not to overlook the human factor, namely the need for skilled entrepreneurs. Many small farmers respond spontaneously to new market opportunities, but the challenges of mastering knowledge-intensive technologies and adjusting to a commercial and competitive business environment require new levels of education and training. Preparing women farmers to become successful entrepreneurs is especially important, as is attracting and retaining farmers in agriculture. These issues are explored by Juma and Spielman in Chapter 12.

Part III of the book addresses the challenge of the many small farms that face particularly difficult social and economic constraints in trying to develop their farms as businesses. Chapter 13 is concerned with strategies for assisting the many small farms that seem likely to remain primarily subsistence oriented or part-time farmers. The chapter draws on IFAD's long experience working with some of the poorest small farmers trapped in these kinds of situations, and reviews that experience.

Many smallholder farmers do not have secure access to land, making it difficult for them to invest in new business opportunities or to farm on a sustainable basis. Conditions vary widely across cultural, economic, and social contexts, but seem particularly challenging in many contexts for women farmers and other disadvantaged groups. In Chapter 14, Deininger reviews recent innovations and experiences in methods of registering and titling land, reformulating national land laws to reconcile overlapping and competing rights between the formal and informal systems, and of ways of strengthening the access and rights of women and other disempowered groups.

In many societies, poor people and especially poor women farmers are disempowered and have limited options for developing new business opportunities. They are often excluded from access to land, water, credit and other financial services, extension advice, and markets. Chapter 15 reviews these problems and shows how they can be overcome, drawing on experiences in Bangladesh.

Climate change is posing new challenges for all types of small farms, but particularly subsistence-oriented farmers and those living in marginal areas. In Chapter 16, Toulmin examines the implications of climate change for small farms and discusses farm and community options for adaptation, and the kinds of public support that are needed.

Many small farms do not have viable futures as businesses or as an adequate source of livelihood, and will either need to sell up or diversify their incomes. Many farm workers will migrate to urban areas as new opportunities arise, but many more will seek full- or part-time employment in the local non-farm economy. In Chapter 17, Wiggins discusses these options and appropriate policies that can promote employment-intensive growth in the rural non-farm economy.

Finally, the book concludes with a chapter in which the editors summarize and synthesize the major findings of the book, and map out a policy agenda for addressing the needs of small farms in the future that is cognizant of their varying agro-climatic, economic, and social contexts.

REFERENCES

Binswanger-Mkhize, H. and A. F. McCalla (2010). 'The changing context and prospects for agricultural and rural development in Africa', in Pingali, Prabhu, and Robert Evenson (eds), *Handbook of Agricultural Economics, Volume 4.* Amsterdam: Elsevier.

Brookfield, H. (2008). 'Family farms are still around: Time to invert the old agrarian question', *Geography Compass*, 2(1), pp. 108–26.

Bryceson, D. F. and V. Jamal (1997). *Farewell to farms: Deagrarianisation and Employment in Africa.* Aldershot: Ashgate.

Cleaver, K. M. and G. A. Schreiber (1994). *Reversing the Spiral: The Population, Agriculture and Environment Nexus in Sub-Saharan Africa.* Washington DC: World Bank.

Collier, P. (2009). 'Africa's organic peasantry: Beyond romanticism', *Harvard International Review*, 31(2), July, pp. 62–5.

Collier, P. and S. Dercon (2009). 'African agriculture in 50 years: smallholders in a rapidly changing world', paper presented at the Expert meeting on How to Feed the World in 2050. Rome: Food and Agriculture Organization of the United Nations.

Eastwood, R., M. Lipton, and A. Newell, (2010). 'Farm size', in Pingali, P. and R. Evenson (eds), *Handbook of Agricultural Economics, Volume 4*. Amsterdam: Elsevier.

Ellis, F. and N. Harris (2004). 'New thinking about urban and rural development', keynote paper prepared for the UK Department for International Development Sustainable Development Retreat.

Fan, Shenggen (ed.) (2008). *Public Expenditures, Growth and Poverty: Lessons from Developing Countries*. Baltimore: Johns Hopkins University Press.

FAO and IFAD (2008). *Water and the Rural Poor*. Rome.

Haggblade S. and P. Hazell (eds) (2010). *Successes in African Agriculture: Lessons for the Future*. Baltimore: Johns Hopkins University Press.

Hayami, Y. (2010). 'Plantations agriculture', in Pingali, P. and R. Evenson (eds), *Handbook of Agricultural Economics, Volume 4*. Amsterdam: Elsevier.

Hazell, P., C. Poulton, S. Wiggins, and A. Dorward (2007). 'The future of small farms for poverty reduction and growth', 2020 Discussion Paper 42. Washington, DC: International Food Policy Research Institute.

Hazell, P. and A. Roell (1983). 'Rural growth linkages: household expenditure patterns in Malaysia and Nigeria', Research Report 41. Washington, DC: International Food Policy Research Institute.

Huang, Y. (1973). 'On some determinants of farm size across countries', *American Journal of Agricultural Economics*, 64(1), pp. 89–92.

IFPRI (2005). *The Future of Small Farms: Proceedings of a Research Workshop*. Washington, DC.

Joshi, P. K., A. Gulati, and R. Cummings Jr (eds) (2007). *Agricultural Diversification and Smallholders in South Asia*. New Delhi: Academic Foundation.

Kenya Central Bureau of Statistics (1977). *Integrated Rural Survey, 1974–75: Basic Report*. Nairobi.

Kherallah, M., C. Delgado, E. Gabre-Madhin, N. Minot, and M. Johnson (2002). *Reforming Agricultural Markets in Africa*. Baltimore: Johns Hopkins University Press.

Lipton, M. (2004). *Crop Science, Poverty, and the Family in a Globalising World*. Plenary Session, Brisbane International Crop Science Conference.

Maxwell, S., I. Urey, and C. Ashley (2001). *Emerging issues in rural development: an issues paper*. London: Overseas Development Institute.

Mellor, J. W. (1976). *The New Economics of Growth: A Strategy for India and the Developing World*. Ithaca, NY: Cornell University Press.

Ravallion, M. and G. Datt (1996). 'How important to India's poor is the sectoral composition of growth?', *World Bank Economic Review*, 10(1), pp. 1–25.

Reardon, T., A. Fall, V. Kelly, C. Delgado, P. Matlon, J. Hopkins, and O. Badiane (1994). 'Is income diversification agriculture-led in the West African Semi-arid Tropics?', in Atsain, A., S. Wangwe, and A. G. Drabek (eds), *Economic Policy Experience in Africa*. Nairobi: African Economic Research Consortium.

Reardon, T., C. P. Timmer, C. Barrett, and J. Berdegue (2003). 'The rise of supermarkets in Africa, Asia, and Latin America', *American Journal of Agricultural Economics*, 85(5), pp. 1140–6.

Rosegrant, M. and P. Hazell (2000). *Transforming the rural Asian economy: The unfinished revolution.* Hong Kong: Oxford University Press.

Spielman, D. J. and R. Pandya-Lorch (2010). *Proven successes in agricultural development: A technical compendium to Millions Fed.* Washington DC: International Food Policy Research Institute.

Start, D. (2001). 'The rise and fall of the rural non-farm economy: Poverty impacts and policy options', *Development Policy Review,* 19(4), pp. 491–505.

Swinnen, J. F. M. and M. Maertens (2007). 'Globalization, privatization, and vertical coordination in food value chains in developing and transition countries', *Agricultural Economics,* 37(1), pp. 89–102.

Tomich, T. P., P. Kilby, and B. F. Johnston (1995). *Transforming Agrarian Economies.* Ithaca, NY: Cornell University Press.

World Bank (2007). *World Development Report 2008: Agriculture for Development,* Washington DC: World Bank.

Part I

The State of Smallholder Agriculture in the Developing World

2 On being a smallholder

GORDON CONWAY

Being a smallholder is a relative term. Having less than 2 hectares (ha) is usually regarded as a smallholding in a developing country. Eighty per cent of all African farms (33 million farms) fall in this category. But in many parts of Latin America a small, predominantly subsistence farm is 10 hectares. Bangladeshis would regard this as a large commercial farm. Clearly 'smallness' is a relative term depending on the resources of the holding, not only the land, but the labour, skills, finances, and technology available. Irrigated rice farmers can produce high yields with two to three crops a year, consuming and selling their harvest even though the farm may be much less than 2 ha in size. In all continents, farms of less than 1 ha and with few resources are usually unable to produce a surplus for sale and cannot provide enough work or substance for the family. Such 'marginal' farms in India comprise 62 per cent of all holdings, and occupy 17 per cent of the farmed land (Nagayets, 2005). As Pingali reminds us, however, the Chinese agricultural revolution was brought about by very small smallholders, with only a *mu* of land, just 1/15th of a hectare (Pingali, 2010).

Very approximately there are 450 million small farms, i.e. under 2 ha, in the world. This implies that some 2 billion people are dependent on smallholding for their livelihoods—a third of the world's population. The great majority of smallholdings, nearly 90 per cent, are in Asia.

Despite the diversity of smallholdings, three facts are not in dispute:

First, the size of land holdings is falling in the developing world, with the fastest decline in Africa (Nagayets, 2005). Average farm holdings for the continents of Asia and Africa are, at the latest 1990 census, 1.5 ha. Despite assertions to the contrary, there is not much extra arable land available for cultivation, and population increase continues to divide land holdings. Smallholders are becoming more numerous and smaller.

Second, land and water are deteriorating. According to the semi-quantitative *Global Assessment of Soil Degradation (GLASOD)* about 300 million hectares (mha), or 5 per cent, of the formerly usable land in developing countries have been lost by severe soil degradation up to the present day, i.e. more than has been brought into production (Oldeman et al., 1991). The current rate of loss is not less than 5 mha per year (Young, 1999). Water, like land, is similarly in short supply and for similar reasons—over-use, inefficient use, and degradation through pollution (IWMI, 2007). Many river basins in the world do not have enough water to meet all the demands: about a fifth of the world's people, more than 1.2 billion, live in areas of physical water scarcity. Rivers are drying up, groundwater levels are declining rapidly, freshwater fisheries are being damaged, and salinization and water pollution are increasing. Large areas of irrigated land in the Middle East, North Africa, and Asia are now maintaining food production through unsustainable extractions of water from rivers or the ground (UNDP, 2006). In China the groundwater overdraft rate exceeds 25 per cent and it is over 56 per cent in parts of northwest India (World Bank, 2006). Both large and small holdings are affected. Arguably the small farms are worst hit.

Third, smallholdings remain of primary importance not only to agriculture, but to rural development in the developing countries. Experience has amply demonstrated the power of agriculture as an engine for economic development (Hazell and Haggblade, 1993). Very few countries have experienced rapid economic growth without preceding or accompanying growth in agriculture (Mellor, 1995). This is not because agriculture has a special capacity for rapid growth (although in certain situations growth can be very fast, e.g. during the Green Revolution) but because of the size of the sector; even modest rates of growth have a considerable multiplier effect. In Africa a US$1 increase in agricultural income leads to an increase in overall income of more than US$2 (UNECA, 2009).

In effect there is a virtuous circle that hinges on agricultural development and, in much of the world, smallholders sit at the centre of this circle:

> As agriculture develops – greater yields and production of subsistence and cash crops – smallholders become more prosperous and the landless also benefit through wage labour. Chronic hunger decreases. The rural economy also grows – through the creation of small rural businesses – providing more employment and improved rural facilities, especially schools and health clinics. Roads and markets develop so that the rural economy connects to the urban economy and to the growing industrial sector. Free trade provides opportunities for greater imports and exports. In particular high value agricultural exports can accelerate agricultural development, further intensifying the virtuous circle... (Conway, 2012).

It has long been recognized that smallholders are in many respects highly efficient (Wiggins, 2009). Small farms produce more per hectare than large farms: many studies have shown there is an inverse relationship between farm size and production per unit of land (however, Paul Collier and Stefan Dercon point out this may be true of small versus large farms, i.e. 1 ha versus 10 ha, but not very large farms, i.e. 1 ha versus hundreds or thousands of ha) (Cornia, 1985; Eastwood et al., 2009; Collier and Dercon, 2009). In the developing countries where labour is relatively cheap and capital relatively expensive, there are few economies of scale. Household labour is the key to smallholder production—usually a family with long experience of the local environment and knowledge of what works and what does not. Because it is 'on the spot' the labour is readily available and motivated, and most important flexible, able to respond immediately to the vagaries of the farming calendar and adaptable to the frequent crises that affect the farm, whether they be pest and disease outbreaks, droughts or floods, or slumps in market prices.

You can see this most clearly in the home garden of a smallholding, a traditional system of agriculture that goes back to the very origins of domestication (probably the first wheats were cultivated when the farmer, most likely a woman, brought seed back from the wild fields to sow on the midden by the dwelling) (Hoogerbrugge and Fresco, 1993). Home or kitchen gardens are particularly well developed on the island of Java in Indonesia, where they are called pekarangan (Soemarwoto and Conway, 1991). Their immediately noticeable characteristic is their great diversity relative to their size: they usually take up little more than half a hectare around the farmer's house. Yet, in one Javanese home garden 56 different species of useful plants were found, some for food, others as condiments and spices, some for medicine, and others as feed for the livestock—a cow and a goat, some chickens or ducks, and fish in the garden pond. Much is for household consumption, but some is bartered with neighbours and some is sold. The plants are grown in intricate relationships with one another: close to the ground are vegetables, sweet potatoes, taro, and spices; in the next layer are bananas, papayas, and other fruits; a couple of metres above are soursop, guava, and cloves, while emerging through the canopy are coconuts and timber trees, such as Albizzia. So dense is the planting that to the casual observer the garden seems like a miniature forest. But it is not a natural ecosystem, it is the product of intimate knowledge and daily care and attention, usually by the woman of the household (Figure 2.1).

For ecologists like me such gardens are a delight. But as Paul Collier warns, we must be careful not to romanticize (Collier, 2009). The high labour input may be capable of producing a large and varied harvest, but the returns on the labour are small. Labour productivity is low and typically insufficient to bring people out of poverty. Quite a large number of countries in Africa have labour productivity in agriculture less than the US$900 needed to bring them out of poverty[1] (Figure 2.2).

[1] This figure is based on Steve Wiggins's (Wiggins, 2009) estimate that labour productivity needs to be over US$700 (in 2004–6) to take them out of poverty. Using agricultural producer price indices of some African countries (as available in FAOSTAT, we estimate that the threshold amount in 2009 should be about US$900.

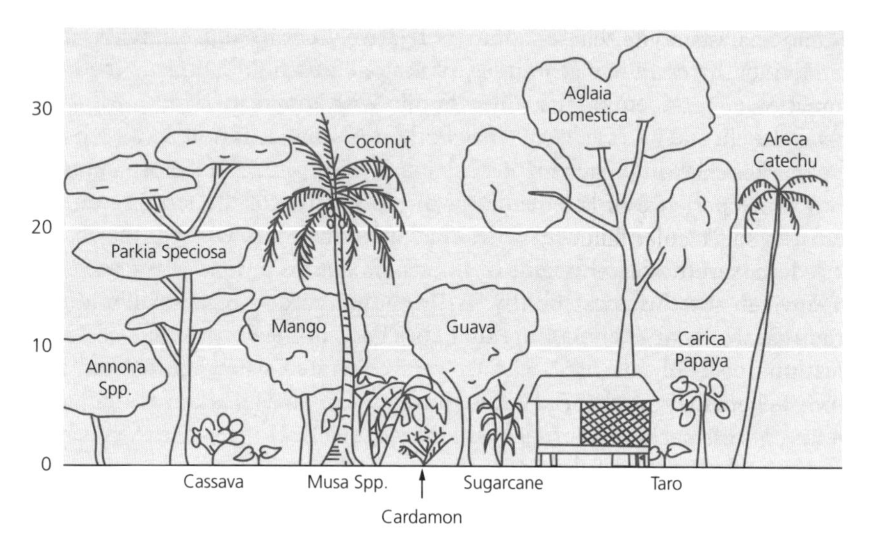

Figure 2.1 The diversity of the Javanese home garden

Source: FAO (1989).

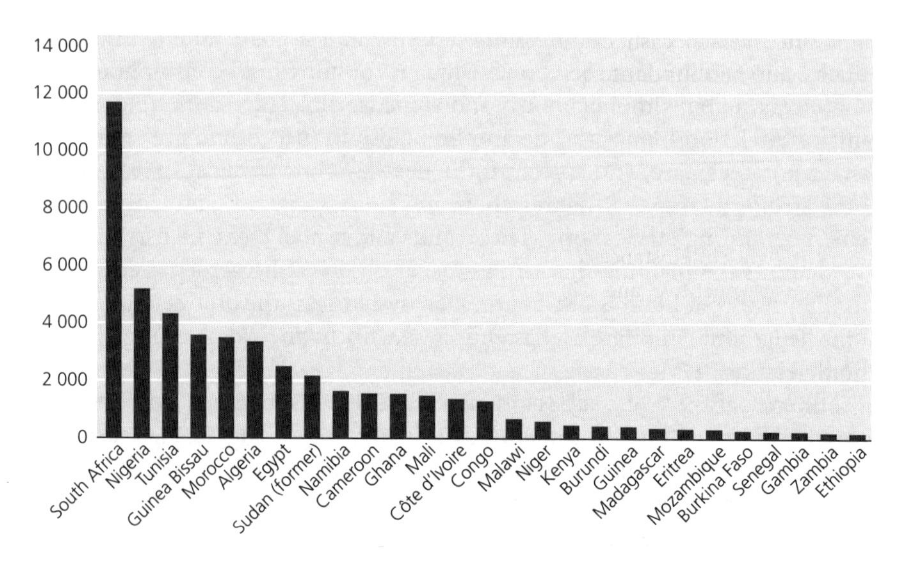

Figure 2.2 Labour productivity in African agriculture

Note: Author's calculations based on data from FAO (2012).

In many respects this challenge lies at the heart of the question: how can we increase the returns so that smallholder farming becomes a route out of poverty? It is not enough to help smallholders achieve subsistence, even if it is sustainable. They also need incomes—not least to pay medical bills and schooling, as well as to pay for food when harvests are poor. The goal is not just sustainable existence, but sustainable development.

Some analysts argue that this can only occur through large-scale farming. The obvious advantages lie in the economies of scale relating to the costs of transactions off the farm—procuring inputs, obtaining credit, and marketing, such as the costs of meeting standards that satisfy buyers from processors and supermarket chains (Wiggins, 2009).

But the experience of large farms in Africa is not all that positive. There have been spectacular failures, especially where inappropriate mechanization has led to severe soil erosion, as in the ill-fated Groundnut Scheme in Tanganyika (now Tanzania) in the 1940s or the export vegetable cultivation in Senegal in the 1970s. Large farms in Africa require experienced management otherwise the costs and the environment can conspire to bring about failure. Where they are appropriate and can be successful is when large capital investments are necessary, for example to support large processing plants that in turn require large-scale production. Examples are farms devoted to high-value and specialist crops such as fruit, vegetables, flowers, intensive pigs, and poultry, and for export crops such as sisal, sugar, tea, rubber, and coffee (Wiggins, 2009).

Contemporary dialogues about agricultural development and food security are plagued by contests between extremes—organic farming versus GM, subsistence versus cash cropping, small-scale irrigation versus large-scale, and so on. But abundant experience tells us that the dialogue should not be about *either/or*, but should focus on *both/and*. Thus the way forward lies not in *either* small *or* large farms, but in making a deliberate choice of *both* small *and* large farms. They both have a role to play in Africa's future agricultural development. In Pingali's (2010) words:

> under the right circumstances smallholders can be just as productive, just as innovative, just as competitive, and just as risk-taking as larger farms.

At the same time, in other circumstances, large farms are appropriate. Large commercial farms can be close to the frontiers of technology, finance, and logistics and hence globally competitive (Collier and Dercon, 2009; Byerlee and Deininger, 2010). This has been the Latin American experience, with farms of more than 10,000 ha growing soybean in the Brazilian Cerrado and over 300,000 ha sugar estates in southern Brazil, many focused on ethanol production. There are also some examples in Africa, including mechanized sorghum and sesame production in Sudan, with farms averaging over 1,000 ha, and some over 20,000 ha.

Although these are impressive developments and it is clear that large-scale farming will have an increasingly significant role in achieving food security, smallholders, because of their sheer numbers and the total land area they occupy, will have to play the dominant development role at least for several decades to come. The question is: can we significantly increase their labour productivity?

Let me now illustrate the challenge with a fairly representative, although fictitious, smallholder in Africa. I will call her Mrs Namarunda (Conway and Toenniessen, 2003). Several years ago, her husband died. Her eldest son inherited the family farm, a single hectare running up one side of a hill near Lake Victoria. The soils are moderately deep and well drained, but they are acidic, highly weathered, and leached. Mrs Namarunda's first son married and moved to Nairobi, where he is an occasional lorry driver and has children of his own.

Mrs Namarunda was left on the farm with four younger children and the responsibility to produce food, fetch water, gather fuel, educate the children, and take care of the family. But shortages of almost everything—land, money, labour, plant nutrients in soil exhausted from many years of continual crop production—mean that she is often unable to provide her family with adequate food. The two youngest children, in particular, suffer from undernourishment and persistent illnesses.

She starts each growing season with a maximum potential harvest of only about 2 tonnes from mixed cropping on her 1 ha of land. To survive, her family requires a harvest of about 1 tonne, so if everything goes right and the maximum harvest is achieved, it would be sufficient to meet their needs and to generate a modest income. But, during the course of every growing season, she faces innumerable threats to her crops that reduce her yields.

Weeds are her most persistent and pervasive problem. Her staple crop, maize, is attacked by the parasitic weed *Striga*, as well as by the streak virus, various boring insects, and a fungus which rots the ears. She has tried growing cassava as an 'insurance crop', but it, too, was attacked, first by mealy bugs and green mites, then it was totally devastated by a new, super-virulent strain of the African cassava mosaic virus. Her banana seedlings are infected with weevils, nematodes, and the fungal disease Black Sigatoka. Her beans, which are intended as a source of protein for the family and nitrogen for the soil, suffer from fungal diseases. She also faces drought at some time during the growing season, which again reduces crop yields. At the end of each season, what she actually harvests is usually less than 1 tonne. She and her children are often hungry, and there is no money for schooling or for health care (see Figure 2.3).

But it does not have to be this way. There are already solutions to many of her pest, disease, and weed problems. New varieties can provide protection and will give greater resilience to drought. Part of the challenge is to ensure that the varieties and other technologies are appropriate, i.e.

- They are productive, in particular they generate high levels of income;
- The production they generate is stable and resilient;
- They are readily accessible and affordable; and
- They do not have significant environmental or human health downsides. (Conway et al., 2009).

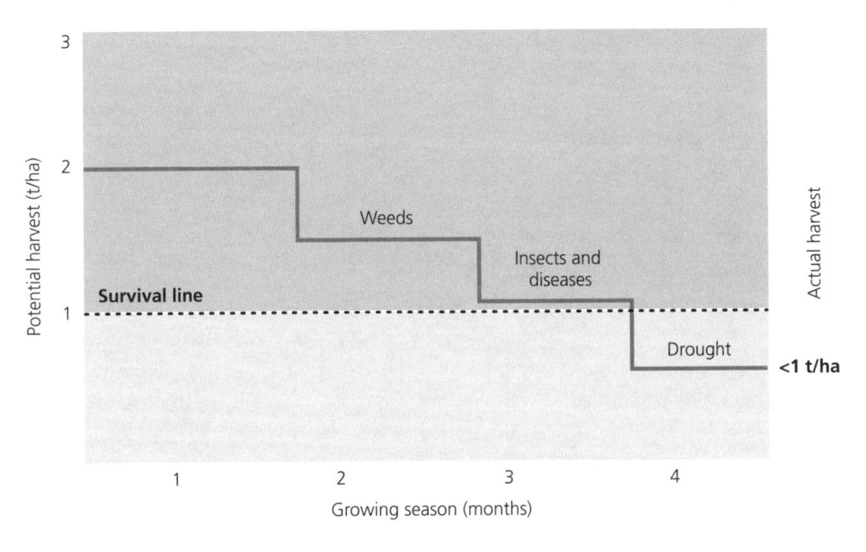

Figure 2.3 An insecure farm in Africa

Source: Conway et al. (2009).

It used to be thought that 'appropriate technologies' for smallholders were in some sense intermediate technologies—lying between traditional technologies and conventional, industrial technologies. But experience suggests that any kind of technology can be appropriate, at least on the criteria described here.

Traditional technologies are approaches to problems that have been used by people for hundreds, if not thousands, of years. They can be thought of as having 'stood the test of time'. Some have clearly worked and still work today. For others there is no scientific evidence that they are effective.

Perhaps one of the most successful of traditional practices is the home garden previously mentioned. Even more ubiquitous is the cultivation of traditional crop varieties, so-called land races, or of 'wild' plants. They may be domesticated to an extent and improved by farmer selection, but usually retain their essential characteristics. Examples include the traditional maizes of Central America, the sticky rices of northern Thailand, and the potatoes of Peru (CIMMYT, 2006; Rerkasem, 2007).

Most traditional technologies, however, are essentially subsistence technologies. They are stable and resilient in their performance, but they rarely increase incomes. To some extent this is also true of those intermediate technologies that combine the best of conventional and traditional technology. There are numerous examples. Some are simple, others more sophisticated. A good example is the development of an affordable and reliable treadle pump (IDE, 2006). For many years, engineers have been developing pumps which allow farmers to replace the arduous task of lifting irrigation water from shallow wells by bucket. The modern treadle pump is ideal in many respects—it is

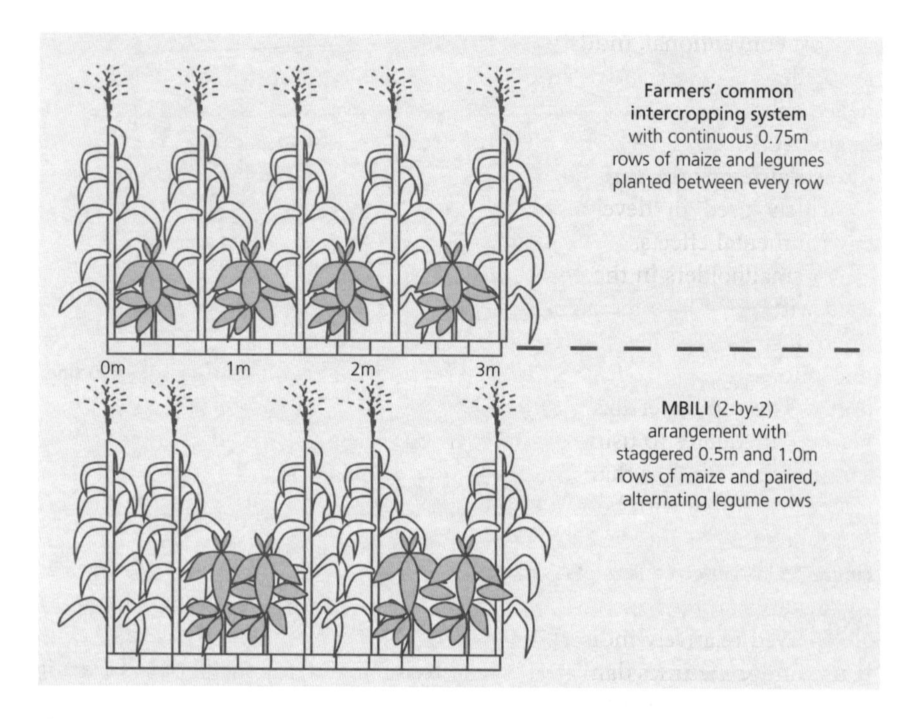

Figure 2.4 Intercropping systems in East Africa

Source: Woomer et al. (2004).

efficient and easy for farmers to use and maintain, and is virtually fool-proof. It is also relatively cheap, as a result of a combination of public subsidies with private manufacture and servicing, and with community involvement. Such pumps can also help to increase incomes if the water is used to irrigate high-value vegetable or horticultural crops.

Another form of intermediate technology is the ecologically based cropping system. An example is the MBILI system developed by an NGO called SACRED (Sustainable Agricultural Centre for Research Extension and Development in Africa) in western Kenya (Woomer et al., 2004). The system was developed on farms with farmers as key partners in its management and experimentation. It consists of intercropping double rows of maize with double rows of higher-value legumes such as beans, green gram, and groundnuts. This allows for better light penetration favouring the legumes. Yields of maize are about 5 tonnes/ha and of legumes about a tonne (Figure 2.4).[2]

[2] A staggered maize–legume intercrop arrangement robustly increases crop yields and economic returns in the highlands of Central Kenya. *Field Crops Research* 115: 132–9, Mucheru-Muna et al., 2009) and innovative maize–legume intercropping results in above- and below-ground competitive advantages for understorey legumes (Woomer et al., 2004).

Most conventional, industrial technologies such as synthetic fertilizers and pesticides and modern irrigation systems provide high income returns and hence enhance labour productivity, as the Green Revolution clearly demonstrated (Conway, 1997). It is why they are so ubiquitous in the farms of the industrialized world. But they tend to be expensive, which is why they are not so widely used in developing countries. They can also have deleterious environmental effects.

For smallholders in the developing world, the answer is to ensure they are used with much greater precision. An example is the technique of fertilizer microdosing developed in Niger. Each microdose consists of a 6-gram mix of phosphorus and nitrogen fertilizer, which just fills the cap of a Coca-Cola bottle. The cap of fertilizer is then poured into each hole before the seed is planted. It equates to using 4 kg/ha of phosphorus, the key limiting nutrient, three to six times less than used in Europe and North America. Microdosing has been credited with boosting millet yields by 50 per cent to 100 per cent in the Sahel, thereby helping to reverse a 50-year trend of declining yields and rising soil degradation (ICRISAT, 2001).

The same principle can be applied to herbicide use. Far too often herbicides are sprayed relatively indiscriminately, killing not only weeds but other wild plants and sometimes damaging the crops themselves. Yet there are extremely serious weed problems that have to be tackled. One such is *Striga*, or witch weed, a devastating parasitic weed in Africa that sucks nutrients from the roots of maize, sorghum, and other crops. *Striga* is also readily controlled by herbicide, imazapyr, but this tends to damage or kill the maize crop. Recently, a mutant gene in maize has been discovered through tissue culture that confers resistance to the herbicide and is being bred into local maize varieties. The maize seed can be dipped into the herbicide before being planted and as the maize plant grows the herbicide will kill the weeds in the ground. Early trials are showing increases in yield from half to over 3 tonnes per hectare (ICRISAT, 2001).

Finally, recent years have seen the development of new scientific 'platforms' for innovation, derived from fundamental discoveries in the physical, chemical, and biological sciences. These have the potential to be developed simultaneously for the needs of the industrialized and the developing world, and in the right circumstances can be appropriate for the needs of developing country smallholder farmers. These new platform technologies include Information and Communication Technology (ICT), Geographic Information Systems (GIS), nanotechnology, and biotechnology (Conway et al., 2009).

In its modern form, crop biotechnology comprises:

- *Marker-aided selection*—particular DNA sequences at specific locations can be used to detect the presence of a gene and so speed up the breeding process.

- *Tissue culture*—which permits the growth of whole plants from a single cell or clump of cells in an artificial medium. This can be used to generate planting materials that are known to be free of disease. Tissue culture has produced new pest- and disease-free bananas in East Africa that can yield up to 50 tonnes/ha (Africa Harvest, 2009).
- *Recombinant DNA* or genetic engineering or modification (GM) technology—which enables the direct transfer of genes from one organism to another.

The first two biotechnologies, in the hands of international and national agricultural research centres, are already delivering improved staple crops to poor farmers. GM technology also has the potential to benefit smallholders in developing countries. At present this largely derives from engineering crops to express a bacterial gene that controls certain insect pests, so reducing the need for harmful synthetic pesticides. Thus *Bt* cotton, which contains a gene producing an insecticidal protein in the plant, has proved to be especially beneficial in a number of countries. Around the world nearly 13 million 'small and resource poor' farmers are growing *Bt* cotton (James, 2009). Burkina Faso, the largest cotton producer in Africa, adopted *Bt* cotton on a commercial scale in 2008 (ISAAA, 2009). In the second season it was being grown on 100,000 ha with yields up to 50 per cent higher than conventional cotton, and the number of sprays reduced from an average of eight to at most two. A World Bank review of the benefits of *Bt* cotton in Argentina, Brazil, Mexico, India, and South Africa showed increases in yields of between 11 and 65 per cent, and increased profits as high as 340 per cent with significantly reduced use of pesticides and pest management costs (Table 2.1) (World Bank, 2007).

Of course, one way of ensuring that technologies are appropriate is to involve smallholders intimately in the development of the technologies (Conway et al., 2009). This means not just involving them in testing new technologies and their local adaptation, but engaging them much earlier in the process, helping to identify objectives for research and development and participating pro-actively in the experimentation. Under the headings of Participatory Rural Appraisal (PRA) and Participatory Learning and Action

Table 2.1 The economic and environmental benefits of growing *Bt* cotton in developing countries

	Argentina	China	India	Mexico	South Africa
Added yield (%)	33	19	26	11	65
Added profit (%)	31	340	47	12	198
Reduced chemical sprays (number)	2.4	–	2.7	2.2	–
Reduced pest management cost (%)	47	67	73	77	58

Source: World Bank (2007).

(PLA), there is a now a formidable array of methods which allow communities to analyse their own situations and, importantly, to engage in productive dialogue with research scientists and extension workers (Chambers, 1997; Scoones and Thompson, 1994).

In the 1990s, farmers in the arid Khanasser Valley in northern Syria started planting olive trees. They did not have a long history of olive cultivation to draw knowledge, and with little water available for irrigation and poor soil quality, yields were far from ideal. The International Center for Agricultural Research in the Dry Areas (ICARDA) came to the area in 2003 and formed a committee of farmers and local extension workers to begin a process of participatory research, evaluation, and innovation. Out of this a number of new technologies were identified and adopted (ICARDA, undated). They include:

- Water Harvesting—the construction of V- or fishbone-shaped, stony-earth bunds around each of the trees to create micro-catchments which can contain water around the tree and control soil erosion.
- Stone Mulching—covering the soil around the tree trunk with stones to reduce evaporation losses from the soil surface. Project experiments showed that basalt stone worked better than the chalky limestone the farmers had previously been using.
- Sub-surface Insert Irrigation—one of the farmers in the valley suggested an irrigation method he had seen being used in Tunisia. A stone pocket or gravel layer is constructed underground around the root zone of each tree. A PVC tube is then inserted vertically into the pocket and water is applied through the tube, so that it goes directly to the deeper roots.

Appropriate technologies of one kind or another have much to offer Mrs Namarunda. They can solve most if not all of her problems, some quickly, some over a longer time. She can then expect to harvest at least 2 tonnes from her hectare. This will allow her to feed her family and leave part of the farm to grow crops for sale and provide the income she desperately needs (Figure 2.5). But, of course, she needs access to these technologies and much else besides.

In the words of Joe De Vries of the Alliance for a Green Revolution for Africa (AGRA), 'Africa's food needs cannot be outsourced. Food needs to be grown here in Africa, and that process starts with getting improved seeds into the hands of farmers'. Some of the new seeds are being produced by the multinational companies. Most of the seeds appropriate to her conditions, however, are coming from partnerships between the international research institutes of the CGIAR and national research institutes, such as the Kenya Agricultural Research Institute (KARI). Attention is now turning to the creation of relatively small, locally based seed companies under AGRA's Programme for Africa's Seed Systems (PASS). So far they have provided start-up capital for 47

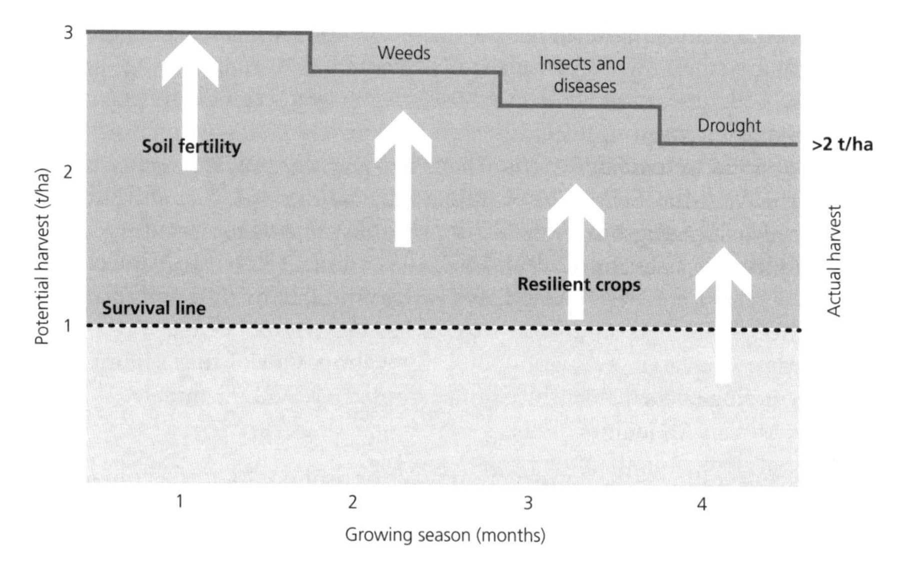

Figure 2.5 A secure farm in Africa

Source: Conway et al. (2009).

African seed companies which have collectively produced 12,807 million tons of certified seed (AGRA, 2011a).

Producing new locally appropriate seeds (and fertilizers) is part of the challenge, but equally important is making them accessible to farmers like Mrs Namarunda. Another of the innovative actions of AGRA (building on the work of the Rockefeller Foundation) is to facilitate the creation of village-level agro-dealers. AGRA has trained and supported over five thousand such agro-dealers in eastern and western Africa. Although the agro-dealer stores are small (what Americans call 'Mom and Pop stores'), they are collectively having a major impact, providing US$45 million worth of improved seeds, fertilizers, and other inputs in 2008. They sell key inputs to farmers in small, affordable quantities and, most significantly, they reduce the distances farmers have to go to get inputs—in one area of Kenya from 17 kilometres in 2004 to 4 kilometres in 2007 (AGRA, 2011b).

Providing physical access to inputs is one challenge, but the new seeds and other inputs need to be affordable. They have to be priced appropriately; a range of seed producers can provide the necessary competitiveness to bring prices down. But smallholders also need subsidies and credit, especially in the early years of new agricultural development. The experience of Malawi has shown what a subsidy programme can do in terms of seed and fertilizer uptake; yields and total production increase dramatically in the first few years (Dorward and Chirwa, 2011). It is questionable, however, whether such high levels of subsidy are sustainable on a national basis.

The medium- to long-term solution is a combination of better targeted, smart subsidies and the provision of microfinance. In recent months there has been much criticism of micro-credit schemes, especially in Asia where a large, private industry has grown up, in many instances charging very high rates of interest. It is worth remembering that the first such schemes back in the 1970s such as SEWA—the Self Employed Women's Association—of Hyderabad were based on small groups of individuals who knew each other and took responsibility for assessing and approving loans. We need to recapture that spirit of lending, and here the agro-dealer network can provide a basis for greater intimacy and locality for credit programmes. As a first step, in Tanzania the National Microfinance Bank is providing US$5 million in loans to agro-dealers to assist them in providing better service to small farmers (African Agriculture, 2008).

The final element in the use of new appropriate technologies is the creation of incentives for adoption. Smallholding is a risky business and farmers need to know they will be fairly rewarded if they take on the risk and also have some recompense if, through no fault of their own, they have a disastrous year (IFAD, 2010). There are three key components: land tenure, insurance schemes, and output markets.

Land tenure is probably the most complicated to resolve. Large numbers of smallholders have insecure rights to their land. In Mrs Namarunda's case her land is owned by her son and presumably he could take it away or sell it, leaving her destitute. Where small-scale land ownership is common and has a legal basis it is imperative to make sure ownership is registered and secure. In one pioneering programme, Rwanda is using satellite technology to resolve disputes and regularize ownership. It is proving cheap and easy to use, with farmers showing considerable skills in using remote sensing images to discuss and resolve boundary disputes.

It is more common, however, for some form of customary tenure to apply, and the challenge is to give this legal status. For example, in Eritrea in 1995 the majority of customary rights were transferred into lifetime rights under guaranteed government protection. Tanzania, Uganda, and Mozambique also all recognize customary tenure, although in different systems, as legally valid (Alden, 2000). The aim in all these instances is to introduce new arrangements and processes that provide sufficient rights to encourage smallholders to invest in technologies and other inputs that will improve their land over the medium to long term.

If they are to improve their lives and livelihoods, smallholders have to take risks. In many instances the downsides of adopting a new technology or borrowing finance can be devastating. A drought or a flood will not only bring starvation, but can wipe out a family's entire savings and force them to sell their assets. In India some 2 million weather index insurance products have been sold through private insurance since 2003, while in Africa a number of similar schemes are currently being piloted (IFAD, 2010).

One such scheme helps small- or medium-scale livestock farmers or pastoralists to insure against the risk of investing in areas of regular drought. It is based on NASA satellite images which measure vegetation and a model which predicts livestock mortality based on the vegetation index.[3] Insurance companies then pay out twice a year for any events that have occurred where the predicted mortality was over a certain threshold. The scheme has multiple benefits: it helps households get loans, prevents them from having to use self-insurance (i.e. having more cattle than they really need), and aids the Kenyan government, which normally must make big payout to farmers after droughts to keep them out of extreme poverty.

Finally, and in some respects most important of all, is the provision of markets that will buy from farmers for a fair price in an open, honest, and non-exploitative fashion. There have been numerous horror stories of farmers investing in new seeds and fertilizers, and obtaining high yields only to find they cannot sell their produce, except at a loss. This will put off even the least risk averse from investing in new technologies.

Part of the answer lies in creating local village grain storage systems, managed by village cooperatives, which will keep some reserves for low harvest season but market the rest at the best prices. This depends on having a countrywide network of small and large markets, and a scheme such as that run by the Kenya Agricultural Commodity Exchange (KACE), a private-sector firm that links sellers and buyers of agricultural commodities, and provides relevant and timely marketing information and intelligence using a mobile phone Short Message Service (SMS) system.

Smallholders are also being encouraged to use Warehouse Receipt Systems. One such system in Tanzania, funded by IFAD, permits farmers to borrow from a Savings and Credit Cooperative up to 70 per cent of the value of the stored grain, but also to sell some of their stock several months after harvest when prices are higher (IFAD, 2010).

So far, I have deliberately talked with a positive optimistic slant; I believe much is possible in both the short and long term. But I have ignored the potential adverse impacts of climate change, which threaten to negate what can be done.

Of course there is a great deal of argument about climate change at present. The deniers are having a field day exploiting inconsistencies and inaccuracies, and exposing questionable actions by some climate scientists. The media delights in such controversy, but we need to be aware that for Mrs Namarunda

[3] Personal communication: Ben Lukuyu, ILRI head of EADD.

climate change is not a fantasy; it is already affecting what happens on her 1-hectare farm.

Further north from where she lives the rainfall pattern in the Sahel has dramatically altered since the 1970s, bringing prolonged drought with periodic heavy rains and devastating flooding on the parched soils. Northern and southern Africa are becoming drier and hotter; the pattern of the Asian and East Asian monsoon is changing; river regimes in Asia and Africa are altering; everywhere the extremes of weather are becoming more frequent and/or more severe. It is not possible to assign any extreme event to anthropogenic climate change, but the trends are consistent with a globe being warmed by increased release of greenhouse gases (Conway, 2009).

It seems that everywhere people are aware of the changes. Ask a group of smallholders whether the climate is changing and they will say, yes of course, and will describe what is happening to their community. Is it affecting your cropping patterns and practices? Yes, they will reply. Asking what they are doing about it will elicit a detailed account of their responses.

In the Atlas Mountains of Morocco, the villagers now cannot grow enough barley to feed themselves, because of the continuing lack of rainfall. They are trying out drip irrigation as a possibility for high-value crops that they can sell in the markets on the coast. They are also harvesting some of the wild, typically drought-tolerant plants growing on the hills around the villages—for example, the Argan tree that produces high-quality oil like olive oil, and the honey from euphorbia. But the women are doing the harvesting, and they are getting little return. The challenge is to process the oil and the honey *in situ* and derive some of the value added in the villages.

Another example is in the village of Nwadhajane in southern Mozambique, the birthplace of the great Mozambique leader, Eduardo Mondlane (Osbahr et al., 2008). The villagers are very aware of climate change affecting them and have already taken significant measures to counteract the worst features. They have two kinds of land—lowland and highland. On the former the crops are very productive, but are washed out by periodic floods; in the highlands they produce good crops in the flood years but poor crops during the droughts. The villagers' response has been to create several farmer associations which have reassigned the land so that each farmer obtains a portion of highland as well as some lowland. The farmer associations are also carrying out experiments with drought-resistant crops.

These are encouraging examples of smallholder experimentation and adaptation, but the solutions also require major government interventions—improvements to irrigation and water storage schemes, coastal protection, river bank strengthening—together with international experimentation on the development of drought- and flood-tolerant crop varieties. Adaptation is going to help Mrs Namarunda if the top-down actions of government are melded with the bottom-up adaptive capacities of rural communities, in particular the millions of smallholders in the world who are already suffering.

The same principle applies to the challenge of mitigating climate change. There is still not enough recognition of the fact that agriculture and deforestation are major producers of greenhouse gases—about 30 per cent of the total (IPPC, 2007). Contributors are: (i) the carbon dioxide resulting from deforestation and the loss of soil carbon in conventional agricultural practice, (ii) the methane emitted by flooded rice and from enteric fermentation in cattle; and finally, (iii) the nitrous oxide from microbial transformation of nitrogen in the soil and in manures. The amounts of methane and nitrous oxide are relatively small, but they are the gases that have the biggest effect on global warming.

It is possible to reduce nitrous oxide and methane by various means: reduced tillage, improved grassland management, restoration of degraded lands, more efficient use of fertilizer, improving water and rice management, planting trees and agroforestry, altering forage and sustainable use of animal genetic diversity, storage and capture technologies for manure, conversion of emissions into biogas, and low emission rice varieties and livestock breeds. But the biggest challenge is to get significant amounts of the carbon back into the soil. Rattan Lal has estimated that we could return at least half of the soil organic carbon that has been lost to the atmosphere since the industrial revolution through a 25- to 50-year programme of soil sequestration (Lal, 2004). In his words,

> soil C sequestration is a truly win-win strategy. It restores degraded soils, enhances biomass production, purifies surface and ground waters, and reduces the rate of enrichment of atmospheric CO_2 by offsetting emissions due to fossil fuel.

But the question for us is: what is in it for Mrs Namarunda? About 70 per cent of the all agriculturally derived greenhouse gases come from developing countries. Moreover, 70 per cent of the mitigation potential is in developing countries, and 90 per cent of that potential is carbon sequestration. So collectively the Mrs Namarundas of this world could make a huge difference. But why should she?

One possibility, of course, is that she seeks out win-win practices. Conservation farming—using minimum or no-till practices—is a good example. It conserves soil and water, increases carbon in the soil, and produces higher yields in drought situations. An interesting agroforestry technology that is already being promoted on a large scale in Africa is the planting of the legume tree, *Faidherbia albida* (Garrity, 2010). This tall tree has the distinctive feature of shedding its leaves in the wet season and putting on green growth in the dry season. Maize can be grown under the trees in the wet season, fertilized by their nitrogen fixation and the leaf mulch. Field trials have shown that 3 tonnes of maize per ha can be produced without added fertilizer, and at the same time the system will return to the soil some 2 to 4 tonnes of C/ha. That is a large amount of carbon to put back.

Perhaps we should take a leaf out of the Reducing Emissions from Deforestation and Forest Degradation (REDD) programme, which is a way of compensating people who live in forests and others for preserving forests (UN REDD, 2009). It is proving to be a successful way of promoting not only forest protection, but also regimes of selective and sustainable logging which can provide incomes for rural households. Agriculture has somehow to produce a similar scheme so that the Mrs Namarundas of this world will be compensated for putting carbon back in the soil. It will not be easy. There are challenges in measuring the level of carbon sequestration, and in finding the sources of the funds required and a fair means of compensation. But in principle it is do-able.

Finally, I want to touch briefly on a key question. In simple terms the question is: If a local community, perhaps aided by the national government or by an aid donor or by an NGO, has been able to build a productive, stable, resilient, and equitable system of technologies and/or processes that works and appears sustainable, can it be replicated on a much larger scale to benefit not just hundreds, but many thousands of smallholders? The challenge facing us is to help take these local systems, especially those that significantly increase farmer and labourer incomes, and to scale them up so that the poor, both farmers and the landless, can benefit in a way that brings about the kind of virtuous circle of rural development I described earlier.

One example of scaling up is the Ugandan Vegetable Oil Development Project, a partnership between IFAD, the government of Uganda, and Bidco, a large private investor. The partnership constructed an oil palm refinery and developed oil palm plantations, disseminated technical expertise and investment, and involved smallholders through the Kalangala Oil Palm Growers Trust, which represents their needs and interests. Once fully established the project will result in 10,000 ha of land being under oil palm production, with one third of this belonging to smallholders (IFAD, 2010).

Another example is the East Africa Dairy Development Project, which started in January 2008 in Kenya, Uganda, and Rwanda, and is a partnership between the International Livestock Research Institute (ILRI) (responsible for monitoring and evaluation), Heifer International (animal husbandry), Technoserve (milk markets and milk policy), and ABS (a US-based company specializing in animal breeding).[4] It has started to help small rural producers gain access to information on best practices for such things as feeding, breeding, and new technologies, and has also established hubs where milk can be collected, cooled, pasteurized, marketed, and sold.[5] To date, 20 Dairy

[4] Personal Communication: Ben Lukuyu, ILRI head of EADD.
[5] Personal Communication: Ben Lukuyu, ILRI head of EADD.

Farmers Business Associations (DFBA) have been formed to manage the business around each hub. There are over sixty-five thousand registered farmers, and ten DFBAs are currently selling a total of 141,000 litres of milk per day to three processors.[6]

Although there is considerable experience of going up in scale there is no simple recipe. However, some principles are beginning to emerge.

The first is that the private sector has much of the necessary experience, skills, and processes to make scaling up work. This is primarily because any agricultural technology or process that significantly increases income is 'marketable' and hence saleable.

Second, the private sector, whether indigenous or foreign, can rarely be left to itself to bring about the scaling-up transformation. In most cases there has to be a public–private partnership. Sometimes this will simply consist of governmental action to provide the right kind of enabling environment for the private sector to operate. In others, more formal public–private–community partnerships are required to harness the different qualities and comparative advantages of the relevant actors.

Third, each value chain is likely to be different. For instance, scaling-up practices will vary for livestock, export high-value crops, and local staples.

Fourth, if equitable benefits are to be derived, the value added needs to be biased to the lower levels of the value chain. Scaling up cannot be seen to benefit only the larger, better-off producers.

Fifth, and related to the previous point, there is likely to be a significant role for farmer associations, cooperatives, and other bodies that will fight to ensure the benefits are widely shared.

Finally, much of the success of scaling up depends on the details of the pathways, processes, and deals struck between the partners.

Some years ago I published a book entitled *The Doubly Green Revolution*, which laid out the argument for a new kind of agricultural revolution that aims to 'repeat the success of the Green Revolution on a global scale in many diverse localities and be equitable, sustainable and environmentally friendly' (Conway, 1997). I believe the concept and its related practices are as relevant today as they were then, if not more so (Conway, 2012).

The elements of a comprehensive framework for support of smallholder farmers in the developing countries draw on the concept of a Doubly Green Revolution. Its aim is to enable national governments, in partnerships with aid agencies, NGOs, and the private sector, to help smallholders achieve food security for themselves and their communities, and at the same time sustainably increase their incomes. I am an optimist, but, more important, my experience over the last 50 years convinces me it can be done.

[6] Personal Communication: Ben Lukuyu, ILRI head of EADD.

REFERENCES

Africa Harvest (2009). *Tissue Culture.* Retrieved from: <http://africaharvest.org/tissue.php>.

African Agriculture (2008). *Tanzanian Microfinance Bank Partners with AGRA to Provide Input Credit to Farmers.* Retrieved from: <http://www.africanagricultureblog.com/2008/02/tanzanian-micro-finance-bank-partners.html>.

AGRA (2011a). *PASS Mid-term Review: Full report 2012.* Retrieved from: <http://www.agra.org/our-results/independent-evaluations/?keywords=africa%26%23039%3Bs+seed+system> (18 July 2013).

AGRA (2011b). *Early Accomplishments.* Retrieved from: <http://www.agra.org/what-we-do/early-accomplishments/?keywords=early+accomplishments>.

Alden Wily, L. (2000). 'Land tenure reform and the balance of power in Eastern and Southern Africa', *Natural Resource Perspectives,* No. 58, June 2000, London: Overseas Development Institute.

Byerlee, D. and K. Deininger (2010). *The Rise of Large Farms: Drivers and Development Outcomes,* World Institute for Development Economic Research, United Nations University. Retrieved from: <http://www.wider.unu.edu/publications/newsletter/articles-2010/en_GB/article-11-12-2010>.

Chambers, R. (1997). *Whose Reality Counts?: Putting the First Last.* London: Intermediate Technology Publications.

Chambers, R. (2005). *Ideas for Development.* Brighton: Earthscan.

CIMMYT (2006). 'Is native maize diversity', *CIMMYT E-News,* 3, no. 11, November 2006. Retrieved from: <http://www.cimmyt.org/en/news-and-updates/item/is-native-maize-diversity>.

Collier, P. (2009). 'Africa's organic peasantry; beyond romanticism', *Harvard International Review,* Summer, pp. 62–5.

Collier, P. and S. Dercon (2009). 'African agriculture in 50 years: smallholders in a rapidly changing world', in *How to Feed the World in 2050.* Proceedings of a technical meeting of experts, 24–26 June 2009, Rome, Italy; FAO09.

Conway, G. (1997). *The Doubly Green Revolution: Food for all in the 21st Century.* London: Penguin Books (also published by Cornell University Press, Ithaca, 1999 and as *Produçãode Alimentos no Século XXI: Biotecnologia e Meio Ambiente,* Estação Liberdade, São Paulo, 2003).

Conway, G. (2009). 'The science of climate change in Africa: impacts and adaptation', Discussion Paper 1, Grantham Institute for Climate Change, Imperial College, London.

Conway, G. (2012). *One Billion Hungry: Can We Feed the World?* Ithaca, NY: Cornell University Press.

Conway, G. and G. Toenniessen (2003). 'Science for African food security', *Science,* 299, pp. 1187–8.

Conway, G., S. Delaney, and J. K. Waage (2009). *Science and Innovation for Development,* Collaborative on Development Sciences, London, UK.

Cornia, G. A. (1985). 'Farm size, land yields and the agricultural production function: An analysis for fifteen developing countries', *World Development,* 13, pp. 513–34.

Dorward, A. and E. Chirwa (2011). 'The Malawi agricultural input subsidy programme: 2005–6 to 2008–9', *International Journal of Agricultural Sustainability,* 9, pp. 232–47.

Eastwood, R., M. Lipton, and A. Newell (2009). 'Farm size', in Pingali, P. and R. Evenson, (eds), *Handbook of Agricultural Economics, Volume 4.* Amsterdam: Elsevier.

FAO (1989): 'Schematic representation of the structural composition of a Javanese home garden', in *Forestry and food security.* Rome: FAO. Retrieved from: <http://www.fao.org/docrep/T0178E/T0178E0g.gif>.

FAO (2012). *FAOSTAT 2012.* The Food and the Agricultural Organisation of the United Nations, Rome, Italy. Retrieved from: faostat.fao.org/site/291/default.aspx.

Garrity, D. (2010). 'Trees, crops and carbon. Creating an evergreen agriculture in Africa', PROFOR Seminar, 10 June 2010. Retrieved from: <http://webcache.googleusercontent.com/search?q=cache:qqcdnHJfuxsJ:www.profor.info/sites/profor.info/files/Evergreen-Agriculture-PROFOR-June2010.pdf+&cd=3&hl=en&ct=clnk&gl=uk&client=firefox-a>.

GKPnet.projects (undated). 'SokoniSMS: Empowering farmers through SMS market in Kenya'.

Hazell, P. and S. Haggblade (1993). 'Farm–nonfarm linkages and the welfare of the poor', in Lipton, M. and J. van der Gaag (eds), *Including the Poor.* Washington, DC: World Bank.

Hoogerbrugge, I. and L. O. Fresco (1993). *Homegarden Systems: Agricultural Characteristics and Challenges.* London: International Institute for Environment and Development, Gatekeeper Series No. 39.

ICARDA (undated). 'Participatory Research: CASE 9: Water and soil management in olive orchards in the Khanasser valley'. Retrieved from: <http://igitur-archive.library.uu.nl/student-theses/2011-0223-200558/EZanden_MScThesis_Final_pdf.pdf>.

ICRISAT (2001). 'Things grow better with coke'. *SATrends,* Issue 2, January. Retrieved from: <http://www.icrisat.org/what-we-do/satrends/01jan/1.htm>.

IDE (2006). *I Am Not Worried About My Children's Future Any More—An IDE Success Story from Bangladesh.* Lakewood, CO: International Development Enterprises.

IFAD (2010). *Rural Poverty Report 2011—New Realities, New Challenges: New Opportunities for Tomorrow's Generation.* International Fund for Agricultural Development, November 2010, Rome, Italy.

IPPC (2007). *Synthesis Report, Summary for Policymakers.* Intergovernmental Panel on Climate Change, UNFCCC.

ISAAA (2009). *Burkina Faso Farmers Gaining from BT Cotton.* 27 November 2009, International Service for the Acquisition of Agri-biotech Applications, Ithaca, NY.

IWMI (2007). Comprehensive Assessment of Water Management in Agriculture, 2007. *Water for Food, Water for Life: A Comprehensive Assessment of Water Management in Agriculture.* International Water Management Institute. London and Colombo: Earthscan.

James, C. (2009). 'Global status of commercialized biotech/GM crops: 2009', *ISAAA Brief* No. 41. ISAAA: Ithaca, NY.

Lal, R. (2004). 'Soil carbon sequestration to mitigate climate change', *Geoderma,* 123, p. 1.

Mellor, J. W. (1995). 'Introduction', in Mellor, J. W. (ed.), *Agriculture on the Road to Industrialization.* Baltimore, Md: John Hopkins University Press.

Mucheru-Muna, M., P. Pypers, D. Mugendi, J. Kung'u, R. Merckx, and B. Vanluawe (2009). 'A staggered maize–legume intercrop arrangement robustly increases crop yields and economic returns in the highlands of Central Kenya', *Field Crops Research,* 115, pp. 132–9.

Nagayets, O. (2005). 'Small farms: current status and key trends', Information Brief, Research Workshop on The Future of Small Farms, Organized by IFPRI, Imperial College and ODI, Wye, June 2005.

Nair, P. K. R. (1988). 'Production systems and production aspects: The International Council for Research in Agroforestry (ICRAF)', Main Paper, presented at the Expert Consultation on Forestry and Food Production/Security, Trivandrum and Bangalore, India, 8–20 February.

Oldeman, L., R. Hakkeling, and W. Sombroek (1991). *World Map of the Status of Human-induced Soil Degradation.* Wageningen: International Soil Reference and Information Centre (ISRIC) and UNEP.

Osbahr, H. et al. (2008). 'Effective livelihood adaptation to climate change disturbance: Scale dimensions of practice in Mozambique', *Geoforum,* 39, pp. 1951–64.

Pingali, P. (2010). *Presentation 'Who is the smallholder farmer?'* Norman E. Borlaug International Symposium, The World Food Prize, 13–15 October 2010, Des Moines, Iowa.

Rerkasem, B. (2007). Having Your Rice and Eating It Too: A View of Thailand's Green Revolution. *Science Asia*, 33, pp. 75–80. Retrieved from: <http://edition.cnn.com/2009/TECH/science/09/04/food.biodiversity/index.html?iref=intlOnlyonCNN>.

Scoones, I. and J. Thompson (1994). *Beyond farmer first: rural people's knowledge, agricultural research and extension practice.* London: Intermediate Technology Development Group.

Scoones, I. and J. Thompson (2009). *Farmer First Revisited, Innovation for Agricultural Research and Development.* Oxford: ITDG Publishing.

Soemarwoto, O. and Conway, G. R. (1991). 'The Javanese homegarden', *Journal for Farming Systems Research and Extension*, 2, pp. 95–117.

UNDP (United Nations Development Program) (2006). *Human Development Report 2006. Beyond Scarcity: Power, Poverty and the Global Water Crisis.* United Nations, New York: Palgrave-Macmillan.

UNECA (2009). *Economic Report on Africa 2009. Challenges to Agricultural Development in Africa.* United Nations Economic Commission for Africa and African Union, Addis Ababa, Ethiopia.

UN REDD (2009). 'UN REDD programme'. Retrieved from: <http://www.un-redd.org/>.

Wiggins, S. (2009). 'Can the smallholder model deliver poverty reduction and food security for a rapidly growing population in Africa?', FAC Working Paper No. 08, Overseas Development Institute, London.

Woomer, P. L., M. Lan'gat, and J. O. Tungani. (2004). 'Innovative maize–legume intercropping results in above- and below-ground competitive advantages for understorey legumes', *West Africa Journal of Applied Ecology*, 6.

World Bank (2006). *Reengaging in Agricultural Water Management: Challenges and Options.* Washington, DC: World Bank.

World Bank (2007). *World Development Report, 2008*, Agriculture for Development. Washington, DC: World Bank.

Young, A. (1999). 'Is there really spare land? A critique of estimates of available cultivable land in developing countries', *Environment, Development and Sustainability*, 1, pp. 3–18.

3 Right place, right time

The state of smallholders in agriculture in sub-Saharan Africa (SSA)

GEOFFREY LIVINGSTON, STEVEN SCHONBERGER, AND SARA DELANEY[1]

1. Introduction

This chapter provides a regional canvas for the broader discussion of the future directions for smallholders in agriculture. A comprehensive overview of all the opportunities and challenges associated with smallholder farming in sub-Saharan agriculture is not attempted; rather, the intention is to communicate the richness and complexity of the continent in comparison with other developing regions, and through assessing the role of smallholder farmers in agricultural growth, focus discussion on some of the key issues which, from the perspectives of the International Fund for Agricultural Development (IFAD) projects in SSA, are particularly relevant for assisting smallholder families to escape poverty through the transition towards 'farming as a business'.

Section 2 provides a brief look at the recent history and current state of agriculture in SSA, including farm size and land distribution, land quality, agricultural production, and trade. Section 3 examines the opportunities for SSA's smallholders, adapting the perspective of IFAD's Rural Poverty Report (IFAD, 2010) to the regional context. We use a risk management lens and focus on business risks, to connect local ecological and market contexts to the specific endeavours of smallholder farmers. In Section 4, the focus is on an issue which merits much greater consideration—the importance of spatial and temporal coordination in reducing risk, increasing returns, and allowing for project success. Some key recommendations on how these ideas can be transformed into an operational approach are provided in the concluding section.

[1] The authors thank Atiqur Rahman for his comments and inputs; they also acknowledge the comments of the anonymous referee for useful comments made on an earlier draft of the chapter.

2. Smallholder farming—sub-Saharan Africa in perspective

2.1 FARM SIZE IN SUB-SAHARAN AFRICA

According to available data, most farmers in sub-Saharan Africa make their living on farms of around 2 hectares (ha) or less, and have been doing so for decades. The average farm size in SSA is estimated to be between 1.6 and 2.4 ha, which is larger than the average for East Asia (1.0 ha) and South Asia (1.6 ha). It is, however, significantly smaller than average holdings in both Central America (10.7 ha) and especially South America (111.7 ha) (Eastwood et al., 2009; Deininger and Byerlee, 2011). And, while Asia is home to by far the greatest number of small farms, with an estimated 87 per cent of all farms worldwide less than 2 ha being in the Asian region, SSA is the region which has the highest percentage of its farms under 2 ha, falling at somewhere between 70 and 85 per cent. This of course varies across countries: for example, 97 per cent of farms in the Democratic Republic of Congo are estimated to be less than 2 ha, 74 per cent in Nigeria, but only 25 per cent in Botswana (Nagayets, 2005; Eastwood et al., 2009; Anriquez and Bonomi, 2007).[2]

Trends in agrarian structure are particularly difficult to ascertain from existing data; however, it appears that there has been little change in average farm sizes over the past 30 years in most countries in SSA. Data on farm sizes in SSA is unfortunately quite sparse, with reliable statistics available for less than half the countries in the region, and many agricultural censuses have not been updated since the 1960s or 1970s (Eastwood et al., 2009). This makes it difficult to draw any detailed conclusions with confidence, or to get a firm grasp on some of the bigger trends in agricultural land tenure over recent years. If one looks at the agricultural land per agricultural population in each country (a reasonable proxy for average farm size, although not perfect), the data shows stagnant or very slightly decreasing cultivated hectares per person since 1980. In some countries, however, agricultural land per person has decreased more dramatically, such as in Zambia and Mozambique where it has dropped from more than 4.5 ha in 1980 to less than 3 ha in 2008 (Figure 3.1). In select cases, largely due to changes in land distribution policies, land per person has increased, such as in South Africa where it grew from 17 ha in 1990 to nearly 21 ha in 2009. Figure 3.1 also shows these estimated trends for all the regions of Africa. The evidence suggests a

[2] As in this book we also use the 2 ha cut-off when citing statistics on smallholder farms in this chapter, as this was the definition used for most available data. We acknowledge that this cut-off is arbitrary, and while appropriate for most countries in SSA, may not capture all smallholders in the region, or be applicable in other regions.

declining trend of agricultural land per capita, both for countries with small land per capita as well as for those with larger land per capita (Mozambique, Zambia).

The equality of land distribution is relatively more equal in SSA than in other developing regions. The majority of agricultural land is cultivated by smallholders, and in most countries land ownership is not overly concentrated in the higher income levels. In DR Congo, for example, 97 per cent of farmers are smallholders and they cultivate 86 per cent of the land. In Ethiopia this changes to 87 per cent smallholders cultivating 60 per cent of land, but there are exceptions like Uganda, where a low 27 per cent of land is cultivated by smallholders who make up 75 per cent of farmers.

A large percentage of these SSA smallholders are women, responsible for key components of household production such as weeding, harvesting, and processing. Further, women often independently grow rice or non-cereal crops for income and are increasingly heading rural households due to male urban migration (Oxfam, 2008).

This compares to more uneven land distribution in Asia and the Middle East and North Africa (MENA). In India, for example, 80 per cent of farmers are smallholders but they cultivate only 36 per cent of the land. In the Latin America and Caribbean (LAC) region, large farms are much more dominant, making the 2 ha cut-off less relevant. However, it can still serve to illustrate the disparity of land holdings in the region—in Brazil, for example, smallholders make up 20 per cent of farmers, but only cultivate 1 per cent of land (Anriquez and Bonomi, 2007).

The size of land holding is of course only one factor of many which come together to determine the productivity and profitability of the land to farmers. As we will discuss, factors such as land quality, location relative to infrastructure and markets, crop and input choice, field and water-management practices, and market demand and price can mean the difference between 2 ha being too little to sustain a small family, and being sufficient to support a thriving enterprise and compete with urban wages.

2.2 AVAILABILITY AND QUALITY OF LAND

While the land available per farmer has remained largely stable in most SSA countries, the amount of land being cultivated has steadily increased along with increases in the agricultural population. In contrast to other regions, increases in agricultural production have occurred largely through expansion of the cultivated area on to the region's relatively abundant land, rather than increases in land productivity (Figure 3.2).

The pursuit of an extensification strategy by SSA's farmers reflects the availability and lower costs of land relative to capital inputs required for intensification, such as credit, fertilizer, and irrigation. There is a further 800 million hectares (Mha) of uncultivated land with rain-fed crop production potential in SSA and 850 Mha in LAC. In comparison, there is almost no available land in South Asia, East Asia, or North Africa (FAO, 2009a).

However, the challenges associated with bringing these areas into production vary considerably across SSA, and some of the land recently brought under cultivation is of marginal quality. The region is home to a large diversity of agro-ecological climates, ranging from the arid drylands of northern Mali, to the humid tropics of the Congo. Figure 3.3 shows SSA divided into six agro-ecological zones, differentiated by the length of the potential growing period for rain-fed agriculture. Rainfall ranges dramatically, from over 2,000 mm/year in central Africa to less than 400 mm/year in arid areas (Bationo et al., 2006).

SSA also has a wide diversity of soil types, differing dramatically in their ability to retain and supply nutrients to plants, to hold or drain water, to withstand erosion or compaction, and to allow for root penetration (Figure 3.3, Bationo et al., 2006).

Many of the soils have suffered severe losses in nutrients, biodiversity, and structure over the years due to unsustainable farming practices. This impacts greatly on the productive capacity of the soils and therefore farmer incomes. The International Fertilizer Development Center (IFDC) has estimated that SSA loses around 8 million tonnes of soil nutrients per year, and that over 95 million hectares of land on the continent has been degraded to the point of greatly reduced productivity. During the 2002–4 cropping season over 80 per cent of countries in Africa[3] were estimated to be losing more than 30kg of nutrients per year, and 40 per cent of countries an astounding 60kg or more per year (Henao and Baanante, 2006).

In addition to land quality factors, the costs of bringing new lands under cultivation in SSA are greatly affected by access to markets, electricity, and communications. Infrastructure in SSA, while improving in some areas, remains a major constraint relative to other regions (Table 3.1). Road condition and density are very low, as we will expand upon in Section 3. Electricity generation capacity has remained stagnant since the 1980s and now averages only 37MW/million people, as compared to an average of 326MW across other low-income regions (Foster and Briceño-Garmendia, 2010). Costs in much of SSA are also significantly higher, averaging 14c/kWh, as compared to 4c/kWh in East Asia and Pacific (EAP) or to 1c/kWh in South Asia, and access is unreliable, with only South Asia experiencing more outages. Many small

[3] In this chapter we focus on sub-Saharan Africa. When the term 'Africa' or 'the continent' is used it is because the statement refers to the continent as a whole, rather than SSA.

Table 3.1 Infrastructure

Region	Paved roads (per cent) (World Bank 2012)	Road density (km^2 of road/surface area) (World Bank, 2009a)	Access to electricity (per cent) (World Bank, 2009a; Teravaninthorn and Raballand, 2008).	Telephone, mobile and fixed (per 100 people) (World Bank, 2009b)
EAP	11	0.72	89*	75
MENA	76	0.33	78	74
LAC	22	0.12	90	99
SAR	57	0.85	52	36
SSA	12	0.13	26	35

*Excluding China. SAR = Southeast Asia Region.

agricultural processing businesses must rely on small diesel-powered generators, with which electricity can cost up to 0.40kWh (Foster and Briceño-Garmendia, 2010). Ten years ago telephone access in SSA was much lower than other developing regions. However, exponential increases in mobile phone use, from 650,000 in 1995 to over 330 million in 2010, have now put the region on a par with South Asia, and on a path for continued expansion in communications connectivity (International Telecommunications Union, 2013).

2.3 AGRICULTURAL PERFORMANCE

Sub-Saharan Africa's rural economy remains strongly based on agriculture. Agriculture in SSA (excluding South Africa) employed 62 per cent of the population and generated 27 per cent of the GDP of these countries in 2005 (Staatz and Dembele, 2007) (Table 3.2). And this production is powered more by smallholder farmers than any other region. Smallholder farms contribute up to 90 per cent of the production in some SSA countries (Nagayets, 2005; Wiggins, 2009). This is similar to many countries in South and East Asia, but much higher overall than in LAC or MENA.

The growth in agricultural GDP in SSA has been relatively strong in recent decades, and was the highest of the developing regions in 2009 (Table 3.3). Overall, increases in agricultural production have kept pace with population growth (FAO, 2012; IFAD, 2010).

Looking at the *net production value per hectare*, which compares the value of production of all agricultural products against a base year (after taking out that used for feed or seed), one can see that while SSA overall has made slower progress compared to others, there is quite a wide degree of variation across the continent. In fact, the more densely populated West Africa region has progressed at a rate slightly greater than that of LAC (Figures 3.4a and 3.4b).

Breaking this down further, there is considerable variation from country to country. For example, Figure 3.5 shows productivity in more crowded Malawi

Table 3.2 Agriculture

	Agriculture value added (% GDP)	Agricultural employment (%)
	2008	2007
SSA	16 (27*)	46 (62*)
APR	13	44
LAC	6.5	12

*Figures excluding South Africa, for 2005

Source: World Bank (2010).

Table 3.3 Growth of agricultural value added (per cent)

	1990s	2000s	2009
SSA	2.7	3.0	4.8
APR	3.6	3.9	3.8
LAC	2.1	3.3	1.5

Source: World Bank (2010).

Table 3.4 Imports of and exports from Africa, 2010

	Imports of Africa by source (% of imports)	Exports of Africa by destination (% of exports)
Europe	42	40
Asia	26	22
North America	7	17
Africa	12	12
Middle East	8	3
Other	5	6

Source: WTO (2010).

climbing since the 1960s, even before the recent fertilizer subsidy push, as compared to slow progress in less densely populated Burkina Faso and Kenya, and stagnation until recently in Senegal.

2.4 TRADE

Africa's share in world trade is proportionally very small, accounting for only about 3 per cent of world exports and imports, shares similar to Latin America, but dwarfed by all other regions. Europe is currently Africa's largest trading partner; however, trade with Asia and other developing regions has been steadily increasing (Table 3.4 and Figure 3.6).

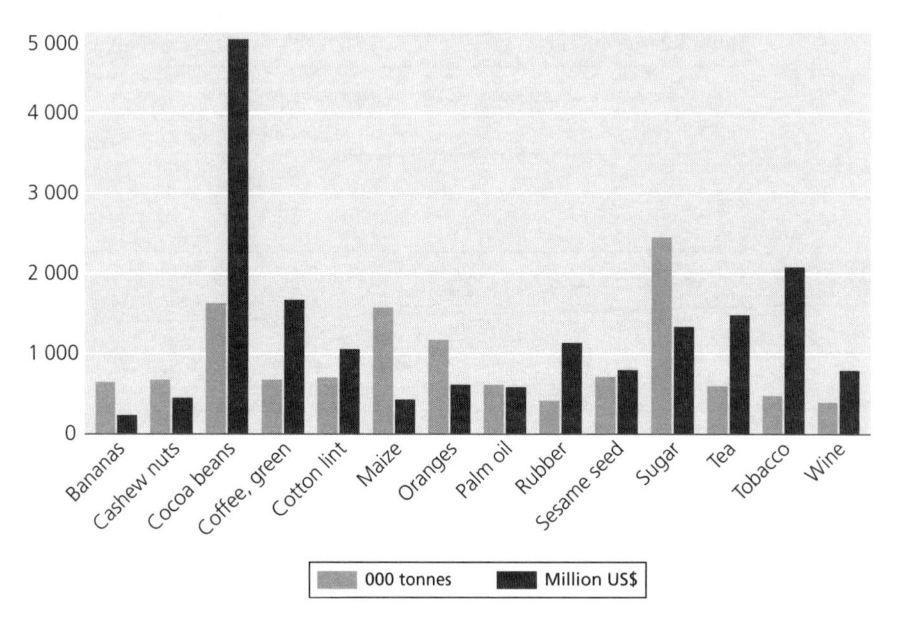

Figure 3.7a Major exports from sub-Saharan African countries: 2010

Source: FAO (2012).

Exports are dominated by fuels (55 per cent), with manufactures (19 per cent), agricultural products (10 per cent), and mining products (9 per cent) making up the remainder (WTO, 2010). Agricultural exports are primarily high-value cash-crops such as cocoa, sugar, coffee, tea, cotton, and oranges (Figure 3.7a). Agricultural imports are primarily basic food items (Figure 3.7b). Wheat is the largest food import to SSA, at 8.6 million tonnes in 2008, owing to the steady increase in demand for bread, particularly in urban areas. Wheat can only be grown on about 1 per cent of the land or 24 Mha in SSA, mostly in the highlands in East Africa (FAO, 1991; Morris and Byerlee, 1993). However, as seen in Figures 3.7a and 3.7b, countries in SSA import large quantities of commodities which could be supplied locally, including rice, maize, sugar, palm oil, and soybeans.

Intra-regional trade is also very low compared to other regions. Total intra-regional trade is only 12 per cent of total trade to and from SSA countries as compared to 52 per cent in Asia and 26 per cent in LAC. The main traded commodities between SSA countries are fuel and mining products, with intra-region agricultural product export totalling only US$8 billion or 18 per cent of officially recorded agricultural exports. However, much informal trade in the region is unrecorded (WTO, 2010).

These comparably lower trade numbers can partially be explained by the difficulty of doing business in most SSA countries. While there has been some

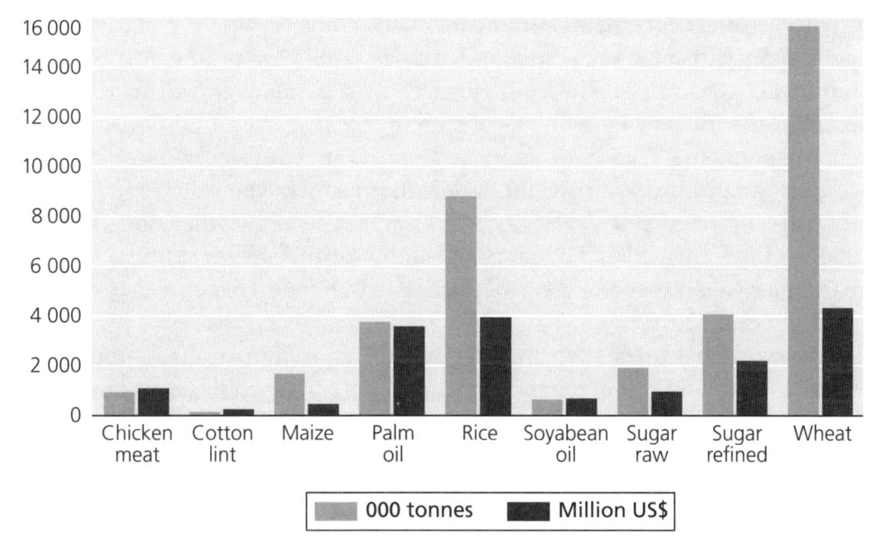

Figure 3.7b Major imports of sub-Saharan African countries: 2010

Source: FAO (2012).

improvement over the years, the countries of SSA account for 9 of the bottom 20 in the Transparency International corruption perception rankings, and 15 of the bottom 20 of the IFC Doing Business Rankings. Starting a business still costs 18 times as much (relative to income per capita), on average, in sub-Saharan Africa as in OECD economies. This encourages firms to remain informal, resulting in disincentives for growth and job creation, and limits the ability of governments to increase their fiscal revenues. At the same time, however, several countries, including Rwanda and Ghana, have improved their standing on both indices substantially, and in 2010 Rwanda was highlighted as a global best performer in improving its business climate (Transparency International, 2010; IFC, 2011).

3. **Opportunities for sub-Saharan Africa's smallholders**

The agricultural sector faces growing global and regional demand for agricultural products for food, feed, industry, and fuel. Continued population and income growth, combined with urbanization, particularly in developing countries, is placing pressure on current food supplies while global productivity increases are levelling off. At the same time, geopolitical and environmental concerns are placing increased emphasis on the replacement of petroleum with renewable sources, such as crops, for production of fuels, lubricants, and fibres. The consequences of this rapid growth in demand, combined with slowing scope for a supply response from traditional producing regions, has resulted in increased sensitivity of agricultural markets to supply variations due to weather and other factors. The resulting tendency towards increased price volatility was clearly observed in the context of the 2007–8 food price crisis. Higher prices and volatility are forecast to continue, particularly because of expected impacts of climate change (for example, Godfray et al., 2010).

The world has turned to sub-Saharan Africa, given its relatively abundant, uncultivated land resources and unrealized potential productivity gains, as a major source of future supply and stability for food and industrial agricultural markets (FAO, 2009a; World Bank, 2009b). At the same time, SSA's governments, recognizing the need to feed an increasingly urbanized population, as well as the opportunity to develop agro-processing industries, are also focused on rapidly increasing agricultural production. This was highlighted at the African Union Congress in 2004 where Heads of State committed to achieving an average 6 per cent annual growth rate in agriculture, supported by a minimum of 10 per cent allocation of public expenditures to the agricultural sector (CAADP, 2011). Allocations have so far been below commitments, however, with the average budget allocation across SSA for agriculture at 9 per cent, and only seven countries in the region meeting the 10 per cent target as of 2009 (RESAKSS, 2010).

While expansion of large, plantation-type operations will likely account for some of SSA's supply response, there remains significant scope for smallholders, and smallholder farmer organizations, to increase their role as commercial suppliers.

The circumstances which favour plantations are theoretically and historically limited to supply chains which require careful timing and rapid transfer of harvests from fields to processing facilities, such as sugar cane, or to areas where governments wish to transfer the onus for significant investments in infrastructure in remote areas to the private sector (Hayami, 2004). However, even in the case of traditional plantation crops such as pineapple, cotton, sugar cane, etc., the labour management challenges of large-scale operations are stimulating the increasing use of alternative institutional approaches which maintain the central role of smallholders, such as nucleus estates and public–private partnerships (Deininger and Byerlee, 2011). Through these partnerships, infrastructure development costs are shared amongst government, farmer organizations, and processors/exporters, often in the context of development projects (cf IFAD Sao Tome). Increasing the number and strength of farmer organizations across the region can help to increase further the scope for these types of partnerships.

SSA's farmers are particularly well positioned to be significant beneficiaries of the institutional innovations which are improving the commercial opportunities for smallholders in agricultural markets. The theoretical efficiency advantages of smallholder production systems for most crops are well known (Binswanger and Rosenzweig, 1986), and these apply particularly in countries with relatively high capital costs relative to labour (Hazell et al., 2007), which is the situation in most parts of SSA. This theoretical advantage of smallholder farming in SSA is confirmed empirically in the World Bank's Awakening Africa's Sleeping Giant study (World Bank, 2009c), which concluded that smallholder production costs at the farm gate for several key crops are competitive with other regions, despite lower productivity, making them competitive suppliers in local markets. For example, Nigerian soybean producers can supply Ibadan markets at 62 per cent of the cost of imports, and Zambian sugar farmers can supply Nakambala markets at 55 per cent of the cost of imports.

3.1 FROM EXTENSIFICATION TOWARDS INTENSIFICATION

While sub-Saharan Africa's relatively abundant, uncultivated arable land suggests significant scope for expansion, this is facing limits, which is increasing the ratio of the cost of land to capital inputs faced by farmers. In particular, efforts to develop these lands through large-scale concessions—particularly by foreign investors—have made it clear that 'uncultivated' does not mean 'unused'.

There are widely varying estimates of the number of hectares transacted and of how much of this is actually being developed for plantation operations. A report from the International Land Coalition indicates that transactions involving nearly 20 million hectares of land in 15 countries across Africa have been documented, with 5 million hectares covered by already signed contracts (Odhiambo, 2011). This reflects the emphasis being given by many African governments to large-scale plantations as a means to rapidly modernize agriculture.

Expansion must take into account impacts on existing economic uses—such as grazing and woodsheds, as well as social and ecological functions whose importance often becomes apparent when efforts are made to convert the land. Importantly, and as discussed in more depth in Section 3, these areas also often suffer from very limited access to infrastructure or market centres (FAO, 2009a).

In order for smallholders to increase production with less additional land and without major increases in labour inputs, they will need to increase productivity through greater capital and technology investments. While there is some scope for increasing the labour intensity of agriculture, given the growing, young population profile, there is little evidence that this can be realised on a broad scale. Youths are much more inclined to seek urban and even rural employment opportunities which offer perceived higher returns for effort than extensive agriculture—particularly given traditional difficulties faced by youth in obtaining land. Attracting youth to agriculture as a livelihood is likely to coincide with increased intensification which improves returns per unit of labour and land. The Sleeping Giant study, along with others, concludes that

> [current farm-level] competitiveness does not represent a sustainable path out of poverty, because at current productivity levels and farm size, agriculture is economically impoverishing and technically unsustainable. The challenge facing African countries is to invest in developing a more sustainable, productivity-driven base for competitive commercial agriculture over the long-run. (World Bank, 2009c; McIntyre et al., 2009)

The smallholder supply response will require increased on-farm investments, such as appropriate seeds and fertilizers, irrigation and mechanization technologies, and reductions in post-harvest losses (PHL). SSA remains well behind other regions in the use of improved seeds and fertilizers. On average, farmers in SSA apply less than 10kg of nutrients/ha, compared to around 140kg/ha in both Latin America and South Asia (World Bank, 2010). Use of high-quality seed is also much lower than it could be, with surveys for staple crops in West Africa indicating improved varieties accounting for only 2–33 per cent of seed, and renewal of seed stock occurring only every 9–13 years. From 1997 to 2007 in West Africa, there was only enough improved maize seed to meet one-third of farmer demand (Ndejeunga and Bantilan, 2002; AGRA, 2011; Adesina, 2010).

Productivity improvements will also require improved water management and greater use of irrigation. While there has been a steady increase in the amount of agricultural land irrigated worldwide in the last 50 years, this has mostly occurred in Asia, where irrigated land has increased from 27 per cent to around 36 per cent. In contrast, only 11 per cent of land is irrigated in LAC, and less than 3 per cent in SSA (IFAD, 2010). Many countries in Asia are now irrigating close to their full potential, while in Latin American countries the percentage of potential used is lower (5–35 per cent) (FAO, 2010). While there is considerable potential to expand irrigation in SSA, opportunities vary greatly across the region, due to differences in rainfall, renewable water resources, and land. While some areas have high irrigation potential, they also receive abundant rainfall, making irrigation less crucial; others receive less rain, but also have less water to draw from. One-third of the potential on the continent is concentrated in two very humid countries: the Congo and Angola (FAO, 2005). The crops that are irrigated in SSA are mainly cash crops; whereas 40 per cent of cash-crop production in SSA comes from irrigated systems, only 15 per cent of cereal production does (Dorosh et al., 2010).

Post-harvest losses represent another key area for improving sectoral efficiency. Once crops are harvested, many farmers in SSA suffer significant losses from grain shattering, spillage during transport, and from bio-deterioration during each step of the supply chain, including storage. Losses in the East and Southern Africa regions, for example, have ranged from 14–17 per cent each year from 2003–9 (weighted average of all cereals) (PHL Network, 2010). However, relatively low-cost storage and transport facilities and protocols are increasingly available in forms and at prices accessible to smallholders based on innovations from Southeast and South Asia (World Bank/FAO, 2010).

Moving from traditional extensification approaches to capital-based intensification requires a greater range of business management skills. Under extensification, the primary resources used are land, labour, and locally available inputs which do not require access to broader markets or credit and which are generally well known to the farmer. Intensification of production requires farmers to make more nuanced decisions on the purchase and use of improved seeds or fertilizers, and the utility of initiating more costly mechanization or irrigation methods. They must also seek credit if necessary and expose themselves to price fluctuations for both inputs and the sale of their produce. SSA's smallholder farmers therefore must become more sophisticated business managers, better able to calculate and manage the risk–return trade-offs of on-farm investments, if they are to seize fully the opportunities of increased global and regional demand for agricultural products.

3.2 THE ROLE OF RISK MANAGEMENT IN FARMING AS A BUSINESS

The ability of SSA's smallholder farmers and farmer organizations to increase on-farm investments in land and labour productivity is constrained by their ability to manage the risk–return trade-offs in moving towards intensified agriculture. IFAD's 2011 Rural Poverty Report (RPR) has highlighted the role of risk in inhibiting smallholders from pursuing commercial agriculture opportunities. The RPR clearly demonstrates that given the precariousness of their livelihood, poor rural families are highly risk-averse and are therefore less inclined than non-poor groups to move up the 'risk–return' ladder towards potential higher incomes, contributing to the growing income disparities we see in developing countries. These risks cover a broad range from social to political to natural resources to business (Figure 3.8).

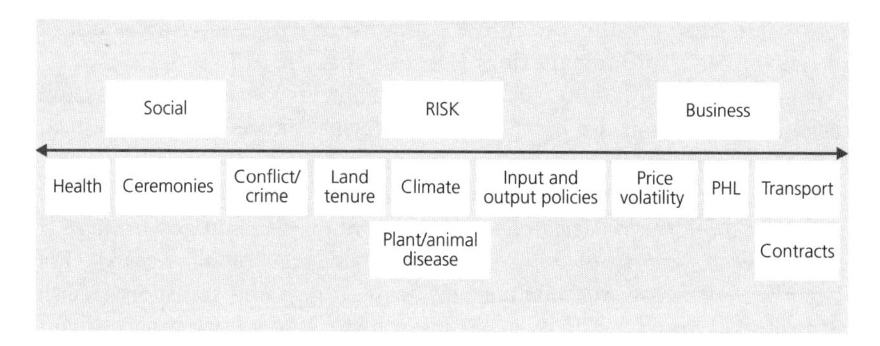

Figure 3.8 Risks faced by smallholders

In the context of SSA, these risks vary significantly between the different agro-ecological zones. As in other regions, it is difficult to generalize risks faced by smallholders given the diversity of farming and marketing systems. In order to capture both the risks and their diversity, we have provided an initial categorization according to agro-ecological systems whose characteristics, along with the marketing systems and market access factors discussed here, drive much of the risk profile faced by smallholders. Table 3.5 shows the principal production zones and associated risks.

While risks associated with health, social obligations, and conflict are fairly consistent across all types of marketing systems, those associated directly with marketing and other aspects of farming as a business are often supply-chain specific, and it is important to focus on the implications these pose to smallholders attempting to move up the risk–return ladder. Table 3.6 summarizes, in broad terms, the different types and characteristics of marketing systems with which smallholders generally work in SSA, and the implications in terms of risks.

Generally speaking, smallholders in dispersed marketing systems are exposed directly to a larger number of business risks, as well as realizing a lower share of returns, while in integrated markets, there is generally a greater level of risk sharing amongst supply-chain actors, as well as a higher share of benefits to producers. While there remains a need for more rigorous evaluation of the relative impacts on livelihoods of participating farmers, experience under IFAD-financed projects which aim to move farmers towards greater market integration (such as for potatoes in Guinea, palm oil in Uganda, cocoa in Sao Tome and Sierra Leone, and coffee in Rwanda) has generally confirmed significant, positive impacts on both the level and stability of incomes of participating smallholders (Raswant and Khanna, 2010).

Table 3.5 Characteristics of agro-ecological zones in sub-Saharan Africa

	Length of growing period (days)	Average rainfall (mm)	Farming systems	Land area (% of SSA)	Agricultural population	Main soil types	Agricultural products	Risks
Arid	<90	0–600	Pastoral	14%	27 million (7%)	Calcisols, arenosols, leptosols	Cattle, camels, sheep, goats.	Health, Social Obligations, Conflict, Crime, Corruption, Drought, Animal Diseases, Prices
Semi-arid	90–179	600–1400	Agro-pastoral	8%	33 million (8%)	Lixisols, arenosols, vertisols	**Sorghum, millet**: with pulses, sesame. Cattle, sheep, goats, poultry.	Health, Social Obligations, Conflict, Crime, Corruption, Land Tenure, Climate, Soils, Animal/Plant Diseases and Pests, Prices, PHL, Transport
Sub-humid	180–269	1400–3000	Mixed cereal/ root-crop	13%	59 million (15%)	Ferralsols, lixisols, plinthosols	**Maize, sorghum, millet**: and **cassava, yams, legumes**, cattle	Health, Social Obligations, Conflict, Crime, Corruption, Land Tenure, Climate, Soils, Animal/Plant Diseases and Pests, Prices, PHL, Transport
			Mixed maize	10%	60 million (15%)	Ferralsols, lixisols, plinthosols	**Maize**: with tobacco, cotton, cattle, goats, and poultry	Health, Social Obligations, Conflict, Crime, Corruption, Land Tenure, Climate, Animal/Plant Diseases and Pests, Soils, Input and Output Policies, Prices, PHL, Transport, Contracts
Humid	>270	3000–4500	Root crops	11%	44 million (11%)	Ferralsols, lixisols, acrisols	**Yams, cassava, legumes**, cattle	Health, Social Obligations, Conflict, Crime, Corruption, Land Tenure, Climate, Animal/Plant Diseases and Pests, Prices, PHL, Transport
			Tree crop	3%	25 million (7%)	Ferralsols, acrisols	**Cocoa, coffee, oil palm, rubber**: with yams, maize	Health, Social Obligations, Conflict, Crime, Corruption, Climate, Plant Diseases and Pests, Input and Output Policies, Prices, PHL, Transport, Contracts

(continued)

Table 3.5 Continued

	Length of growing period (days)	Average rainfall (mm)	Farming systems	Land area (% of SSA)	Agricultural population	Main soil types	Agricultural products	Risks
			Forest-based	11%	28 million (7%)	Ferralsols, acrisols	**Cassava**: with maize, sorghum, beans, and cocoyams.	Health, Social Obligations, Conflict, Crime, Corruption, Climate, Plant Diseases and Pests, Input Policies, Prices, PHL, Transport
Highlands	180–>270	1400–4500	Highland Perennial	1%	30 million (8%)	Andosols, cambisols	**Banana, plantain, enset, coffee**: with cassava, sweet potato, beans, cereals. **Cattle**.	Health, Social Obligations, Conflict, Crime, Corruption, Land Tenure, Climate, Animal/Plant Diseases and Pests, Input Policies, Prices, PHL, Transport, Contracts
			Highland Temperate	2%	28 million (7%)	Andosols, cambisols	**Wheat and Barley**: with peas, lentils, broad beans, rape, tef, and potatoes. **Cattle**.	Health, Social Obligations, Conflict, Crime, Corruption, Climate, Animal/Plant Diseases and Pests, Input Policies, Prices, PHL, Transport, Contracts

Note: Prepared by the authors based on Dixon et al., (2001) and Bationo et al. (2006)

Table 3.6 Indicative risks associated with smallholder supply chains

Marketing system	Typical products	Characteristics of supply chain	Risks for smallholder
Highly integrated	Exports of high value for processing in specialized markets Ex: (organic and fair trade cocoa, coffee, oil palm, cotton, honey) and in fresh markets (flowers, fruits, vegetables)	Producer share: High (producers receive 60 to 80% of export/processor price) Structure: Highly structured/integrated: Lead firm directly manages chain back to individual producer or co-op to ensure quality requirements and certification; usually involves contract with international trade, value-added and/or quality standards specified. High level of interlinking.	Failure to meet quality standards Input and trade policy Climate
Integrated with intermediary	High value for domestic/ regional fresh and processed markets Ex: (dairy, eggs, fruits, vegetables, meat) and exports for processing (conventional coffee, cocoa, oil palm, cotton)	Producer share: Medium (producers receive 40 to 75% of export price) Structure: Structured with one or two local aggregators who transmit/enforce quality standards between producers and lead firm—often contract and/or informal credit. Some interlinking. Some cases of co-ops integrating chain.	Failure to meet quality standards Contract enforcement Price volatility Input and trade policy Climate
Dispersed	Low value domestic/regional staple crops and biofuel stocks Ex: (cassava, rice, corn, millet, sorghum)	Producer share: Low (producers receive 15 to 50% of price to processor or consumer) Structure: Absence of lead firm—unstructured, spot market transactions with multiple channels and numerous intermediary transporters/aggregators—limited or no transmission of quality standards to participants in supply chain. Absence of long-term investments or inter-linked market relationships (weak access to private input markets)	Post harvest losses Transport delays and costs Price volatility Input and trade policy Climate

Note: Prepared by the authors based on Pingali and Rosengrant, 1995, Swinnen et al. (2007), and USAID (2009).

3.3 HELPING SSA'S SMALLHOLDERS MANAGE THE RISKS OF FARMING AS A BUSINESS

The potential benefits for smallholders of participation in more integrated supply chains are well recognized by sectoral stakeholders. Farmer organizations, such as the Network of Farmers' and Agricultural Producers' Organisations of West Africa (ROPPA), are now focused on helping members to become competitive suppliers of high-value products for domestic markets, such as rice, meat, dairy, and vegetables. African Governments have placed increased attention on revitalizing the role of smallholders in production of export crops and domestic import substitutes, as reflected in the Comprehensive Africa Agricultural Development Programme (CAADP) investment plans. Support for improved supply-chain integration, market access, and commercialization is prominent in 11 out of the 15 currently completed plans (CAADP, 2011).

Donor financing is increasingly focused on helping smallholders gain capacity to participate in more integrated supply chains through investments in irrigation, roads, rural finance, research, weather insurance, inputs, and farmer organization (OECD, 2011). These investments are intended to reduce both the costs and the risks faced by smallholders with intensification of production so as to become more efficient suppliers of markets.

However, evidence on the ground is highlighting that these investments, of themselves, are often not producing the level of results expected. This is seen in the case of irrigation and roads in SSA, where assessments have demonstrated lower returns relative to projects in other regions (Inoncencio et al., 2007; World Bank, 2009a), and in lower adoption rates for improved seeds and fertilizer (Ndejeunga and Bantilan, 2002). While there is a myriad of explanations offered, most emphasizing the weakness of institutions or governance, we suggest that a more focused and practical element merits greater attention: effective coordination in terms of place and timing of development support, as discussed in the next section.

4. Right place, right time—doing a better job of lowering business risk

In seeking to lower risks—particularly business risks—and increase opportunities for smallholder farmers, there seems to be a pronounced tendency to search for new technical solutions while often ignoring the potential to strengthen the impact of existing investments. There is a predisposition among development practitioners, be they academic researchers, officers of international financial institutions, or staff from non-governmental organizations, to look for new solutions to rural poverty alleviation, food security, and income generation. Improved seed varieties, micro-dosing of fertilizers, more crop per drop irrigation schemes, sustainable agronomic practices, innovative financial instruments, and the like all certainly have a role to play in improving the lives of the rural poor in Africa. However, there are potentially enormous gains that can be achieved through improving the spatial and temporal coordination of existing development interventions by simply responding to the two proverbial 'elephants in the room':

> More than one-third of all sub-Saharan rural Africans are so geographically and economically isolated from market towns that, at present, they are virtually condemned to a life of subsistence agriculture, regardless of their access to modern inputs, irrigation infrastructure or financial services.

The majority of development programmes and projects do not deliver goods, services, and works in a timely manner, and this has had an extremely negative impact on productivity and increased revenues of project clients. Successful agricultural campaigns are all about planning and timing. Without rigorous project management, farmers will not gain the full benefits of new technological solutions.

4.1 RIGHT PLACE

Increasing access to markets figures prominently in rural poverty reduction strategies. But the extreme degree of geographical isolation of SSA smallholders is not widely apprehended. According to a spatial analysis undertaken in 2007 (Sebastian, 2008), 34 per cent of the rural population in sub-Saharan Africa live more than five hours from a market town of 5,000 people. As can be seen from Table 3.7, SSA has the greatest percentage of population and the second greatest number of people living five hours or more from a market town.

The density of SSA's road network is low compared to other developing regions and is, in fact, regressing, as there are fewer kilometres of roads today in SSA than there were 30 years ago. There are only around 200 kilometres of road, only 12 per cent of which is paved, per 1,000 square kilometres of land area. Only 34 per cent of Africans live within 2 kilometres of an all-season road, compared to 65 per cent in other developing regions (Dorosh et al., 2010; Foster and Briceño-Garmendia, 2010).

These sobering sub-Saharan statistics mask significant regional differences. The East Africa region has a much lower overall population density, smaller local markets, and lower road connectivity than West Africa. The average travel time to a major city (of 100,000 people) is 2.2 times greater than in West Africa. This large inter-regional difference is attributable to a much less dense secondary and tertiary road network in East Africa, where distances to these smaller roads are 1.8 times further than in West Africa (Dorosh et al., 2010).

Table 3.7 Access to market towns in different regions and SSA

	People living more than 5 hours from a market town of 5,000 people or more		People living less than 1 hour from a market town of 5,000 people or more	
	%	Number in millions	%	Number in millions
SAR	5	45	56	512
EAP	17	188	33	366
LAC	20	26	46	61
MENA	31	23	26	19
Central Asia	32	32	26	26
SSA	34	131	21	81

Source: Teravaninthorn and Raballand, 2008.

So what is the impact of remoteness to market on agricultural productivity?

If a farmer cannot profitably market her surplus, then there is no logical reason to produce more than her family can store and/or consume. There is thus no motivation to adopt productivity-enhancing technologies, particularly those external inputs which are costly and, in any event, are not likely to be available. This intuitive conclusion is borne out by the Africa Infrastructure Country Diagnostic (AICD) study 'Transport Prices and Costs in Africa: A Review of the Main International Corridors' (Teravaninthorn and Raballand, 2008), which estimates that an average farmer is producing at 45 per cent of the theoretical agronomic potential when located four hours from a major city. This percentage drops to 20 per cent when the farmer is six hours away and 12 per cent when eight hours away. So, fully one-third of the rural population in sub-Saharan Africa is sitting on the side-lines with no real chance of participating in the market economy nor improving their economic condition.

What about the other two-thirds of rural sub-Saharan Africans who theoretically are in sufficient proximity to market towns to sell surplus production? The 2007 spatial analysis (Sebastian, 2008) categorized access to market as high (<1 hour), medium (2–4 hours), and low access (>5 hours). There is a significantly lower percentage of rural people in SSA than other regions with high access to market towns, as can be seen from Table 3.7.

The number of rural sub-Saharan Africans who enjoy medium access to market towns is, at 46 per cent, higher than all other regions with the exception of East Asia. However, the significance of distance to market is primarily one of transport costs to receive inputs and deliver products. And here, the relative benefits of medium access to market evaporate in the face of high transport costs. It can be seen from Figure 3.9 that travelling the same distance can cost up to five times in parts of Africa as it does in Pakistan. This, in effect, means that from a cost perspective, a SSA farmer located one hour away from a market town (high access) pays the same for transport as a Pakistani farmer located five hours from a market town (low access). The extremely high cost of transport in Africa dramatically increases the impact of distance to market.

The importance of widely differing transportation costs comes into clear focus when considering the relative cost of agricultural inputs and the opportunities for agricultural import substitution in SSA.

A concrete example is the impact of transport costs on fertilizer prices: differences in transport costs are the most important contributor to the higher prices in SSA countries as compared to Thailand (Figure 3.10). One notes that a tonne of fertilizer is, on average, 80 per cent more expensive in

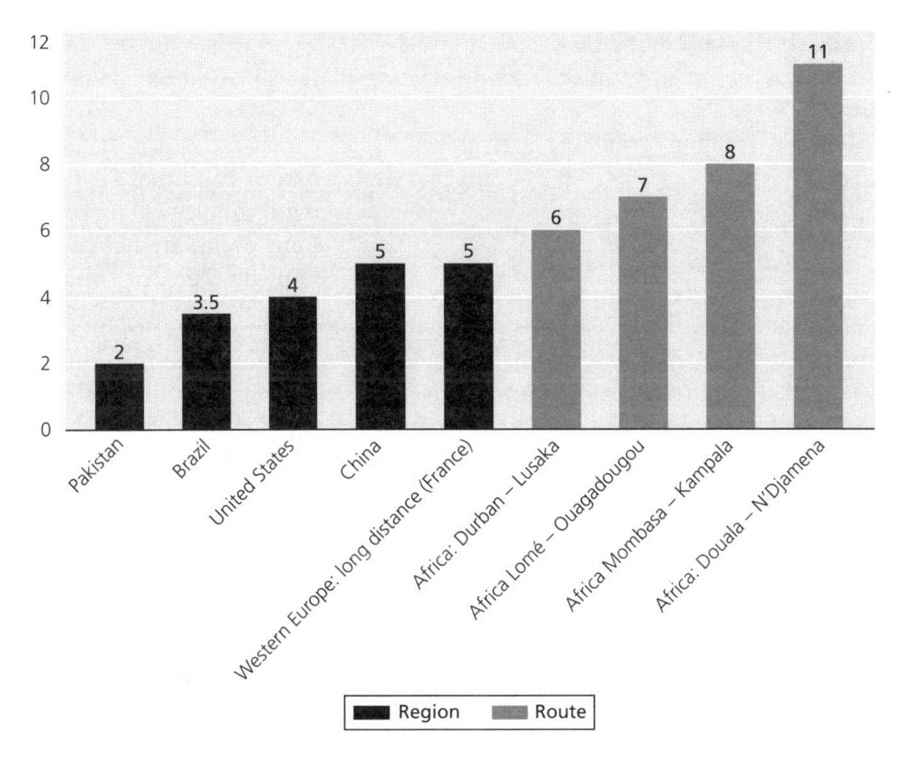

Figure 3.9 Average transport prices: A global comparison in 2007

Source: Teravaninthorn and Raballand (2008).

Mali than in Thailand (Bumb, 2009). Landlocked countries typically must absorb US$50–100 per tonne in additional transport costs to have goods delivered from the nearest port to their own border. The problem is made worse by small market size, fragmentation, and unnecessary product differentiation, making it difficult to achieve the economies of scale needed for more efficient production or import (World Bank, undated).

As can be seen from Figure 3.7b (Section 2), SSA countries import more than 7 million tonnes of rice per year (whole and broken), more than 90 per cent of which is produced in Thailand, China, and Pakistan by smallholder farmers. Rice imports represent almost 40 per cent of total consumption in SSA. In much of the region, locally produced rice cannot compete on a cost basis with product grown in remote rural areas of Asia and transported several thousand kilometres to market, and this, despite similar or lower smallholder farm-gate production costs in Africa (Africa Rice Center, 2008).

The *Sleeping Giant* study identifies very promising opportunities for smallholder import substitution in domestic and regional markets for rice, as well as soybeans, sugar, and maize. These opportunities, however, can only be realized through dramatic decreases in SSA's transportation costs to render them cost competitive in primary destination markets (World Bank, 2009c).

Reasons for high transport prices

The high price of road transport in SSA is widely recognized but the reasons for this are much less widely understood. High prices have generally been attributed to the poor state of the road network and, secondarily, to inefficient logistics and endemic corruption[4]. Several donors and in particular the World Bank have, since the 1970s, invested heavily in improving SSA's transportation infrastructure, particularly along key trade corridors. These investments have facilitated road transport and have, in fact, reduced the costs of transport.

According to the 2008 transport price study, reduced transport costs have not translated into reduced transport prices for farmers and shippers in much of Africa. On the contrary, particularly in West and Central Africa (WCA), the reduction in transport costs due to improved road conditions has served principally to increase profit margins for trucking companies (Teravaninthorn and Raballand, 2008).

The study calls into question much conventional wisdom regarding transport in Africa, and the role that donors and governments must play to facilitate access to markets. For example:

- The cost of transport along major corridors is not significantly higher in SSA than in European countries. The report estimates that average costs per vehicle kilometre in Central and East Africa (US$1.87 and US$1.33) were similar to those in France (US$1.52) and Germany (US$1.71), and were, in fact, lower than in Poland (US$2.18). Although variable costs, primarily for vehicle maintenance and fuel, were significantly higher, fixed costs for truck acquisition and drivers' salaries were much lower.
- The trucking industry in Pakistan shares many similarities with SSA (purchase of old used trucks, very poor condition of transport infrastructure, low wage levels for drivers), yet transport prices in Pakistan are but a fraction of what they are in most of sub-Saharan Africa. The prevailing low prices for truck transport in Pakistan enable Pakistani rice to be competitive on African markets.
- Estimated profit margins registered by trucking companies are very high, particularly in Central and West Africa. Estimates range from 118 per cent on the Ngaoundëré–N'Djamema corridor, 80 per cent on the

[4] Inefficient and expensive maritime port operations are also a significant barrier to increasing the competitiveness of smallholder farmers in SSA, impacting the cost of imported inputs and prices achieved for agricultural exports. Inadequate and antiquated infrastructure to meet the rapidly expanding volume of maritime traffic, and cumbersome institutional and regulatory frameworks, result in much higher operating costs in Africa than elsewhere. Average costs to unload containers in West African ports (US$320) are, on average, twice the amount of other continents.

Tema–Bamako corridor, 86 per cent on the Mombassa–Kampala corridor to a low of 18 per cent on the Lusaka–Johannesburg corridor.
- Under present industry conditions, rehabilitation of major road networks, reduction in border crossing times, decreases in fuel prices, and bribes will have *no* impact on transport prices for smallholder farmers in WCA.

The study highlights the very different prevailing conditions in East and Southern Africa (ESA), as opposed to those in WCA, in particular the much lower transport costs and prices in ESA. The main reason for this is competition within the sector.

In WCA the trucking industry is highly regulated. Freight allocation is controlled by trucking cartels, which erect barriers to entry, promulgate, and enforce regulations which limit competition, and create conditions which favour bribery to increase market share under the 'tour de role' freight allocation system. The structure of the industry and the policy frameworks do not provide incentives to invest in newer lorries or maximize existing trucking capacity through more efficient use of vehicles. It is no coincidence that trucking firms in these regions are often owned by the politically well connected.

By contrast, in ESA, direct contracting between shipper and transporter is the norm. This has fostered a more open, competitive environment which is reflected in more efficient services, newer truck fleets, and overall lower costs and prices in the region. The changes have not been absolute however, with corruption and power relations still inflating prices and reducing quality on some routes in the region.

The 2008 transport price study examined the impact of the 1994 deregulation on the trucking industry in Rwanda (the only such case of deregulation in SSA, according to the study) following the genocide. Deregulation of the parastatal trucking industry resulted in a decline in prices by more than 30 per cent in nominal terms and almost 75 per cent in real terms. Moreover, lower prices also led to an increase in the size of the Rwandan fleet. The positive impact of deregulation on transport prices in Rwanda is similar to the experiences in other countries. Morocco, Mexico, Indonesia, France, as well as a handful of central European countries, have deregulated their trucking industries during the last 30 years, and, in all instances, prices have decreased while service has improved.

While not identified as a main contributor to transport costs on the principal international corridors analysed in the AICD Study (Teravaninthorn and Raballand, 2008), illicit payments to government agents, associated delays, and subsequent spoilage weigh heavily on the cost of local transport of agricultural products in SSA. These additional transport costs are distributed to producers in the form of lower farm-gate prices and to consumers through higher retail prices. Producers of perishable goods, who are predominantly women smallholders, are disproportionately affected by these issues because of significant losses incurred during transport. The World Bank (2009a) argued that illicit payments act as a major barrier to agricultural growth. Figure 3.11, for example, shows the number of checkpoints, time of delay, and the average amount of bribes for a section of the roads in West Africa—an illustration which makes starkly clear the challenge faced by local producers and traders.

So, for much of sub-Saharan Africa, revising the policy environment for the trucking industry to make it more competitive would effectively shorten the 'economic distance' from farm to market, decrease the cost of purchased inputs, improve profit margins for smallholder marketable surpluses, and level the playing field for millions of SSA smallholders so that they may better seize opportunities for import substitution. Farmer organizations have been working in concert with regional economic communities to draw attention to these issues through collective action, but much more needs to be done.

4.2 RIGHT TIME

> Observe due measure, for right timing is in all things the most important factor. (Heriod, Greek Poet, *circa* 800 BC)

Development stakeholders pay insufficient attention to the critical importance of temporal coordination—planning and management of delivery of inputs and outputs—as a factor in rural poverty reduction. Discussions concerning improved productivity, greater food security, and increased revenues centre on the new silver bullet—the breakthrough technology, the new 'it'. Talking about the impact of improved planning, prompt disbursements, and timely delivery of goods, works, and services on poverty reduction and economic growth is not very exciting. It does not generate research grants or fellowships, or provide the raw material for refereed papers in prestigious journals. Perhaps this is why there is so little data on the impact of poor management and late delivery of inputs on economic growth.

But anyone with even a passing knowledge of hands-on development practice knows that SSA agriculture, as the engine of economic growth, is running on perhaps four of its six cylinders, and this is because inputs, be they financial, agronomic, or technical, are all too often not being delivered in a timely manner. This may be the fault of a donor institution, ministry of agriculture, the project implementation unit, the bank or micro-finance institution, or the agricultural cooperative. Of course, no one knows the overall impact of untimely delivery of goods, services, and funding, but there is widespread recognition that the problem is pervasive. It is also correctable.

Agricultural development projects mostly provide the same services: financing, technical expertise, and physical inputs, but they achieve varying levels of success. And the difference, more often than not, comes down to the delivery of the required goods and services in a timely manner.

The history of fertilizer subsidy programmes in SSA is a case in point. Late fertilizer deliveries to farmers' fields have been a salient characteristic of fertilizer subsidy programmes. Dorward (2009) examined ten fertilizer subsidy programmes and found that in nine of the ten cases, fertilizer was delivered late to a substantial proportion of the beneficiaries. In Ghana, Yawson et al. (2010) reported that 82 per cent of respondents indicated that subsidized fertilizer was not available at planting time. A similar situation was described by Xu et al. (2006).

It is commonly recognized that late application of fertilizers has a negative effect on yield response (Kabambe et al., 1998; Minde et al., 2008; Xu et al., 2006) and the lowered yield response can often discourage farmers from adopting improved input packages. In Zambia, Xu found that late application of fertilizer on maize in several provinces had a significant negative impact on yield, compared to those farmers who applied the recommended doses at planting time. Of the 21 sample groups examined, yields were significantly higher in all cases among those farmers who had access to fertilizer in a timely manner. On-time fertilizer applications registered value/cost ratios of >2 in 8 of the 21 cases, compared to none among those who received fertilizer late. It is generally admitted that value/cost ratios of >2 are a condition for farmers to adopt fertilizer in their production systems.

To our knowledge, no multi-country study examines the larger impact of late fertilizer delivery on profitability and adoption, but the empirical evidence seems quite clear: subsidized fertilizers are frequently delivered late; late delivery and application decrease yields compared to 'right time' application and deter adoption of a technology which could improve food security and livelihoods. The increased use of inorganic fertilizers is not, of course, a panacea. Their use, even when the appropriate blend is employed in a timely fashion, may not always be financially justifiable. Moreover, inorganic fertilizers need to be associated with better agronomic practices and integrated soil fertility management (McIntyre et al., 2009). But in cases where marginal returns are positive, improved management of fertilizer procurement could have a significant impact on food security, smallholder revenues, and adoption rates.

Another glaring case where delayed delivery of inputs has a pronounced negative impact on smallholders is the tardy delivery of marketing loans to cooperatives for raw material acquisition. Over the past decade, Rwanda has invested tens of millions of dollars in the construction of coffee wet mills, new coffee trees, and training to improve coffee quality and capitalize on the increased world-wide demand for premium coffees. A significant portion of this development has been through cooperatively owned processing facilities. Cooperatives depend on marketing loans to finance coffee cherry[5] purchases from cooperative members and non-members. Generally, commercial bank loans are provided in tranches during the marketing season against warehoused inventories of parchment[6] coffee. The history of coffee cooperatives in Rwanda is characterized by late payment of marketing loans, which in turn interrupts cherry purchases as farmers will not sell their cherries on credit to the cooperative and instead sell to neighbouring privately owned processing facilities or other coffee cooperatives. Interrupted cherry deliveries mean that cooperatives cannot process adequate amounts of coffee to pay off long-term capital loans for processing facility construction. The result is frequent cooperative insolvency, cessation of activity, and outstanding bad loans which weaken banks' financial positions and discourage further investment in the coffee sector.[7]

The reasons for delayed provision of marketing loans are varied. It may be due to late or incomplete submission of loan applications by cooperatives, or tardiness on the part of bank loan officers, but the upshot is still the same: the lack of 'right time' management is a serious impediment to capitalizing fully on existing investments and is a significant brake on increasing smallholder food security and incomes.

Multilateral and bilateral donors share a significant part of the responsibility for this pervasive lack of urgency in the implementation of development initiatives. Lag times between donor project approval and the start of programme implementation in terms of direct services to farmers commonly exceed 12 months. Processing of requests for disbursement of loan and grant funds sent to donors from borrowing governments are nowhere near as expedient as they should be. Donor programme evaluations are often published so long after fieldwork that their recommendations are outdated. Delays between the publication of vacancy announcements and the recruitment of project staff are frequently so lengthy that candidates get discouraged and take other positions.

The impact of consistently slow response time on the part of donors is far reaching. Not only does it retard programme implementation, but, perhaps more importantly, it sends exactly the wrong message to client governments, agricultural cooperatives, private-sector partners, and smallholders that timing is not important. It also undermines donor credibility. How can donors credibly advocate more expedient government procurement processes when their own administrative processes exhibit the same weaknesses?

[5] Coffee cherries are the fruit of the coffee tree. The fruit is harvested and the beans are extracted in processing.

[6] Dried but unhulled coffee beans.

[7] Personal communication from Reiner, C., IFAD's Country Portfolio Manager.

IFAD, like other multilateral and bilateral donors, recognizes the contribution which public–private partnerships (PPPs) can play in improving food security and income generation for smallholders. To be a viable partner to the private sector, however, 'right time' responses are vital. For example, in Sierra Leone a very successful PPP has been developed with smallholder cocoa growers and high-quality exporters, facilitated by an IFAD-funded project. However, an otherwise positive relationship amongst farmers, marketers, and the government project is strained by consistent delays in transferring reimbursable funds to the companies and cooperatives who either have to delay disbursement to their members for work accomplished or have to pre-finance from their own resources, calling into question the advantages of a PPP. In this case, the project is addressing the issue, but it highlights how getting the timing right is critical to the ability of public organizations to be attractive partners in the expanding context of PPPs.

So why have development stakeholders not done a better job of getting the timing right? The reasons are multiple. Some involve complex issues such as bureaucratic accountability and competing stakeholder political agendas while others are simpler. First, in a non-profit environment, there are few penalties for not delivering on time, in contrast to the private sector which must rely on customer satisfaction for profit and success. Second, improved operational planning, as mentioned in the introduction to this section, is not a very glamorous subject and has not received the importance which it is due; realistic, detailed chronological planning and expedient turnaround times are at the heart of successful agricultural development programmes. Because they have not received their due, insufficient attention has been paid to management training of project coordinators, transparent monitoring of metrics which gauge response time, and the establishment of lapse-of-time contracting mechanisms (which provide for automatic approval unless specific concerns are raised within a specified time period). These would serve to heighten the sense of accountability and promote the timely provision of goods, works, and services.

In the final section on conclusions and recommendations, we define the broad changes required to put temporal and spatial coordination at the forefront of programme design and implementation. We also provide a number of concrete recommended actions for key stakeholders which are essential for improving the effective delivery of smallholder development initiatives at the right place and at the right time.

5. Conclusions and recommendations

Sub-Saharan Africa's unique resource endowment and political geography provide a challenging context which has traditionally translated into relatively weak social and economic progress compared to other regions. However, while tremendous challenges, particularly in infrastructure and governance, remain, the benefits of a growing, better-educated, and healthier population and improving economic management are gradually translating into stronger economic and agricultural sector growth. In fact, several SSA countries are emerging as the continent's 'lions', with globally leading growth rates realized and anticipated over the coming years (Economist, 2011).

Sub-Saharan Africa's women and men smallholder farmers are poised to be a key driver of future economic growth and poverty reduction, provided the challenges leading to high levels of risk can be overcome. Growing global demand for agricultural products is translating into improved production incentives for SSA's farmers. Given the high proportion of smallholders and the potential productivity advantages, even in the context of current constraints, SSA's smallholders should be major beneficiaries of these opportunities. For this to happen, smallholder farming in SSA needs to become more business-oriented, through the introduction of investments which increase efficiency and income, and the strengthening of farmer-led organizations which increase bargaining power and market-integration.

As highlighted in IFAD's Rural Poverty Report, farmers' willingness to undertake these investments is dependent on reducing the actual and perceived risks which trap many rural households in a subsistence equilibrium around the poverty line. Governments, donors, NGOs, and others are investing increasingly in technology, infrastructure, and other efforts to reduce risks and increase returns, and to facilitate smallholders' participation in more integrated supply chains. However, the effectiveness of these efforts in terms of sustainability and adoption by farmers, and their impact in terms of economic returns, is not currently reaching its full potential.

Efforts to boost the impact of investments in smallholder agriculture have not paid sufficient attention to issues of spatial and temporal coordination. The tendency is often to address under-performance with the search for new technologies or paradigms when the very practical constraints of spatial and economic isolation and lack of timeliness in delivery of development assistance are staring stakeholders in the face. The impact of geographic isolation, high transport costs, time lost to road blocks, as well as failure to deliver on time to producers, or for producers to deliver products into the market when prices are more favourable, can completely undermine the returns on investment in irrigation, inputs, credit, insurance, post-harvest storage, training, etc. The result is a likely abandonment of these efforts by producers who are understandably cautious in investing their own time and scarce cash in the necessary sustaining activities. This can disproportionately impact women who generally devote a greater share of their agricultural effort on perishable market production relative to staple crops. Hence there is a need for much more focus on both 'right place' and 'right time'.

How do we transform right place, right time from a slogan or analytical framework into an operational approach?

Operationalizing 'right place' requires a more realistic assessment of marketing opportunities through the implementation of spatially aware value-chain assessments which serve as the basis for designing investment plans, programmes, and projects in the context of the CAADP framework. This entails assessing not only the physical distance from markets and the quality of roads, but also the economic distance in terms of freight charges, and official and unofficial controls and tariffs—something which is rarely addressed, at least in the context of IFAD-financed projects. In addition, stakeholders will need to work together systematically to bring transparency to the official and unofficial policy distortions which undermine smallholder competitiveness vis-à-vis imports—particularly in meeting the opportunities of growing, regional urban markets. Finally, co-financing schemes which involve both transport and agricultural development, so frequently characterized by piecemeal approaches focused on physical distance and road quality alone, need to give way to joint planning and implementation based on assessment of impacts on market competitiveness if the essential synergies are to be realized.

Operationalizing 'right time' requires much more attention to valuing the timing as well as the act of delivery of goods and services to farmers and to markets. Again, in the first instance this requires analysis of the timing requirements for production and marketing, as well as assessment of the roles of actors to achieve this. In terms of implementation, the issue of timing needs to be at the forefront of planning of activities and needs to be a key driver of management oversight. For example, in the context of IFAD-financed projects, annual work plans and budgets are prepared and discussed between the projects' management units, governments, IFAD, and other partners. Although the timing of activities is presented, these are rarely highlighted relative to the timing needs of farmer production cycles, or marketing windows.

Making progress in this area will be challenging, as it will require not only a change in processes and priorities on the part of implementing institutions, but will also require an improvement in the interface between the business practices of donors and national governments and those of smallholder farmers and other actors in the supply chain.

In summary, our recommendations for each of the actors involved are:

National Governments:

- Prioritize farmer-focused infrastructure.
- Acknowledge the importance of reducing transport costs and take steps to open up the national transport sector, increase transparency, and reduce unnecessary road blocks.

- Review transport-related taxation policies to ensure that they promote renewal of trucking fleets, greater vehicle utilization, and consequent lower cost structures.
- Put in place conducive trade and market policies to support local production and decrease transaction costs.

Farmer organizations:

- Bring attention to the importance of issues such as reliable and affordable transport, reliable and timely delivery of services, and to other factors which constrain production.
- Lobby for positive changes and provide evidence of the impacts of these factors on sectoral competitiveness.

NGOs:

- Continue work in agronomic knowledge-sharing and extension services to complement government activities.
- Facilitate farmer capacity strengthening through programmes which help farmers form cohesive and informed organizations.
- Offer entrepreneurial and business management training in rural areas to enable participants to assess business opportunities and more effectively manage both on-farm and off-farm income-generating activities.

Research institutions:

- Research and document the importance of place in agricultural production strategies and highlight opportunities appropriate to specific locations.
- Research and document the importance of timely access and utilization of inputs and services such as seeds, fertilizer, pesticides, market information, and credit, and also the negative effects of late delivery or application.

Private sector:

- Increase use of institutional innovations which expand client-base and product range to include a greater number of SSA smallholder farmers, working with public partners as appropriate to reduce costs and business risks.
- Contribute business marketing, production, and delivery expertise to facilitate timely delivery of desired inputs.

Donor institutions:

- Work with farmer organizations and others to draw attention to the importance of spatial and temporal issues in agriculture through CAADP and other agricultural-sector policy dialogue fora.
- Promote greater synergies among co-financiers by adopting joint design and implementation.

- Streamline internal processes to expedite responsiveness to client governments, and regularly monitor and communicate progress to concerned stakeholders.
- Use value-chain assessments to identify key constraints to competitiveness and as a tool to better determine the probable returns to investments at the level of producer groups and individual smallholder farms. Ensure that critical timing aspects are identified, agreed with, and understood by all project stakeholders and staff. Work timing aspects into project work plans and budgets, and include in staff orientation, training, and evaluation.
- Include specific indicators for critical transport cost and timing aspects into implementation and monitoring framework.

The analysis presented in the chapter indicates the vast opportunities to improve spatial and temporal coordination in support of sub-Saharan Africa's smallholder, and it is expected that this chapter will stimulate attention to this issue and begin a process of discussion and action towards these recommendations.

REFERENCES

Adesina, A. (2010). 'Global food and financial crises: lessons and imperatives for accelerating food production in Africa', Presidential Lecture delivered at the 3rd Conference of the African Association of Agricultural Economists, 19–23 September 2010.

Africa Rice Center (2008). *Africa Rice Trends, 2007.* Cotonou: Africa Rice Center.

AGRA (2011). 'Facts and figures about Africa's seed systems'. Retrieved from: <http://www.agra.org/what-we-do/seed/>.

Anriquez, G. and G. Bonomi (2007). 'Long-term farming and rural demographic trends', background paper for the WDR 2008.

Bationo, A., A. Hartemink, O. Lungu, M. Naimi, P. Okoth, E. Smaling, and L. Thoimbiano (2006). 'African soils—their productivity and profitability of fertilizer use', background paper for the Africa Fertilizer Summit, June 2006, Nigeria.

Binswanger, H. and M. Rosenzweig (1986). 'Behavioral and material determinants of production relations in agriculture', *Journal of Development Studies*, 2223 (April), pp. 503–39.

Bumb, B. (2009). 'Fertilizer supply chain in Africa', presentation at a COMESA fertilizer policy training event, Zambia, 2009 by IFDC.

CAADP (2011). 'The Comprehensive Africa Agriculture Development Programme'. Retrieved from: <http://www.caadp.net>.

Deininger, K. and D. Byerlee (2011). 'The rise of large-scale farms in land-abundant developing countries. Does it have a future?', World Bank Policy Research Working Paper 5588. Washington, DC: World Bank.

Dixon, J. and A. Gulliver, with D. Gibbon (2001). 'Chapter 2: Sub-Saharan Africa', in *Farming Systems and Poverty: Improving Farmers' Livelihoods in a Changing World.* Rome and Washington, DC: FAO and World Bank.

Dorosh, P., H. Wang, L. You, et al. (2010). 'Crop production and road connectivity in sub-Saharan Africa a spatial analysis', World Bank Policy Research Working Paper Series, July 2010 (no. 5385).

Dorward, A. (2009). *Rethinking Agricultural Input Subsidy Programmes in a Changing World.* London: SOAS.

Eastwood, R., M. Lipton, and A. Newel (2009). 'Chapter 65: Farm Size', *Handbook of Agricultural Economics*, Vol. 4, 2010, pp. 3323–97.

Economist (2011). 'The lion kings? Africa is now one of the world's fastest-growing regions', 6 January. Retrieved from: <http://www.economist.com/node/17853324>.

FAO (1991). 'How good the earth? Quantifying land resources in developing countries: FAO's agro-ecological zones studies'. Rome: FAO.

FAO (2005). *Irrigation in Africa in figures, Aquastat Survey 2005.* Edited by K. Franken, FAO Water Reports 2005. Rome: FAO.

FAO (2009a). 'How to feed the World in 2050', Forum Background document, 12–13 October, Rome. Retrieved from: <http://www.fao.org/wsfs/forum2050/wsfs-background-documents/hlef-issues-briefs/en/>.

FAO (2009b). 'How to feed the World in 2050—the special challenge for sub-Saharan Africa', Forum Background document, 12–13 October, Rome. Retrieved from: <http://www.fao.org/wsfs/forum2050/wsfs-background-documents/hlef-issues-briefs/en/>.

FAO (2010). *State of Food Insecurity in the World 2010—Addressing food insecurity in protracted crises.* Rome: FAO.

FAO (2010a). *FAO Aquastat*, the Food and the Agricultural Organisation of the United Nations, Rome. Retrieved from: <http://www.fao.org/nr/water/aquastat/main/index.stm>.

FAO (2012). *FAOSTAT*, the Food and the Agricultural Organisation of the United Nations, Rome. Retrieved from: <http://faostat.fao.org/site/291/default.aspx>.

Fischer, G., M. Shah, H. van Velthuizen, and F. O. Nachtergaele (2002). *Global Agro-ecological Assessment for Agriculture in the 21st century: Methodology and Results.* Vienna: IISA; Rome: FAO.

Foster, V. and C. Briceño-Garmendia (eds) (2010). *Africa's Infrastructure: A Time for Transformation.* AICD Flagship Report.

Godfray, H. C. J., I. R. Crute, L. Haddad, D. Lawrence, J. F. Muir, N. Nisbett, J. Pretty, S. Robinson, C. Toulmin, and R. Whiteley (2010). 'The future of the global food system', *Philosophical Transactions of the Royal Society: B*, 365, pp. 2769–77.

Hayami, Y. (2004). 'Family Farms and Plantations in Tropical Development', *Asian Development Review*, 19(2), pp. 67–89.

Hazell, P., C. Poulton, S. Wiggins, and A. Dorward (2007). 'The future of small farms for poverty education and growth', 2020 Discussion Paper 42. Washington, DC: IFPRI.

Henao, J. and C. Baanante, (2006). *Agricultural Production and Soil Nutrient Mining in Africa—Implications for Resource Conservation and Policy Development.* Alabama: IFDC, Muscle Shoals.

IFAD (2010). *Rural Poverty Report 2011—New Realities, New Challenges: New Opportunities for Tomorrow's Generation.* Rome: IFAD.

IFC (2011). *Doing Business 2011—Making a Difference for Entrepreneurs.* Washington DC: IFC.

IMF (2010). *Regional Economic Outlook—sub-Saharan Africa—April 2010.* Washington DC: IMF.

Inoncencio, A., M. Kikuchi, M. Tonosaki, A. Maruyama, D. Merry, H. Sally, and I. de Jong. (2007). 'Costs and performance of irrigation projects: a comparison of sub-Saharan Africa and other developing regions', IWMI Research Report 109, IWMI, Colombo, Sri Lanka.

International Telecommunication Union (2013). *World Telecom/ICT Indicators Database, 2013*, 17th edition, ITU, Geneva. Retrieved from: <http://www.itu.int/en/ITU-D/Statistics/Pages/stat/default.aspx>.

Kabambe, V., J. Kumwenda, W. Sakala, and R. Ganunga (1998). 'The effects of time of applying basal dressing fertilizers on maize yield under varying rainfall regimes and inherent soil fertility in Malawi', Sixth Eastern and Southern Africa Regional Maize Conference, 21 September, pp. 290–2.

McIntyre B., H. Herren, J. Wakhungu, and R. Watson (eds) (2009). *International Assessment of Agricultural Knowledge, Science and Technology for Development (IAASTD), Global Report*. Washington, DC: Island Press.

Minde, I. J., T. S. Jayne, E. W. Crawford, J. Ariga, and J. Govereh (2008). 'Promoting fertilizer use in Africa: current issues and empirical evidence from Malawi, Zambia, and Kenya', RESAKSS Working Paper No. 13.

Montpellier Panel (2010). *The Montpellier Panel Report, Africa and Europe: Partnerships for Agricultural Development, Agriculture for Impact*. London: Imperial College. Retrieved from: <http://www3.imperial.ac.uk/africanagriculturaldevelopment/themontpellierpanel/panelreport>.

Morris, M. L. and D. Byerlee (1993). 'Narrowing the wheat gap in sub-Saharan Africa: a review of consumption and production', *Economic Development and Cultural Change*, 41(4), pp. 737–61, 29.

Nagayets, O. (2005). 'Small farms: current status and key trends', information brief prepared for the Future of Small Farms Research Workshop, 26–29 June, Wye College, London.

Ndejeunga, J. and M. Bantilan (2002). 'Uptake of improved technologies in the semi-arid tropics of West Africa: why is agricultural transformation lagging behind?', paper presented at International Conference on 'Green Revolution in Asia and transferability to Africa', Tokyo.

Odhiambo, M. O. (2011). *Commercial Pressures on Land in Africa, a Regional Overview of Opportunities, Challenges and Impacts*. Rome: International Land Coalition.

OECD (2011). *OECD Statistics*. Retrieved from: <http://stats.oecd.org>.

Oxfam (2008). *Double-Edged Prices*, Oxfam Briefing Paper 121. Oxford: Oxfam International. Retrieved from: <http://www.oxfam.org.uk/resources/policy/conflict_disasters/downloads/bp121_food_price_crisis.pd>.

PHL Network (2010). *Estimated Post-Harvest Losses (2003–2009)*. Post Harvest Losses Network. Retrieved from: <http://www.aphlis.net/index.php?form=losses_estimates>.

Pingali, P. and M. Rosengrant (1995). 'Agricultural commercialization and diversification: processes and policies', *Food Policy*, Vol. 20, No. 3, pp. 171–85.

Raswant, V. and R. Khanna, with T. Nicodeme (2010). *Pro-poor Rural Value-chain Development: Thematic Study*. Rome: IFAD.

RESAKSS (2010). 'Regional strategic analysis and knowledge support system', facilitated by IFPRI. Retrieved from: <http://www.resakss.org>.

Sebastian, K. (2008). *GIS/Spatial analysis, contribution to 2008 WDR*, Background Paper for the World Development Report 2008, Market Access Dataset. Retrieved from: <http://siteresources.worldbank.org/INTWDR2008/Resources/2795087-1191427986785/SebastianK_ch2_GIS_input_report.pdf> (20 July 2013).

Staatz, J. and N. Dembele, with A. Mabiso (2007). 'Agriculture for development in sub-Saharan Africa', Background paper for the World Development Report 2008.

Swinnen, J., A. Vandeplas, and M. Maertens (2007). 'Governance and surplus distribution in commodity value chains in Africa', paper prepared for 106th Seminar of EAAE, Montpellier.

Teravaninthorn, S. and G. Raballand (2008). 'Transport prices and costs in Africa: a review of the main international corridors', Africa Infrastructure Country Diagnostic, Working Paper 14.

Transparency International (2010). *Corruptions Perception Index 2010*. Berlin: Transparency International.

UNICEF (2007). Nutrition table 2000–2006, in *The State of the World's Children 2008—Child Survival.* Retrieved from: <http://www.unicef.org/sowc08/docs/sowc08_table_2.pdf>.

USAID (2009). 'Briefing paper: value chain governance', Accelerated Microenterprise Advancement Project. Washington, DC: USAID.

West Africa Trade Hub (2007). *Report on the First Results of the Improved Road-Transport Governance (IRTG) Initiative on Interstate Highways.* Retrieved from: <http://www.watradehub.com/sites/default/files/resourcefiles/aug09/report20on20first20irtg20results20english20jw.pdf> (20 July 2013).

Wiggins, S. (2009). 'Can the smallholder model deliver poverty reduction and food security for a rapidly growing population in Africa?', paper for the Expert Meeting on How to Feed the World in 2050, Rome.

Wiggins, S. (2010). 'African agriculture: a time for cautious optimism?', presentation at ODI, 14 July, ODI, London.

World Bank (undated). 'Why are fertilizer prices higher in Africa? World Bank fertilizer toolkit'. Retrieved from: <http://www.worldbank.org/afr/fertilizer_tk/bpractices/HighPrices.htm>.

World Bank (2009a). *World Development Report 2009.* Washington, DC: World Bank.

World Bank (2009b). *World Development Indicators 2009.* Washington, DC: World Bank.

World Bank (2009c). *Awakening Africa's Sleeping Giant—Prospects for Commercial Agriculture in the Guinea Savannah Zone and Beyond.* Washington, DC: World Bank.

World Bank (2010). *World Development Indicators 2010.* Washington, DC: World Bank.

World Bank (2012). *World Development Indicators 2012.* Washington, DC: World Bank.

World Bank/FAO (2010). 'Missing food: the case of post harvest losses in sub-Saharan Africa', Working Paper, Washington, DC: World Bank; Rome: FAO.

WTO (2010). *World Trade Developments, International Trade Statistics 2010.* Geneva: WTO.

Yawson, D. O., F. A. Armah, E. K. A. Afrifa, and S. K. N. Dadzie (2010). 'Ghana's fertilizer subsidy policy: early field lessons from farmers in the central region', *Journal of Sustainable Development in Africa*, 12(3).

Xu, Z., J. Govereh, J. R. Black, and T. S. Jayne (2006). 'Maize yield response to fertilizer and profitability of fertilizer use among small-scale maize producers in Zambia', paper prepared for the International Association of Agricultural Economists Conference, 12–18 August, Australia.

4 Smallholder farming in Asia and the Pacific: challenges and opportunities

GANESH THAPA AND RAGHAV GAIHA[1]

1. Introduction

Small farms have been defined in a variety of ways. The most common measure is farm size: many sources define small farms as those with less than 2 hectares (ha) of crop land. Others describe small farms as those depending on household members for most of the labour or those with a subsistence orientation, where the primary aim of the farm is to produce the bulk of the household's consumption of staple foods (Hazell et al., 2007). Yet others define small farms as those with limited resources, including land, capital, skills, and labour. The World Bank's Rural Development Strategy defines smallholders as those with a low asset base, operating less than 2 ha of crop land (World Bank, 2003). An FAO study defines smallholders as farmers with limited resources relative to other farmers in the sector (Dixon et al., 2003). In this chapter, small farms have been defined as those with less than 2 ha of land area and those depending on household members for most of the labour.[2] Whether inadequacy of farm income supplemented by wages necessarily implies that they are mostly net food buyers is not self-evident or generally corroborated.

It is estimated that about 87 per cent of the world's 450 million small farms (less than 2 ha) are in Asia and the Pacific region (IFPRI, 2007). China and India alone account for 193 million and 93 million small farms, respectively.

[1] The authors wish to thank Nidhi Kaicker and Manoj Pandey for meticulous research assistance. The revision has benefited from the comments of the participants at the IFAD conference on New Directions for Smallholder Agriculture. The authors also wish to thank the anonymous referee and the editors of this volume for their helpful comments.

[2] With the exception of the analysis of size, marketed surplus and price in India, where land owned is measured in terms of acres instead of hectares. This was done because of the small number of observations in the size interval >2 ha.

Three other Asian countries with a large number of small farms are Indonesia (17 million), Bangladesh (17 million), and Vietnam (10 million).

Often a sharp dichotomy is drawn between smallholders and labourers, ruling out any overlap between their interests. An important point is that, since smallholders eke out a bare subsistence if the soil quality is poor, suffer from outdated technology and limited access to markets, and product mix is un-remunerative, they are forced to pursue other livelihood options. Typically, household income is supplemented by working on neighbouring farms and elsewhere. The National Sample Survey (NSS) data for India in 1993–2004 corroborate this. Among those in the lower land interval of 0–1 ha, for example, about 28 per cent were self-employed in agriculture in 1993, and a slightly lower proportion (over 24 per cent) in 2004. A little under half of households in this interval (about 48 per cent) worked as labourers in 1993 and a slightly lower proportion (about 46 per cent) did so in 2004. Of agricultural labourers, three-fourths operated/owned land in the interval 0–2 ha in 1993 and 98 per cent in 2004.

So the overlap between smallholders and labourers is large, suggesting that drawing a sharp dichotomy between them runs the risk of a false separation. Some policies designed to enhance the welfare of one would benefit the other too. Higher productivity, for example, would enhance the welfare of smallholders as well as agricultural labourers.

Agriculture in Asia is characterized by smallholders cultivating small plots of land[3]. The average size of operational holdings (actual area cultivated) in Bangladesh is only 0.5 ha, in Nepal and Sri Lanka 0.8 ha, in India 1.4 ha and in Pakistan 3 ha. About 81 per cent of farms in India have land holdings of less than 2 ha, whereas their share in total cultivated area is about 44 per cent (NCEUS, 2008). In China 95 per cent of farms are smaller than 2 ha. In Nepal 93 per cent of cultivated land is operated by small farmers (<2 hectares) covering 69 per cent of the cultivated area. In Bangladesh, small farms account for 96 per cent of operational holdings with a share of 69 per cent of cultivated area. Pakistan is an exception, with a relatively high concentration of large landholdings. Fifty eight per cent of farms in Pakistan are of less than 2 ha, but they operate only 16 per cent of the farm area. In contrast, farms of more than 10 hectares occupy 37 per cent of total farm area.

The overall trend in Asia has been that of declining farm size. For example, in China farm size decreased from 0.56 ha in 1980 to 0.4 ha in 1999 (Fan and Chan-Kang, 2003); in Pakistan it declined from 5.3 ha in 1971/73 to 3.1 ha in 2000; in the Philippines the average farm size fell from 3.6 ha in 1971 to 2 ha in 1991; and in India it declined from 2.2 ha in 1950 to 1.8 ha in 1980, to 1.4 ha in 1995–6 and to 1.33 ha in 2000–1 (Nagayets, 2005; Government of India,

[3] For a comprehensive review, see Thapa (2010).

Table 4.1 Changes in farm size and land distribution in selected Asian and Latin American countries

Country	Period	Land distribution (Gini)		Average farm size (hectares)		Change in total number of farms (%)	Change in total area (%)
		Start	End	Start	End		
Smaller farm size, more inequality							
Bangladesh	1977–96	43.1	48.3	1.4	0.6	103	−13
Pakistan	1990–2000	53.5	54.0	3.8	3.1	31	6
Thailand	1978–93	43.5	46.7	3.8	3.4	42	27
Smaller farm size, less inequality							
India	1990–5	46.6	44.8	1.6	1.4	8	--5

Sources: World Bank (2007); Anriquez and Bonomi (2007).

2008). In Bangladesh, the average farm size declined from 1.4 ha in 1977 to 0.6 ha in 1996, whereas in Thailand, it declined from 3.8 to 3.4 ha between 1978 and 1993 (Table 4.1).[4]

In Asia, the Gini coefficient of land distribution is declining in India, whereas it is increasing in other countries such as Bangladesh, Pakistan, and Thailand. In many countries of Asia and the Pacific, unequal access to land is perpetuated through social mechanisms, which leave many households belonging to indigenous peoples or ethnic minorities without land or with land plots too small to meet their needs.

The number of small farms and their share in total cultivated area has been increasing over time in some Asian countries. For example, in India, small farms accounted for almost 81 per cent of operational holdings in 2002–3 compared to about 62 per cent in 1960–1 (Table 4.2). Correspondingly, the area operated by small farms increased from about 19 per cent to 44 per cent during this period (NCEUS, 2008). The distribution of landownership in India has become less skewed. The share of land area owned by small farms increased from 20 per cent in 1961–2 to 43.5 per cent in 2003. Also, the trend towards landlessness appears to have been arrested, with the percentage of landless between 1971–2 and 2003 remaining approximately at 10 per cent. In India, the distribution of operational holdings closely mirrors the distribution of land owned.[5]

Smallholders' contribution to the total value of agricultural output is also significant in many countries of Asia. For example, in India, their

[4] For a persuasive explanation of persistence of small farms on efficiency grounds, see Lipton (2006).

[5] Further investigation of this similarity requires data on leasing-in and leasing-out of data by size class of land owned to which we did not have access.

Table 4.2 Changes in percentage distribution of operated area by size of operational holdings in India, 1960–1 to 2002–3

	% distribution of operational holdings				% distribution of operated area			
Land class	60–61	81–82	91–92	02–03	60–61	81–82	91–92	02–03
Small	61.7	68.2	75.3	80.6	19.2	28.1	34.3	43.5
Medium	33.8	28.8	24.8	18.1	51.9	53.7	50.5	44.7
Large	4.5	3.1	1.9	1.3	29.0	18.2	15.2	11.8

Note: Small: <2 ha; medium: 2–10 ha; large: >10 ha.

Source: Computed from: NCEUS (2008).

contribution to total farm output exceeds 50 per cent although they cultivate only 44 per cent of land. Many studies have also confirmed the inverse relationship between farm size and productivity per hectare. Small farmers are characterized by smaller applications of capital but higher use of labour and other family-owned inputs, and a generally higher index of cropping intensity and diversification. The inverse relationship between farm size and productivity is a powerful rationale for land reform policies, including land redistribution for both efficiency and equity gains.[6] Small farms tend to grow a wide variety of cultivars, many of which are landraces. These landraces are genetically more heterogeneous than modern varieties, offer greater resilience against drought, disease, and pests, and so enhance harvest security (Clawson, 1985).

More recent evidence from India confirming but elaborating the inverse size–productivity relation in agriculture is given in Section 3.

Our recent analyses with household data in Laos and Cambodia suggest that proportions of poor (including those below the cut-off of $1.25) are highest in the lowest size interval <about 2 ha, but there are regional variations.[7] Although there are plausible grounds for asserting that this is the most poverty-prone group because of lack of access to credit, technology, and markets, firm evidence is lacking except for a few countries in the Asia-Pacific region.

[6] Small farmers are not just more productive (per ha of land) but also exhibit higher returns per unit of investment (Lipton, 2006). More interestingly, Foster and Rosenzweig (2010), based on an all-India household survey in 2006, confirm (i) that returns on investment decline rapidly with landholding size; (ii) the profits per acre peak at a little over 4 acres (1 ha = 2.47 acres). The important point is that even at small landholding size the profits per acre are high despite credit and other constraints; (iii) finally, the marginal returns on fertilizer fall as landholding size increases, further confirming efficient use of scarce inputs. Lipton (2006) also emphasizes the key role of smallholders in poverty reduction. Briefly, in most areas with widespread poverty, anti-poverty paths must enhance the physical assets of the poor, their employment income, and food entitlements.

[7] For details, see APR Report (2011), Gaiha and Annim (2010), and Gaiha and Azam (2011).

Experience has shown that Asian countries such as India that promoted small farms were able to launch the Green Revolution. Countries such as China started supporting smallholder farming after collective farms could not provide adequate incentives to increase production and productivity.

This paper assesses the challenges and opportunities faced by small or family farming in Asia and the Pacific region in sustainable agricultural production and productivity enhancement, and in diversifying into high-value commodities. It first gives a brief account of the transformation of the agriculture sector in the region from the mid-1960s to the mid-1990s, which was characterized by a dramatic increase in agricultural production and productivity through major breakthroughs in technological innovations, and the more recent transformation, which is characterized by significant changes in diets brought about by increases in incomes, urbanization, and globalization, and the resulting changes in production of high-value commodities and major transformation in the agrifood industry.

The chapter then discusses the challenges faced by smallholders in addressing the problems related to the sustainability of food production as well as agricultural diversification. Of particular importance in this context is responsiveness of marketed surplus of food commodities to prices. As an illustration, based on a recent household survey in India, new light is thrown on whether smallholders are constrained in marketing their outputs of these commodities. Also, two interrelated issues are examined: (i) whether large-scale investments in agriculture—especially in some of the poorest countries in Asia and the Pacific—are justified on efficiency grounds; and (ii) whether complementarities between large investors and smallholders could be better exploited. Following this, the chapter highlights some of the technological and institutional innovations that have been tested to address such challenges. It then discusses the policy and programme support provided by selected countries in the region to small or family farms in enhancing productivity and in benefiting from emerging markets in high-value commodities. Finally, it identifies some measures that the governments, the private sector, and international development partners can take to support small farmers in dealing with emerging challenges, and in sharing experiences and learning from one another.

2. Transformation of agriculture

This section briefly discusses two important transformations in the agriculture sector, which have had a profound impact on the small or family farms of the two regions. In the first one, small farms played an important role, particularly in Asia, in raising food production and incomes based on biological, chemical, and mechanical innovations. The second transformation, linked to dietary transformation and high-value chains, is more recent and presents a considerable challenge as well as opportunity for these farmers to benefit from new agriculture.

2.1 THE GREEN REVOLUTION

The Green Revolution in Asia, which mainly comprised a dramatic increase in the production of three important cereal crops—rice, maize, and wheat—between 1965 and 1990, was driven by rapid advances in the sciences and substantial public investment in and policy support for agriculture (Hazell, 2009). This represented the first major transformation of the agriculture sector in Asia in modern history. Cereal production more than doubled in Asia between 1970 and 1995, from 313 to 650 million tonnes per year (Table 4.3). As a result, per capita calorie availability increased by about 30 per cent and real prices of wheat and rice decreased. Higher production of all three major cereal crops was realized mainly through yield growth. Between 1965 and 1982, average rice, maize, and wheat yields increased by 2.54 per cent, 3.48 per cent, and 4.07 per cent per year, respectively. During the same period, cultivated area expanded by only 0.7 per cent, 1.09 per cent, and 1.3 per cent, respectively.

The success of the Green Revolution in raising food production and productivity, broadening economic growth and reducing poverty has been impressive. Nevertheless, in recent years agricultural production has experienced a number of challenges that have cast doubts on the sustainability of past gains.[8]

Table 4.3 Changes in cereal yield and production in Asia, 1970 and 1995

	India	Other S. Asia	China	SE Asia	Developing Asia
Cereal yield (mt/ha)					
1970	0.93	1.20	1.77	1.35	1.32
1995	1.74	1.85	4.01	2.24	2.63
% change	88.4	54.2	126.5	65.6	99.5
Cereal production (million mt)					
1970	92.8	25.4	161.1	33.8	313.2
1995	174.6	48.1	353.3	73.6	649.6
% change	88.1	89.3	119.3	117.8	107.4

Source: Hazell (2009).

[8] For further elaboration, see Section 3.

2.2 RECENT TRANSFORMATION IN AGRICULTURE

Growth in consumption and production of high-value commodities

Rapid economic and income growth, urbanization, and globalization are leading to a significant shift in diet in Asia and the Pacific region, away from staples and increasingly towards livestock and dairy products, fruits and vegetables, and fats and oils. Rapid income growth is a key factor in the rising demand for high-value agricultural products. In most Asian countries urbanization is increasing rapidly and studies have shown that urban households spend more on meat, fish, and sugar and less on rice than rural households, even after taking into account income and household characteristics (Minot et al., 2003).

Urbanization, rapid growth in per capita incomes, and the increase in the opportunity cost of women's time as a result of their entry into the workforce, led to greater demand for non-staples, particularly perishables and processed foods in Latin American countries (Reardon et al., 2002). On the supply side, trade liberalization since the early 1980s made it easier and cheaper to import food and non-food products.

Trade liberalization has also contributed to the growth of high-value agriculture. The reduction in import barriers in industrialized countries has favoured the growth of high-value exports such as fish and seafood products. Likewise, foreign direct investment has also facilitated the transformation of agricultural production in developing countries. It has facilitated the expansion of food processing, animal feed production, exports, and food retailing. The entry of foreign companies into the agriculture sector has put competitive pressure on domestic agribusiness companies (Gulati et al., 2005).

A recent study by the International Food Policy Research Institute (IFPRI) analysed the growth of high-value agriculture in Asia and its implications on the restructuring of the agricultural supply chain, and on the role of small farmers (Gulati et al., 2006). These countries include the largest and most important transforming countries of Asia—Bangladesh, India, and Pakistan in South Asia; Indonesia, the Philippines, Thailand, and Vietnam in South-East Asia; and China in East Asia.

The study documented a clear shift in food consumption from grains and other starchy staple crops such as cassava and sweet potatoes to meat, milk, eggs, fish, fruits, and vegetables mainly due to income increases (Table 4.4). In these countries, per capita grain consumption either increased very slowly or even decreased between 1990 and 2000. In contrast, per capita demand

Table 4.4 Average annual percentage growth in per capita consumption of selected foods in selected Asian countries, 1990–2000

	B'desh	India	Pak	Indo	Phil	Thai	Viet	China
Cereals	0.2	−0.4	0.0	0.9	0.1	0.2	1.2	−1.3
Veg	0.2	2.1	2.2	3.3	0.0	0.5	4.9	8.5
Fruits	−1.5	2.9	0.5	1.9	0.2	0.3	1.7	10.0
Milk	0.2	1.9	3.0	5.9	1.5	5.0	13.5	5.0
Meat	1.0	0.9	0.2	0.4	4.7	1.5	4.3	6.8
Eggs	4.6	1.9	1.9	3.7	1.6	−0.4	5.8	9.7
Fish	4.7	2.0	1.6	3.2	−1.4	3.9	3.7	8.4

Source: Gulati et al., 2006 (based on FAO Food Balance Database).

for vegetables, fruits, and animal products increased substantially in all countries.[9]

In addition to rising domestic demand, these high-value commodities have also experienced high export demand. High-value products such as fruits, vegetables, livestock products, and fish constitute a rapidly growing share of international trade in agricultural products. In these countries as a group, the share of high-value exports in total agricultural exports increased from 47 per cent to 53 per cent.

Due mainly to the high growth in domestic demand and, to some extent, an increase in exports, production of high-value commodities in many Asian countries has grown more rapidly than that of foodgrains. The production of foodgrains in the eight countries under study increased by 1.3 per cent per year during the 1990s, slightly below the population growth rate of 1.5 per cent. In contrast, the production of high-value commodities grew much more rapidly during this period (Table 4.5). For example, fruit and vegetable production increased by 7.7 per cent in these eight countries. China, in particular, achieved a very high growth rate in the production of fruits and vegetables. Between 1980 and 2004, 58 per cent of the increase in global horticulture production came from China, 38 per cent from all other developing countries, and the remaining 4 per cent from developed countries (Ali, 2006). India, Indonesia, Pakistan, and Vietnam also recorded an annual growth rate of more than 3 per cent in the production of fruits and vegetables in the 1990s.

The production of livestock products also increased impressively in many Asian countries during the 1990s. Milk production grew by 4.6 per cent per

[9] The dietary changes have significant nutritional implications. In 2000, 56 per cent of all the calories consumed in developing countries were obtained from cereals and 20 per cent from meats, dairy, and vegetable oils. By 2050, the contribution of cereals will drop to 46 per cent and that of meat, dairy, and fats will rise to 29 per cent (*The Economist*, Special Report: Feeding the World, 26 February 2011). For a similar shift in India see, Gaiha et al. (2010) and Kaicker et al. (2011).

Table 4.5 Average annual percentage growth in production of foodgrains and high-value commodities in selected Asian countries, 1990–2000

	B'desh	India	Pak	Indo	Phil	Thai	Viet	China
Grains	3.6	1.9	3.8	1.7	1.4	3.7	5.7	0.1
Fruits & Veg	1.7	4.3	3.8	4.1	2.1	2.1	4.7	10.2
Milk	3.0	4.2	5.7	2.8	−6.5	14.8	3.5	5.8
Eggs	6.4	4.2	4.6	4.9	3.4	1.1	6.7	10.8
Meat	3.4	3.0	2.8	1.6	5.6	3.6	6.3	7.6
Fish	7.0	4.0	2.7	5.0	0.4	3.0	7.6	11.3

Source: Gulati et al., 2006 (based on FAO Agricultural and Fisheries Production databases).

year in these eight countries during this period. Most countries also achieved high growth rates in the production of eggs, meat, and fish.

Transformation of agrifood industry

The growth in domestic consumption and production of high-value agricultural commodities in Asia and the Pacific was accompanied by a transformation of the agrifood industry, which includes processing, wholesale, and retail. Governments contributed to this mainly through investment in municipal wholesale markets, parastatal processing firms, and state-run retail chains. However, the main new developments are private-sector investment in and consolidation of processing and retail (Reardon et al., 2009; and Timmer, 2009).

An important element of this transformation is the restructuring of the wholesale sector, which started with the public investment phase in the 1970s–1980s in many parts of Asia and in the 1990s in China. This phase was characterized by public investment in the expansion and upgrading of wholesale markets, and investment in market information systems to reduce transaction costs for small farmers to gain access to growing urban markets. In the 1990s and 2000s, more attention was paid to deregulation of wholesale markets to allow greater entry and competition.

The second element of this transformation is the restructuring of the processing sector. In the 1990s, private small- and medium-sized processing companies grew due to liberalization in the processing sector. This growth was facilitated by a rapid increase in the consumption of processed foods spurred by rising incomes and urbanization, and a concomitant increase in the number of women working outside their homes.

The third element is the restructuring of the retail sector, which is mainly characterized by the supermarket revolution and a rapid spread of fast-food chains in many countries of the region. The growth in supermarkets, which started in the early to mid-1990s, was driven by a massive flow of foreign direct investment and competitive domestic private investment, privatization of retail parastatals, rising incomes and urbanization, and procurement system change (Reardon et al., 2009; Timmer, 2009; and Gaiha and Thapa, 2008). The spread of modern retail took place in three waves, first in East Asia outside China, then in South-East Asia, and finally in China, India, and Vietnam. Within a given country, supermarkets first sold processed products, then semi-processed, and recently fresh produce.[10]

3. Challenges faced by small farms

Farmers are facing a number of challenges in producing food in a sustainable manner as well as in diversifying from their dependence on cereal production to the production of high-value commodities. Although some of these challenges affect both large and small farms, there is evidence that they apply more strongly to small farms. For example, small farmers cannot take advantage of higher food prices by expanding production if they have difficulty in accessing services and credit. Similarly, when new technologies require higher capital inputs or mechanization, small farmers may be at a disadvantage unless they are helped in reducing their transaction costs to access inputs, credit, and marketing facilities.

In recent years, productivity growth of major food crops has declined quite significantly.[11] However, funding has shifted from public to private research, particularly in biotechnology. This change is reportedly disadvantageous to small farmers because private research companies lack incentives to address small farmers' concerns (Pingali and Traxler, 2002). Also, the impacts of both environmental degradation and climate change are usually more severe for small farmers than for large farmers because small farmers have less access to human, social, and financial capital and information than large farmers (Hazell et al, 2007; APR Report, 2011).

An important insight relates to supply response to higher food prices. The slowdown in growth rate of agricultural capital formation was in part a consequence of a long spell of unfavourable prices facing producers, resulting in capital moving out of agriculture. The incentives offered by spiralling food prices are likely to accelerate agricultural growth and dampen food price inflation (Imai et al., 2011).

[10] In a perceptive comment, Timmer (2009) points to a significant feature of the food policy agenda in this context. He observes that the food system is more consumer-driven than before, the marketing system is far more crucial as the efficient system for transmitting consumer preferences to farmers' opportunities, but there are fewer players in the system.

[11] But this is subject to a caveat, as emphasized later in this section.

Attractive investment opportunities have opened up in agriculture, leading to large-scale investments and competition for land (e.g. rubber plantations in Cambodia, palm oil production in Indonesia, cereals in Kazakhstan) (Deininger and Byerlee, 2010). New sources of economies of scale have emerged as a result of technical change (zero tillage and GMOs), new markets (contracts with supermarket chains for large continuous and uniform deliveries), and institutional changes (e.g. access to international finance).[12] However, frequently the large-farm advantage is due to market failures (e.g. credit), institutional gaps (e.g. weak extension services), and policy distortions (e.g. minimum support prices). Elimination of such biases against smallholders would enhance their competitiveness. State interventions and collective action by producers' organizations would make a significant difference.

Box 4.1 on Cambodia offers illustrative evidence on the limited benefits of land concessions to the host country.

BOX 4.1 LAND CONCESSIONS IN RURAL CAMBODIA

In the case of Cambodia, the evidence on the benefits of land concessions to the poor is mixed, and serious concerns have been raised whether such concessions are a more effective way of helping the rural poor than the alternative of granting small plots of land to the landless and enhancing their access to credit and technology (CEA, 2010; and ADB, 2009). Going by the official records, 60 companies have been awarded economic land concessions, involving a total land area of 1 million ha. Most of these are in non-flooded areas and degraded forests, and cover a range of crops such as rubber, palm oil, sugar cane, cashew, coffee, and forest plantations. Most of these land concessions are frequently disputed and remain unresolved for long periods, resulting in overlapping land claims of local villagers and the affluent. These land disputes are a result of weak and patchy environmental impact assessment, inefficient implementation of business plans, fewer instances of consultation with the local communities prior to the approvals, and, finally, the disruption of traditional livelihoods of the disadvantaged indigenous groups living in remote areas.

The reservoir rice cultivation in the plains of the Tonle Sap Lake yielded substantial benefits to the local community, including the poor. In one district, the reservoir owners recorded high returns amounting to 92 per cent of the reservoir investment cost in one year. On rented farms, the tenants made an attractive profit, $285 per ha. However, in late 2008–9, as prices crashed and yields decreased, 80 per cent incurred large losses, of $245 per ha. High fertilizer costs, rents, and interest to buy fertilizer were the main reasons why barely 10 per cent of the surveyed households had rented the rice fields for the following year. In another case, for the investment project of Sofcin-KCD, land was cleared without prior notice, resulting in conflicts

(*continued*)

[12] Many land concessions in Lao PDR and Cambodia—two of the poorest countries in Asia and the Pacific—were withdrawn either because there was lack of transparency in granting them and/or because no investment was made. For details, see Gaiha and Annim (2010), Gaiha and Azam (2011), and APR Report (2011).

BOX 4.1 CONTINUED

and violence. Subsequently, the company agreed to offer compensation, at the direction of the Land Conflict Resolution Committee, in both cash and land exchange. By contrast, the Dak Lal Company negotiated with landowners before developing the land for rubber plantations (50 per cent agreed upon). On the villagers' share of land (remaining 50 per cent), the company developed the land and planted rubber trees, and trained villagers on how to take care of them. People can harvest the rubber latex in seven years and sell it to the company at a guaranteed price of 80 per cent of the international price. This model that allows large and small individual plantations to coexist evoked enthusiastic support from the villagers.

These cases have important policy implications: (i) granted concession land should not overlap with local people's land; (ii) land reserves for exchange should be cultivable and not far from the village; (iii) better understanding of options offered and freedom to choose; and (iv) above all, monitoring and regulation of land concessions strictly in accordance with agricultural priorities.

Source: Gaiha and Azam (2011).

A feasible option is to explore mutually beneficial complementarities between large and small farms. In cooperatives, for example, large farmers could be cast in an entrepreneurial role that enables small farmers to access technology and markets.

In what follows, we throw new light on how constrained smallholders are marketing their produce, based on a recent nationwide household survey in India in 2006. This analysis is essentially illustrative. Structured around a well-known and insightful model of marketed surplus, new evidence is reported on why responsiveness of marketed surplus varies by size class of land owned, own price, cross-prices of other food crops, and access to markets, among others. Although limited to India, these findings are of considerable significance given diverse agro-climatic conditions.[13]

3.1 SIZE, MARKETED SURPLUS, AND PRICE

To serve as a backdrop to our analysis, a distillation of available evidence on market arrivals and size of holdings in India is given below. Many of the important contributions were based on farm management studies and cost of cultivation surveys carried out by Krishna (1995a, 1995b, 1995c), Bardhan (2003), and Bardhan and Bardhan (2003), among others. A notable recent addition is Kanwar (2006). The insights from these studies are highly relevant in the context of rising food and oil prices, and their implications for the rural poor.

[13] Ideally, similar applications to other countries would have been more helpful, but data and time constraints precluded it.

One important finding relates to the price response of marketed surplus of foodgrains. Bardhan and Bardhan (2003) first specify a theoretical model of farmers' foodgrain marketing decisions, positing that in the production decision the relevant prices are those of foodgrains relative to competing crops and agricultural inputs, whereas in the consumption decision the relevant prices are those of foodgrains relative to competing consumer good(s)—including manufactured consumables. They conclude that the marketed surplus of grains is higher when the relative cereal price is higher, and it is lower when the relative price of commercial crops is higher. The intuition underlying these results is that, when the relative cereal price is high, more is marketed as less is consumed; and when the relative price of commercial crops is high, marketed surplus of grains is lower because of switching of acreage.

The analysis given here builds on this literature by using a recent all-India survey (Rural Economic and Demographic Survey [REDS]) conducted by the National Council for Applied Economic Research (NCAER, 2006) in 17 states of India in 2006.[14] As the household and village data are being subjected to consistency checks, our results are not to be treated as definitive. The sample consists of 5,695 households in the 17 states. We have worked with smaller samples, as outliers had to be eliminated.

Our focus is on marketed surplus (amount marketed divided by crop output) by size of land owned. As the entire land data are in acres, for analytical convenience we have grouped households into cultivating <2 acres (small), between 2 and 5 acres (medium) and >5 acres (large).[15] Since there are inter- and intra-group variations in soil conditions and use of irrigation, our grouping is essentially a first order of approximation. Although recent cross-country evidence confirms robustly a positive supply response of food commodities to prices, the present analysis seeks to extend it by analysing the responsiveness of market surplus of various commodities to their own prices by size of land owned.[16]

Another contribution of this analysis is that food commodities are disaggregated into four groups: cereals, pulses, oil seeds, and vegetables. As the consumption basket has changed in recent years, as illustrated earlier, it is worthwhile examining whether smallholders are responding to the high-value chains (e.g. by producing and marketing more of high-value commodities such as oil seeds and vegetables in response to market prices).[17]

[14] The states include Tamil Nadu, Kerala, Karnataka, Maharashtra, Gujarat, Rajasthan, Punjab, Haryana, Uttar Pradesh, Bihar, Jharkhand, West Bengal, Orissa, Chhattisgarh, Madhya Pradesh, Himachal Pradesh, and Andhra Pradesh.

[15] As the observations on large farms were limited, it was not feasible to use a classification in ha.

[16] For details of the cross-country evidence, see Imai et al. (2011).

[17] In fact, evidence has accumulated pointing to a dietary transition in India. For details, see Kulkarni and Gaiha (2010) and Gaiha et al. (2010).

In a broadbrush treatment, let us consider distribution of farming households into small, medium, and large, shares of land irrigated, proportions using fertilizers, and access of sample villages to rural infrastructure.

About three-fourths of the sample households are small landholders, about 15 per cent are medium, and just under 10 per cent are large. About 57 per cent of land of smallholders is irrigated, with slightly lower shares of land of medium and large landholders. However, out of the total land irrigated, more than half belongs to large landholders and less than one-fifth to smallholders.

Given the cost of fertilizer, it is not surprising that the fraction of farmers *not* using fertilizers is highest among smallholders—in fact, it is nearly three times higher among smallholders than large landholders.[18] Two striking features with respect to the educational attainment of household heads are: (i) the proportion of illiterate heads is highest among smallholders and lowest among large landholders; and (ii) the proportion with more than ten years of schooling is lowest among smallholders and highest among large landholders. As access to new technology and markets with more remunerative prices is positively linked to educational attainments—though admittedly these links have weakened somewhat with advances in ICT—smallholders are at a disadvantage (Byerlee et al., 2010).

Table 4.6 describes access to different forms of rural infrastructure. Unfortunately, access to these is in relation to a village and not a household. Hence we are unable to capture inequity in access by size of holding. Subject to this caveat, we note that village access varies enormously depending on the type of infrastructure. For example, about 72 per cent of the villages had a *pacca* road, and about 70 per cent had a telephone facility; by contrast, more than half the villages had access to a wholesale agricultural product market at a distance of more than 10 km; about 48 per cent of the villages had access to an input store at a distance exceeding 5 km while about 35 per cent of the villages had access within <5 km; about 41 per cent of the villages had access to banks within the narrow range <5 km while about 33 per cent had access within 5–10 km; as access to the nearest town makes a difference to marketing of output and purchase of input options, it is of some concern that the nearest town for over 43 per cent of the villages was at a distance >10 km.

Investment in rural transportation and other facilities (e.g. banking, communication, storage) is likely to make agricultural markets more efficient as well as benefit the poor more. Evidence for other Asia and Pacific countries points in the same direction (Gaiha et al., 2009).

[18] This should not be taken to imply that the use of fertilizer does not vary by crop.

Table 4.6 Distribution (%) of villages by access to infrastructure and markets

Distance ranges	Nearest Wholesale Agriculture Product Market	Nearest *Pacca* Road	Nearest Agricultural Input Store	Nearest Bank	Nearest District headquarters	Nearest Town	Nearest Telephone facility
0 kms	3.42	72.27	16.17	14.49	0.00	2.54	69.66
0–5 kms	18.8	22.27	34.89	41.12	0.84	21.61	22.65
5–10 kms	27.35	2.52	25.11	32.71	5.88	32.63	5.13
above 10 kms	50.43	2.94	23.83	11.68	93.28	43.22	2.56
Total	100.00	100.00	100.00	100.00	100.00	100.00	100.00

Note: Authors' calculations based on REDS of NCAER (2006).

3.2 CROP YIELDS BY SIZE

As a *descriptive* technique, we approximate distributions of crop yields by size using kernel density functions (see the graphs in Annex A: Figures A4.1 through A4.4). Relative to histograms, these are smoother and not influenced by the end points of bins.[19]

Figure A4.1 shows that kernel densities of cereal yields among smallholders are unimodal, with a cluster around moderately high values; the densities among medium landholders are unimodal too, with the cluster at slightly lower yields than among smallholders; and the densities among large landholders are bimodal with clusters at low and moderately high yields.

Figure A4.2 illustrates that kernel densities of pulses are bimodal among both large and smallholders with very different clusters of yields; among the latter, yields cluster around very low and large values, while among the former the clusters are around very low and slightly larger values; in striking contrast are the unimodal densities among medium landholders, with a cluster around low yields, but skewed to the right, implying that many obtain low yields while others obtain moderate to high yields.

Figure A4.3 illustrates that vegetable kernel densities are bimodal among all three size groups. Among smallholders clusters of yields occur at moderate or high values, with a few obtaining very high yields; among medium landholders clusters of yields occur at moderate or high values; in striking contrast are the yield densities among large landholders, with a cluster at low values and another at moderately high values.

Figure A4.4 depicts yet another striking contrast in oilseed yields by size. The densities are unimodal among large landholders, with a cluster of yields at low values. The densities among medium landholders, by contrast, are bimodal, with clusters at low and moderately high yields; and, while the

[19] The underlying distribution is Gaussian. For a lucid exposition of why kernel densities are to be preferred to histograms, see Deaton (1997).

kernel densities are bimodal among smallholders too, the clusters occur at low and very high values.

In sum, while the generalization that has dominated the size–productivity debate, with rich and fascinating explanations of why smallholders are more productive, is confirmed, our descriptive analysis suggests that this relation varies with food commodity group (not-so-strong, for example, in cereals). Another point that emerges is that, while much lower fractions of smallholders are concentrated in lower ranges of yields compared with medium and large landholders, segments of smallholders also obtain very low yields (for example, in oilseeds).

3.3 DETERMINANTS OF MARKETED SUPPLY[20]

The main findings of the tobit results on the marketed surplus of cereals, pulses, vegetables, and oilseeds are given in Tables 4.7 and 4.8. In all these cases a quadratic relation is indicated between marketed surplus and land size (see Figures B4.1 through B4.4 in Annex B).

(a) Cereals[21]

The main findings of the tobit results on the marketed surplus of cereals (Table 4.7) are[22]:

1. The higher the household head's schooling, the higher was the marketed surplus of cereals.
2. Lower caste households (the Scheduled Castes [SCs], and Other Backward Castes [OBCs]) marketed lower fractions relative to Others (the omitted group), presumably because of discriminatory practices in output and credit markets.
3. Controlling for these and other effects, small landholders marketed significantly lower proportions than large landholders (the omitted group), and these proportions were substantially lower.
4. The higher the price of cereals, the larger was the marketed surplus. The elasticity of marketed surplus of cereals to its own village price is about 0.39, implying that a 1 per cent higher price is likely to induce a 0.39 per cent larger marketed surplus.
5. The cross-price effects are statistically significant, but not economically. The higher the price of oilseeds, the higher the marketed surplus of cereals.

[20] The specification used and the results are given in Annex B.

[21] Out of 5,694 observations in the sample, the uncensored were 2,791. The use of a tobit is thus justified, taking the size of land owned as predetermined.

[22] As the log of a variable has a monotonic relation to the values of the variable, we avoid use of log for expositional convenience.

Table 4.7 Factors affecting marketed surplus of cereals and pulses: Tobit estimates per acre of land for different farm size groups

	Cereals		Pulses	
Explanatory Variables	Coefficient (t-statistic)	Elasticity (z-statistic)	Coefficient (t-statistic)	Elasticity (z-statistics)
Log of household head's years of schooling	0.20***(3.65)	0.03***(3.65)	−0.11(0.70)	−0.0072(0.70)
Caste dummy: SC	−2.13***(−11.59)	−0.04***(−11.63)	−2.38***(4.14)	−0.0184***(4.18)
Caste dummy: ST	0.31(1.42)	0.003(1.42)	−2.33***(2.96)	−0.0089***(2.98)
Caste dummy: OBC	−0.39***(−3.18)	−0.02***(−3.18)	−0.07(0.20)	−0.0018(0.20)
Land-owned dummy: Small	−2.02***(−11.39)	−0.20***(−11.40)	−3.11***(6.92)	−0.1234***(7.05)
Land-owned dummy: Medium	−0.05(−0.23)	−0.001(−0.23)	−1.88***(3.51)	−0.0145***(3.53)
Log of village-level trader's price for cereals	0.49***(13.13)	0.3942***(13.23)	−0.49***(5.78)	−0.1591***(5.88)
Log of village-level trader's price for oilseeds	0.06***(3.24)	0.0266***(3.24)	−0.28***(6.65)	−0.0542***(6.79)
Log of village-level trader's price for vegetables	0.02(0.85)	0.01(0.85)		
Log of village-level trader's price for pulses			1.82***(15.85)	0.3466***(23.64)
Interaction of log of village-level trader's price for oilseeds and vegetables	−0.01**(−2.45)	−0.01**(−2.45)		
Constant	−0.92***(−2.86)		−10.67***(9.77)	
/sigma	3.63		5.9603	
Number of observations	5694		5694	
Left-censored observations at dep. variable = 0	29		5060	
Uncensored observations	2791		634	
LR chi-square (10)	726.11***		1199.87***	
PseudoR-square	0.0365		0.1716	
Log likelihood	−9588.2501		−2896.5451	

Note: Log of market surplus of cereals is the dependent variable. *** refer to significance at the 1% level and ** at the 5% level.
The elasticities are based on the uncensored observations. After controlling for interaction terms, the marginal effects of log of village-level trader's price for oilseeds and log of village-level trader's price for vegetables at the means were −0.0049 and −0.0550, respectively.

Table 4.8 Factors affecting market surplus of vegetables: Tobit estimates per acre of land for different farm size groups

Explanatory Variables	Vegetables		Oilseeds	
	Coefficient (t-statistic)	Elasticity (z-statistics)	Coefficient (t-statistic)	Elasticity (z-statistics)
Log of household head's years of schooling	0.8826***(3.58)	0.0466***(3.64)	−0.0101(−0.06)	−0.0007(−0.06)
Caste dummy: SC	−0.6093(−0.82)	−0.0037(−0.83)	−1.4538***(−2.62)	−0.0111***(−2.63)
Caste dummy: ST	−6.2826***(−3.03)	−0.0188***(−3.09)	1.4485**(2.13)	0.0054**(2.13)
Caste dummy: OBC	−0.7312(−1.36)	−0.0142(−1.36)	0.0369(0.10)	0.0009(0.10)
Land-owned dummy: Small	−2.2901***(−3.23)	−0.0715***(−3.27)	−5.1216***(−11.23)	−0.1993***(−11.95)
Land-owned dummy: Medium	−0.8137(−0.96)	−0.0049(−0.96)	−2.3031***(−4.45)	−0.0175***(−4.49)
Log of village-level trader's price for pulses	−0.1050*(−1.66)	−0.0158*(−1.67)		
Log of village-level trader's price for cereals	−0.1316(−0.71)	−0.0340(−0.71)	−0.4379***(−3.52)	−0.1410***(−3.56)
Log of village-level trader's price for vegetables	1.5609***(14.48)	0.1381***(22.84)	0.0493(1.09)	0.0054(1.09)
Log of village-level trader's price for oilseeds			1.4809***(16.73)	0.2834***(23.36)
Constant	−15.7938***(−9.23)		−9.5779***(−9.03)	
/sigma	7.2989		6.2926	
Number of observations	5694		5694	
Left-censored observations at dep. variable = 0	5393		5093	
Uncensored observations	301		601	
LR chi-square(9)	644.43***		1041.03***	
PseudoR-square	0.1692		0.1550	
Log likelihood	−1582.4242		−2838.4201	

Note: Log of market surplus of vegetables is the dependent variable. *** and * refer to significance at the 1% and 10% levels, respectively. The elasticities are based on the uncensored observations.

However, the elasticity is low (0.02). The interaction effect of prices of oilseeds and vegetables is negative, but with a low elasticity (–0.01).

(b) Pulses[23]

1. The head's schooling does not have a significant positive effect on marketed surplus of pulses.
2. However, the caste affiliations matter, as both Scheduled Castes (SCs) and Scheduled Tribes (STs) market lower proportions of pulses produced.
3. Smallholders market significantly lower proportions, as also medium landholders, than large landholders.
4. Controlling for these effects, the price of pulses and marketed surplus are positively related and the elasticity is 0.35. This implies that if the cereal price rises by 1 per cent, the marketed surplus rises by 0.35 per cent. This elasticity is slightly lower than that of cereals.
5. The cross-price effects of cereals and oilseeds are negative, with larger (absolute) elasticity of cereals (–0.16) relative to that of oilseeds (–0.05). These results imply that there are significant production substitutions and lower marketed surplus of pulses in response to changes in relative prices.

(c) Vegetables

The sample of households that grew vegetables was small (283). The main findings for vegetables (Table 4.8) are:

1. Head's schooling and marketed surplus of vegetables are positively related.
2. ST households marketed significantly lower fractions than Others.
3. Smallholders marketed a significantly lower proportion of vegetables than large landholders.
4. Controlling for these effects, the price has a robust effect on marketed surplus. The elasticity is 0.14, implying that a 1 per cent higher price induces a 0.14 per cent higher marketed surplus.
5. The higher price of pulses, however, induces a lower marketed surplus, but the (absolute) value of the elasticity is low (-0.016).
6. In another specification without price of cereals, the longer the distance to a wholesale market, the lower is the marketed surplus. This is highly plausible as, given lack of cold storage facilities, vegetables cannot be marketed over long distances. The elasticity is 0.09, implying a 1 per cent increase in distance to the nearest market resulted in a 0.09 per cent lower marketed surplus.

[23] The uncensored observations were 634.

(d) Oilseeds

The tobit results for oilseeds (sample of 601) given in Table 4.8 are:

1. Somewhat surprisingly, the head's education is not linked to marketed surplus of oilseeds.
2. While SCs market lower fractions, STs market higher fractions (relative to Others).
3. Both small and medium landholders market lower fractions of their output than large landholders—especially the former.
4. Controlling for these effects, there is a significant positive price effect on marketed surplus of oilseeds. The elasticity is 0.28, implying that a 1 per cent higher price induces a 0.28 per cent higher marketed surplus.
5. The cross-price effect of cereals is significant with a negative sign and a moderately large (absolute) elasticity (–0.14). This further corroborates production substitution and lower marketed surplus.

In sum, our analysis confirms the important own-price effects on marketed surplus of each of the four food commodity groups: cereals, pulses, vegetables, and oilseeds. However, elasticities with respect to own price vary, with the highest for cereals, followed by pulses and then oilseeds. For vegetables, easier access to markets matters, given lack of cold storage facilities. There is also evidence of cross-price effects, implying production substitutions and changes in marketed surplus. Education of household head matters too in two commodity groups. To the extent that education facilitates access to new technology and market prices, it is also positively related to marketed surplus. In all four cases, smallholders are associated with lower marketed surplus. Our analysis, however, could not throw light on whether smallholders marketed lower fractions because they received lower farm-gate prices and/or because their access to markets was more constrained[24].

3.4 DECLINING PRODUCTIVITY GROWTH

A number of studies have confirmed a slowdown in productivity growth in cereal crops such as rice and wheat in major irrigated areas of Asia, such as the Indo-Gangetic plain and East Asia (Bhandari et al., 2003; Pingali et al., 1997). For example, rice yield growth in irrigated areas of Asia declined from 2.31 per cent per annum in 1970–90 to 0.79 per cent in 1990–2000 (Hossain, 2006). The major reasons for this decline in yield growth include: the displacement of cereals on better lands by more profitable crops; diminishing returns on modern varieties when irrigation and fertilizer use are already at

[24] For a list of variables and crop compositions, see Tables B4.1 and B4.2 in Annex B.

high levels; and the recent low price of cereals relative to input costs, making additional intensification less profitable (Hazell, 2009). In intensive mono-crop systems such as the rice–wheat system of the Indo-Gangetic plains, deteriorating soil and water quality is an important problem; degradation of soils and build-up of toxins have been reported in intensive paddy systems in several Asian countries (Pingali et al., 1997; Ali and Byerlee, 2002).

Researchers have documented stagnating or even declining levels of total factor productivity in some of these production conditions (Janaiah et al., 2005). An analysis of data from long-term yield trials in several countries of South Asia found stagnating or declining yield trends in rice and wheat when input use was held constant (Ladha et al., 2003). One of the reported reasons for slow yield growth is the pest and disease resistance to chemical pesticides[25].

While these findings are informative, an important recent contribution offers a new perspective, based on careful calculation of TFP growth for a large sample of countries and over a very long period (Fuglie, 2010). Briefly, in developing regions, productivity growth has accelerated since the 1980s. Input growth slowed but remained positive. China sustained exceptionally high total factor productivity (TFP) growth rates since the 1980s. Few other countries and sub-regions in Asia and Pacific also performed well. TFP performance is strongly correlated with national investments in 'technology capital'—a measure of a country's ability to develop and extend improved technology to farmers.[26]

3.5 ENVIRONMENTAL PROBLEMS

Poor water management in many countries of Asia has resulted in land degradation in irrigated areas through salinization and waterlogging. It is estimated that almost 40 per cent of irrigated land in dry areas of Asia are affected by salinization (Millennium Ecosystem Assessment, 2005).

Inappropriate use of fertilizers and pesticides has led to water pollution and damage to larger ecosystems, where excess nitrates from agriculture enter water systems. Fertilizer nutrient run-off from agriculture has become a major problem in intensive systems of Asia, causing algal bloom and destroying wetlands and wildlife habitats.

[25] But this is not as worrying as it may seem, as population growth is slowing and yields of some crops, notably maize, are still rising at a modest pace. More importantly, as a recent study (Fuglie, 2010) demonstrates, growth in yields has slowed, as farmers are cutting inputs for environmental reasons or simply because they are more concerned about quality than quantity. So farmers' productivity is still rising at a healthy 1.4 per cent per annum. However, it needs to rise to 1.75 per cent per annum (*The Economist*, 26 February 2011).

[26] For further details, see APR Report (2011) and Fuglie (2010).

Serious soil and water degradation has taken place in the rice–wheat system of India and Pakistan due to intensive and continuous monoculture of rice in summer and wheat in winter (Ali and Byerlee, 2002; World Bank, 2007). The effects of soil nutrient mining, salinization, and declining organic matter have been exacerbated by depletion of groundwater aquifers and build-up of pest and weed populations and resistance to pesticides.

3.6 LAND AND TENURE SECURITY

In many countries of the region, marginalization is linked to the lack of access to land and land-use rights. Improving poor people's access to land is important to improve equity as well as production, as small farms tend to be more productive than large farms (Lipton, 1993, 2006). The political prospects for redistributive land reform are not bright for many developing countries. Also, land scarcity has become acute, and rapid urbanization is reducing the area available for agriculture (Cassman et al., 2003). Crop land per capita of agricultural population is only 0.23 ha in East Asia and the Pacific and 0.27 ha in South Asia, compared to 0.48 ha in sub-Saharan Africa, 0.74 ha in the Middle East and North Africa, 1.55 ha in Latin America and the Caribbean, and 3.53 ha in Europe and Central Asia.

Some aspects of land reform, such as the extension of tenurial security, may be less difficult to implement than other aspects, such as land ceilings. IFAD-supported tribal development projects in India provide examples illustrating the importance of security of tenure. For example, the Orissa Tribal Development Project provided titles to land above ten degrees in slope to tribal groups. Land occupied by tribals became transferable to women in the form of inheritable land titles in perpetuity. Such land titling led to major improvements in natural resource management, creating incentives derived from clear property rights.

In socialist countries like China and Vietnam, land-tenure reform has led to significant increases in agricultural production and rural poverty reduction. In Vietnam under the *Doi Moi* reform process, in 1988 agricultural collectives were converted to contract land to households for 15 years for annual crops and 40 years for perennial crops (Kirk and Nguyen, 2009). This reform, together with the relaxation of price controls and the opening up of domestic and international trade, promoted entrepreneurship and productivity. Vietnam passed a Land Law in 1993 that extended land tenure to 20 years for annual crops and 50 years for perennial crops. These reforms generated strong incentives to invest in agriculture, which led to greater food security and better nutrition. Land transactions increased greatly as a result of tenure reforms. There is an active land market in the country, with the percentage of households participating in land transactions increasing from 3.8 per cent in 1993 to 15.5 per cent in 1998. Although land sales are not allowed, with more secure land rights many farmers have diversified their production into aquaculture, livestock, and perennial crops such as coffee and cashew. Land titles in Cambodia raised rice productivity and reduced rural poverty (Gaiha and Azam, 2011). In China, land rentals have contributed to rural diversification and income growth.

An analysis of land reforms in India by Deininger et al. (2009) using a 20-year panel (1981–99) of household data for rural India yields useful insights into their effects. First, by allowing households to increase investment, land reforms had a positive impact on accumulation of assets, both human and physical. Partly through this channel, land reforms promoted growth.[27] Second, the benefits to the poor were disproportionately large, implying a positive impact on equity. Third, the impact of reforms declined with time—land transfers have come to a virtual standstill in recent years—emphasizing the need for more imaginative approaches that take note of existing opportunities to access land, the obstacles preventing such access, and the potential economic returns from land compared to the alternatives.

3.7 WATER SHORTAGES

In much of Asia, the demand for water for both agricultural and non-agricultural uses is rising and water scarcity is becoming acute, thus limiting the future expansion of irrigation.[28] Irrigated food production in large areas of China and South Asia is being maintained through unsustainable extraction of water from rivers or the ground (UNDP, 2006). The expansion of tubewell irrigation in South Asia has resulted in serious overdrawing of groundwater and falling watertables. In the agriculturally advanced states of India—Haryana, Punjab, Rajasthan, and Tamil Nadu—more than one-fifth of groundwater aquifers are overexploited (World Bank, 2007). As a result, water pumping has become difficult and too costly.[29] The most affected are small farmers, who have little access to expensive pumps and often have insecure water rights.

In Asia in general, and South Asia in particular, the area of land irrigated by large-scale surface schemes has been declining since the early 1990s. For example, between 1994 and 2001, India and Pakistan together lost more

[27] Tenancy reforms and ceilings have significant and positive effects on income, consumption, and assets, with the former yielding stronger effects (Deininger et al., 2009).

[28] What may aggravate water shortage is dietary shift towards meat. It takes 1,150–2,000 litres of water to produce 1 kg of wheat, but about 16,000 litres of water for 1 kg of beef. As more people eat more meat, rising demand by farmers will be pitted against contracting water supplies (*The Economist*, 26 February 2011).

[29] In Punjab (an Indian state), for example, the watertable has plummeted from a couple of metres below the surface to, in parts, hundreds of metres down (*The Economist*, 26 February 2011).

than 5.5 million ha of canal-irrigated areas, despite very large investments in rehabilitation and new projects (Mukherji et al., 2009). Some of these areas were lost due to irrigation-induced soil salinity and waterlogging.

3.8 DIVERSIFICATION

Small farmers have the potential to raise their incomes by switching from grain-based production systems to high-value agriculture. Although the production of high-value agriculture is labour-intensive and thus more suitable for smallholders, they face a number of constraints. Since high-value agricultural commodities are perishable and their markets are fragmented, there is high volatility in their prices, and thus high market risk. In addition, small farmers have low volumes of marketable surplus and the land they cultivate is mostly located in remote areas with poorly developed infrastructure. As a result, smallholders face high transaction costs and risks in production and marketing of such commodities. They also face poor access to credit, and stringent food safety and quality standards.[30]

While growth of urbanization and rising incomes fuelled the growth of a diversified agricultural sector and integration into high-value chains linked to supermarkets in some parts of Asia and the Pacific region, following the food crisis, there is evidence of erosion of trust in markets allocating food supplies in countries worst affected, and heightened concerns for self-sufficiency in food staples. Manifestation of such concerns (reflected in protectionist policies towards rice in particular) runs the risk of slowing down diversification of agriculture.

3.9 IMPACT OF CLIMATE CHANGE

Researchers have predicted that climate change will have serious consequences for agriculture, particularly for smallholders in poor developing countries. In tropical countries even moderate warming (1 degree C for wheat and maize, and 2 degrees C for rice) can reduce yields significantly because many crops are already at the limit of their heat tolerance (World Bank, 2007). In parts of Asia and Central America, wheat and maize yields could decrease by 20 to 40 per cent as the temperature rises by 3 to 4 degrees, even if farm-level adjustments are made to accommodate higher average temperatures, such as changing the date of seeding or planting drought-resistant

[30] As noted in Section 3, although yields of food crops are higher among smallholders, the fact that they market substantially lower fractions of their outputs suggests that lack of easy access to credit and markets are major impediments.

varieties (Long et al., 2007).[31] Rice yields would also decline, although less than wheat and maize yields.

In low-lying areas, agriculture will be adversely affected by flooding and salinization due to sea level rise and saltwater intrusion in groundwater aquifers. Water scarcity will increase in areas such as Nepal and parts of China and India due to decreasing snow cover over time, where glacial melt is an important source of irrigation water.

In an unpleasant taste of what climate change may do, during the 2010 summer the jet stream (air currents of 7,000–12,000 m above sea level which affect the winds and weather) changed its course. This was linked to catastrophic floods in Pakistan and forest fires in Russia, resulting in spiralling food prices later in 2010 (*The Economist*, 26 February 2011).

Both mitigation and adaptation measures are necessary, with greater emphasis on the latter. As the 'world's appetite for emissions reduction has been revealed to be chronically weak', it is imperative 'to find ways of adapting to many possible future climates' (*The Economist*, 25 November 2010).

Adaptation calls for not just expanded research into improved crop yields and tolerance of temperature and water scarcity, but also research into management of pests, soil conservation, and cropping patterns that enhance their resilience (Gaiha and Mathur, 2010). There is also a case for weather insurance, which will pay not when crops fail but when specific climatic events occur (e.g. rainfall below a set level) (Gaiha and Thapa, 2006).

Strategies of adaptation by smallholders raise specific concerns. They are likely to suffer impacts of climate change that are locally specific and hard to predict. The variety of crop and livestock species produced by them, and the importance of non-market relations will increase the complexity both of the impacts and the subsequent adaptations, relative to commercial farms with more restricted ranges of crops. While small farm sizes, low technology, low capitalization, and diverse non-climate stressors (e.g. population-driven land fragmentation, limited access to markets) add to their vulnerability, their existing patterns of diversification away from agriculture and the store of indigenous knowledge impart greater resilience (Morton, 2007).

3.10 RISK AND VULNERABILITY

Smallholders face a number of individual risks such as disease, injury, and death of animals, as well as common or aggregate risks such as drought, epidemics, and economy-wide shocks, affecting everyone in the locality. The consequences of these risks can be extremely severe, potentially leading to

[31] Some researchers point to the fact that the projections of crop yield losses made by different climate change models may be overestimated, as they tend to be based on cereal mono-crops with high rates of chemical fertilizer use.

malnutrition, disease, starvation, or even death. As a result, managing and coping with risks are an integral part of the daily lives of poor rural people.

In addition, there has been concern that the recent successes of market-oriented policy reforms (e.g. in India and China) or the advance of globalization may have further increased the degree of potential income fluctuations, thereby exacerbating the already precarious position of poor rural people, comprising principally landless and small farmers (Dercon, 2005). Evidence points to high vulnerability of small farmers in the semi-arid region of south India to crop shocks. What is worse, occasionally they are subject to a series of such shocks, making it harder for them to escape persistent poverty (Gaiha and Imai, 2004). Other evidence comes from the Philippines, Bangladesh, and Cambodia confirming the significant effects of natural hazards (e.g. *El Nino* in the Philippines, floods in Bangladesh, and droughts, floods, and windstorms in Cambodia) on various indices of poverty and anthropometric measures of under-nutrition.[32] Disasters often disrupt food production, resulting in loss of livelihoods and higher food prices. Finally, not only do poor rural people lose assets, but they also lack access to risk-sharing mechanisms such as insurance. It is therefore not surprising that disasters substantially increase poverty levels (e.g. 50 per cent of the increase in the incidence of poverty in the Philippines during the 1998 crisis was due to *El Nino*). Although the devastation is seldom confined to the poorer segments—including small farmers—in the absence of easy access to credit and insurance they find it harder to recover their previous standard of living (Jalan and Ravallion, 2001).

Although there is overlap between poverty and vulnerability to poverty, with a diverse pattern both *within* and *between* countries for which evidence exists, a useful insight is that poverty and vulnerability are *distinct*. Thus interventions designed to target the latter must differ from those designed for the former. Specifically, more careful attention must be given to risk mitigation and dealing with vulnerability to poverty—especially in rural areas.

4. Opportunities for higher productivity, higher incomes, and sustainability

This section discusses technological as well as institutional innovations that can enable small or family farms sustainably to raise agricultural productivity and to increase incomes by accessing emerging markets for high-value commodities.

[32] See, for example, Gaiha and Azam (2011) for a robust confirmation of how natural hazards aggravate rural poverty in Cambodia.

4.1 TECHNOLOGICAL INNOVATIONS TO ADDRESS ENVIRONMENTAL PROBLEMS AND YIELD GROWTH

To address concerns about the sustainability of Green Revolution technologies and their ability to benefit poor farmers, particularly in less-favoured areas, many advocate new technological approaches (e.g. Pender, 2008). These include low external input and sustainable agriculture approaches based on ecological principles of farming; organic agriculture based on a similar set of agro-ecological principles but without the use of artificial chemical fertilizers, pesticides or genetically modified organisms; and biotechnology. Although biotechnology and agro-ecological approaches seem to be in opposition to one another, both approaches focus on biologically based rather than chemically based technologies, and there may be potential for realising complementarities between these approaches. In fact, it has been argued that a combination of ecological and biotechnology approaches is needed to bring about a 'Doubly Green Revolution' (Conway, 1997). Others have argued that integrated agricultural and natural resource management innovations are needed that combine improved germ plasm (using both conventional methods and biotechnology) and improved and integrated management of soils, water, biodiversity, and other natural resources (CGIAR, 2005).

Conservation agriculture/zero tillage

To address the declining productivity growth of the rice-wheat system in the Indo-Gangetic plain, zero tillage has been promoted by the Rice-Wheat Consortium, a partnership of the Consultative Group on International Agricultural Research centres and national agricultural research and extension system, with the support of IFAD and other development partners. This technology involves planting wheat immediately after rice, without tillage, so that wheat seedlings germinate using the residual soil moisture from the previous rice crop. Zero tillage has been reported to have many advantages over conventional tillage in the rice-wheat system. It saves labour, fertilizer, and energy, minimises planting delays between crops, conserves soil, reduces irrigation water needs, increases tolerance to drought, and reduces greenhouse gas emissions (Erenstein et al., 2007).

Organic agriculture

Organic agriculture is a specific type of low external input whose requirements are more restrictive—no use of chemicals or genetically modified organisms. Based on certification, price premiums of 10 to 50 per cent are common for developing country exports of organic products (IFAD, 2005).

Organic farming has increased rapidly in many Asian countries in the last few years. In 2000–02, there were about 60,000 farms producing certified organic products on about 600,000 ha. This increased to more than 90,000 farms on more than 3.8 million ha in 2005–06 (Pender, 2008). China, India, and Indonesia are the major organic producers in Asia.

Several studies have shown favourable impacts of organic agriculture on the costs of production and yields (IFAD, 2005; Reunglertpanyakul, 2001). However, there are several constraints to the adoption of organic farming. Profit margins usually diminish due to increased competition, and organic producers may face greater market risks as the sector grows. Perhaps the most important concern among smallholder farmers relates to the costs of certification and assuring compliance with organic standards. These problems can be addressed by developing farmer organisations at the local level and through efforts by outside agencies to develop local capacities and facilitate linkages to markets.

Biotechnology

Broadly defined, biotechnology includes a wide variety of techniques, from traditional methods such as conventional plant and animal breeding to more modern techniques such as tissue culture, embryo transfer, cloning, breeding using marker-assisted selection, genetic engineering of plants or animals, and genomics (ADB, 2001). In current literature, the term biotechnology is used to refer to modern agricultural biotechnology, and it is also used synonymously with genetic engineering. Biotechnology is reported to have the potential of incorporating many traits in crop varieties that can address problems faced by smallholders, such as drought resistance, disease and pest resistance, yield improvement, and quality improvement.

Since 1996, there has been a rapid adoption of a few genetically modified (GM) crops globally. Among Asian countries, an estimated 6.4 million small farmers in China (on an average area of 0.5 ha) and one million small farmers in India (on an average area of 1.3 ha) were growing *Bt*[33] cotton by 2005, while more than 50,000 farmers in the Philippines (on an average area of 2 ha) were growing *Bt* maize (Pender, 2008). Studies have shown that *Bt* cotton has contributed to increasing yields, reducing costs of production, increasing farmer incomes and reducing negative health and environmental effects of high pesticide use, particularly in China (Smale et al., 2006; Huang et al.,

[33] Bt stands for Bacillus thuringiensis, which is a soil-dwelling bacterium. It is used as a biological pesticide in genetically modified crops such as cotton and maize using Bt genes.

2002). Other studies conducted in India have also reported reduced pesticide use and increased yields (Bennett et al., 2006; Qaim et al., 2006).

GM cotton has been adopted by large numbers of smallholders in China and India, indicating that the technology can be adopted equally by large and small farmers. It further confirms the ability of smallholders to adopt new technologies, although there may be lags in adoption due to considerations of costs and risks. The dissemination of biotechnology to developing countries is inhibited by intellectual property rights issues, the lack of interest of multinational corporations in investing in the development of GM crops in poor countries and less-favoured areas, difficulties in establishing public-private partnerships, and the lack of investment and leadership in biotechnology by international agricultural research centres (Pender, 2008).

4.2 INSTITUTIONAL INNOVATIONS FOR PRODUCTIVITY ENHANCEMENT AND DIVERSIFICATION

Although smallholders face formidable challenges, a number of innovative institutional models are emerging that can help small farmers benefit from the 'new agriculture' dominated by value chains. These include: the development of farmer/producer organisations for marketing; the promotion of contract farming; the development of supply chains for high-value exports through an appropriate mix of private- and public-sector initiatives; facilitating private-sector provision of market information through telecommunication; and directing fiscal stimulus to rural areas.

Farmer/producer organisations

To overcome challenges related to high transaction costs, small farmers in many countries have formed producer organisations. These organisations are of various kinds, including cooperatives, associations, and societies. They support smallholders in gaining access to markets and public services, and for advocacy. One of the most well-known producer organisations in Asia is the Indian Dairy Cooperative, which in 2005 had a network of more than 100,000 village-level dairy cooperatives with 12.3 million members and which accounts for 22 per cent of milk produced in the country (NDDB, 2006). Sixty per cent of members are landless or smallholders; women make up 25 per cent of the membership. This cooperative model was replicated with the brand name 'Safal' for fruit and vegetables to meet growing demand in the Indian capital Delhi.

Contract farming

Contract farming has been promoted in many Asian countries as a potential means to incorporate small farmers into growing markets for high-value commodities. Since contracts often include the provision of seed, fertilizer, and technical assistance for accessing credit and a guaranteed price at harvest, this form of 'vertical coordination' has the potential to address many constraints to small-farm productivity. In this sense, it has been viewed as an institutional solution to the problems of market failure for credit, insurance, and information (see Box 4.2).

BOX 4.2 COOPERATIVE MODEL FOR VEGETABLE AND FRUIT MARKETING, INDIA

To meet the growing demand for fresh fruits and vegetables in the Indian capital city Delhi, the Mother Dairy Fruit and Vegetables Limited (MDFVL) was established in 1988 as a subsidiary of the National Dairy Development Board, which has brought about a milk revolution in India through farmer cooperatives. MDFVL sells 250 metric tonnes of fresh vegetables and fruits to about seventy-five thousand customers every day. It sources fruits and vegetables from over one hundred and fifty producer associations comprising 18,000 farmers. (Note: these figures are from Birthal et al., 2005, Singh may have used another reference.) These associations are informal cooperatives or self-help groups and are not governed by the State Cooperative Act. MDFVL helps producer associations procure improved seed varieties, fertilizer, and chemicals and also provides extension services. It links producers with input dealers for the supply of production inputs at wholesale rates. It also organizes training programmes for farmers on good agronomic practices to increase production and minimize the use of chemicals. MDFVL has established quality standards for fruit and vegetables, and the produce is graded and priced as per agreed norms. Farmers are paid every two weeks through their associations.

Source: Birthal et al. (2005).

Several studies have assessed to what degree smallholder farmers have participated in contract farming in Asia, and the evidence has been mixed. A recent study of contract and non-contract growers of apples and green onions in Shandong province of China found no bias toward large farmers in contract farming schemes (Miyata et al., 2009). In contrast, another study found that small farmers were less likely to participate in contract farming than larger farmers (Guo et al., 2005). Singh (2002) identifies several problems associated with contract vegetable production in Punjab—power imbalance between farmers and companies, violation of the terms of the agreements, social differentiation, and environmental unsustainability.

Most studies indicate positive impacts of contract farming on incomes. For example, Birthal et al. (2005) found that the gross margins for contract dairy farmers in India were almost double those of independent dairy farmers, largely because contract farmers had lower production and marketing costs. Miyata et al. (2009) also found that contract farmers earned more than non-contract farmers even after controlling for household labour availability, education, farm size, share of land irrigated, and proximity to the village leader. Major factors for this difference included higher yields obtained by contract growers due to the technical assistance and specialised inputs provided by the packers, and higher prices received.

Two challenges are: (i) to achieve discipline in collective action for the producer organisation to meet the terms of the contract and at the same time ensure that members resist the temptation of side-sales, particularly when prices are rising and local markets exist for the contracted product; and (ii) ensure that the commercial partner, often with monopsony power, does not renege on the contractual arrangement when the crop is ready, by offering lower prices or imposing higher quality standard (Byerlee et al., 2010).

Supply chains and supermarkets

Several researchers have argued that smallholders enjoy several advantages over large commercial farmers in supplying supermarkets. The first advantage is linked to production technologies and the associated labour requirements. Thai Fresh United, for example, has a portfolio of 140 herbs, spices, vegetables, and fruits, each of which has stringent quality requirements (Gaiha and Thapa, 2008). Smallholders, especially women, are able to give the careful attention that such crops require. Small producers supplying Hortico in Zimbabwe, for example, had lower rejection rates for certain non-traditional vegetables relative to large farmers. Second, the traditional agro-economic and production practices of smallholders are more amenable to the requirements of supermarkets. For example, in Thailand, Tops in 2003 has found that smallholders adapt more easily to organic production through crop rotation and selection among resistant varieties (Gaiha and Thapa 2008).

However, smallholders need support for intermediation and internalisation to be able to integrate into the supply chains (Gaiha and Thapa, 2008; Lipton, 2006; Swinnen, et al., 2006). Intermediation can take different forms involving the cooperation of public and private agencies. For example, food safety standards might be laid down by national governments, and private agencies might help smallholders implement them; rural infrastructure might be strengthened by the public sector through private financing; suppliers might help finance the provision of inputs and provide extension. Internalisation involves organisations of producers, especially small producers, which negotiate production and marketing arrangements with supermarkets or their suppliers.

A study sponsored by IFAD found the prospects for the expansion of supermarkets to be promising in most Asian countries (Gaiha and Thapa, 2008). It also saw good potential for the integration of smallholders in a rapidly transforming food and agricultural sector, provided they receive adequate support from the public and private sectors.

Information and communication technology

Information and communication technologies can reduce information asymmetries by providing information to smallholders on weather, input and output prices, and production technologies. Many successful examples of smallholders benefiting from ICT are emerging (Box 4.3).

Fiscal stimulus

Although the contagion of the financial crisis did not dampen growth in Asia and the Pacific region as much as initially feared, the projected reductions in growth rates are 2 per cent or more in 2009. This is largely due to the resilience of China and, to a lesser extent, India (ADB, 2009a). In anticipation of such losses, and to minimise them, fiscal stimulus was undertaken by many countries in the region, ranging from 0.5 per cent of gross domestic product to more than 5 per cent (ADB, 2009b). A study undertaken by IFAD's Asia and the Pacific Division (Gaiha et al., 2009) demonstrates the potential of fiscal stimulus in accelerating overall growth through agricultural growth. If mechanisms are put in place to direct the fiscal stimulus to rural areas where both physical and social infrastructure are inadequate to sustain the growth impulse, substantial increases in yields and revenues from agriculture are likely. Various studies have confirmed the vital role of rural roads, transportation and market access in enabling small farmers and others to reap greater benefits from higher prices (Fan and Rao, 2008; Gaiha et al., 2009). Of particular significance are the findings of a study by Shilpi and Deininger (2008), focusing not only on distance to a market in the Indian state of Tamil Nadu, but also on the facilities available in that market. Their analysis shows that additional investments in market facilities are indeed pro-poor, since the sales by poorer farmers increase more than those by wealthy farmers. In other words, while the wealthier farmers capture the benefits of existing facilities better than the poorer farmers, the marginal benefit from an improvement of market facilities is substantially greater for small (poorer) farmers.

Sustainability of the fiscal stimulus, however, seems doubtful amidst fears of inflation in emerging Asian countries (notably China and India).

BOX 4.3 MARKETING SUPPORT TO SMALLHOLDERS THROUGH INFORMATION AND COMMUNICATION TECHNOLOGY: THE CASE OF E-CHOUPAL IN INDIA

The e-Choupal initiative of the Indian Tobacco Company (ITC) is changing the lives of thousands of farmers in India. Between 2000 and 2007, the agribusiness division of ITC set up 6,400 Internet kiosks called e-Choupals in nine Indian states, reaching about thirty-eight thousand villages and 4 million farmers. ITC establishes an Internet facility in a village and appoints and trains an operator (*sanchalak*) from among the farmers in the village. The *sanchalak* operates the computer to enable farmers to get free information on local and global market prices, weather, and farming practices. The e-Choupal also allows farmers to buy a range of consumer goods and agricultural inputs and services (sourced from other companies).

The e-Choupal serves as a purchase centre for ITC for 13 agricultural commodities, with the *sanchalak* acting as the commission agent in purchasing the produce and organizing its delivery to ITC. In 2006/7 ITC purchased about 2 million tonnes of wheat, soybeans, coffee, shrimp, and pulses valued at $400 million through the e-Choupal network. This direct purchasing cuts marketing costs for both farmers and ITC. It improves price transparency and allows better grading of produce. It also allows farmers to realize a bigger share of the final price.

Source: World Bank (2008): Agriculture for Development.

5. **Enabling policy and programme support to small farms—an example from Asia and the Pacific**

There are powerful efficiency and equity reasons to support small farms in Asia and the Pacific. They are economically more efficient relative to large farms, can create large amounts of productive employment, reduce rural poverty and food insecurity, support a more vibrant rural non-farm economy, and help to contain rural-urban migration (Hazell, 2005). The Green Revolution experience showed strong commitment of many Asian governments to agriculture, which led to significant investments in technologies and rural infrastructure as well as major policy and institutional reforms in support of agriculture. In countries such as China and India, public interventions such as land policies, agricultural marketing and support services, and agricultural research and extension benefited commercially oriented small farms. In China, small farms were supported after collective farms could not provide adequate incentives to increase production and productivity.

The reform of the rural economic system in China in 1978 laid an institutional foundation for rural development and poverty reduction. The main

BOX 4.4 POLICY SUPPORT TO SMALL FARMERS IN CHINA

In recent years the government has implemented a series of policies to strengthen the agriculture sector and to benefit small farmers. First, the government has significantly improved resource allocation to agriculture to benefit small farmers in rural areas: from RMB 432 billion in 2007 to RMB 596 billion in 2008 and to RMB 716 billion in 2009. Second, since 2006 the government has abolished agricultural tax and other taxes and fees, which has changed the age-old distribution relationship between the state and farmers. Third, the government has implemented the policy of minimum procurement price for grains to protect farmers' interest and national food security. Fourth, more resources have been allocated to build rural infrastructure and to improve rural production and living conditions. Fifth, since 2007 China has exempted tuition and fees for students in rural elementary and secondary schools, which has benefited over 148 million rural children. The government has also established a new rural cooperative medical system covering 815 million farmers.

Although smallholder farming has contributed significantly to enhance agricultural production and to reduce rural poverty during the past 30 years, it is experiencing new challenges due to globalization and trade liberalization. These include the inability to achieve economies of scale, ineffectiveness in the dissemination of new technologies, and difficulties in risk prevention and control. The government has taken a number of steps to deal with the challenge of declining farm size. Although farmers had land-use contracts for 15 years, administrative reallocation was regularly practised in response to population growth or to make land available for non-agricultural purposes. With the rapid rise in rural–urban migration, decentralized land rentals have complemented and eventually replaced administrative reallocations. Such land rentals are reported to have had favourable impacts on land productivity, occupational structures, and welfare (World Bank, 2007). Net revenue on rented land increased by about 60 per cent, as land was transferred from those with low ability or interest in agriculture to better farmers. Net income increased both for renters and landlords by 25 per cent and 45 per cent, respectively. Land rentals also transformed the occupational structure in rural areas. Almost 60 per cent of farmers, who rented out their land, depended on agriculture as their main source of income before entering land rental markets. Their number declined to 17 per cent following land rentals, with 55 per cent migrating and 29 per cent engaging in local non-farm activity. This shows that, in a context of strong non-farm growth and migration, a well-functioning land rental market can contribute to productivity growth as well as welfare. However, there is a need to continue efforts to strengthen farmers' property rights and to reduce the discretionary powers of officials.

element of the reform was to change the agricultural production model from centralised planning to household contract farming. This reform significantly boosted farmers' incentives to produce more and promoted agricultural development.

Box 4.4 gives the highlights of the current programme of the Chinese government in support of small farmers.

6. Concluding remarks

Small farms have proved resilient over time and they continue to contribute significantly to agricultural production, food security, rural poverty reduction, and biodiversity conservation in Asia and the Pacific region despite the challenges they continue to face with respect to the access to productive resources and service delivery. They are now facing new challenges of integration into a new agriculture dominated by value chains, adaptation to climate change, and management of market volatility and other risks and vulnerability.

However, they have also shown their ability to integrate into the emerging value chains, if they are given support through intermediation and internalisation. Intermediation may take a variety of forms whereby public and private agencies cooperate (e.g. food safety standards might be laid down by governments, and private agencies might help smallholders implement them; rural infrastructure might be strengthened by the public sector through private financing; suppliers might help finance the provision of inputs and provide extension). Internalisation involves organisations of producers, especially small producers, which negotiate production and marketing arrangements with supermarkets or their suppliers.

In the wake of the food price crisis, attractive investment opportunities have opened up in agriculture, leading to large-scale investments and competition for land. However, frequently the large farm advantage is due to market failures (e.g. credit), institutional gaps (e.g. weak extension services) and policy distortions (e.g. minimum support prices). Elimination of such biases against smallholders would enhance their competitiveness.

Institutional innovations can play an important role in the provision of inputs and services to small or family farmers when there are market failures. In some cases, the private sector has adequate incentives to innovate (as discussed above in the sections on contract farming and supermarkets). However, the government should play an active role in coordinating the delivery of input, financial, technical, and output marketing services to small farms. Support will also be needed to enable small farmers to face emerging challenges related to climate change impacts and market volatility.

Annex A

KERNEL DENSITY FUNCTIONS OF YIELDS

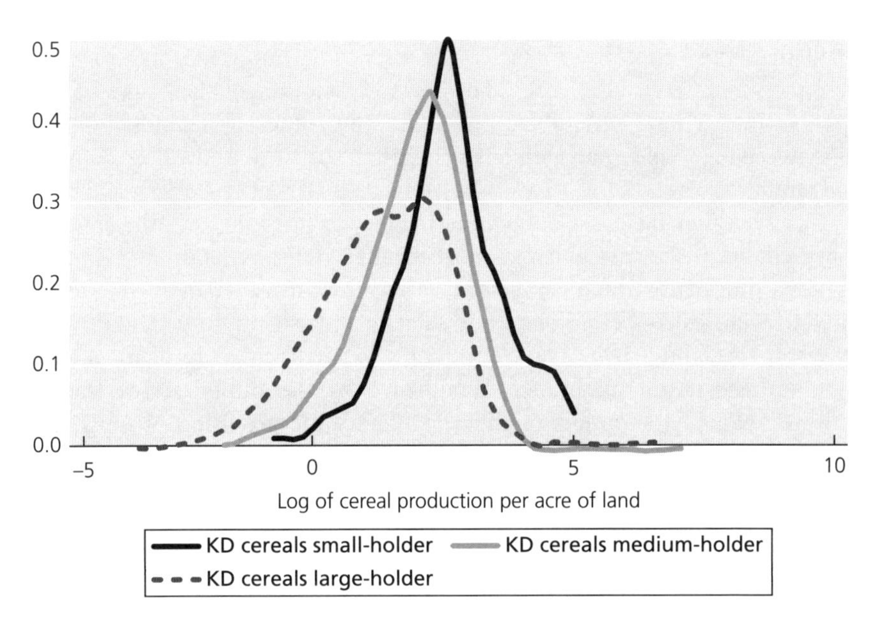

Figure A4.1 Kernel density function for log of cereals

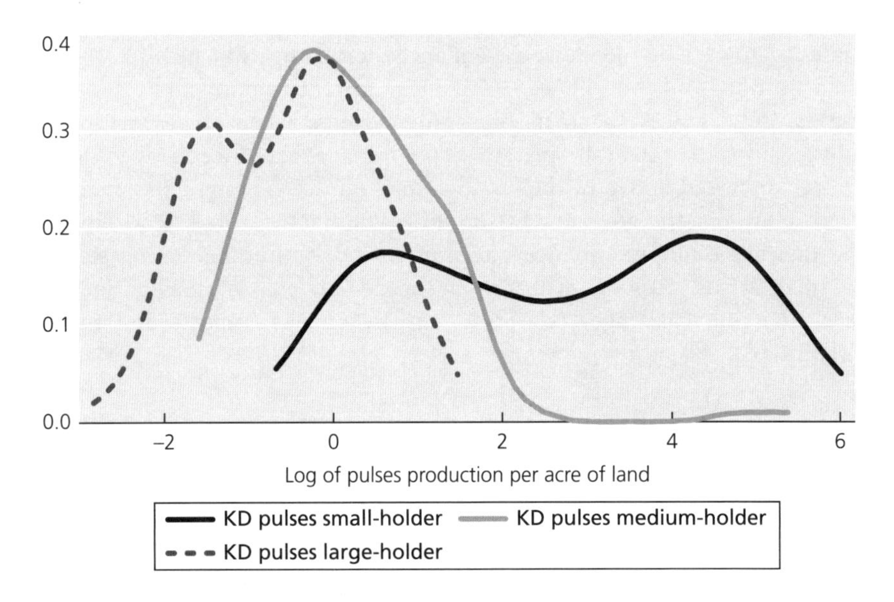

Figure A4.2 Kernel density function for log of pulses

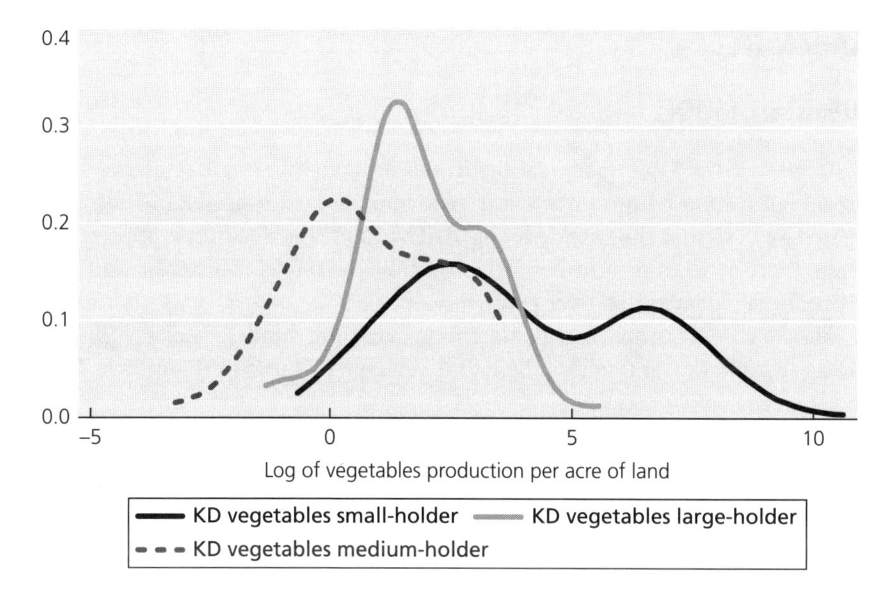

Figure A4.3 Kernel density function for log of vegetables

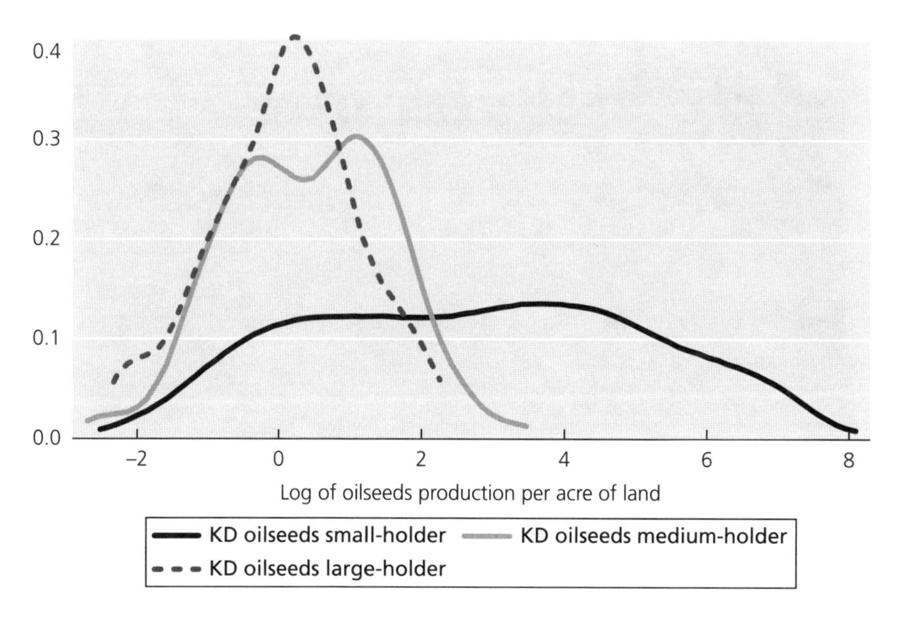

Figure A4.4 Kernel density function for log of oilseeds

Annex B

THE TOBIT MODEL

We have used a tobit specification in which (positive) values of marketed surplus of a food commodity are transformed logarithmically and zeros are treated as 1 (so that the natural log is 0). The tobit specification is appropriate when there is a large number of zeros for a variable of interest and it is continuously distributed over positive values.[34]

The censored normal regression model, or tobit model, is one with censoring from below at 0 where the latent variable is linear in regressors. Thus

$$y^* = \beta_0 + x\beta + \mu, \mu|x \sim \text{Normal}(0, \sigma^2) \quad (1)$$

$$y = \max(0, y^*) \quad (2)$$

The latent variable y^* satisfies the classical linear model assumptions: in particular, it has a normal, homoscedastic distribution with a linear conditional mean. Equation (2) implies that the observed variable, y, equals y^*

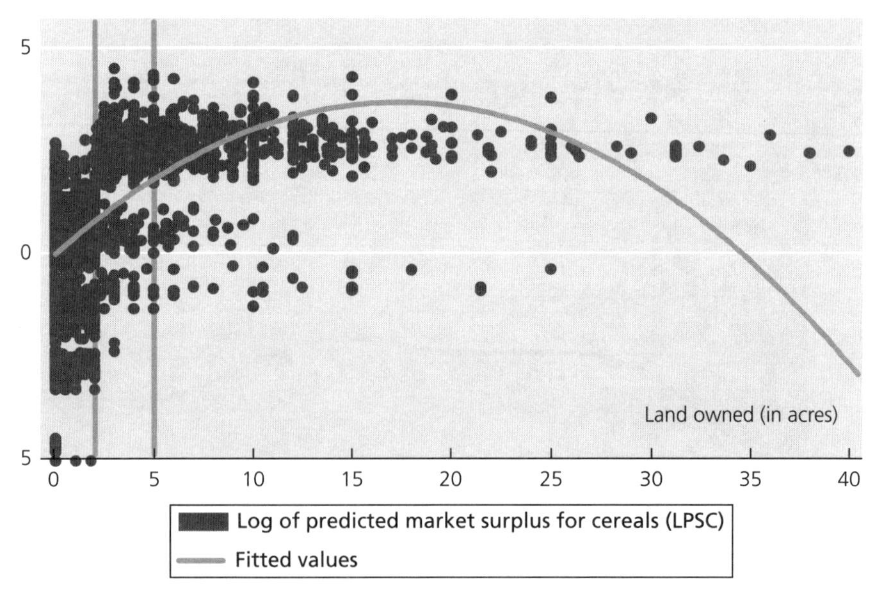

Figure B4.1 Log of predicted market surplus for cereals by landholdings

Note: Vertical lines are drawn at 2 acres and 5 acres, respectively. The lines separate small, medium, and large landholders.

[34] An assumption here is that size distribution of land is predetermined.

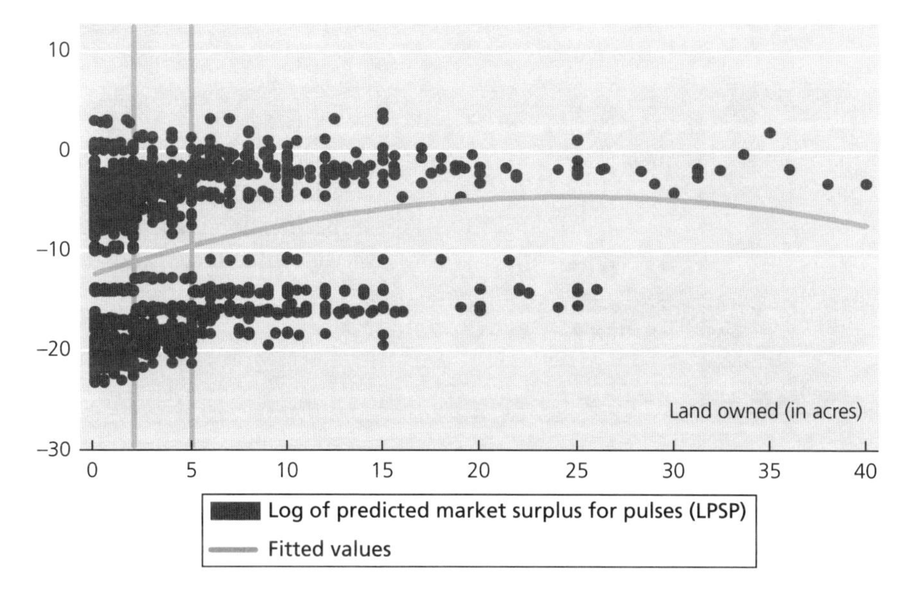

Figure B4.2 Log of predicted market surplus for pulses by landholdings

Note: Vertical lines are drawn at 2 acres and 5 acres, respectively. The lines separate small, medium, and large landholders.

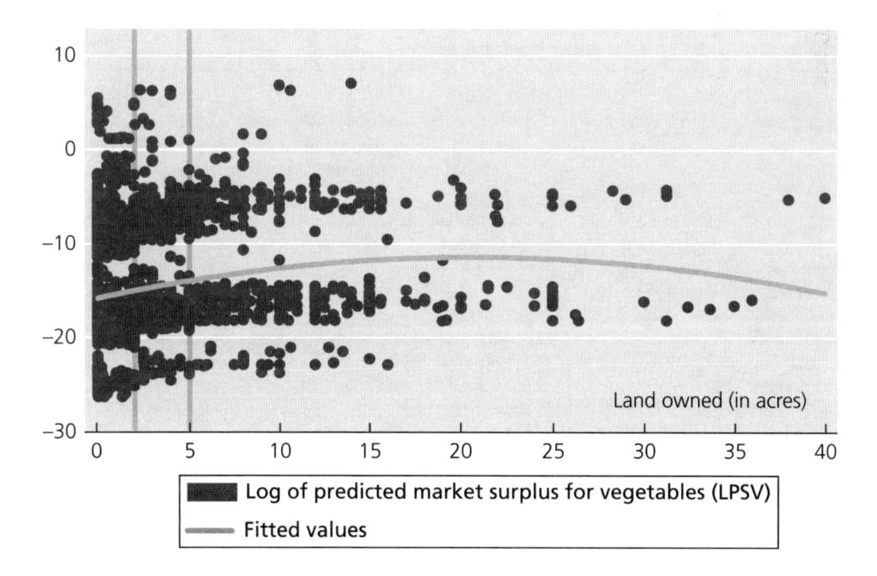

Figure B4.3 Log of predicted market surplus for vegetables by landholdings

Note: Vertical lines are drawn at 2 acres and 5 acres, respectively. The lines separate small, medium, and large landholders.

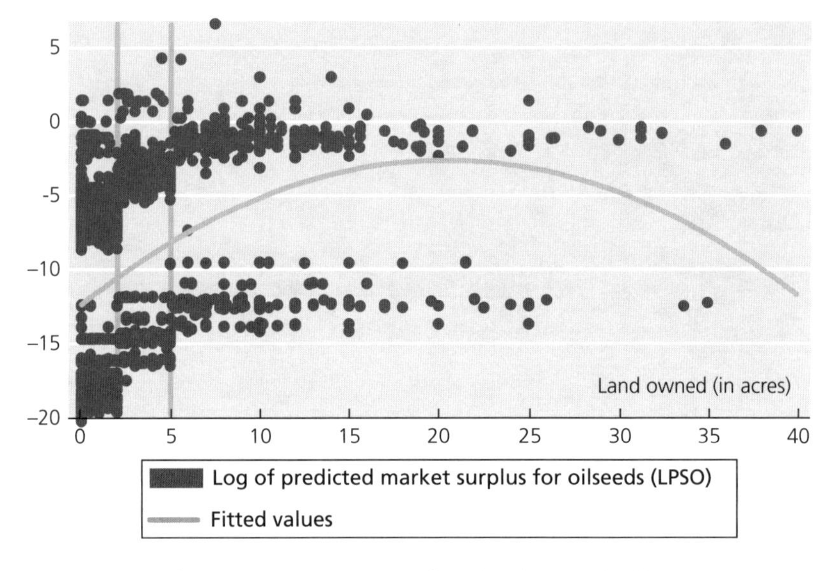

Figure B4.4 Log of predicted market surplus for oilseeds by landholdings

Note: Vertical lines are drawn at 2 acres and 5 acres, respectively. The lines separate small, medium, and large landholders.

Table B4.1 Definitions of variables used in regression analysis

Variables	Definitions
Dependent Variables	
Log of market surplus for commodity i	Log of market surplus for commodity i = log (total sale quantity of commodity i multiplied by 100 and divided by total production quantity of that commodity). i = cereals or pulses or vegetables or oilseeds
Log of production for commodity i	Log of total quantity produced (in quintals) for commodity i. i = cereals
Explanatory Variables	
Log of household head's years of schooling	Log of household head's years of schooling
Caste dummy: SC	= 1 if household is Scheduled Caste; 0 otherwise
Caste dummy: ST	= 1 if household is Scheduled Tribes; 0 otherwise
Caste dummy: OBC	= 1 if household is Other Backward Caste; 0 otherwise
Caste dummy: Others	Reference group
Land-owned dummy: Small	= 1 if land owned by household is less than 2 acres; 0 otherwise
Land-owned dummy: Medium	= 1 if land owned by household is greater than 2 acres but less than or equal to 5 acres; 0 otherwise
Land-owned dummy: Large	Reference group
Log of village-level trader's price for cereals	Log of village-level price (in Rs/quintal) on which produced cereals were sold to traders
Log of village-level trader's price for pulses	Log of village-level price (in Rs/quintal) on which produced pulses were sold to traders
Square of log of village-level trader's price for pulses	Square of log of village-level trader's price for pulses
Log of village-level trader's price for vegetables	Log of village-level price (in Rs/quintal) on which produced vegetables were sold to traders

Log of village-level trader's price for oilseeds	Log of village-level price (in Rs/quintal) on which produced oilseeds were sold to traders
Log of village-level price of chemical fertilizer	Log of village-level price (in Rs/kg) of chemical fertilizer
Log of distance of wholesale agricultural market from the village	Log of distance of wholesale agricultural market from the village
Distance of *pacca* road from village	Distance of *pacca* road from village
Square of distance of *pacca* road from village	Square of distance of *pacca* road from village

Table B4.2 Crops included under cereals, pulses, and vegetables

Commodity	Crops included
Cereals	Paddy, Wheat, Barley, Maize, Jawar, Bajra, Ragi, other cereals and millets
Pulses	Black Gram (Urd), Green Gram (Moong), Pigeon Pea (Arhar, Tur), Horse Gram (Kulthi), Cowpea (Lobia), Kidney Bean (Moth), Lentil (Masoor), Fiels Pea (Matar), Bengal Gram (Chana), and other pulses
Vegetables	Ash Gourd (Kohla), Beet Root (Chukandar), Bitter Gourd (Kerela), Bottle Gourd (Louki), Brinjal, Eggplant (Baingan), Broad Bean (Baakla), Cabbage (Pattagobby), Carrot (Gajat), Cauliflower (Phool Gobby), Cluster Bean (Guvar Ki Fali), Cowpea (Lobia), Cress, Garden Cress (Pani Dhleem), Cucumber (Khera), Double Bean, Drum Stick (Sejana), Elephant Ear, Edible Arum (Akhi, Arvi), Elephant Foot (Gimmy Kand), French Bean (Jungli Sem, Frans Bean), Garden Pea, Pea (Matar), Goose Foot (Bathua), Indian Bean (Sem), Knolknol (Gaath Gobhi), Lady's Finger (Bhindi), Lettuce (Salad), Lime Bean, Little Gourd (Kundroo, Tindora), Mountain Spinach (Pahari Palak), Musk Melon (Kharbooja), Onion (Piaz), Pointed Gourd (Parwal), Potato (Aaloo), Pumpkin (Petha), Radish (Mooli), Red Pumpkin (Sitaphal, Kaddu), Ridge Gourd (Tori), Round Gourd (Tinda), Smooth Gourd (Kali Tori), Snake Guard (Chachera, Chachinda), Spinach (Palak), Sword Bean, Sweet Potato (Sakar Kandi), Tomato (Tamattar), Turnip (Saljam), Velet Bean (Khamch, Tohar Sem), Water Melon (Tarbooj), Yam (Tataaloo), and other vegetables
Oilseeds	Sesamam (Til), Groundnut, Castor, Sunflower, Niger (Ramtil), Soybean, Safflower (Kusum, Kardi), Rapeseed/Mustard (Sarsoan), Indian Mustard (Rai), Linseed (Alsi), other oilseeds

when $y^* \geq 0$, but $y = 0$ when $y^* < 0$. Since y^* is normally distributed, y has a continuous distribution over strictly positive values.

In the estimating equation, the dependent variable, y, represents marketed surplus of food, **x** is a vector of independent variables, $\boldsymbol{\beta}$ is a vector of unknown coefficients, and μ is an independently distributed error term assumed to be normally distributed with 0 mean and variance.

In the tobit, two expectations are of particular interest: $E(y|y>0, x)$, which is sometimes called the 'conditional expectation' because it is conditional on $y>0$, and $E(y|x)$, which is unfortunately called the 'unconditional expectation'. (Both expectations are conditional on the explanatory variables.)[35] We have used the former.

[35] For further details, see Wooldridge (2006).

REFERENCES

Ali, M. (2006). 'Horticultural Revolution for the Poor: Nature, Challenges and Opportunities', background paper for the WDR 2008.

Ali, M. and D. Byerlee (2002). 'Productivity growth and resource degradation in Pakistan's Punjab: A decomposition analysis', *Economic Development and Cultural Change*, 50(4), pp. 839–63.

Anriquez, G. and G. Bonomi (2007). 'Long-term farming and rural demographic trends', background paper for the World Development Report 2008, World Bank.

ADB (2001). *Diagnostic and Analysis of Binding Constraints to Rural Development and Poverty Reduction in Cambodia.* Manila: Asian Development Bank (ADB).

ADB (2001a). *Agricultural Biotechnology, Poverty Reduction, and Food Security.* Manila: Asian Development Bank (ADB).

ADB (2009). *Sector Assistance Programme Evaluation for the Agriculture and Rural Development Sector in Cambodia.* Manila: Asian Development Bank (ADB).

ADB (2009a). *Global financial turmoil and emerging market economics: Major contagion and a shocking loss of wealth.* Manila: Asian Development Bank (ADB).

ADB (2009b). *The global economic crisis: Challenges for developing Asia and ADB's response.* Manila: Asian Development Bank (ADB).

APR Report (Asia and the Pacific Division, IFAD) (2011). 'Agriculture-pathways to prosperity in Asia and the Pacific', Rome: IFAD, (mimeo).

Bhandari, A., R. Amin, C. Yadav, E. Bhattarai, S. Das, H. Aggarwal, R. Gupta, and P. Hobbs (2003). 'How extensive are yield declines in long-term rice-wheat experiments in Asia?', *Field Crops Research*, 81.

Bardhan, P. (2003). 'Size, productivity and returns to scale: an analysis of farm-level data in Indian agriculture', in P. Bardhan (ed.), *Poverty, Agrarian Structure, & Political Economy in India* (pp. 227–47). New Delhi: Oxford University Press.

Bardhan, P. and K. Bardhan (2003). 'Price response of marketed surplus of foodgrains: an analysis of Indian time-series data', in P. Bardhan (ed.), *Poverty, Agrarian Structure, & Political Economy in India* (pp. 248–61). New Delhi: Oxford University Press.

Bennett, R., U. Kambhampati, S. Morse, and Y. Ismael (2006). 'Farm-level economic performance of genetically modified cotton in Maharashtra, India', *Review of Agricultural Economics*, 28 (1).

Birthal, P., P. K. Joshi, and A. Gulati (2005). 'Vertical coordination in high-value commodities—implications for smallholders', MTID Discussion Paper 85, International Food Policy Research Institute, Washington DC, USA.

Byerlee, D., A. de Janvry, and E. Sadoulet (2010). 'Agriculture for development—revisited', Conference on Agriculture for Development, University of California, Berkeley.

Cassman, K., A. Doberman, D. Walters, and H. Yan (2003). 'Meeting cereal demand while protecting natural resources and improving environmental quality', *Annual Review of Environmental Resources*, 28, pp. 315–58.

CEA (Cambodian Economic Association) (2010). 'Does large-scale agricultural investment benefit the poor?', Phnom Penh (mimeo).

CGIAR Science Council (2005). *System Priorities for CGIAR Research 2005–2015*, Science Council Secretariat: Rome, Italy.

Clawson, D. L. (1985). 'Harvest security and intraspecific diversity in traditional tropical agriculture', *Economic Botany*, 39, pp. 56–67.

Conway, G. (1997). *The Double Green Revolution: Food for All in the Twenty-First Century.* Ithaca: Comstock Publishing Associates.

Deaton, A. S. (1997). *The Analysis of Household Surveys: A Microeconometric Approach to Development Policy.* Baltimore: The Johns Hopkins University Press.

Deininger, K. and D. Byerlee (2010). 'The rise of large-scale farms in land-abundant developing countries: does it have a future?', Conference on Agriculture for Development Revisited, University of California, Berkeley.

Deininger, K., S. Jin, and H. K. Nagarajan (2009).'Land reforms, poverty reduction and economic growth: evidence from India', (mimeo).

Dercon, S. (2005). 'Risk, insurance and poverty', in Dercon, S. (ed.), *Insurance Against Poverty.* Oxford: Oxford University Press.

Dixon, J., K. Taniguchi, and H. Wattenbach (eds) (2003). 'Approaches to assessing the impact of globalization on African smallholders: Household and village economy modelling', *Proceedings of a working session on Globalization and the African Smallholder Study.* Rome: FAO.

Erenstein, O., U. Farook, R. Malik, and M. Sharif (2007). 'Adoption and impacts of zero tillage as a resource conserving technology in the irrigated plains of South Asia', Comprehensive Assessment Research Report 19, International Water Management Institute (IWMI), Colombo.

Fan, S. and C. Chan-Kang (2003). 'Is small beautiful? Farm size, productivity and poverty in Asian Agriculture', plenary paper prepared for the 25th International Conference of Agricultural Economists, 17 July, Durban.

Fan, S. and N. Rao (2008). *Public Expenditure, Growth and Poverty in Developing Countries.* New Delhi: Oxford University Press.

Foster, A. D. and M. Rosenzweig (2010). 'Barriers to farm profitability in India: mechanisation, scale and credit markets', Department of Economics, Brown University (mimeo).

Fuglie, K. O. (2010). 'Total factor productivity in the global agricultural economy: Evidence from FAO data', in Alston, J., B. Babcock, and P. Pardey (eds), *The Shifting Patterns of Agricultural Production and Productivity Worldwide.* Midwest Agribusiness Trade and Research Information Centre.

Gaiha, R. and S. Annim (2010). 'Agriculture, GDP and prospects of MDG 1 in Lao PDR', School of Economics Discussion Paper 1012, University of Manchester (mimeo).

Gaiha, R. and Md. Azam (2011). 'Agriculture, GDP and prospects MDGI: a study of Cambodia in transition', APR, IFAD.

Gaiha, R. and K. Imai (2004). 'Vulnerability, persistence of poverty, and shocks-estimates for semi-arid rural India', *Oxford Development Studies*, vol. 32, no. 2.

Gaiha, R., K. Imai, G. Thapa, and W. Kang (2009). 'Fiscal stimulus, agricultural growth and poverty in Asia and the Pacific region: evidence from panel data', Economics Discussion Paper Series Number, EDP-0919, University of Manchester, Manchester.

Gaiha, R., R. Jha, and Vani S. Kulkarni (2010). 'Diets, malnutrition and poverty: the Indian experience', ASARC Working Paper 2010/20, Australian National University, Australia.

Gaiha, R. and S. Mathur (2010). Commentary on 'Does research reduce poverty? Assessing the impacts of policy-oriented research in agriculture', *IDS Bulletin*, vol. 41, no. 6.

Gaiha, R. and G. Thapa (2006). 'Issues in crop and weather insurance', APR, IFAD.

Gaiha, R. and G. Thapa (2008). 'Supermarkets, smallholders and livelihood prospects in selected developing countries', in R. Jha (ed.) (2008), *The Indian Economy Sixty Years After Independence.* Basingstoke: Palgrave Macmillan.

Government of India (2008). *Agricultural Statistics at a Glance.* Directorate of Economics and Statistics, Department of Agriculture and Cooperation, Ministry of Agriculture, New Delhi.

Gulati, A., N. Minot, C. Delgado, and S. Bora (2005). 'Growth in high value agriculture in Asia and the emergence of vertical links with farmers', paper presented at the symposium, *Toward High-Value Agriculture and Vertical Coordination: Implications for Agribusiness and Smallholders*. National Agricultural Science Centre, Pusa, New Delhi.

Gulati, A., N. Minot, C. Delgado, and S. Bora (2006). 'Growth in high-value agriculture in Asia and the emergence of vertical links with farmers'. Forthcoming chapter in Swinnen, J. (ed.), *Global Supply Chains, Standards, and Poor Farmers*. London: CABI Press.

Guo, H., R. W. Jolly, and J. Zhu (2005). 'Contract farming in China: supply chain or ball and chain?', paper presented at Minnesota International Economic Development Conference, University of Minnesota, 29–30 April.

Hazell, P. (2005). 'Is there a future for small farms?', in Colman, D. and N. Vink (eds), *Reshaping Agricultural Contributions to Society*, Proceedings of the 25th International Conference of Agricultural Economists (ICAE), 16–22 August (USA: Blackwell).

Hazell, P. (2009). 'The Asian Green Revolution', IFPRI Discussion Paper 00911, November 2009, Washington DC: IFPRI.

Hazell, P., C. Poulton, S. Wiggins, and A. Dorward (2007). 'The future of small farms for poverty reduction and growth', International Food Policy Research Institute (IFPRI) 2020 Discussion Paper 42, May 2007. Washington DC: IFPRI.

Hossain, M. (2006). 'Rice technology for poverty reduction in unfavourable areas', PowerPoint Presentation at the IFAD Workshop on Evaluation of Asia/Pacific Regional Strategy for Rural Poverty Reduction, June 2006. Manila: ADB.

Huang, J., S. Rozelle, C. Pray, and Q. Wang (2002). 'Plant biotechnology in China', *Science*, 295.

IFAD (2005). 'Organic agriculture and poverty reduction in Asia: China and India Focus: Thematic evaluation', Report No. 1664, Rome.

IFPRI (2007). *The Future of Small Farms: Proceedings of a Research Workshop*, Wye, UK, 26–29 June 2005, jointly organized by International Food Policy Research Institute (IFPRI)/2020 Vision Initiative Overseas Development Institute (ODI) Imperial College, London.

Imai, K., R. Gaiha, G. Thapa, and A. Ali (2011). *A Re-examination of Supply Response to Changes in Food Commodity Prices in Asian Countries*, to be published by the International Economic Association.

Jalan, J. and M. Ravallion (2001). 'Household income dynamics in rural China', draft.

Janaiah, A., K. Otsuka, and M. Hossain (2005). 'Is the productivity impact of the green revolution in rice vanishing?', *Economic and Political Weekly*, 40(53), pp. 5596–600.

Kaicker, N., Vani S. Kulkarni, and R. Gaiha (2011). 'Dietary transition in India: an analysis based on NSS data for 1993 and 2004', Faculty of Management Studies, University of Delhi, mimeo.

Kanwar, S. (2006). 'Relative profitability, supply shifters and dynamic output response in a developing economy', *Journal of Policy Modelling*, 28, pp. 67–88.

Kirk, M. and T. Nguyen (2009). 'Land-tenure policy reforms: decollectivization and the Doi Moi system in Vietnam', International Food Policy Research Institute (IFPRI) Discussion Paper. Washington, DC: IFPRI.

Krishna, R. (1995a). 'Farm supply response in India-Pakistan: a case study of the Punjab region', in V. Krishna (ed.), *Selected Writings: Raj Krishna* (pp. 7–19). New Delhi: Oxford University Press.

Krishna, R. (1995b). 'The marketed surplus function for a subsistence crop: an analysis with Indian data', in V. Krishna (ed.), *Selected Writings: Raj Krishna* (pp. 20–36). New Delhi: Oxford University Press.

Krishna, R. (1995c). 'Agricultural price policy and economic development', in V. Krishna (ed.), *Selected Writings: Raj Krishna* (pp. 49–90). New Delhi: Oxford University Press.

Kulkarni, V. and R. Gaiha (2010). *Dietary Transition in India*, Centre for the Advanced Study of India, University of Pennsylvania.

Ladha, J., D. Dawe, H. Pathak, A. Padre, R. Yadav, B. Singh, Yadvinder Singh, Y. Singh, P. Singh, A. Kundu, R. Sakal, N. Ram, A. Regmi, S. Gami, A. Bhandari, R. Amin, C. Yadav, E. Bhattarai, S. Das, H. Aggrawal, R. Gupta, and P. Hobbs (2003). 'How extensive are yield declines in long-term rice-wheat experiments in Asia', *Field Crops Research*, 81, pp. 159–80.

Lipton, M. (1993). 'Land reform as commenced business: The evidence against stopping', *World Development*, 21(4), pp. 641–57.

Lipton, M. (2006). 'Can small farmers survive, prosper, or be the key channel to cut mass poverty?', *Electronic Journal of Agricultural and Development Economics*, vol. 3, no. 1.

Long, S., E. Ainsworth, A. Leakey, J. Nosberger, and D. Ort (2007). 'Food for thought: lower-than-expected crop yield stimulation with rising CO_2 concentrations', *Science*, 312(5782), pp. 1918–21.

Millennium Ecosystem Assessment (2005). *Current State and Trends Assessment*. Washington, DC: Island Press.

Minot, N., M. Epprecht, T. Anh, and L. Trung (2003). 'Income Diversification and Poverty in the Northern Uplands of Vietnam', report prepared for the Japan Bank for International Cooperation, Hanoi, Vietnam.

Miyata, S., N. Minot, and D. Hu (2009). 'Impact of contract farming on income: linking small farmers, packers, and supermarkets in China', *World Development*, vol. 37, no. 11.

Morton, J. F. (2007). 'The Impact of climate change on smallholder and subsistence agriculture', Proceedings of the National Academy of Sciences, 11 December.

Mukherji, A., T. Facon, J. Burke, C. de Fraiture, J. M. Faures, B. Fuleki, M. Giordano, D. Molden, and T. Shah (2009). *Revitalizing Asia's Irrigation, to Sustainably Meet Tomorrow's Food Needs*. Colombo, Sri Lanka: International Water Management Institute; Rome: Food and Agriculture Organization of the United Nations.

Nagayets, O. (2005). 'Small farms: Current status and key trends', in *The future of small farms: Proceedings of a research workshop*, Wye, UK, 26–29 June 2005. Washington DC: IFPRI.

NCAER (National Council of Applied Economic Research) (2006). *Rural Economic and Demographic Survey (REDS)*. New Delhi.

NCEUS (2008). *A Special Programme for Marginal and Small Farmers*, National Commission for Enterprises in the Unorganised Sector (NCEUS), New Delhi, India.

NDDB (2006). *Annual Report, 2005–2006*, National Dairy Development Board, Anand, India.

Pender, J. (2008). 'Agricultural technology choices for poor farmers in less favoured areas of South and East Asia', Occasional Paper 5, APR, IFAD, Rome.

Pingali, P., M. Hossain, and R. Gerpacio (1997). *Asian Rice Bowls: The Returning Crisis*. Wallingford: CABI.

Pingali, P. and G. Traxler (2002). 'Changing locus of agricultural research: will the poor benefit from biotechnology and privatization trends?', *Food Policy*, 27(3), pp. 223–38.

Qaim, M., A. Subramanian, G. Naik, and D. Zilberman (2006). 'Adoption of *Bt* cotton and impact variability: insights from India', *Review of Agricultural Economics*, 28(1).

Reardon, T., C. Barrett, J. Berdegue, and J. Swinnen (2009). 'Agrifood industry transformation and small farmers in developing countries', *World Development*, 37, pp. 1717–27.

Reardon, T., J. Berdegue, and J. Farington (2002). *Supermarkets and Farming in Latin America: Pointing Directions Elsewhere?*, ODI Natural Resource Perspectives No. 81, December 2002. London: The Overseas Development Institute.

Reunglertpanyakul, V. (2001), 'National study: Thailand', paper on organic farming in Asia presented at the United Nations Economic and Social Commission for Asia and the Pacific regional workshop, *Exploring the Potential of Organic Agriculture for Rural Poverty Alleviation in Asia and the Pacific*, Chiang Mai, Thailand, 26–29 November.

Shilpi, F. and D. Umali-Deininger (2008). 'Market facilities and agricultural marketing: evidence from Tamil Nadu, India', Washington, DC: World Bank (mimeo).

Singh, S. (2002). 'Contracting out solutions: political economy of contract farming in the Indian Punjab', *World Development*, vol. 30, no. 9.

Smale, M., P. Zambrano, and M. Cartel (2006). 'Bales and balance: a review of the methods used to assess the economic impact of *Bt* Cotton on farmers in developing economies', *AgBio-Forum*, 9(3).

Swinnen, J. M., L. Dries, N. Noeva, and E. Germenjia (2006). 'Foreign investment, supermarkets, and the restructuring of supply chains: evidence from Eastern European dairy sectors', LICOS Discussion Papers 16506, LICOS—Centre for Institutions and Economic Performance, KU Leuven.

Thapa, G. (2010). 'Smallholder farming in transforming economies of Asia and the Pacific: challenges and opportunities', paper presented at the Roundtable on the role of smallholder agriculture and family farming in Asia and Latin America and options for South–South cooperation organized by the International Fund for Agricultural Development (IFAD) on 18 February 2010 in Rome.

The Economist (2010). 'Adapting to Climate Change: Facing the Consequences', 25 November.

The Economist (2011). 'Feeding the world', 26 February 2011.

Timmer, C. P. (2009). 'Do supermarkets change the food policy agenda?', special issue on Agrifood industry transformation and small farmers in Developing countries, *World Development*, vol. 37, no. 11.

UNDP (2006). 'Beyond scarcity: power, poverty and the global water crisis', *Human Development Report 2006*, United Nations Development Programme, New York, Palgrave MacMillan.

Wooldridge, J. (2006). *Introductory Econometrics: A Modern Approach*. Cincinnati, OH: South-Western College Publishing.

World Bank (2003). 'Reaching the rural poor: A renewed strategy for rural development', Washington, DC.

World Bank (2007). 'World Development Report 2008: Agriculture for development', Washington, DC.

5 The state of smallholders in agriculture in Latin America

JULIO A. BERDEGUÉ AND RICARDO FUENTEALBA[1]

1. Introduction

Latin America (LAC[2]) has undergone fundamental economic, social, cultural, and political changes since 1980 and in many respects, it is truly a very different place than a generation ago.

What has not changed to the same degree, unfortunately, is the poverty and the dismal distribution of income. In the early 1980s there were 124 million rural inhabitants in LAC[3], 74 million of whom were poor and, of these, 41 million could not even meet their food needs. Thirty years later, the numbers are 119 million, 62 million, and 35 million, respectively (Berdegué, 2009; CEPAL, 2010). This mediocre performance in rural poverty reduction is even more disappointing if one considers that in the same period GDP per capita increased by over 25 per cent in real terms.

Inequality is in large part the reason why economic growth and, indeed, the rapid transformation of LAC societies, have not resulted in a more substantial poverty reduction. If adjusted by inequality, the Human Development Indexes (HDI) of 18 LAC countries for which there is data, drop below the HDI for Africa (4 countries) or Asia (11 countries)[4]. The richest 20 per cent of the rural population earn between 10 and 50 times more than the poorest 20 per cent ranges (CEPAL, 2010); in 9 of 16 countries for which there is data, this measure of income distribution is worsening (Berdegué, 2009). The majority of the countries for which there is data have Gini coefficients of rural income that are higher than 0.5, thus confirming rural LAC as the most unequal rural sector in the world (Schejtman and Berdegué, 2009). Inequality

[1] The authors wish to thank Ms Felicity Proctor and an anonymous referee for comments/suggestions that helped improve the chapter.

[2] Here defined to include the 19 Spanish- and Portuguese-speaking countries South of the Mexico–US border, including two Caribbean countries (Dominican Republic and Cuba).

[3] Using the official definition of 'rural', which in all LAC countries significantly under-represents the true size of the rural population, by as much as 100 per cent or more (de Ferranti et al., 2005).

[4] Data taken from <http://hdrstats.undp.org/en/tables/default.html>.

of access to land is even worse, with a Gini of 0.78, compared with Africa's 0.62 (Justino et al., 2003).

Because of this deep inequality, *average* national incomes give a very distorted image of the reality of LAC's rural people, portraying them as 'middle income' when, in fact, many are very poor. For example, while Mexico's GDP per capita is $8,920, the average income of the poorest 40 per cent of the rural population is $652 per year, and that of the poorest 20 per cent is of $456 per year (equivalent to the GDP per capita of Tanzania).

Rural households whose head declares him or herself to be primarily 'self-employed in agriculture' have seen a deterioration in their welfare over the past twenty years or so. In 10 out of 15 countries analysed, there has been a growing gap in poverty rates between this category and the rural average: Costa Rica (gap grows by 22 percentage points), Panama (15 points), Mexico (14 points), Chile (10 points), El Salvador (9 points), Guatemala (7 points), Nicaragua (4 points), Honduras (3 points), Paraguay (2 points), and Bolivia (1 percentage point). Peru remains stable, while there are improvements (narrowing of the gap) in Dominican Republic (12 percentage points), Colombia (10 points), Brazil (5 points), and Venezuela (1 point) (Berdegué et al., 2006). However, according to Modrego et al. (2006), there has been a significant reduction in the gaps in services such as education of household members over 15 years of age and access to electricity, between households headed by 'self-employed in agriculture' and those headed by 'employers in agriculture'.

This chapter explores the state of family farmers in Latin America as a (diverse) social group caught between these two realities: a rapidly changing context that creates new incentives and new opportunities, and the dead weight of structural inequalities that constrain many from participating in and taking advantage of development processes.

After this section we propose an operational definition of smallholder to overcome the limitations of alternative definitions based on the single criterion of farm size, which is particularly useless in reflecting the reality of the smallholder sector in LAC. This leads us to a typology based on the asset endowment of the farm household and the conditions of the proximate environment in which the household must operate; three large groups or 'types' of smallholders are identified.

In the third section, using this typology we review the available data, referring to secondary information and published sources. We recognize that this approach is far from ideal, as many publications group smallholders in ways different to the typology that we use.

In the fourth section we turn again to the typology, this time discussing the second of the classification variables—the proximate environment in which smallholders operate. While we could discuss numerous characteristics of what is clearly a multi-dimensional environment, we focus on two aspects:

local trends of growth and poverty reduction, and changes in agrifood systems. We argue that these contextual conditions affect the potential and performance of smallholders, very often to a far greater extent than the size of the farm.

The fifth section presents four normative conclusions: (a) it is necessary and possible to differentiate strategies and policies according to the three categories of family farms; (b) policies ought to focus not only on developing the assets and capabilities of farmers, but also on changing the contexts in which they operate; (c) greater attention needs to be paid to domestic food markets, with an emphasis on commodities; and (d) it is necessary to emphasize the development of public services and public goods that can work at the scale of 15 million family farms, in contrast with programmes that have given greater priority to transferring private assets to, inevitably, only a small proportion of family farmers.

2. **Concepts**

Since Alexander Chayanov published his theory of the peasant economy in 1925, there has been a recognition that the key feature of the agricultural smallholder sector is its reliance on family labour, that leads to linking the operation of the family farm to the family's consumption, labour circumstances, and demographic cycles (Chayanov, 1986).

More recently, authors as diverse as CEPAL (1982), Lipton (2005), or the World Bank (2007) have forged a strong consensus that the definitional characteristics of this type of agriculture include: small farms, family operated farms, no or limited non-family hired labour. There is less agreement on whether other factors ought to be included in the definition, the most important of which (for both analytical and for policy and even political purposes) is the ability of the household or family to sustain its livelihood on the basis of its self-employment in its own farm. But regardless of the ongoing debates, the above-mentioned characteristics are the crux of the matter.

Yet, when it comes to making use of the concept, we have somehow ended with a working definition of smallholder agriculture as 'farms of 2 hectares or less' (Nagayets, 2005; Wiggins et al., 2010; Hazell et al., 2010; IFAD, 2010). The author of the most recent analysis that uses the 2 hectares definition recognizes the limitations of this approach,

> given that it fails to properly account for the quality of resources, the types of crops grown, or disparities across regions... The size-based definition also precludes

analysis or comparison of institutional or market arrangements available to farmers . . . as well as their access to key social services . . . further, the size-based definition does not shed light on a farm's labour arrangements . . . which can also have substantial implications for the farms' efficiency and productivity (Nagayets, 2005, p. 355).

In short, the 2 hectare definition is a measure of our ignorance and not of our understanding of smallholder farming, nor of what is needed for well-designed strategies and policies.

In this chapter, smallholder or family based agriculture is defined as a social and economic sector made up of farms that are operated by farm families, using largely their own labour. We therefore include in the smallholder or family based agriculture sector two categories which could be controversial. First, a large sub-sector often referred to as 'subsistence farmers', who derive a large fraction of their household income from non-farm sources, including non-farm employment, remittances, as well as cash and in kind social welfare support. Second, a sub-sector that is smaller in number of farms, but of much greater importance when it comes to economic participation; these are commercial family farmers who may employ one or two permanent non-family workers, but where much of the farm work and farm management is done by family members.

Furthermore, we want to add another dimension to the discussion of the smallholder sector. While the farm size and family labour criteria are very important, they miss a key point: the actual performance and potential of a family farm is conditioned not only by the availability and use of its assets, but by the characteristics of its proximate environment, socio-economic as well as bio-physical (de Janvry and Sadoulet, 2000).

Assets and context are used by Berdegué and Escobar (2002) to propose a framework as depicted in Figure 5.1. On the vertical axis, Figure 5.1 recognizes variation in land, labour, and/or capital asset endowment, with farms with more/better assets on top of the graph. Environments or contexts vary along the horizontal axis, with more favourable ones on the right and unfavourable areas on the left. With this framework, the agricultural sector can be categorized in three large groups, each of which requires specific strategies and policies in order to make the best possible use of their assets in the context in which they operate.

The first category, group A in Figure 5.1, includes some (a few) smallholders that have a relatively important asset endowment (land, labour, and/or access to capital) and are located in places (territories, regions) where the productivity of those assets is high. These farmers are usually fully integrated in a market economy and make a substantial contribution to the production of food for domestic and international markets. Their future depends at least as much on market trends and developments as on the specificities of public policy.

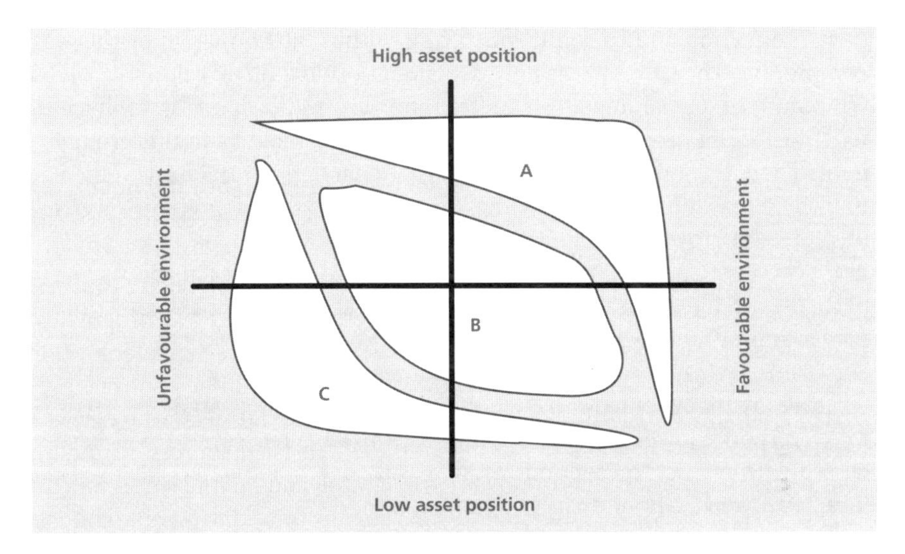

Figure 5.1 Types of family farms according to asset endowment and context

Source: Berdegué and Escobar, (2002). .

The second category, group B in Figure 5.1, is often ignored in polarized policy debates, that is, those reduced to dichotomous categories (large vs small farms; commercial vs subsistence; 'viable' vs 'unviable'; poor vs rich). The transformation of LAC rural societies has led to a significant sector of family farms that have some assets but often lack a few critical elements to make the difference (e.g. they have land, but lack enough credit or are not part of an effective producers' organization). Or they may be located in places (territories or regions) where the bio-physical and socio-economic conditions are 'good enough', but are not fast-moving regions of the highly competitive, globalized agribusiness. This sector falls through the cracks of institutional failures and state weaknesses, often ignored by policymakers focused on promoting agricultural exports *and* by those whose mandate is to target the poorest—not eligible for many public programmes targeting the most disadvantaged, nor strong enough to make it on their own in open and competitive markets. This is a sector made invisible by the 2 hectare limit, and yet it represents the best opportunity in LAC for strategies and public policies aimed at revitalizing rural societies and for promoting socially inclusive economic growth.

The third category, group C in Figure 5.1, is the other extreme of the distribution. It is composed of resource-poor farmers located in places where conditions are adverse not only for agriculture, but often for other economic activities. Many, perhaps the majority of these farm families, derive only a fraction of their income from agriculture; they trade only a fraction of their production and consume much of it within the household or the local community. They are increasingly dependent on low-quality non-farm jobs, temporary or partial migration and remittances, and cash and in kind transfers from social programmes. The majority of smallholders in this group are poor. There is significant agreement in LAC that agricultural development is unlikely to be sufficient to move these smallholders out of poverty. Nevertheless, there is growing recognition that agricultural income is an essential, if not dominant, part of the livelihood strategies of these smallholders, in particular if the focus is on reducing vulnerability to shocks and on the prevention or amelioration of crises that can push these families to levels of abject misery from which they may never recover.

There are many dimensions to both 'context' and 'assets', and one needs to make certain choices. For the purpose of this chapter, context will be discussed with respect to economic growth and poverty reduction trends at the local level, and agrifood market trends.

In conclusion, a simplification of the heterogeneity of smallholder agriculture that is useful for the purpose of designing and implementing development strategies, policies, and programmes is as follows:

- Asset-poor smallholders in territorial and regional contexts that are not conducive to economic growth and social development.
- Smallholder agriculture with some limitations of assets in territorial and regional contexts where there is a measure of economic growth and social development.
- Asset-rich smallholders in territorial and regional contexts that are very conducive to economic growth and social development.

3. The smallholder sector in LAC

Using the definition of the smallholder sector as that comprised by farms of less than 2 hectares, Nagayets (2005) finds that there are about 5 million small farms in the Americas. This estimate, adopted by other authors (e.g. Wiggins et al., 2010; Hazell et al., 2010) as well as by IFAD in the background concept note for the conference 'New Directions for Smallholder Agriculture', is patently wrong. While a limit of 2 hectares perhaps fits the distribution of land holdings in Asia[5], it certainly does not in LAC. There, this procedure distorts our understanding of smallholder agriculture, and misguides the design of public strategies and policies, as it reduces the smallholder to a

[5] According to the Nagayets' estimate (2005), 87 per cent of the world's 400–500 million small farms in the world are in Asia, and Africa has about 40 million small farms.

fraction of its real size, particularly if measured in terms of economic and social contributions.

Based on agricultural census data, Chiriboga (1999) estimated that in 15 LAC countries there were about half a million corporate farms, controlling roughly 55 per cent of the farm land in LAC. The rest, the smallholder sector according to this analyst, would be made up of about 6 million commercial family farms (42 per cent of the land) and 11 million subsistence farms (3 per cent of the land). Distribution of farms according to this classification varies by sub-region, with a higher proportion of commercial family farms in the Southern Cone (Argentina, Chile, Uruguay, and South of Brazil), than in the Andes and in Central America.

Schejtman and Berdegué (2009), using different sources to those of Chiriboga (1999), estimate that there are 7.3 million type B farms (see Figure 5.1) in Brazil, Chile, Colombia, Honduras, Mexico, Paraguay, and Peru, representing between 14 per cent and 53 per cent of the total number of farms in those countries.

Soto Baquero et al. (2007) studied six countries (Brazil, Chile, Colombia, Ecuador, Mexico, and Nicaragua) and concluded that the family based agricultural sector was made up of 11 million units (and 50 million people), and that they controlled between 30 per cent and 60 per cent of the agricultural land in those countries (including forests). These authors differentiate three strata: subsistence farms (equivalent to type C in Figure 5.1), in transition (type B), and consolidated (partly in type B and partly in type A). Subsistence farms number 7 million (and 63 million hectares), the group in transition adds up to 3 million farms (43 million hectares), and the consolidated family farms number 1 million (29 million hectares), all according to Soto Baquero et al. (2007).

Schejtman (2008) added data for six countries (Argentina, Bolivia, Guatemala, Paraguay, Peru, and Uruguay) to the study by Soto Baquero et al. (2007), reaching a total of 14 million farms, of which 60 per cent correspond to subsistence smallholders, 28 per cent to small farmers in transition, and 12 per cent to consolidated family farms.

All of the classifications cited here are based on estimates of how much land approximately correlates with the potential to have a production surplus to sell in the marketplace.

Recent evidence has established that the smallholder sector is not decreasing in numbers in LAC. Under different scenarios of economic growth and urban–rural wage differentials, Bezemer and Hazell (2006) projected 'little change in Latin America and the Caribbean' (p. 13). Modrego et al. (2006) analysed household surveys for nine countries and found slow annual changes in the proportion of rural households that they defined as 'self-employed in agriculture' (Table 5.1); this is taken directly from the questions asked in the household survey and generally refer to the position of the head of household.

Table 5.1 Land holdings in Argentina

Size (hectares)	Number	Area (hectares)
Less than 5	40,957	105,895
5.1–10	22,664	177,973
10.1–25	39,833	714,584
25.1–50	33,787	1,290,129
50.1–100	34,881	2,660,005
100.1–200	34,614	5,150,390
200.1–500	40,211	13,113,229
500.1–1000	21,441	15,261.566
1000.1–2500	16,621	26,489,560
2500.1–5000	6256	22,525,345
5000.1–7500	2088	12,962,493
7500.1–10,000	1285	11,546,633
10,000.1–20,000	1851	27,296,370
20,000 and over	936	35,514,388
Total	297,425	174,808,564

Source: Agricultural Census 2002.

In fact, in four of the nine countries (Chile, Colombia, Guatemala, and Honduras), their participation increased over time.

Now we will look at more detailed data for several countries in the region. Missing from the analysis are a few countries where smallholder agriculture is important, but for which there is a lack of recent and reliable analyses of the issues of interest to this chapter. The absence of data for Mexico, Peru, and Bolivia is particularly noteworthy; Mexico and Peru have recent Agricultural Censi, and Bolivia is scheduled to do one in 2011, and therefore it should be possible to close this important gap.

3.1 ARGENTINA

The distribution of agricultural holdings by size is shown in Table 5.2, according to data from the 2002 Agricultural Census. We can see that while smallholder farms of less than 5 hectares comprise the largest category in number (14 per cent of the total), they account for less than one-tenth of 1 per cent in terms of land access; the standard 2 hectare definition of smallholders is useless in Argentina.

The Inter-American Institute for Cooperation on Agriculture has defined an Argentinean family farm as one operated by the farm family, with no permanent salaried workers, and with a size limit that varies by macro-regions and provinces (Scheinkerman, 2009); this upper limit is estimated from the 2002 Agricultural Census data to range between 500 hectares (in the Corrientes and Misiones provinces) to 2,500 hectares in Patagonia, and even to 5,000

Table 5.2 Rural households by economic activity of head of household in Latin America (per cent)

		Group								
Country	Year	Self-employed ag	Employers ag	Employees ag	Agriculturally based HH	Self-employed non ag	Employer non ag	Employee non ag	Not agriculturally based HH	Unemployed/Not in the labour force
Chile	1990	18.35	2.79	33.66	**54.81**	5.14	0.46	11.66	17.26	**27.93**
	2003	19.48	1.74	27.54	**48.76**	5.70	0.74	12.24	18.67	**32.56**
	D	1.13	−1.05	−6.13	**−6.05**	0.56	0.28	0.58	1.41	**4.64**
Colombia	1995	19.66	5.63	23.78	**49.07**	16.12	2.03	16.64	34.79	**16.14**
	2000	24.73	5.42	18.07	**48.22**	16.72	1.87	14.78	33.37	**18.42**
	D	5.08	−0.21	−5.71	**−0.85**	0.60	−0.16	−1.87	−1.42	**2.27**
Costa Rica	1995	10.44	3.18	18.55	**32.17**	11.61	3.56	32.05	47.22	**20.61**
	2001	9.76	4.67	17.04	**31.47**	12.00	4.65	30.11	46.76	**21.92**
	D	−0.68	1.49	−1.52	**−0.70**	0.39	1.09	−1.94	−0.46	**1.31**
Guatemala	1989	9.42	0.52	6.89	**16.83**	19.42	3.47	43.75	66.64	**16.53**
	2002	34.09	5.19	16.71	**55.99**	9.27	3.88	15.03	28.18	**15.84**
	D	24.67	4.67	9.82	**39.16**	−10.16	0.41	−28.71	−38.46	**−0.69**
Honduras	1995	34.44	1.95	14.23	**50.62**	14.97	2.25	14.66	31.88	**17.50**
	2003	40.94	1.29	16.34	**58.57**	12.29	0.51	11.44	24.24	**17.19**
	D	6.50	−0.66	2.12	**7.96**	−2.68	−1.74	−3.21	−7.64	**−0.32**
Mexico	1994	29.36	5.05	24.36	**58.77**	8.94	1.05	17.27	27.27	**13.96**
	2002	26.10	4.10	22.31	**52.50**	9.82	0.98	20.69	31.49	**16.01**
	D	−3.27	−0.96	−2.05	**−6.27**	0.88	−0.07	3.42	4.22	**2.05**
Nicaragua	1993	36.40	0.13	15.08	**51.62**	8.57	0.08	14.56	23.21	**25.18**
	2001	34.17	8.74	16.52	**59.43**	7.69	1.45	14.35	23.50	**17.07**
	D	−2.23	8.60	1.44	**7.82**	−0.88	1.37	−0.21	0.29	**−8.11**
Paraguay	1995	52.65	0.00	8.73	**61.38**	13.11	2.21	13.49	28.82	**9.80**
	2001	43.31	3.64	8.28	**55.23**	10.81	2.42	16.45	29.68	**15.09**
	D	−9.35	3.64	−0.44	**−6.15**	−2.31	0.21	2.96	0.86	**5.28**
Peru	1994	61.78	0.00	11.62	**73.40**	10.88	0.00	10.18	21.06	**5.54**
	2002	53.63	9.10	8.24	**70.97**	9.23	1.13	12.49	22.86	**6.18**
	D	−8.15	9.10	−3.38	**−2.43**	−1.65	1.13	2.32	1.80	**0.64**

*Indicates difference not statistically significant at 5%

Source: Modrego et al. (2006).

hectares in the southern limits of the Patagonia. According to the IICA study (Scheinkerman, 2009), 75 per cent of Argentinean farms are family farms, and they control 31 million hectares of farm land (18 per cent of the total).

Scheinkerman (2009) classifies Argentinean family farms in four groups:

- Type A: does not own a tractor, holds less than fifty equivalent cattle units, has less than two irrigated hectares, is not a fruit producer, and does not practise greenhouse farming. This group has 113,234 members and 5.9 million hectares of farmland.
- Type B: tractors are more than 15 years old, has between 51 and 100 equivalent cattle units or between 2 and 5 irrigated hectares, of which up to 0.5 hectares may have fruit trees. This group has 58,602 members and 6.3 million hectares of farmland.
- Type C: tractors are less than fifteen years old, has more than one hundred equivalent cattle units, or more than five irrigated hectares or more than 0.5 hectares with greenhouses or fruit trees. This group has 47,032 members and 11.4 million hectares of farmland.
- Type D: while this category shares some of the characteristics of Type C farmers, this group also employs one or two wage workers. This group has 32,248 members and 7.4 million hectares of farmland.

According to the same study, the Pampas with 72,000 family farms (158 hectares per farm on average) and the north-eastern provinces with 71,000 family farms (91 hectares per farm on average) account for 57 per cent of all family farms in Argentina. Those in the Pampas, however, face a much more favourable environment than those in the north-east.

Another study is that of Carmagnani (2008), which sets an upper limit for family farms at 500 hectares, but also excludes those of 50 hectares as belonging to a social category whose livelihood is mainly based on non-farm sources of income, including agricultural and non-agricultural wages, and social policy cash and non-cash transfers. Carmagnani calls this group of less than 50 hectare farms 'sub-family farms', meaning that income from the farm activity is insufficient to sustain a family. He also defines an upper category of family farms that in several ways resemble corporate farms, for example, by employing permanent wage workers. Table 5.3 categorizes Argentinean farms according to Carmagnani definitions and 2002 Agricultural Census data.

Hence, according to Carmagnani (2008), family farms, strictly speaking, in Argentina would be 35 per cent of the total and they hold about 9 per cent of the farmland.

An IICA study published in 2007 (Scheinkerman et al., 2007) finds that 40 per cent (or 132,272) of households registered in the 2002 Agricultural census are poor. Poverty rates are highest in the north-west (63 per cent), the north-east (59 per cent), and Patagonia (48 per cent), and significantly lower in the

Table 5.3 Classification of family farms in Argentina

Type of family farm	Number (000)	Size (hectares)	Area (000 hectares)	Total labour (000)	Workers per farm
Sub-family	103	0–25	999	183	1.8
Family	103	25–200	9,101	183	1.8
Corporate-like	91	200–max	164,710	416	4.6

Source: Carmagnani (2008) and Agricultural Census 2002.

Cuyo (27 per cent) and Patagonia (21 per cent) regions, the latter being the one with the highest number of family farms.

3.2 BRAZIL

In Brazil, a family farm is defined by the Family Farming Law (Law 11,326) based on four criteria:

- Does not have, under any tenure regime, an area of more than four fiscal modules.
- Predominantly relies on its own family labour.
- The household income predominantly originates in the family farm.
- The family operates the farm.

Bollinger and Olivera (2010), however, have criticized the legal definition of family farming on the basis that it excludes 'hundreds of thousands' of households whose main income is not from the family farm. This is particularly important in a country that has seen a notorious expansion of the non-farm rural economy, of Conditional Cash Transfer programmes, and of a very successful universal pension scheme that includes rural households.

As shown in Table 5.4, if one were to use the 2 hectare criterion, there are slightly more than 1 million smallholder units in Brazil (20 per cent of the total), holding 829,000 hectares (about two-tenths of 1 per cent). This concept grossly misrepresents the Brazilian smallholder sector as defined in the international literature and by Brazilian law.

The 2006 Agricultural Census in Brazil included a number of questions that allows for a direct measurement of the family farm sector, according to the legal definition in place in the country that is consistent with the international literature. In 2006, there were 4.3 million family farms (84 per cent) and 808,000 non-family farms in Brazil. The family farm sector controlled 80 million farmland hectares (24 per cent).

According to a recent study (Soto Baquero et al., 2007), almost half of all Brazilian family farms are in the north-eastern region, followed by the south

Table 5.4 Land holdings in Brazil

Size (hectares)	Number of farms (000)	Area (000 hectares)
Less than 2	1,049	829
2–10	1,428	6970
10–50	1,581	36,410
50–100	391	26,483
100–500	371	75,738
500–1000	54	36,958
1000–10,000	45	105,845
10,000 and more	2	40,708
No land	255	–
Total	5,176	329,941

Source: Agricultural Census 2006 (IBGE, 2009).

Table 5.5 Classification of family farms in Brazil

Type	Number (000)	Area (000 hectares)
A (consolidated)	406	24,141
B (transitional)	994	33,810
C (subsistence)	2739	49,817
Total	**4,139**	**107,768**

Source: Soto Baquero et al. (2007).

(22 per cent) and the south-east (15 per cent). These are two contrasting environments: a more favourable one in the south and south-east, and one that is notoriously unfavourable in the north-east (as will be discussed in greater detail later in this chapter). Again, the issue is not only one of how many smallholder farms there are and how much land and other assets each one has on average, but of the context in which those families will use those assets and, hence, what is the productivity potential of those assets and what are the transaction costs that will be faced.

Soto Baquero et al. (2007) categorized family farms using 2006 Brazilian Agricultural Census data, as shown in Table 5.5. According to the authors, two-thirds of the 4.1 million family farms in Brazil, with a total of almost 50 million hectares (18 hectares per farm, on average), belong to the subsistence group. The intermediate or transitional group includes 24 per cent of family farms that control 34 million hectares (31 per cent), with an average farm size of 34 hectares. Finally, there is a category of 40 thousand consolidated family farms that control about 7 per cent of the farmland of the family farming sector, with average holdings of 59 hectares.

Carmagnani (2008) implicitly coincides with the legal definition, by concluding that land holdings of less than 5 hectares should not be considered part of the family farming group, stating that 'less than 6 per cent of their income is cash farm income' and, hence, their livelihood is based on wage labour and other non-farm sources. According to this author, family farming is limited to holding between 5 and 100 hectares, but he states that even the 5–20 hectare group faces great difficulties in making a living out of a family farm. Carmagnani (2008) concludes that only 35 per cent of the units officially categorized as family farms, with a total of 54.4 million hectares, would fit his more stringent definition of family farms that can derive most or all of their income from the farm and not from non-farm sources.

3.3 CHILE

According to the 2007 Agricultural Census, 43 per cent of the agricultural land holdings in Chile have 5 hectares or less of land, but they control less than 1 per cent of the total farm land in the country (Table 5.6).

A study by Echenique and Romero (2009) defines family agriculture according to the legal and operational criteria used by the Ministry of Agriculture of Chile, with an upper limit of 12 'basic irrigated hectares' (BIH). The study concludes that those farms below 2 BIH derive most of their income from non-farm sources, so that the relevant group are those families that farm between 2 and 12 BIH.

After converting land to irrigated land equivalents (BIH), Echenique and Romero (2009) conclude that there are about 195 thousand farms (70 per cent) with less than 2 BIH, holding about 1.2 million hectares (6 per cent of the total). This group of subsistence farmers derives a large proportion of their income from non-farm sources. The next category in Table 5.7 is that of

Table 5.6 Land holdings in Chile

Size (hectares)	Number (000)	Area (000 hectares)
1 or less	34,699	18
1–5	84,975	210
5–10	46,139	325
10–20	42,611	596
20–50	36,965	1,145
50–100	14,911	1,028
100–200	8149	1,125
200–500	5,677	1,737
500–1000	2056	1,414
1000–2000	1048	1,441
2000 and more	1430	20,742
Total	278,660	29,782

Source: INE (2007).

Table 5.7 Classification of farms in Chile according to farm size standardized according to biological yield potential

Type	Number of holdings	Area (000 hectares)
Less than 2 BIH[(a)] (subsistence farms)	195,309	1,205
2–12 BIH (family farms)	67,795	2,599
12–60 BIH	19,351	3,307
60 and more BIH	5331	13,335
Total	287,786	20,446

(a) BIH = basic irrigated hectares, and equivalent to the biological yield potential of one irrigated hectare in the Maipo valley
Source: Echenique and Romero (2009).

Table 5.8 Classification of family farms in Chile, 2007 Census

Type	Number of farms
Self-consumption rural household	74,459
Pluri-active rural household	30,224
Peasant family farms	58,379
Non-peasant family farms	1569
Semi-commercial non-family farms	14,189
Corporate agriculture	15,706
Total	194,526

Source: Jara et al. (2009).

family farms that represent about 23 per cent of the total number of agricultural holdings with approximately 13 per cent of the total farmland, and an average farm size of 38 BIH.

Jara et al. (2009) use 2007 Agricultural Census data on farm size, farm labour, percentage of the family income derived from work on the farm, and access to markets to classify Chilean farms in six categories (Table 5.8). Three of them correspond to the group that is relevant to smallholder policies. The first category, subsistence rural households, includes 38 per cent of all farms in Chile; those in this group rely to a significant degree on self-consumption of agricultural produce, and public cash and non-cash social subsidies. The second group, with 16 per cent of all farms, is similar to the previous one, except that here non-farm income is more prominent. The third category is that of 'peasant family farms', referring to farm families who derive their livelihood predominantly from their own work on the farm; this group includes about 30 per cent of all farms in Chile.

Soto Baquero et al. (2007) revisit the 1997 Agricultural Census data and classify Chilean farms in the three categories that we have already described (Table 5.9). A first result is that family farms represent 87 per cent of all

Table 5.9 Classification of family farms in Chile, 1997 Census

Type	Number	Area (1000 hectares)	Average Area (hectares)
Subsistence	154,820	2,656	17.2
Transitional	120,626	3,214	26.7
Consolidated	8,942	5,89	66.0
Total family farms	284,388	6,460	22.7
Non-family farms	41,127	44,840	1,090.3
Total	325,515	51,300	155,7

Source: Soto Baquero et al. (2007), based on 1997 Agricultural Census data.

Chilean farms, controlling only 13 per cent of the farmland; the average farm size of a family farm in Chile is 23 hectares. The subsistence group in Soto Baquero's et al. classification is the largest, with about 54 per cent of all family farms, and 48 per cent of all farms, but with only 5 per cent of the total farmland, and an average farm size of 17 hectares. The group of transitional family farms includes 120,000 units, about 37 per cent of all farms in Chile, while controlling 6 per cent of all farmland, with an average farm size of 27 hectares. Finally, consolidated family farms are only 3 per cent of all Chilean farms, and they only control 1 per cent of the farmland; the average farm size of this group is of 66 hectares.

Unfortunately, the studies of Soto Baquero et al. (2007) and of Jara et al. (2009) that use, respectively, the 1997 and the 2007 Agricultural Censi of Chile, are not comparable because they used different criteria and assumptions in their classifications. It is a pending task to know if any of these groups are growing in relative importance.

3.4 COLOMBIA

According to the 2001 Agricultural Census of Colombia, there are about four hundred thousand farms that fit the 2 hectare smallholder definition; they control less than one-half of 1 per cent of the total land. As in other countries, this definition is of little use (Table 5.10).

Forero and Galeano (2010) have estimated the number of smallholders per department of Colombia on the basis of the 2004 National Agricultural Survey. They generate two estimates, one with a 10 hectare cut-off point and another one with 20 hectares (Table 5.11). The sum of both groups gives a total of 1.7 million farms of less than 20 hectares. Four departments (Boyacá, Cauca, Cundinamarca, and Nariño, in that order), have two-thirds of all farms of less than 10 hectares. When it comes to farms between 10 and 20 hectares, Cauca and Nariño are replaced in the ranking by Santander and Antioquia, and these two departments plus Boyacá and Cundinamarca account for almost 40 per cent of all farms in this class.

Table 5.10 Land holdings in Colombia

Size	Number	Area (hectares)
1 or less	366,244	191,820
1–3	465,025	844,523
3–5	236,633	899,925
5–10	291,752	2,042,050
10–20	225,238	3,127,283
20–50	219,912	6,884,453
50–100	108,715	7,487,517
100–200	55,906	7,566,533
200–500	40,797	11,598,122
500 and more	11,669	10,063,221
Total	2,021,891	50,705,447

Source: FAO (2010).

Table 5.11 Land holdings of less than 20 hectares in Colombia by department

Department	Less than 10 hectares		10–20 hectares	
	Number	Percentage in the department	Number	Percentage in the department
Antioquía	74,571	55	19,838	15
Atlántico	14,368	100	0	0
Bolívar	12,133	39	0	0
Boyacá	392,259	91	22,726	6
Caldas	20,339	66	3805	13
Cauca	188,176	86	17,237	8
Córdoba	22,641	42	8872	16
Cundinamarca	180,175	80	22,496	10
Huila	58,576	69	11,010	13
La Guajira	7666	48	0	0
Magdalena	7259	24	4477	16
Meta	13,310	27	5817	12
Nariño	171,201	91	10,632	6
Nte de Santander	29,657	47	13,277	21
Quindío	9448	73	1722	13
Risaralda	21,296	82	2294	9
Santander	51,628	52	19,918	19
Sucre	16,851	48	7281	21
Tolima	58,412	60	16,780	17
Valle del Cauca	29,421	60	0	0
Casanare	10,112	47	0	0
Other Departments	40,063	21	35,868	18
Total	1,429,562	69	224,050	11

Source: National Agricultural Survey (2004), in Forero and Galeano (2010).

In 12 of the 22 departments reported by Forero and Galeano (2010), smallholdings of less than 10 hectares make up 50 per cent or more of the total number of farms. If one considers the 20 hectare cut-off point, another three departments join the group of those with a majority of smallholders.

Of the 'self-employed in agriculture' households, almost 44 per cent are in the Western region (according to the regional classification of the Regional Economic and Social Planning Council), 26 per cent in the centre–east region, and 25 per cent in the Atlantic region.

The self-employed in agricultural Colombian households are further classified by Soto Baquero et al. (2007) into three distinct groups according to their total household incomes. Unfortunately, we have no information about land holdings, but we do have estimates of monthly household incomes (Table 5.12).

3.5 ECUADOR

According to the 2 hectare definition, smallholder agriculture in Ecuador includes 29 per cent of all farms, but controls only 0.7 per cent of the land (Table 5.13). Clearly, this definition does not serve any useful policy or analytical purpose in this country.

Forero and Galeano (2010) estimate the distribution of smallholdings of less than 10 hectares and of 20 hectares in Ecuador, with data from the 2000 Agricultural Census (Table 5.14). They conclude that there are 635,000 family

Table 5.12 Classification of family farms in Colombia

Type	PPP US dollars monthly	Number
Subsistence	< 579.2	585,540
Transitional	579.2–1158.4	95,316
Consolidated	>1158.4	57,093
Total	–	737,949

Source: Soto Baquero et al. (2007).

Table 5.13 Land holdings in Ecuador

Size (hectares)	Number	Area (hectares)
Less than 1	248,398	95,834
1–2	117,660	156,016
2–3	78,850	183,354
3–5	90,401	339,021
5–10	101,066	688,987
10–20	75,660	1,017,807
20–50	76,792	2,372,027
50–100	34,498	2,242,409
100–200	12,941	1,666,879
200 and more	6616	3,593,496
TOTAL	842,882	12,355,831

Source: National Census of Agriculture 2000.

Table 5.14 Land holdings of less than 20 hectares in Ecuador by province

Department	Less than 10 hectares		1,020 hectares	
	Number	Percentage in the province	Number	Percentage in the province
Azuay	80.128	80	9537	10
Bolívar	22.402	58	6450	16
Cañar y el Piedrero	25.188	78	3357	10
Carchi	7171	56	2582	20
Chimborazo	68.289	84	7352	9
Cotopaxi	54.319	80	5856	9
El Oro	10.050	45	3586	17
Esmeraldas, Golondrinas y Concordia	2689	12	1852	10
Guayas	34.602	53	12.893	20
Imbabura	26.228	78	2743	8
Loja	33.754	51	12.016	19
Los Ríos	19.596	47	8931	21
Manabi y Manga del Cura	36.474	48	11.542	15
Morona Santiago	2046	12	1186	7
Napo	596	12	401	7
Orellana	459	8	511	8
Pastaza	1008	19	158	3
Pichincha	41.418	65	6330	10
Sucumbios	577	7	585	8
Tungurahua	67.069	94	2282	3
Zamora Chinchipe	700	8	342	4
Total	534.763	63	100.492	12

Source: Agricultural census (2000), in Forero and Galeano (2010).

Table 5.15 Classification of family farms in Ecuador

Type	Number (000)	Area (000 hectares)
Subsistence	456	2,510
Transitional	274	1,932
Consolidated	10	640
Total family farms	740	5,083
Non-family farms	103	7,272
Total	843	12,355

Source: Soto Baquero et al. (2007).

farms in the country, of which 84 per cent are of less than 10 hectares. The provinces of Azuay, Chimborazo, Tungurahua, Cotopaxi, and Pichincha, in that order, contain 58 per cent of all farms of less than 10 hectares. In 16 of the 22 provinces reported by Forero and Galeano (2010), smallholders represent 50 per cent or more of the total number of farms.

Soto Baquero et al. (2007) analyse the 2000 Agricultural Census and find that family agriculture includes 88 per cent of all farms and 41 per cent of the farmland (Table 5.15). Using the same types already reported for other countries, they conclude that the 'subsistence' group included almost half a million farms (54 per cent), with about 2.5 million hectares of land (20 per cent), and an average farm size of 5.5 hectares. The second category, 'transitional' farms, includes 33 per cent of all farms, 15 per cent of the farmland, with an average farm size of 7 hectares. Finally, the group of 'consolidated' family farms includes slightly less than ten thousand units that control 5 per cent of the farmland, with an average farm size of 66 hectares. It should be noted that the average farm sizes of 'consolidated family farms' and 'non-family farms' are not too different (66 and 77 hectares, respectively), again driving home the message that in this day and age, land is important but hardly sufficient to determine the potential of a farm.

Table 5.16 shows the regional distribution of the different types of family farms in the Soto Baquero et al. (2007) classification. First of all, a healthy 36 per cent of family farms are located in the Coast region, which generally speaking is a better or more conducive environment for agricultural development; it is not a surprise that there are fewer subsistence farms in the Coast than in the Andean highlands. Transitional family farms are more or less equally distributed in the Coast and in the Highlands, while consolidated family farms also have a strong presence in the Amazon Basin, probably linked to coffee, cocoa, and cattle production.

An interesting piece of information comes from the Soto Baquero et al. (2007) study, which helps to put the issue of farms size (assets) vs farm context in perspective. On average, a subsistence farm in the Coastal plains is twice as large as one in the Andean Highlands, while one in the Amazon Basin is eight times larger. At the same time, a farm of 4.5 hectares in the Andes is already 'transitional', while one of 25 hectares in the Amazon Basin is still in the 'subsistence' group. Clearly, using farm size alone as a criterion will lead to wrong strategic and policy decisions.

Carmagnani (2008) takes a more restrictive approach, since he considers that family agriculture is distinct from subsistence agriculture, where the family depends to a large extent on non-farm income. Using this classification, he estimates that there are over half a million subsistence farms (63 per cent of the total), holding about 6 per cent of the land. Family agriculture includes a

Table 5.16 Regional distribution of different types of family farms (per cent)

Type	Subsistence	Transitional	Consolidated	Total family farms
Coastal plains	31	44	43	36
Highlands	62	51	36	58
Amazon Basin	7	5	20	6
Total	100	100	100	100

Source: Soto Baquero et al. (2007).

quarter of a million farms, with one-third of the land. Finally, corporate farms are only 6 per cent of the total, but they control 61 per cent of the total farmland. According to this analyst, the size of subsistence farms is on average 1.44 hectares, while family farms are 11 times larger (16 hectares on average).

3.6 GUATEMALA

The 2003 Guatemalan Agricultural Census found 830,000 farms in the country, of which around 76 per cent were smallholders according to the 2 hectare definition (Table 5.17). These farms control about 13 per cent of the 3.7 million hectares recorded in the census, a very high proportion for LAC standards. Yet this definition leaves out a sizable proportion of what would constitute smallholders or family farmers under the international consensus of farms operated by farm families with no or little wage labour.

Fradejas and Gauster (2006) classify Guatemalan farms in four groups (Table 5.18). The first two categories, which they call 'infra-subsistence' (household members cannot possibly survive based on farm production and income alone) and 'subsistence' account for a staggering 92 per cent of

Table 5.17 Land holdings in Guatemala

Size (hectares)	Number	Area (hectares)
0.4–0.7	375,708	121,655
0.7–1.4	185,196	170,976
1.4–3.5	157,681	317,124
3.5–7.1	46,099	210,296
7.1–22.6	39,599	475,998
22.6–45.2	10,929	332,138
45.2–452	14,593	1,299,209
452–903.2	610	361,983
903.2–2258	222	284,784
2258–4516	37	114,187
4516–9032	9	50,973
9032 and more	1	11,530
Total	830,684	3,750,853

Source: FAO (2010).

Table 5.18 Classification of family farms in Guatemala

Size (hectares)	Number	Area (hectares)	Average area (hectares)
< 0.7 (less than subsistence)	375,708	172,413	0.46
0.7–7 (subsistence)	388,976	989,791	2.5
7–45 (surplus)	50,528	1,145,318	22.7
> 45 (commercial)	15,472	3,008,316	194.4
Total	830,684	5,315,838	6.4

Source: Fradejas and Gauster (2006).

all farms, but they only control 22 per cent of the land. The situation of the first group is worth noting: 45 per cent of the farms with less than 3 per cent of the land, or half a hectare per household. Clearly it is not possible to expect that this group of Guatemalan families will base their livelihood strategies on self-employment in agriculture.

The second and third groups of the Fradejas and Gauster (2006) classification would likely be prioritized by smallholder policies. Together they account for over half of all Guatemalan farms, and they control 41 per cent of the land, with an average farm size of slightly less than 5 hectares.

3.7 NICARAGUA

According to the 2001 Nicaraguan Agricultural Census, there are about 40,000 farms of less than 2 hectares, controlling 77 thousand hectares, that is, just over 1 per cent of the total farmland (Table 5.19). If the 2 hectare criterion for defining what constitutes a smallholder had any real meaning, the past 30 years or so of Nicaraguan history, including a revolution, an agrarian reform, and a civil war, would have to be described as examples of major conflicts over nothing.

At the same time, it is not reasonable to think that under the conditions of the Nicaraguan rural sector, a family can sustain its livelihood on the basis of an average farm size of 1 hectare (for the three first lines in Table 5.19). For this reason Carmagnani (2008) argues that the bottom limit of Nicaraguan family farming is at around 5.6 hectares, while the upper limit is at about 50 hectares.

If Carmagnani is right, then the family farming sector in Nicaragua is made up of about 90,000 farms (45 per cent of the total), with about 1.8 million hectares (29 per cent of the total farmland), and average farm sizes of about 20

Table 5.19 Land holdings in Nicaragua

Size (hectares)	Number	Area (hectares)
Less than 0.4	7337	1936
0.4–0.7	10,745	7146
0.7–1.8	21,379	28,389
1.8–3.5	26,517	72,808
3.5–7	28,576	159,300
7–14	27,022	298,717
14–35	38,780	982,308
35–70	21,684	1,172,423
70–140	10,746	1,139,997
140–350	5169	1,153,030
350 and more	1594	1,238,462
Total	199,549	6,254,516

Source: FAO (2010).

hectares. Below that there is a sector of about 80,000 subsistence farmers with about 160,000 hectares (2 hectares per farm). Soto Baquero et al. (2007) categorize Nicaraguan family farmers into the three groups with which we have seen in other countries in this chapter. They base their report on the 2001 National Household Survey, and not on the Agricultural Census. However, their analysis needs to be considered with much care since the authors—after expanding the survey data—arrive at 293,000 farms in the country, that is, 50 per cent more than the number of farms accounted for in the 2001 Agricultural Census. In the absence of the Soto Baquero et al. (2007) classification, we lack any other source to group Nicaraguan smallholders within a good typology.

3.8 URUGUAY

Uruguay is a small country with only 57,000 farms according to the 2000 Agricultural Census (Table 5.20). Farms of less than 2 hectares are less than 3 or 4 per cent, and they control about one-tenth of 1 per cent of the land. This category is therefore meaningless for any analytical or policy purpose.

Carmagnani (2008) proposes that all units of more than 100 hectares are corporate farms that do not belong to the family farm sector. At the same time, he argues that the lower boundary of this sector is of about 10 hectares, below which a family has to rely on non-farm sources of income. Projecting this analysis on to the 2000 Census data, one would conclude that with this restricted definition, the family farm sector in Uruguay is made up of about 23,000 farms (40 per cent of the total), that control about 856,000 hectares, with an average farm size of 37 hectares. If one considers the subsistence sector it would add 13,000 farms but only 65,000 hectares, thus reducing the average farm size significantly to 25 hectares.

Table 5.20 Land holdings in Uruguay

Size (hectares)	Number	Area (000 hectares)
1–4	6260	17
5–9	7086	48
10–19	7118	98
20–49	8934	285
50–99	6647	473
100–199	6382	910
200–499	6783	2163
500–999	3887	2726
1000–2499	2912	4,442
2500–4999	838	2837
5000–9999	228	1505
10,000 and more	56	918
Total	57,131	16,420

Source: National Census of Agriculture 2000.

3.9 IN SUMMARY

In summary, a detailed reading of the best estimates of the size of smallholder agriculture in LAC, allows us to conclude that it is made up of around 15 million farms. About 65 per cent correspond to a category of smallholders that rely significantly and perhaps increasingly on non-farm sources of income to sustain their livelihoods; for these farmers, agriculture complements other activities, remittances, and cash and in kind social transfers and supports are of great importance. Still, this group owns or controls well over 100 million hectares. Even if small, the income derived from this land is absolutely critical for their survival and to reduce their vulnerability to shocks of all kinds. Many in this group would be considered poor. Yet an agriculture-based or agriculture-led development strategy would miss the fundamentals in the case of this group.

A second category is those family farmers that indisputably and most clearly meet the criteria considered by most authors. Their livelihood depends predominantly on the operation of their farms, they hire little or no non-family labour, and therefore operate and manage their farms with the members of the farm family. They are integrated in agricultural markets, but face significant challenges derived from the limits of their own households and farm assets, and because of the imperfections of factor and product markets, and the gaps and limitations of institutional frameworks of all kinds. This group is made up of about 4 million small farms, which control around 200 million hectares of farmland. The contribution that this group makes to feeding Latin America and, increasingly, other regions of the world, cannot be underestimated. Because they are deeply embedded in the local economies, their agriculture-based development has production and consumption linkages that make them important local and regional players. This is a group made invisible by the definition of smallholders according to the 2 hectare criterion, but at least in LAC, we believe that they represent the best bet for the revitalization of rural societies.

The third and final component of the smallholder sector in LAC, are farms that are at the border between the family farm and the corporate agriculture sectors. The key factor that distinguishes this group from the previous group is that these farmers routinely hire non-family labour to help with the farm operations. Yet at least some of the family members continue to be engaged in the operation of the family farm, and certainly in its management functions. Of course, these are fully commercial farms, many highly competitive, and are behind many of the recent booms that have put Latin American agriculture in the global map of food production. There are probably slightly more than 1 million of these farms (about 8 per cent of the total smallholder sector), and they control about 100 million highly productive hectares. As in the case of the second group, because of the forward and backward linkages with other sectors of the local and regional economy, and also because of the labour they hire, these farmers are crucial players in the rural economies of Latin America.

4. The proximate context of smallholders

A key proposition of this chapter is that the performance and the development potential of smallholders in LAC depends to a very significant degree on the characteristics of the proximate context in which they make decisions. While this statement is not controversial, it is nevertheless true that the vast majority of the smallholder development programmes and policies are aimed at improving the assets of the farm, of the farm family, or of the farmer, with little effort to changing or influencing contexts.

We will argue that, unfortunately, vast areas of LAC currently present contexts that are quite unfavourable for the development of smallholder agriculture. Under such conditions, investing in farm, farm family, or farmer assets is unlikely to yield the results intended by policies and programmes.

In our opinion, this is a major factor that helps explain why the agricultural boom experienced in many LAC countries since the late 1990s and early 2000s, did not translate into higher rates of rural poverty reduction; that is, why agriculture did very well but a very large contingent of farmers did not (da Silva et al., 2009). It should be noted that the agricultural boom of the early 2000s was reinforced by the strong reduction of the net implicit taxation that affected agriculture and farmers' incomes until the 1980s (Anderson and Valdés, 2008). That is, both the macro-economic context and the agricultural sector's trends were very favourable, and yet the evidence is that in several countries, smallholders failed to benefit and that rural poverty reduction was mainly due to other sources of employment and of income.

4.1 LOCAL ECONOMIC GROWTH AND POVERTY REDUCTION

What have we learned about the interaction between proximate context and the dynamism of smallholder agriculture?

Petrolina-Juazeiro, an area of 53,000 square kilometres and 510,000 inhabitants in the states of Bahia and Pernambuco in Brazil, was no different than most of the rural areas in the north-eastern region. Its economy was based on a stagnant agriculture, dominated by the production of cotton, livestock, and subsistence crops. From the mid-1990s to 2006, a public corporation implemented six projects with close to forty-six thousand hectares of irrigated lands that led to the emergence of more than two hundred agricultural firms producing high-value crops for export, about two thousand two hundred small farmers, and more than one hundred thousand wage workers (40 per cent of which were women) with incomes and wages way above the region's average (Damiani, 2006).

The Salinas district (in the Bolivar Department of Ecuador in the Andes, where the majority of the population are Quechua Indians) is today nationally renowned for the high quality of its cheese production, having 22 small and medium cheese factories linked to 28 savings and loan cooperatives that cover the whole territory. Marketing is done through their own outlays, as well as through supermarkets and pizza parlours (exhibiting their trademark 'El Salinerito') and exports are made through Camari, a development NGO. A series of other local manufacturing activities were derived from its development, including hams and cold meats, and toys from native woods.[6]

One of the main lessons learned from this type of successful agricultural development experience with smallholders in a prominent position, is that economic growth with social inclusion is not only about what happens in the farms of smallholders, or even in their communities and organizations. Such development outcomes involve whole territories, with a multitude of inter-linked actors (poor and non-poor, agrarian and not, urban and rural, private and public) that mobilize complementary assets and capabilities. We submit the hypothesis that in Latin America today, the condition of this proximate context is a far more important determinant of the performance and opportunities of smallholders than what happens on the farm, or the peasant community, or the producers' organization.

Dynamic agricultural systems are increasingly characterized by strong linkages with services and industry that lead to income and employment diversification (at the territorial and household levels). Backwards and forwards production linkages to suppliers of inputs and equipment and to processors and traders tend to increase in importance and complexity as agriculture develops. Of particular relevance is the accelerated transformation of the agri-food systems in developing countries, where supermarkets and other large national and FDI-based industries are becoming the main destiny of agricultural products. The impact of these trends on local incomes and employment will depend on the degree of openness and competitive capacity of the economy and on the relative weight of the modern retail systems.

Production and consumption linkages to agriculture can also originate in other sectors of the economy, as when an agro-industry creates new procurement options (like in contract agriculture) or when non-farm but rural-based activities (e.g. tourism) generate or increase the demand for locally produced

[6] Manuel Chiriboga, personal communication.

food. Labour market linkages derived from the seasonal nature of agricultural employment can stimulate rural non-farm activities in the slack periods.

Rural–urban linkages are a consequence of the mostly urban location of industrial processing and related services. The nature and intensity of the links between the agricultural hinterland and urban nuclei are critical to the development of dynamic agri-food systems. Cities have been an important source of generation and dissemination of agricultural technology. Since capital, inputs, labour, and product markets tend to be less imperfect in the urban environment, spillover effects to neighbouring agricultural areas can lead to increased smallholder farm productivity, better prices for smallholder products, and higher wages in the agricultural labour markets.

In short, the opportunities that smallholders will be able to take advantage of, as well as the challenges they will face, depend to a very large extent on the dynamics of economic growth, with or without social inclusion, in their proximate geographic context.

A series of publications has recently documented changes in the past decade or so in different indicators of economic and social development at the local (municipal) level in 11 LAC countries.[7] The authors of these publications have collectively processed census and household survey data for 400 million persons, in 10,000 municipalities; the studies cover 80 per cent of the population of LAC. These studies are based on outcomes and not on the determinants of those outcomes; hence, we do not know why a certain place has economic growth or not, for example. Such determinants could, of course, make a big difference in terms of the opportunities or constraints to smallholders (e.g. it would not be the same if the localized economic growth is due to drug trafficking or a new irrigation system leading to increases in farm productivity; or if poverty reduction is due to better social safety nets or to more and better jobs being made available).

Table 5.21 summarizes data from these 11 studies, for 2 important dimensions: changes between the late 1980s or early 1990s and the early–mid 2000s in average per capita income (a proxy of economic growth), and incidence of poverty (or headcount poverty), both at the level of municipalities (except for the study of Peru, which looks at provincial districts, and for Ecuador, which looks at sub-municipal entities called *Parroquias*). The first four rows of Table 5.21 show that the majority (66 per cent) of the population of these eleven countries live in over six thousand municipalities (59 per cent of the total) that have not experienced economic growth, as indicated by the lack of positive and statistically significant change over this period of time.

[7] See Damianović et al., 2009, for El Salvador; Escobal y Ponce, 2008, Peru; Favareto y Abramovay, 2009, Brazil; Fernández et al., 2009, Colombia; Flores et al., 2009, Honduras; Gómez et al., 2009, Nicaragua; Hinojosa et al., 2009, Bolivia; Larrea et al., 2008, Ecuador; Modrego et al., 2008, Chile; Romero y Zapil, 2009, Guatemala, and; Yúnez-Naude et al., 2009, Mexico.

Table 5.21 Changes in per capita income or consumption and in incidence of poverty

Context	Population		Municipalities	
	Numbers in 1,000s	Percentage	Numbers	Percentage
Changes in average per capita income or consumption				
Significant increase	133,952	34	4245	41
No positive change	265,556	66	6176	59
Total	399,509	100	10,421	100
Changes in headcount poverty				
Significant decrease	136,127	34	4818	46
No positive change	263,381	66	5603	54
Total	399,509	100	10,421	100
Combined changes				
Positive change in average income/ consumption and headcount poverty	95,730	24	3389	33
Positive change in average income/ consumption but not in headcount poverty	38,221	10	856	8
Positive change in headcount poverty but not in average income/consumption	40,396	10	1429	14
No positive changes in either indicator	225,160	56	4747	46
Total	399,509	100	10,421	100

The second group of four rows in Table 5.21 show similar results for changes in the incidence of poverty: two-thirds of the population live in slightly more than half of the municipalities that have seen no significant poverty reduction over the period covered by these studies.

Finally, the group of five rows at the bottom of Table 5.21 look at the combinations of both indicators of development. Only one-fourth of the population of these 11 countries (that is, about 20 per cent of the population of LAC) live in places that have experienced growth *with* poverty reduction in the past decade or so. What is distressing is that around 50 per cent of the population (and also of the municipalities) are in a lose-lose position, that is, no economic growth and no poverty reduction either.

One could tentatively conclude that most smallholders will experience unfavourable environments where there is no economic growth and/or such growth is not socially inclusive or 'pro-poor'. However, smallholders are not randomly distributed in space, and it could well be that they could be mostly concentrated in regions where conditions are generally more favourable. Unfortunately, there are no studies that systematically tell us *where* smallholders are located in each country; we can only give an approximate answer, looking in greater detail at some of the data of the studies summarized in Table 5.21. We can explore this issue in the case of Brazil, thanks to the data of the Favareto and Abramovay (2009).

Table 5.22 presents data for three macro-regions of Brazil. The Brazilian north-east is a region with very large numbers of smallholders, many of them poor; the centre–west region is one of large-scale corporate agriculture; finally, the south is a rich region with many fully commercial and very competitive family owned and operated farms. The numbers in Table 5.22 confirm that the smallholders in the north-east would face much more challenging proximate contexts, compared with the other two regions. In the north-east there are fewer municipalities experiencing economic growth (44 per cent), poverty reduction (71 per cent), or growth with poverty reduction (39 per cent), compared with the centre–west (51 per cent, 77 per cent, and 42 per cent, respectively), and the south (63 per cent, 86 per cent, and 56 per cent, respectively). That is, it is very likely that the same smallholder development policy, with the same instruments, and the same level of investment per farmer, would yield less positive results in the north-east where family farmers have to fight against a more adverse context, than in the south where family farmers can ride the wave of a vibrant territorial context.

However, what is perhaps more important is that even within the adverse north-east region, there are 564 municipalities (39 per cent of the total in the macro-region), clustered in around forty to fifty territories, where smallholders face a favourable proximate context of economic growth with poverty reduction. Conversely, in the favourable and rich southern region of Brazil

Table 5.22 Changes in per capita income or consumption and in incidence of poverty in regions of Brazil
(Percentage of municipalities)

Context	Northeast	Centre–West	South
Changes in average per capita income or consumption			
Significant increase	44	51	63
No positive change	56	49	37
Total	100	100	100
Changes in headcount poverty			
Significant decrease	71	77	86
No positive change	29	23	14
Total	100	100	100
Combined changes			
Positive change in average income/consumption and headcount poverty	39	42	56
Positive change in average income/consumption but not in headcount poverty	5	10	7
Positive change in headcount poverty but not in average income/consumption	32	36	30
No positive changes in either indicator	25	13	7
Total	100	100	100

Source: Authors, with data from Favareto and Abramovay (2009).

there are over one hundred and twenty municipalities, clustered in about fifteen or so territories, where smallholders would face a context of economic and social stagnation.

4.2 TRENDS IN AGRI-FOOD MARKETS

For many good reasons, much of the political and policy emphasis in the region since after the end of the structural adjustment processes has been placed on creating favourable conditions and capacities to access global markets, with a special interest on the promotion of non-traditional exports (NTEXs). This interest has been spurred on by the numerous trade agreements signed by a majority of the countries in the region. However, one could argue that this interest has often been accompanied by an unwarranted neglect of policies to improve and exploit the domestic food markets.

In the case of 16 Latin American countries that collectively represent more than 80 per cent of the regional agricultural GDP, the domestic market consumes 73 per cent of the agricultural output; the figure is 46 per cent in the case of agro-industrial products in 9 countries (Figure 5.2). Even in the case of fresh fruits and vegetables, where the non-traditional export market receives much attention from international agencies and national policy-makers, it has been estimated that the sales of supermarkets in domestic markets represent over 1.5 times the value of the fresh fruit and vegetable exports from the region (Reardon and Berdegué 2002). In 2003, domestic food sales by modern retailers in LAC amounted to over US$169 billion (Reardon and Berdegué 2006).

Driven by demographic growth, urbanization, and dietary transitions, markets within developing countries for agricultural products are growing at a faster rate than those in the industrialized countries. Between 1998 and 2002, sales of Nestlé and Unilever respectively grew by 7 and 3.2 per cent in Europe and by 29.8 and 8.3 per cent in Latin America, and sales of packaged products in 1996–2002 grew by 29 per cent in lower- to middle-income countries, compared with only 3 per cent in high-income countries (Wilkinson and Rocha, 2006).

In short, the domestic market in LAC as a region and in most of its countries individually, is the largest and the fastest growing market for agricultural products. This creates important opportunities for agricultural growth. The domestic markets in all LAC countries are also rapidly changing in their structure and in the way they work. Figure 5.3 illustrates the changing consumption patterns of the Latin American and Caribbean people; not only do they eat 22 per cent more food per capita than 30 years ago, they also eat differently (in particular, more meat, dairy products, fresh fruits and vegetables, and vegetable oils).

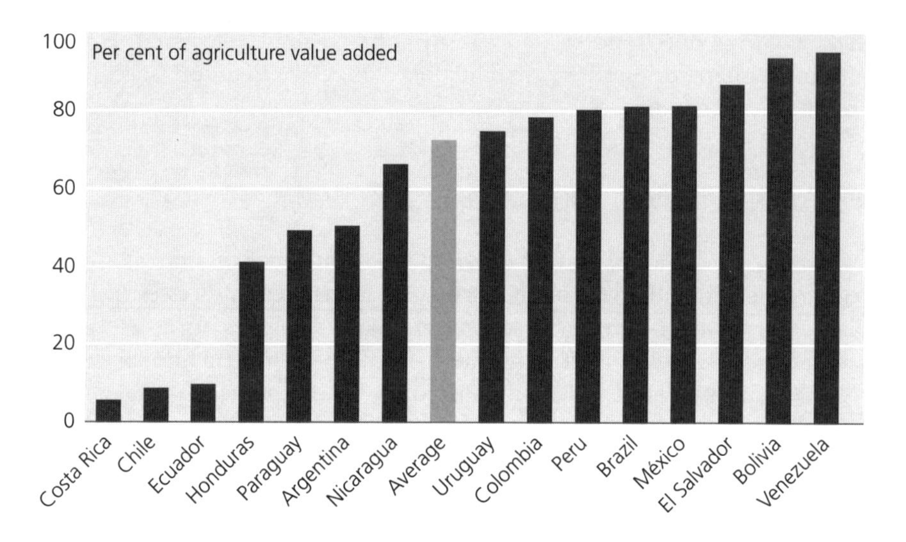

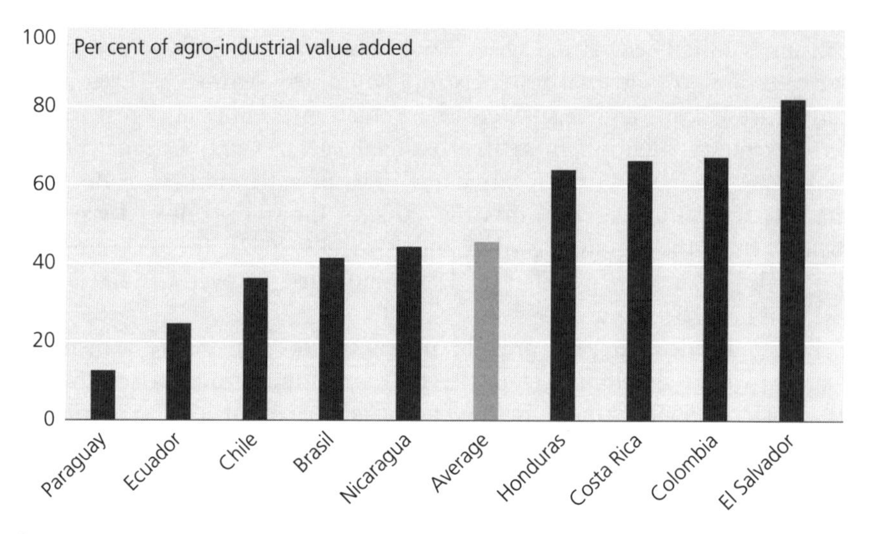

Figure 5.2 Domestic market share of agricultural and agro-industrial products in 2002 (%)

Source: Berdegué et al. (2006).

Due to liberalization of Foreign Direct Investment policies and to trade agreements, it is becoming more difficult for LAC farmers to compete and to meet the conditions of the rapidly changing domestic markets. Long gone are the days when domestic markets were those of inefficient wholesale markets and informal intermediaries. Today, the new domestic food markets in LAC are dominated by supermarket chains, giving rise to four major trends in food

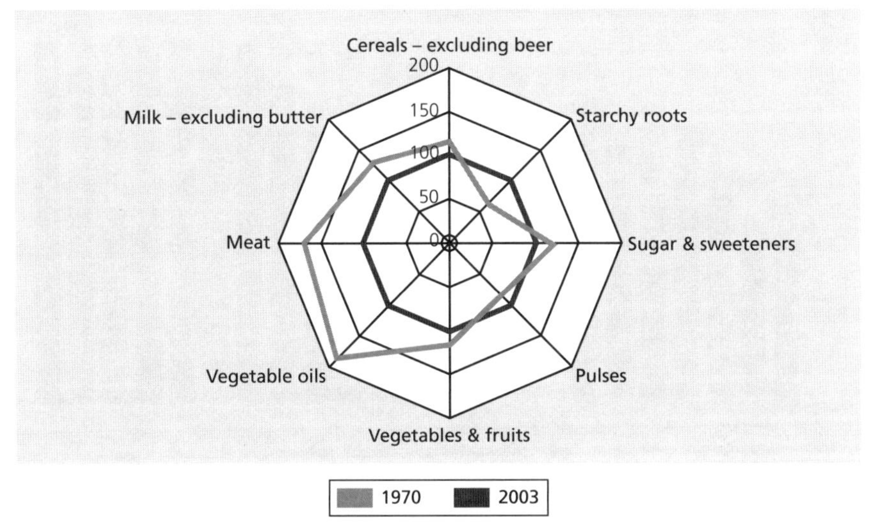

Base 100 = consumption levels at 1970: Cereals = 114.7, Starchy roots = 80.5, Sugar and sweeteners = 40.0, Pulses = 13.6, Vegetables and fruits = 129.2, Vegetable oils = 6.14, Meat = 34.7, Milk exc. butter = 84.6

Figure 5.3 Changing patterns of food consumption in Latin America and the Caribbean, 1970–2003 (Food per capita per year in kg, base 1970 = 100).

Source: FAOSTAT (Food Balance Sheets).

procurement systems: extension and integration of catchment areas, reliance on specialized wholesalers and modern logistics firms, greater vertical coordination and rapid emergence of a variety of contractual arrangements that are displacing spot markets, and a growing importance of private quality standards and of private enforcement of public standards. More and more, domestic and global markets converge in their dynamics, organizational forms, and institutional settings (Reardon and Berdegué, 2006).

While production for the export market tends to be concentrated in capitalized farms and agri-businesses, a large percentage—probably the majority—of medium and small family farms and agri-processors tend to focus on the domestic market. This creates a potential for direct and indirect impacts of agricultural growth on the reduction of rural poverty and inequality. The case of Chile is particularly illustrative of this point; despite the fact that this is one of the most export-oriented countries (Figure 5.4), there are 11 times more farmers engaged in the domestic market than those dedicated primarily to the export sector. Of those Chilean farmers who produce food for the domestic market, 89 per cent are commercially oriented small and medium family farmers. Two-thirds of Chilean commercially oriented small farmers produce for the domestic market (ODEPA 2002). These trends are likely to be augmented in countries with a large domestic agriculture market and high

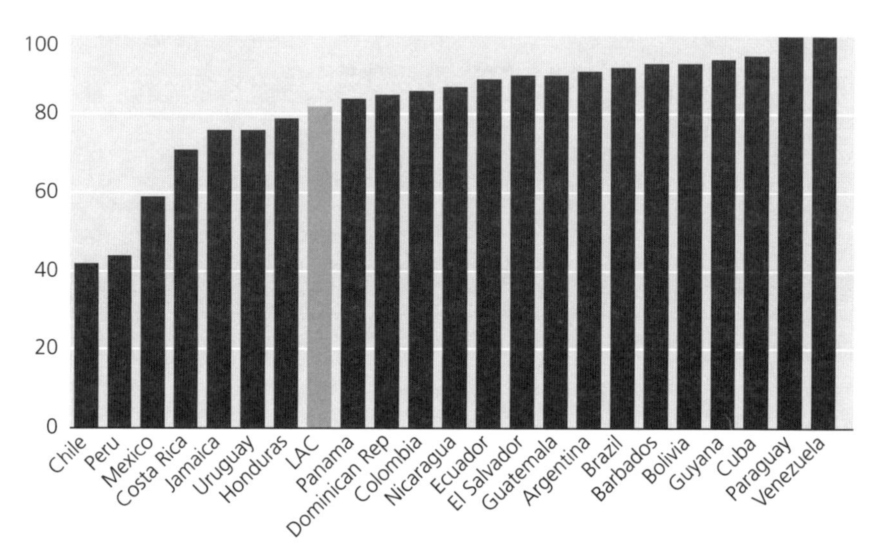

Figure 5.4 Traditional commodities as percentage of LAC food exports.

Source: CEPAL (2006) and FAO (2004).

proportions of small-scale farmers, such as, for example, Bolivia, Brazil, Colombia, Guatemala, Mexico, or Peru.

Agricultural exports account for 11 per cent of total LAC exports, but in about half the countries of the region this contribution is of 20 per cent or greater (Piñeiro, 2005). Despite the significant efforts made in promoting higher-value and non-traditional exports, the traditional commodities that have been important for a long time, such as coffee, cocoa, cereals, banana, vegetable oils, and meat, still account for over 80 per cent of regional exports (Figure 5.4).

There are six LAC countries with a significant participation[8] in global exports of higher-value food products: Argentina, Brazil, Chile, Costa Rica, Ecuador, and Mexico. As explained by Henson (2006), exports of higher-value products tend to be dominated by middle-income countries because of the very substantial private and public investments and the well-developed institutional contexts that are required to be successful in these markets. In addition, global markets of each product tend to be dominated by a small number of early entrants. While this does not rule out the possibility that other LAC countries can gain a foothold in global markets of higher-value agricultural products, it does mean that for most countries, there are clear limits to what can be expected.

[8] Defined as equal to or greater than 1 per cent of global exports.

Higher-value export markets tend to be the domain of a relatively small number of capitalized farmers, urban entrepreneurs who invest in agriculture, processing, and trading firms. However, family farmers—in some instances including poor households—have been able to achieve a significant participation in some niche markets, notably organic coffee where 13 LAC countries provide almost half the global planted area (Henson, 2006), and the Fair Trade markets for banana, coffee, fresh fruits and vegetables, honey, fruit juice, and sugar where LAC accounts for two-thirds or more of the certified producers in the world (Farnworth and Goodman, 2006; Lyon, 2006). While these examples are limited in scope, they do show that given the right incentives, effective producers' organizations, and availability of financial and technical support services, small family farms can rapidly innovate and participate successfully in very dynamic and competitive markets.

5. The challenges

There are four normative messages arising from this chapter: (a) there is an urgent need to differentiate development strategies and policies according to the three categories of family farms; (b) policies ought to focus not only on developing the assets and capabilities of farmers, farm households, farms, and farmers' organizations, but on the territorial contexts in which they operate[9]; (c) we ought to give much greater attention to domestic food markets, with an emphasis on commodities. The fourth conclusion is a corollary of the previous ones: (d) we need to emphasize the development of public services and public goods that can work at the scale of 15 million family farms, in contrast with programmes that have given greater priority to transferring private assets to, inevitably, only a small proportion of family farmers.

Despite many declarations, public strategies and policies continue to fail to internalize the heterogeneity of smallholder agriculture in LAC. In large part, this is due to the weakness of the public sector organizations charged with agricultural development, as differentiated strategies and policies are far more challenging to design, implement, and manage than 'one-size-fits-all' approaches.

Recent policy practice in LAC has resulted in a rural development toolkit with four major types of policy objectives and instruments: (a) guaranteeing minimum living standards; (b) reducing vulnerability; (c) improving the

[9] Of course, there are other highly important contexts above the territorial scale: regional conditions, national conditions, and global conditions, all of which impact on the potential and the performance of smallholders.

contexts in which decisions are made and assets utilized; and (d) strengthening assets and developing capabilities.

We want to argue that guaranteeing minimum living standards and reducing vulnerability need to be leading strategies to support the more than 9 million subsistence smallholders, whose livelihoods are already dependent on non-farm sources of income. Under these circumstances agricultural development (that is, strengthening the on-farm components of the household's livelihood strategies) ought to emphasize an objective of reducing vulnerability. That is, agricultural development already is a safety-net more than the engine of these families' livelihood strategies. Policy ought to support that approach.

On the contrary, when it comes to the remaining 6 million family farmers, improving the contexts in which decisions are made and assets utilized, and strengthening their assets and capabilities, become paramount objectives. Here the plea is for strategies and policies to recognize this dual pillar: contexts, and assets and capabilities. The scale is tilted heavily in favour of targeting the farmer and the farm, and not enough consideration is paid to the proximate context in which those farmers and farms work. The results of this imbalance are agricultural booms that miss too many smallholders.

The vast majority of smallholders, surely more than 90 per cent, work for and depend on domestic markets, and, more precisely, domestic commodity markets. Yet, agricultural development agencies, international and domestic, are heavily under-investing in improving access to and participation in those markets, and in the past ten years or so have turned their energy and their most creative attention to export markets and, within them, to higher-value and niche markets. This is a terrible mistake, and one which is particularly costly to most of the 9 million subsistence farmers and to the majority of the 4 million family farmers in the intermediate group discussed in previous pages.

Policies that support the improvement of public services and the provision of public goods are a good part of the solution to these challenges. Unfortunately, they are not in vogue. In all but three or four countries, policy has been reduced to a collection of targeted projects, and a disproportionate share of the public effort is focused on delivering private goods and services to groups of farmers. While this may be satisfactory from the perspective of donors who want to measure impact as fast as possible and certainly within three years, it surely spells disaster for the majority of smallholders who don't happen to be included among the lucky few.

A return to public policy focused on the provision of public goods is, in our opinion, the key to future progress. Since the 1980s, public policies for public goods in most countries have been abandoned in favour of a collection of development projects that largely deliver private goods to an important but inevitably small number of beneficiaries (relative to the 15 million smallholders in the region). While these projects may and often do improve the asset position of the beneficiaries, they can do very little to change the second axis in our typology, that is, the contextual conditions. The task is demanding, as it requires serious rethinking, retooling, and investing in strengthening a public sector that is capable of meeting the challenges of the coming 30 or 50 years (Piñeiro, 2009).

REFERENCES

Anderson, K. and A. Valdés (eds) (2008). *Distortions to Agricultural Incentives in Latin America*, Washington, DC: The World Bank.

Berdegué, J. A. (2009). 'Estrategias y programas de reducción de la pobreza rural', in M. Piñeiro (ed.), *La Institucionalidad Agropecuaria en América Latina: Estado Actual y Nuevos Desafíos*, Rome: FAO, pp. 420–46.

Berdegué, J. A. and G. Escobar (2002). 'Rural diversity, agricultural innovation policies, and poverty reduction', *AgREN Network Paper*, No. 122, London: Overseas Development Institute.

Berdegué, J. A., A. Schejtman, M. Chiriboga, F. Modrego, R. Charnay, and J. Ortega (2006). 'Towards regional and national "Agriculture for Development" agendas in Latin America and the Caribbean', background paper for the *World Development Report 2008*. Santiago, Chile: Rimisp–Latin American Center for Rural Development.

Bezemer, D. and P. Hazell (2006). 'The agricultural exit problem: an empirical assessment', background paper for the *World Development Report 2008*. Santiago, Chile: Rimisp–Latin American Center for Rural Development.

Bollinger, F. and O. Olivera (2010). 'Brazilian Agriculture: A Changing Structure', selected paper prepared for presentation at the Agricultural & Applied Economics Association's 2010 AAEA, CAES & WAEA Joint Annual Meeting, Denver, Colorado, 25–27 July 2010. IBGE—Brazilian Institute of Geography and Statistics.

Carmagnani, M. (2008). 'La agricultura familiar en América Latina. Problemas del Desarrollo', *Revista Latinoamericana de Economía*, Vol. 39, núm. 153, abril–junio 2008.

CEPAL (Comisión Económica para América Latina) (1982). *Economía Campesina y Agricultura Empresarial. Tipología de Productores del Agro Mexicano.* México: Siglo XXI Editores.

CEPAL (Comisión Económica de las Naciones Unidas para América Latina y el Caribe) (2006). Anuario Estadístico de America Latina y el Caribe. Santiago: CEPAL.

CEPAL (Comisión Económica para América Latina) (2010). *Panorama Social de América Latina 2009.* Santiago: CEPAL.

Chayanov, A. V. (1986). *The Theory of Peasant Economy.* Madison, Wisconsin: The University of Wisconsin Press.

Chiriboga, M. (1999). 'Desafíos de la pequeña agricultura familiar frente a la globalización', in L. Martínez (ed.), *El Desarrollo Sostenible en el Medio Rural.* Quito, Ecuador: FLACSO.

Damiani, O. (2006). 'Rural development from a territorial perspective. Case studies in Asia and Latin America', background paper for the *World Development Report 2008.* Santiago, Chile: Rimisp–Latin American Center for Rural Development.

Damianović, N., Valenzuela, R., and Vera, S. (2009). 'Dinámicas de la desigualdad en El Salvador: hogares y pobreza en cifras en el período 1992–2007'. Documento de Trabajo No. 52. *Programa Dinámicas Territoriales Rurales.* Rimisp–Centro Latinoamericano para el Desarrollo Rural, Santiago, Chile.

da Silva, J. G., S. Gómez, and R. Castañeda (eds) (2009). *Boom Agrícola y Persistencia de la Pobreza Rural. Estudio de Ocho Casos.* Rome: FAO.

de Ferranti, D., G. E. Perry, W. Foster, D. Lederman, and A. Valdés (2005). *Beyond the City. The Rural Contribution to Development.* Washington, DC: The World Bank.

de Janvry, A. and E. Sadoulet (2000). 'Rural poverty in Latin America. Determinants and exit paths', *Food Policy*, 25, pp. 389–409.

Echenique, J. and L. Romero (2009). *Evolución de la Agricultura Familiar en Chile en el Período 1997–2007.* Corporación Agraria para el Desarrollo—Oficinal Regional de la FAO para América Latina y el Caribe. Santiago, Chile.

Escobal, J. and C. Ponce. (2008). 'Dinámicas provinciales de pobreza en el Perú 1993–2005'. Documento de Trabajo No. 11. *Programa Dinámicas Territoriales Rurales.* Rimisp–Centro Latinoamericano para el Desarrollo Rural, Santiago, Chile.

FAO (2010). *2000 World Census of Agriculture. Main results and Metadata by Country (1996–2005).* Food and Agriculture Organization of the United Nations, Rome.

FAO (2004). 'Tendencias y desafíos en la agricultura, los montes y la pesca en América Latina y el Caribe'. *Oficinal Regional de la FAO para América Latina y el Caribe*, Santiago, Chile.

Farnworth, C. and M. Goodman (2006). 'Growing ethical networks: the Fair Trade market for raw and processed agricultural products'. Background Paper for the *World Development Report 2008.* Rimisp–Latin American Center for Rural Development, Santiago, Chile.

Favareto, A. and R. Abramovay (2009). 'O surpreendente desempenho do Brasil rural nos anos 1990'. Documento de Trabajo No. 32. *Programa Dinámicas Territoriales Rurales.* Rimisp–Centro Latinoamericano para el Desarrollo Rural, Santiago, Chile.

Fernández, M., C. Hernández, A. M. Ibáñez, and C. Jaramillo (2009). 'Dinámicas departamentales de pobreza en Colombia 1993–2005', Documento de Trabajo No. 33. *Programa Dinámicas Territoriales Rurales.* Rimisp–Centro Latinoamericano para el Desarrollo Rural, Santiago, Chile.

Flores, M., H. Lovo, W. Reyes, and M. Campos (2009). 'Cambios en la pobreza y concentración del ingreso en los municipios de Honduras: desde 1988 a 2001'. Documento de Trabajo No. 50. *Programa Dinámicas Territoriales Rurales.* Rimisp–Centro Latinoamericano para el Desarrollo Rural, Santiago, Chile.

Forero, J. and Galeano, J. (2010). 'Cálculo del número de productores campesinos en Colombia, Ecuador y Perú'. Unpublished manuscript.

Fradejas, A. and S. Gauster (2006). 'Perspectivas para la Agricultura Familiar Campesina de Guatemala en un contexto DR-CAFTA'. Guatemala: Coordinación de ONG y Cooperativas, CONGCOOP. Manuscript.

Gómez, L., B. Martínez, F. Modrego, and H. M. Ravnborg (2009). 'Mapeo de cambios en municipios de Nicaragua: consumo de los hogares, pobreza y equidad 1998–2005'. Documento de Trabajo No. 12. *Programa Dinámicas Territoriales Rurales.* Rimisp–Centro Latinoamericano para el Desarrollo Rural, Santiago, Chile.

Hazell, P., C. Poulton, S. Wiggins, and A. Dorward (2010). 'The future of small farms: Trajectories and policy priorities', *World Development*, 38(10), pp. 1349–61.

Henson, S. (2006). 'New markets and their supporting institutions: opportunities and constraints for demand growth', background paper for the *World Development Report 2008.* Rimisp–Latin American Center for Rural Development, Santiago, Chile.

Hinojosa, L., J. P. Chumacero, and M. Chumacero (2009). 'Dinámicas provinciales de bienestar en Bolivia'. Documento de Trabajo No. 49. *Programa Dinámicas Territoriales Rurales.* Rimisp–Centro Latinoamericano para el Desarrollo Rural, Santiago, Chile.

IBGE (Brazilian Institute of Geography and Statistics) (2009). *Censo agropecuário 2006.* Brasil, Grandes Regiões e Unidades da Federação, Rio de Janeiro, Brasil.

IFAD (2010). *Rural Poverty Report 2011.* International Fund for Agricultural Development, Rome.

INDEC (2002). *Censo Nacional Agropecuario 2002.* Argentinian Institute of Statistics and Census. Retrieved from: <http://www.indec.gov.ar/>.

INE (2007). *VII Censo Agropecuario.* Chilean Institute of Statistics. Retrieved from: <http://www.censoagropecuario.cl>.

Jara, E., F. Modrego, J. A. Berdegué, and X. Celis (2009). 'Empresas agrícolas en Chile: Caracterización e implicancias para las políticas de innovación y competitividad en el sector agroalimentario'. *Informe a la Fundación para la Innovación Agraria.* Santiago, Chile: Rimisp–Centro Latinoamericano para el Desarrollo Rural.

Justino, P., J. Litchfield, and L. Whitehead (2003). 'The impact of inequality in Latin America', Working Paper No. 21, Poverty Research Unit. Sussex: University of Sussex.

Larrea, C., R. Landín, A. I. Larrea, W. Wrborich, and R. Fraga (2008). 'Mapas de pobreza, consumo por habitante y desigualdad social en el Ecuador: 1995–2006. Metodología y resultados'. Documento de Trabajo No. 13. *Programa Dinámicas Territoriales Rurales.* Rimisp–Centro Latinoamericano para el Desarrollo Rural, Santiago, Chile.

Lipton, M. (2005). 'The family farm in a globalizing world: the role of crop science in alleviating poverty', 2020 Discussion Paper No. 40. Washington, DC: International Food Policy Research Institute.

Lyon, S. (2006). 'Fair Trade in Latin America'. Background paper for the *World Development Report 2008.* Rimisp–Latin American Center for Rural Development, Santiago, Chile.

Modrego, F., R. Charnay, E. Jara, H. Contreras, and C. Rodríguez (2006). 'Small Farmers in Developing Countries: Some Results of Household Surveys Data Analysis', background paper for the *World Development Report 2008.* Santiago, Chile: Rimisp–Latin American Center for Rural Development.

Modrego, F., E. Ramírez, and A. Tartakowsky (2008). 'La heterogeneidad espacial del desarrollo económico en Chile: Radiografía a los cambios en bienestar durante la década de los 90 por estimaciones en áreas pequeñas'. Documento de Trabajo No. 9. *Programa Dinámicas Territoriales Rurales.* Rimisp–Centro Latinoamericano para el Desarrollo Rural, Santiago, Chile.

Nagayets, O. (2005). 'Small farms: Current status and key trends', in *The Future of Small Farms,* Proceedings of a Research Workshop, Wye, UK, 26–29 June. Washington, DC: International Food Policy Research Institute.

ODEPA (2002). 'Agricultura Chilena. Rubros según tipo de productor y localización geográfica'. Documento de Trabajo No. 8. Oficina de Estudios y Políticas Agrarias. Santiago, Chile.

Piñeiro, M. (2005). 'Rural development in Latin America: Trends and policies'. Background document for the European Commission-Inter-American Development Bank Rural Development Dialogue, Brussels, February 2005. Manuscript.

Piñeiro, M. (2009). *La institucionalidad agropecuaria en América Latina: Estado actual y nuevos desafíos.* Santiago: FAO.

Reardon, T. and J. A. Berdegué (2002). 'The rapid rise of supermarkets in Latin America. Challenges and opportunities for development'. *Development Policy Review,* 20(4), pp. 371–88.

Reardon, T. and J. A. Berdegué (2006). 'The retail-led transformation of agrifood systems and its implications for development policies'. Background paper for the *World Development Report 2008.* Rimisp–Latin American Center for Rural Development, Santiago, Chile.

Romero, W. and P. Zapil (2009). 'Dinámica territorial del consumo, la pobreza y la desigualdad en Guatemala: 1998–2006'. Documento de Trabajo No. 51. *Programa Dinámicas Territoriales Rurales.* Rimisp–Centro Latinoamericano para el Desarrollo Rural, Santiago, Chile.

Scheinkerman de Obschatko, E. (2009). 'Las explotaciones agropecuarias familiares en la República Argentina: un análisis a partir de los datos del Censo Nacional agropecuario 2002'. Ministerio de Agricultura, Ganadería y Pesca de la Nación: Instituto de Cooperación para la Agricultura-Argentina. *Estudios e Investigaciones*: 23.

Scheinkerman de Obschatko, E., M. Foti, and M. Román (2007). 'Los pequeños productores en la República Argentina: importancia en la producción agropecuaria y en el empleo en base al censo nacional agropecuario 2002'. Secretaría Agricultura, Ganadería, Pesca y Alimentos. Instituto de Cooperación para la Agricultura-Argentina. *Estudios e investigaciones*: 10.

Schejtman, A. (2008). 'Alcances sobre la agricultura familiar en América Latina'. Documento de Trabajo No. 21. *Programa Dinámicas Territoriales Rurales.* Santiago, Chile: Rimisp–Centro Latinoamericano para el Desarrollo Rural.

Schejtman, A. and J. A. Berdegué (2009). 'The social impact of regional integration in rural Latin America', in P. Giordano (ed.), *Trade and Poverty in Latin America.* Washington DC: Inter-American Development Bank, pp. 249–322.

Soto Baquero, F., M. Rodríguez Fazzone, and C. Falconi (2007). *Políticas para la Agricultura Familiar en América Latina y el Caribe.* Santiago, Chile: FAO-BID.

Wiggins, S., J. Kirsten, and L. Llambí (2010). 'The future of small farms', *World Development*, 38 (10), pp. 1341–8.

Wilkinson, J. and R. Rocha (2006). 'Agri-processing and developing countries'. Background paper for the *World Development Report 2008*, Rimisp–Latin American Center for Rural Development, Santiago, Chile.

World Bank (2007). *World Development Report 2009. Agriculture for Development.* Washington DC: The World Bank.

Yúnez-Naude, A., J. Arellano González, and J. Méndez Navarro (2009). 'México: Consumo, pobreza y desigualdad a nivel municipal 1990–2005'. Documento de Trabajo No. 31 *Programa Dinámicas Territoriales Rurales.* RimispCentro Latinoamericano para el Desarrollo Rural, Santiago, Chile.

Part II

The Business Agenda for Smallholders

6 Targeting investments to link farmers to markets: a framework for capturing the heterogeneity of smallholder farmers

MAXIMO TORERO[1]

1. Introduction

With roughly three-quarters of the world's poor living in rural areas, addressing global poverty requires paying special attention to rural populations in developing countries, especially smallholder farmers. In South Asia, sub-Saharan Africa, and East Asia and the Pacific, the rural population represents two-thirds or more of the total population; together, these three regions hold about 1.1 billion poor people (living on less than a dollar a day)—roughly 90 per cent of the world's poor. While many poor people in these three regions do live in urban areas, the critical challenges clearly lie with rural populations.

The share of agriculture in developing countries' gross domestic product (GDP) has fallen over time—in Latin America and the Caribbean (LAC), agriculture now represents less than 8 per cent of GDP. In the Least Developed Countries (LDCs), on the other hand, agriculture still represents about one-third of GDP. However, agriculture in LACs is more than ten times more productive in terms of value added per worker than agriculture in the LDCs. Similarly, agricultural exports represent about one-third of all merchandise exports from LDCs, while agricultural exports from LAC countries make up about 30 per cent of merchandise exports (Orden et al., 2004).

Though many agricultural exports in LAC are produced on very large farms, smallholders are responsible for over 90 per cent of the agricultural goods produced in sub-Saharan Africa. In South Asia, for example, out of 125

[1] The author acknowledges with thanks the comments and suggestions of anonymous referees.

million farm holdings, more than 80 per cent have an average size of 0.6 hectares; farmers with less than 2 hectares account for 40 per cent of total food grain production. In sub-Saharan Africa, more than two-thirds of the region's farm holdings have an average size of less than 1 hectare and account for over 90 per cent of agricultural output. In Latin America, there is also a huge inequality in the distribution of land. FAO estimates that the largest 7 per cent of land holdings in the region (those above 100 hectares) account for 77 per cent of the land, while the smallest 60 per cent account for only 4 per cent of the land. Most of these smallholders either practise subsistence farming or operate largely in local markets due to a lack of connectivity to more lucrative markets at provincial, national, or global levels. As a result, incentives remain weak and investments remain low, as does the level of technology adoption and productivity, resulting in a low-level equilibrium poverty trap.

How can smallholders escape this poverty trap? Two instruments appear critical to break this deadlock: (i) physical infrastructure, such as roads, ports, and information technologies, that connects smallholders to markets, and (ii) accompanying institutions that can reduce the marketing risks and transaction costs associated with the process of exchange between producers and consumers. Smallholders are generally exposed to a higher degree of risk and higher transaction costs due to their small production surpluses. Thus, any innovative institutions that link farms to markets, reduce transaction costs, and minimize risk will help smallholders participate in markets.

However, the exact nature of the infrastructure and institutions that can enable smallholders to progress from subsistence farming in a village economy to active participation in provincial, national, and international markets varies from country to country and even from region to region within a country. This chapter attempts to develop a framework that can capture smallholder heterogeneity, and therefore identify and prioritize the types of institutions and infrastructure that would help link different types of smallholders to markets. It also aims to improve knowledge about the impact that complementary investments in rural institutions and infrastructure, both capital-intensive infrastructure (roads, electricity, potable water and drainage water for irrigation, and telecommunications) and post-harvest technologies (storage services, processing infrastructure, etc.), may have on market development and poverty reduction.

The next section details the conditions necessary to advance smallholders by capturing their heterogeneity. The third section details the proposed methodology used to capture smallholders' heterogeneity followed by in the fourth the way the methodology is implemented for Mozambique, Armenia, and Honduras. Finally, conclusions are presented.

2. Capturing the heterogeneity of smallholders through a typology of micro-regions

Large inequalities exist within rural populations; farmers can range from large- and medium-sized modern farmers with good access to markets and services, good land quality, and access to water and roads, to subsistence smallholders with very low-quality land and limited access to markets, roads, and water. Rural households are extremely diverse in their economic characteristics due to: (i) heterogeneity in the quantity and quality of their assets (land, water, human capital); (ii) access to technologies; (iii) transaction costs in markets for outputs and inputs; (iv) credit and financial constraints; (v) access to public goods and services; and (vi) local agro-ecological and biophysical conditions.

Based on these differences, we propose the development of a methodology to identify a typology that incorporates production efficiency, linkage to markets and income generation, geographical interdependence, and the bottlenecks associated with the livelihoods of the rural poor (Torero et al., 2009). Based on these criteria, it is possible to identify two key dimensions to construct a typology of smallholders: (i) the potential of each household and micro-region conditional on geographic location, biophysical conditions and markets; and (ii) the degree of existing potential agricultural efficiency given smallholders' access to technology, inputs, public services, infrastructure, and human capital.

This typology of micro-regions is an alternative way to classify and analyse very small rural areas within a country where smallholders are located. Unlike other classification methods, such as poverty maps or cluster analyses, this typology allows us to justify the resulting classification on economic criteria such as productive potential and efficiency in resource management. Such a typology will allow policy interventions to be targeted at those regions and households with the most potential for productive efficiency gains, thus leading to higher returns on those interventions, which could be oriented to productive development, market creation (agricultural or non-agricultural), or even welfarism.

The identification of productive potential and efficiency is achieved through the estimation of an econometric stochastic profit frontier model that takes into account indicators of socioeconomic and market conditions as well as biophysical and accessibility factors. These indicators explain a big portion of the heterogeneity among rural households; therefore, their inclusion in any policy analysis is fundamental. Each indicator's level of importance as a determinant of productive potential and efficiency is strictly determined by economic theory and empirical evidence.

The stochastic frontier approach provides an ideal framework to build a typology of micro-regions. Conceptually, it is developed from a theory of producer behaviour in which the motivation is the standard optimization criteria (minimize costs or maximize profits), but in which success is not guaranteed. The associated estimation procedures allow for failures in efforts to optimize and for different degrees of success among producers. This opens up the possibility of analysing the determinants of variation in the efficiency with which producers pursue their objectives. The stochastic frontier approach accounts for the fact that, conditional on geographic location, elements such as prices, biophysical conditions, and (in the short run) fixed inputs are exogenous to the farmer's decision process. Hence, by holding these factors fixed, there exists an optimal technology and production plan that generates the maximum profit a farmer can obtain. With this approach, it is possible to identify where the profit frontier lies, as well as how much of the difference between the profit frontier and actual observed profits (i.e. profit loss) can be explained by the farmer making choices that result in profit inefficiencies (an inefficiency will be the gap between where the farmer's current production and subsequent profits are and where they could be in the profit possibility frontier, i.e. the maximum potential profits that can be obtained if resources are used in the most efficient way). Adding the stochastic component allows for a better fit to the farm production process, which is very sensitive to unpredictable changes in exogenous conditions such as weather or international prices. In this context, profit inefficiency is defined as the monetary loss which results from not operating at the profit frontier given the prices and levels of fixed production factors faced by the farm.

As shown in Figure 6.1 the typology of micro-regions, once built, can be combined with other relevant information if the goal is also to target interventions based on criteria such as malnutrition and/or poverty. Table 6.1 provides an example of the classifications that can be obtained by mixing potential and efficiency with poverty. For example, we could identify areas with high levels of poverty (dark red and dark green areas of Table 6.1) and combine them with efficiency and profit potential. If those areas were of low agricultural potential, independent of their level of efficiency (red portion of Table 6.1), broad rural development programmes, conditional cash transfers, and nutritional programmes would be recommended. However, if those areas were of high/medium agricultural potential (dark green portion of Table 6.1), agricultural development strategies with nutritional programmes would be promoted, according to the micro-region's level of efficiency.[2]

[2] It is possible to obtain a more detailed characterization of each area in order to recommend ad-hoc policies for each particular reality.

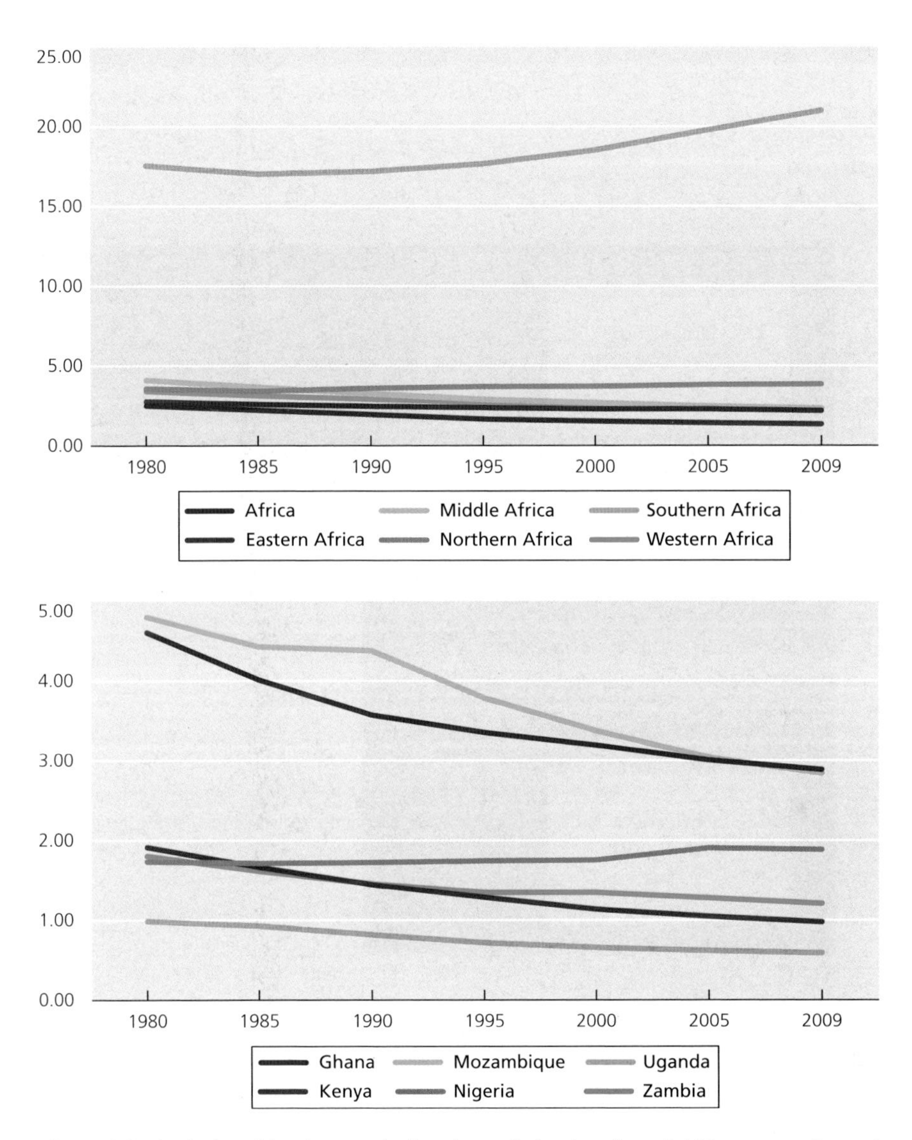

Figure 3.1 Agricultural land per agricultural population in selected African countries and regions

Note: Authors' calculations based on FAO (2012).

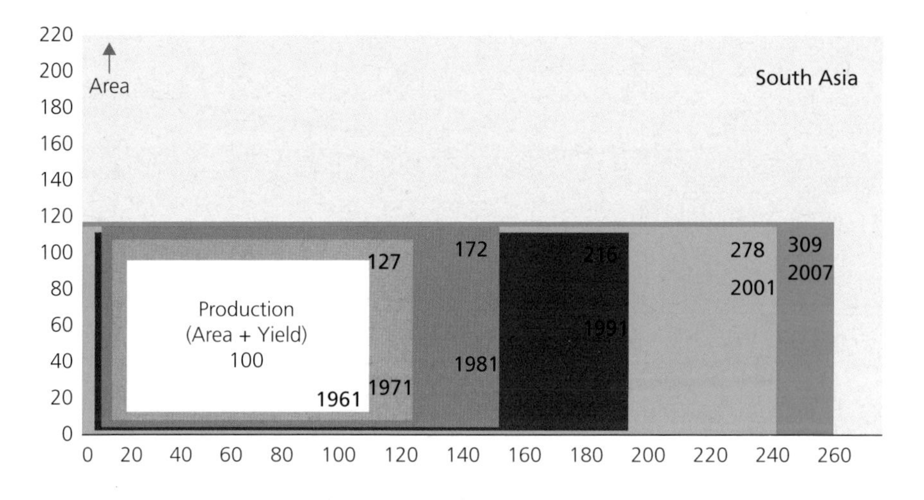

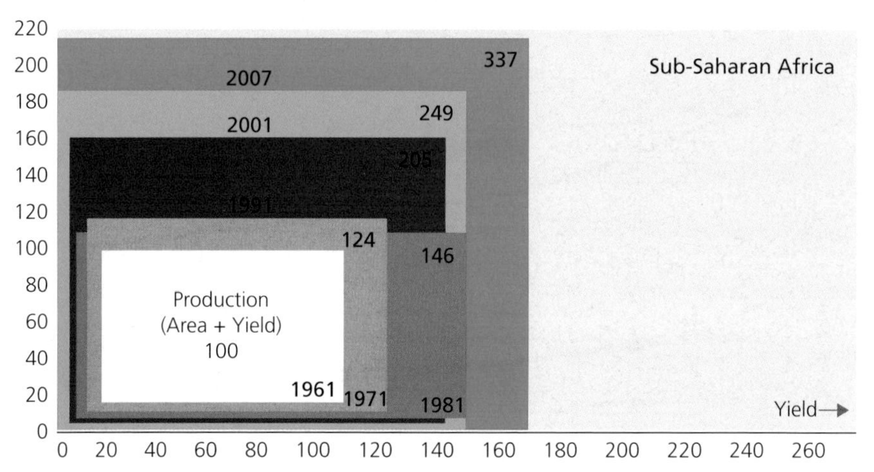

Figure 3.2 Increases in cereal production in South Asia and sub-Saharan Africa

Source: Henao and Baanante (2006).

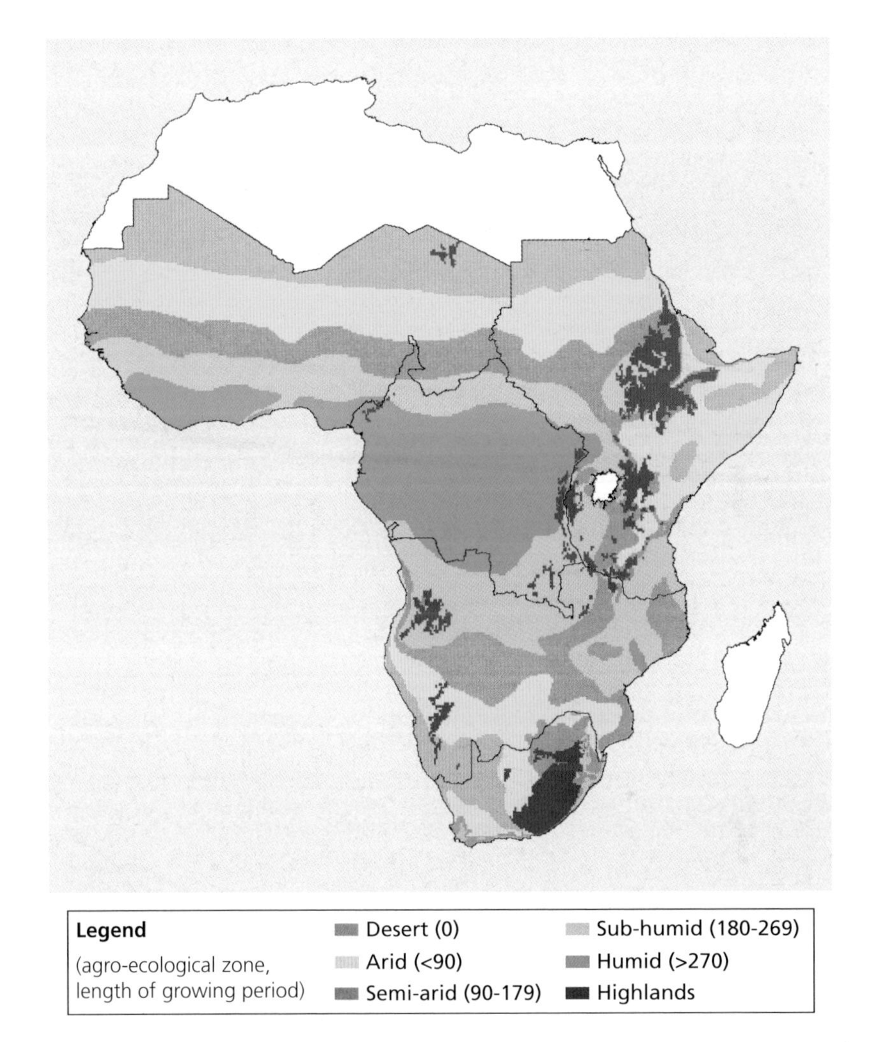

Figure 3.3 Agro-ecological zones in sub-Saharan Africa

Source: Fischer, G. et al. (2002).

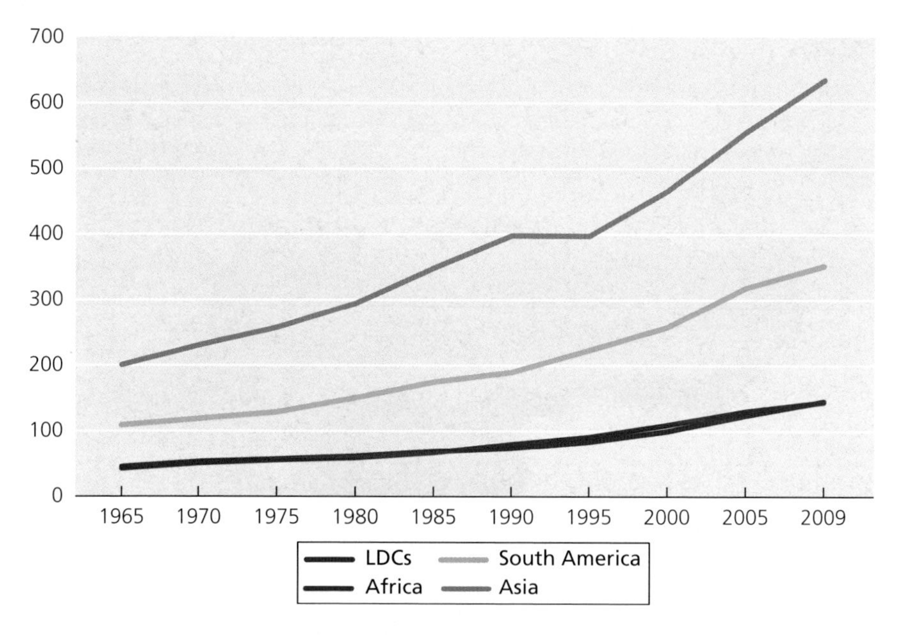

Figure 3.4a Net value of agricultural production in Africa and other regions

Authors' calculations based on FAO (2012)

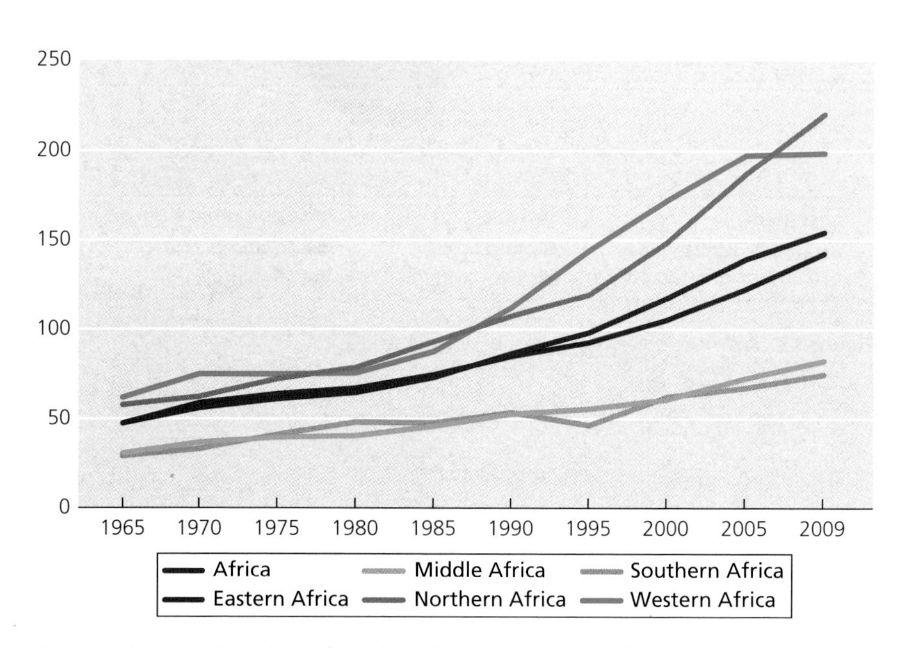

Figure 3.4b Net value of agricultural production in different African regions

Source: FAO (2012).

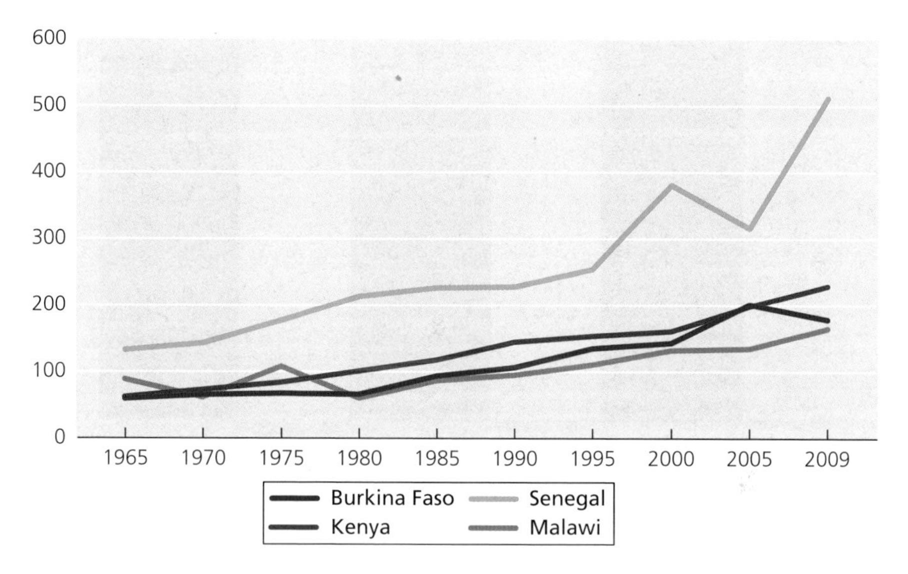

Figure 3.5 Net value of agricultural production in some African countries

Source: FAO (2012).

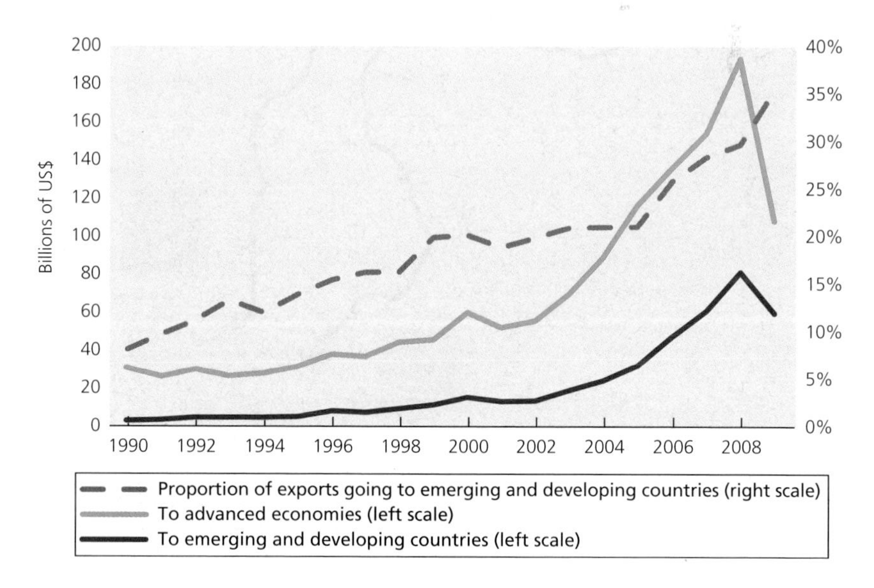

Figure 3.6 Destinations of exports from Africa: 1990–2008

Source: IMF (2010).

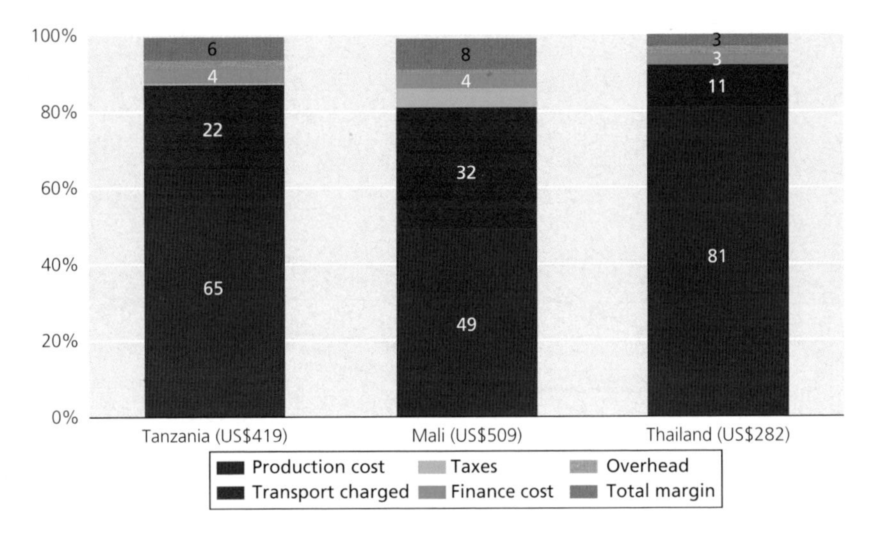

Figure 3.10 Fertilizer price formation in Thailand, Tanzania, and Mali in 2006

Source: Bumb (2009).

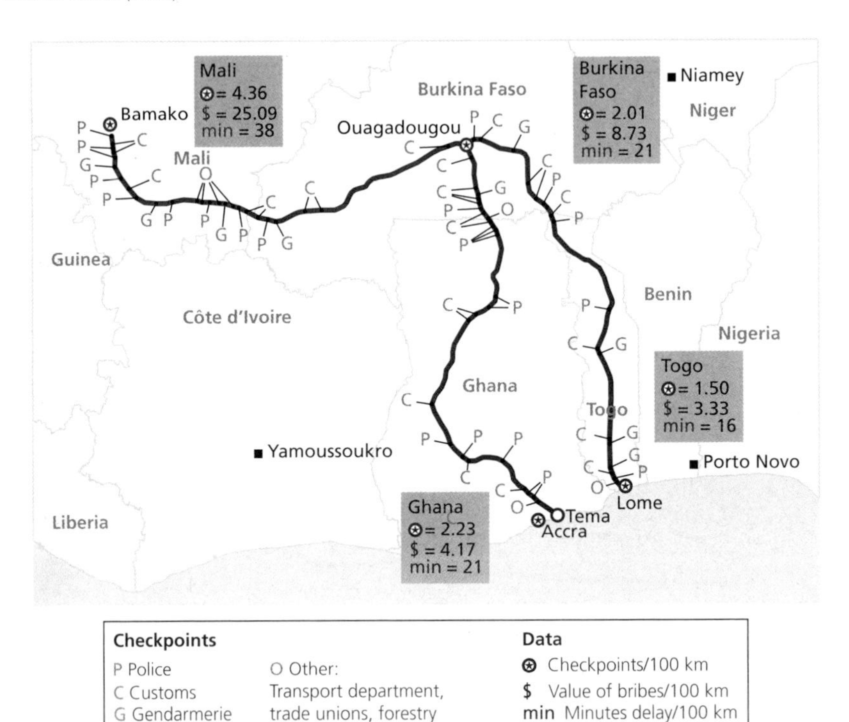

Figure 3.11 Checkpoints, bribes, and delays on corridors in West Africa

Source: West African Trade Hub (2007).

Micro-regions	Poverty	Potential	Efficiency
Critical, lacking, agricultural potential	High	Low	High-medium-low
Medium priority, no agricultural opportunities	Medium	Low	High-medium-low
Low priority	Low	Low	High-medium-low
High priority	High	Medium-high	High-medium-low
Medium priority with agricultural opportunities	Medium	Medium-high	Medium-low
Low priority with agricultural opportunities	Low	Medium-high	Medium-low
High performance	Low	Medium-high	High

Table 6.1 Example of a three-dimensional classification

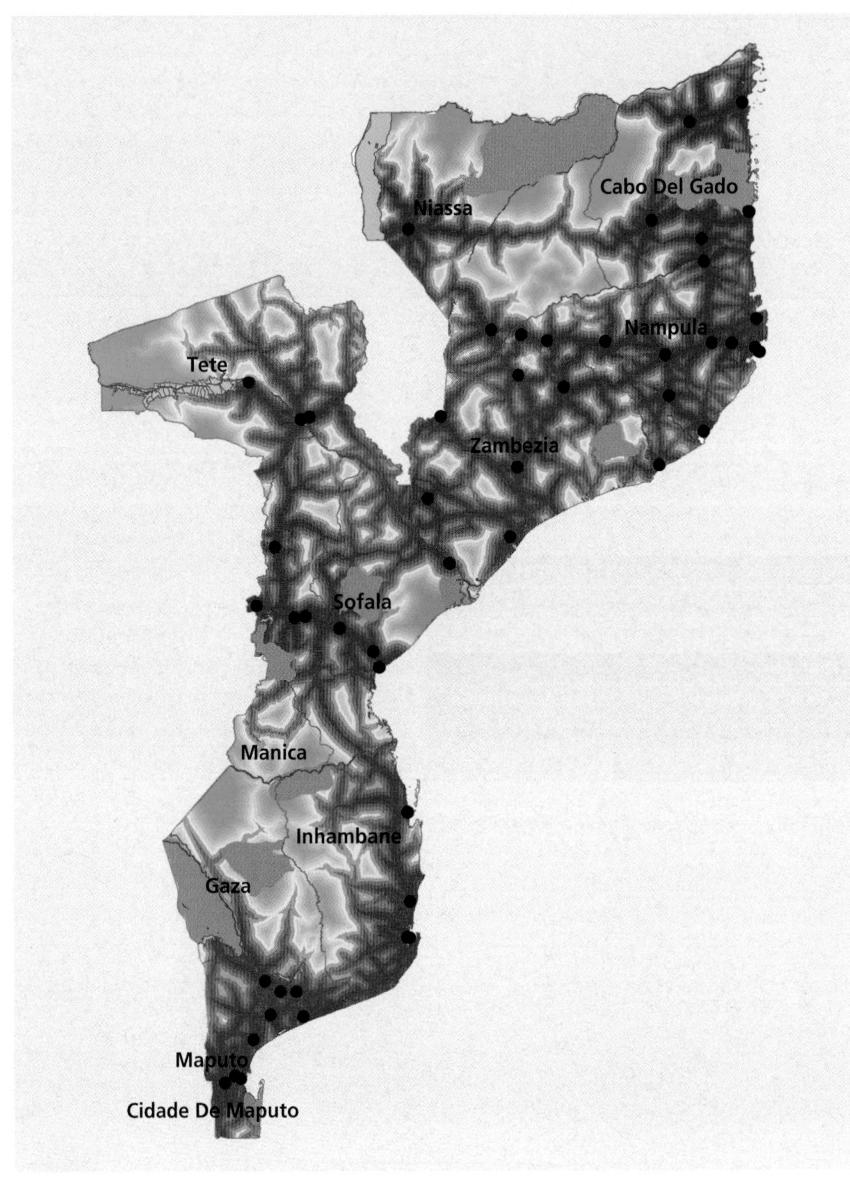

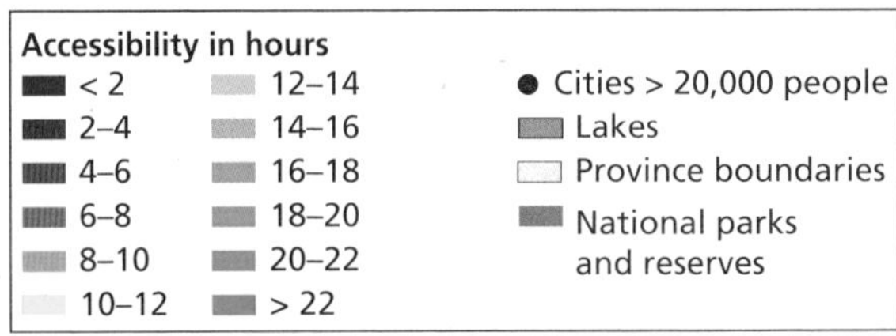

Figure 6.4 Mozambique—market access

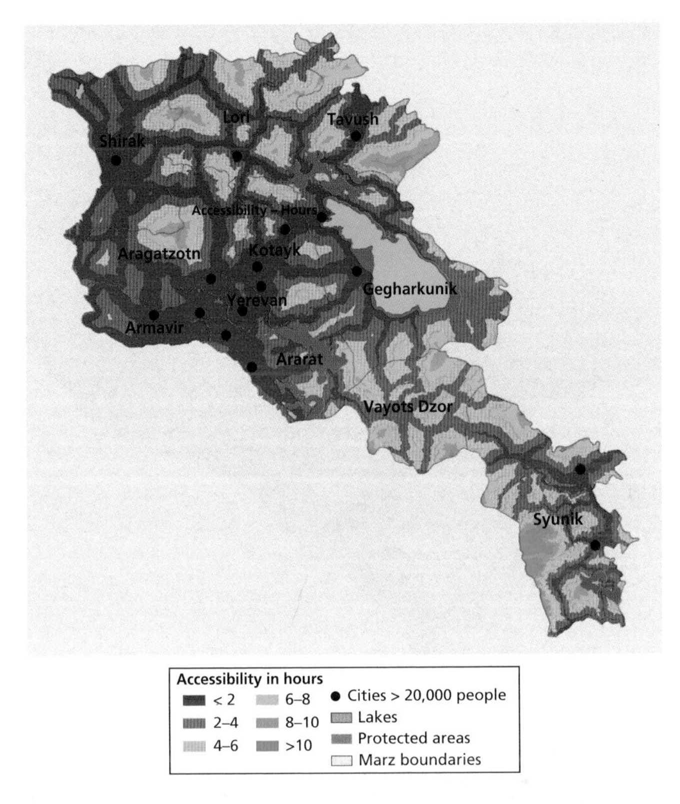

Figure 6.5 Armenia—market access

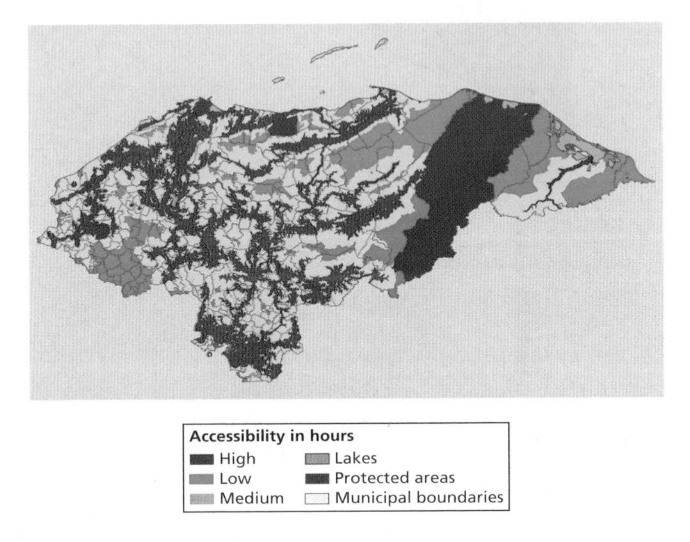

Figure 6.6 Honduras—market access

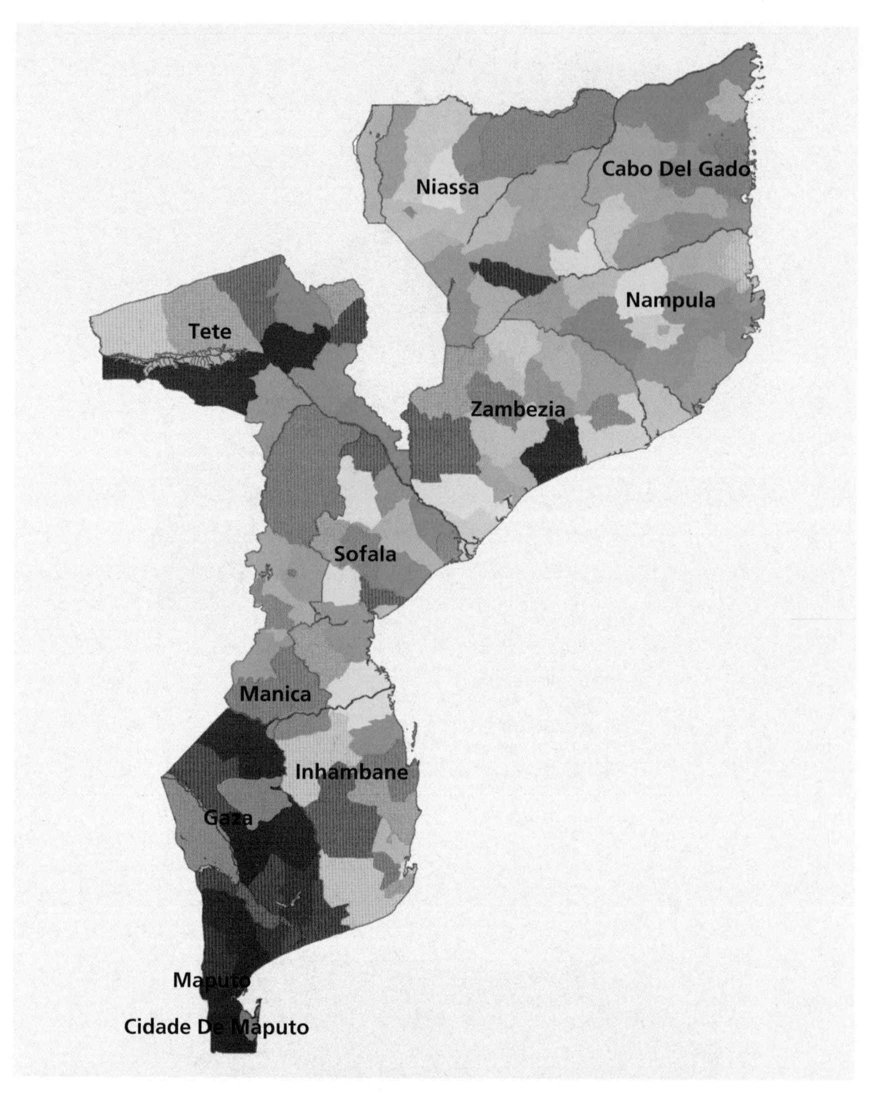

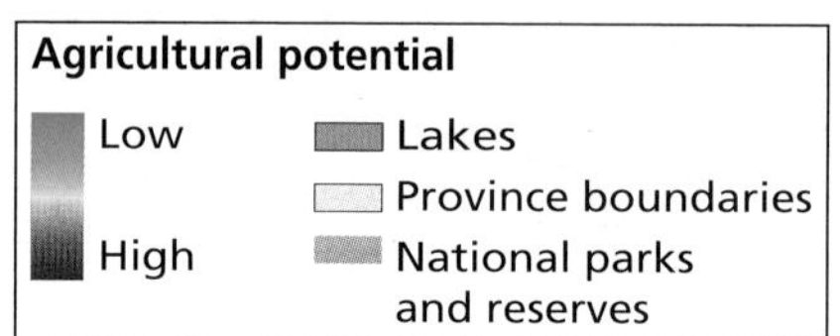

Figure 6.7 Mozambique—agricultural potential

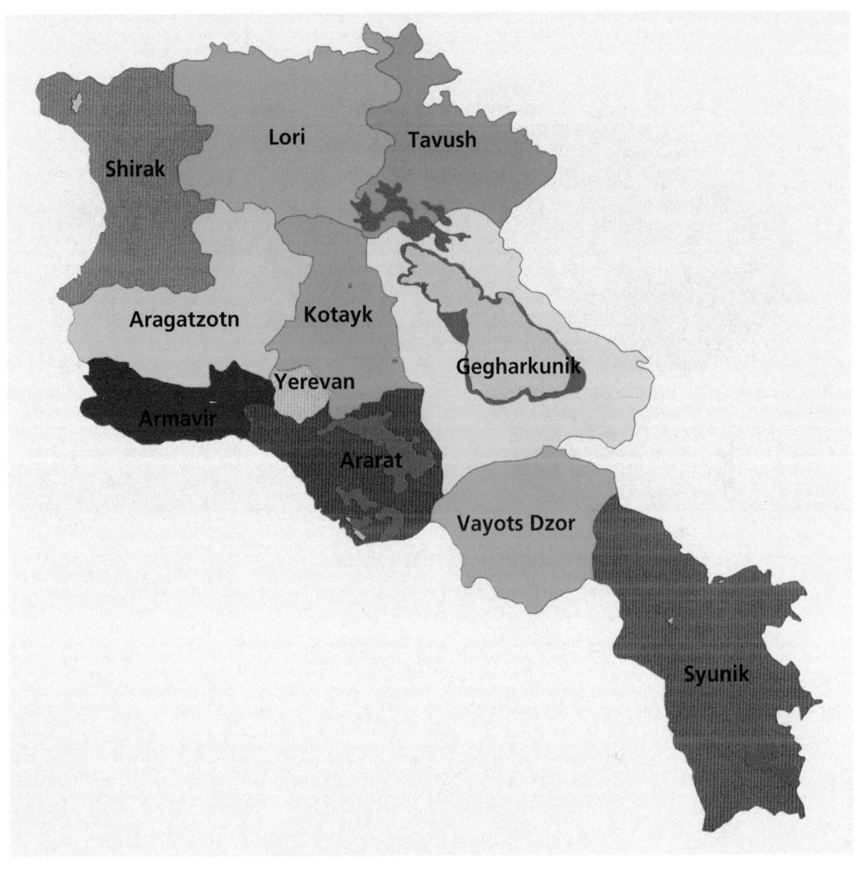

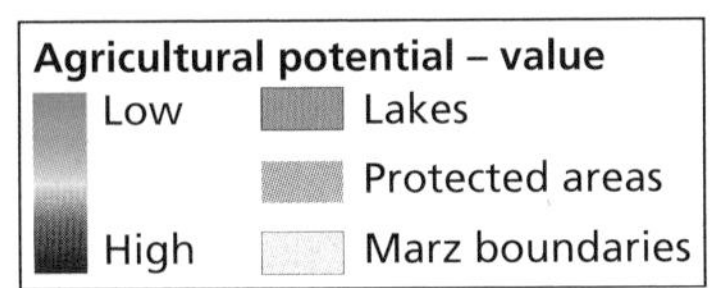

Figure 6.8 Armenia—agricultural potential

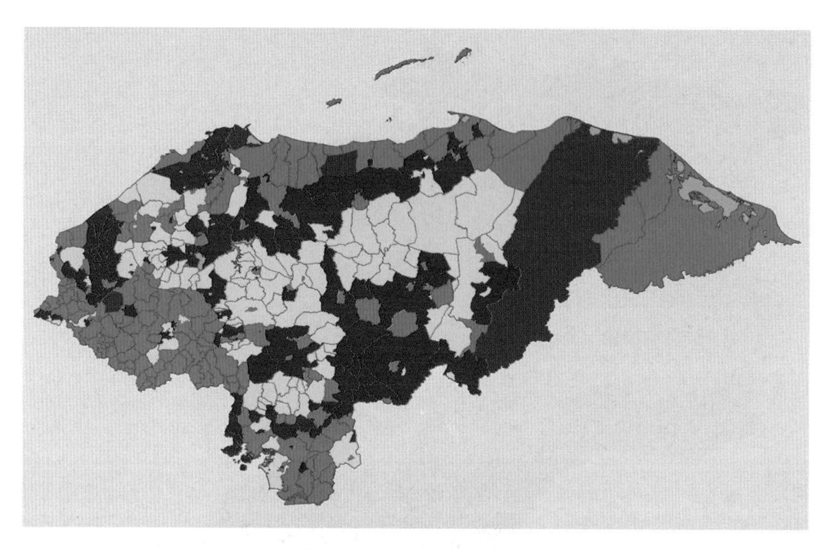

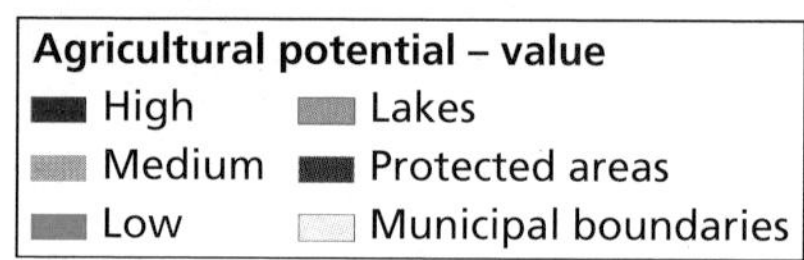

Figure 6.9 Honduras—agricultural potential

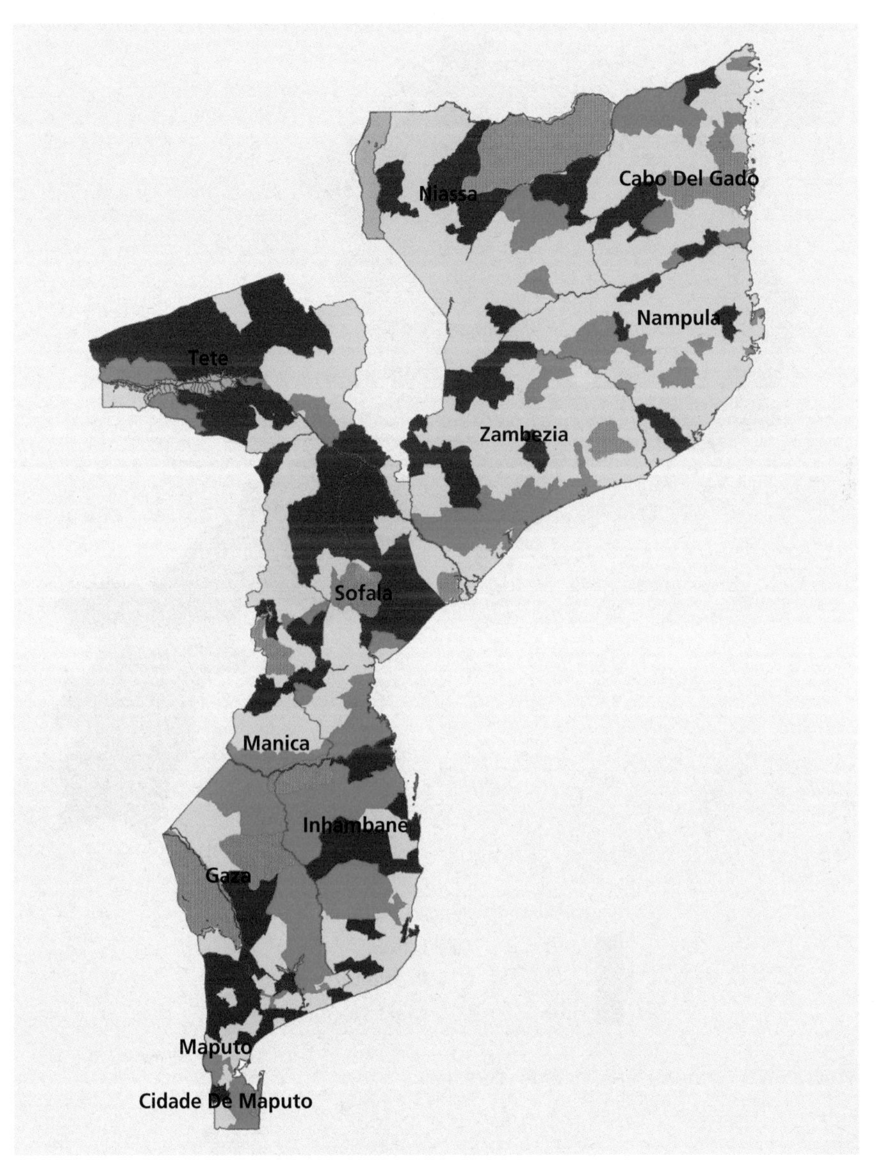

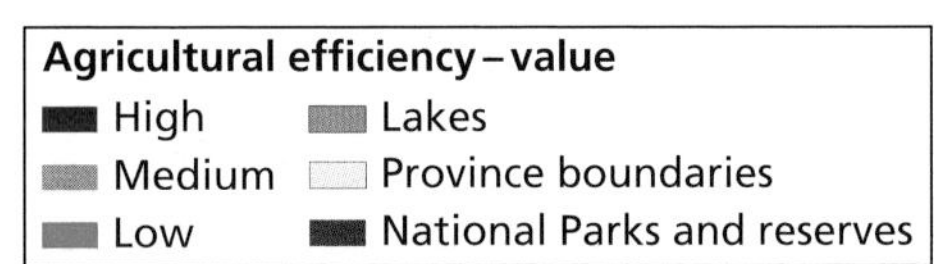

Figure 6.10 Mozambique—agricultural efficiency

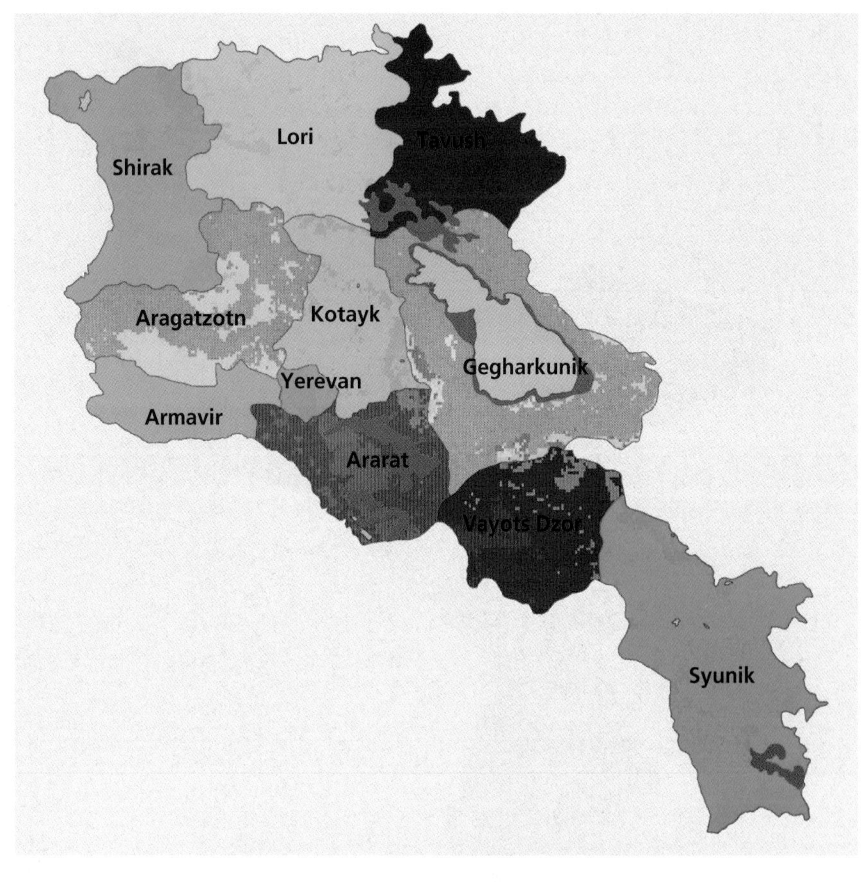

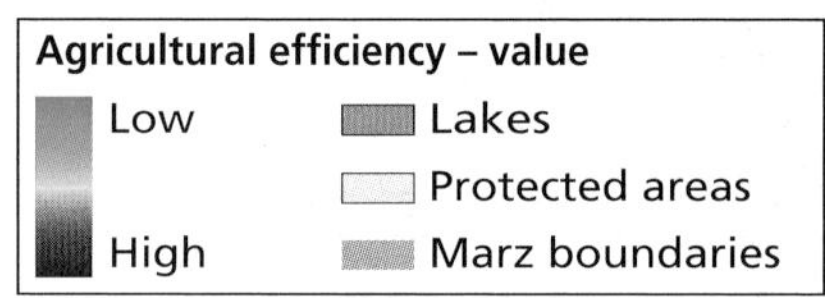

Figure 6.11 Armenia—agricultural efficiency

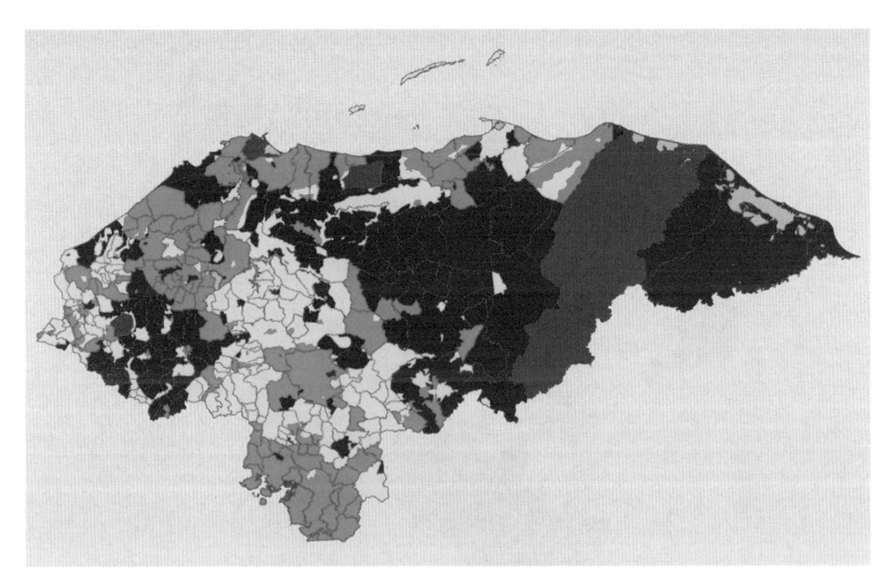

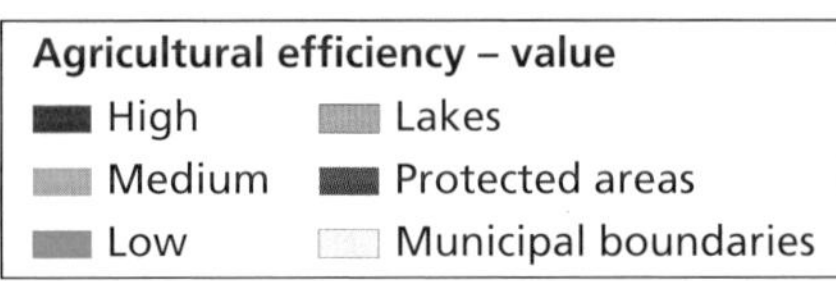

Figure 6.12 Honduras—agricultural efficiency

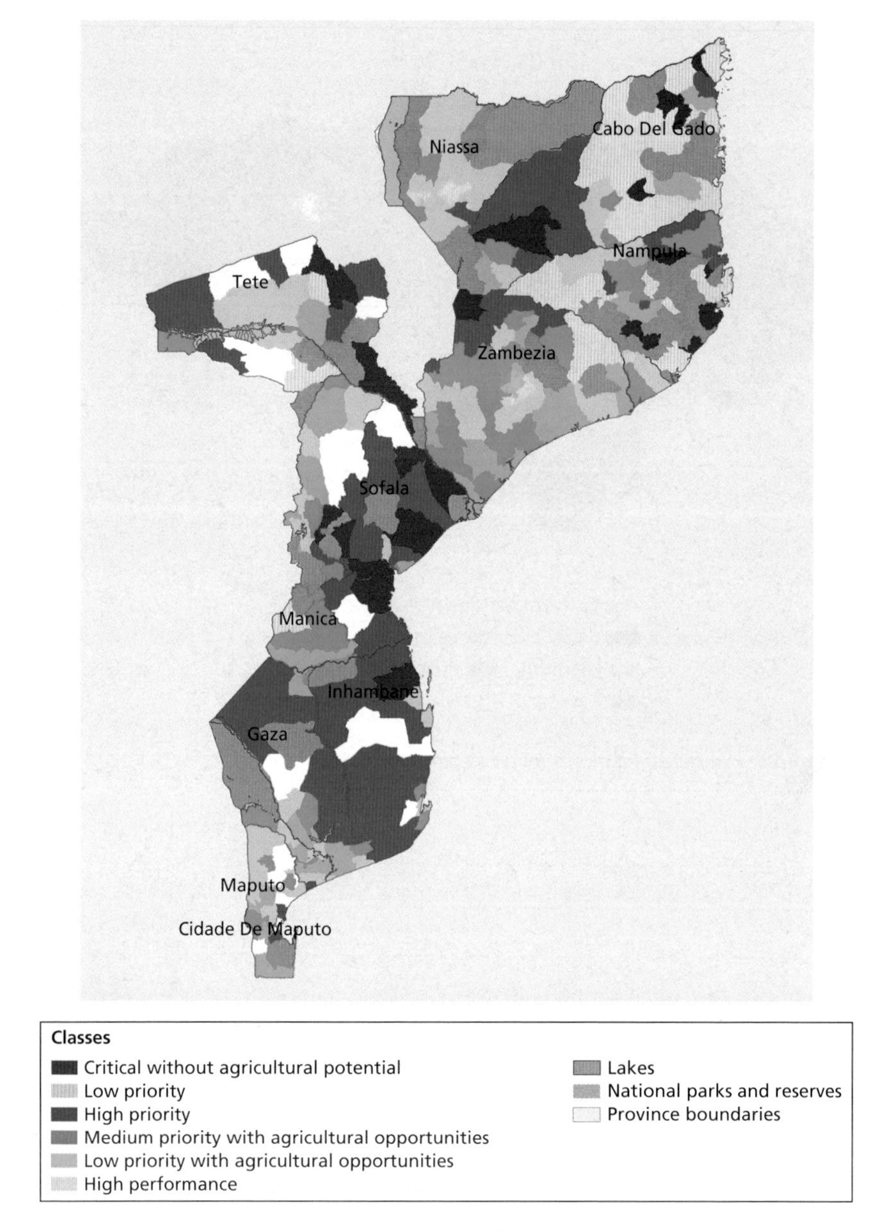

Figure 6.13 Mozambique—typology combining efficiency, potential, and poverty in seven categories

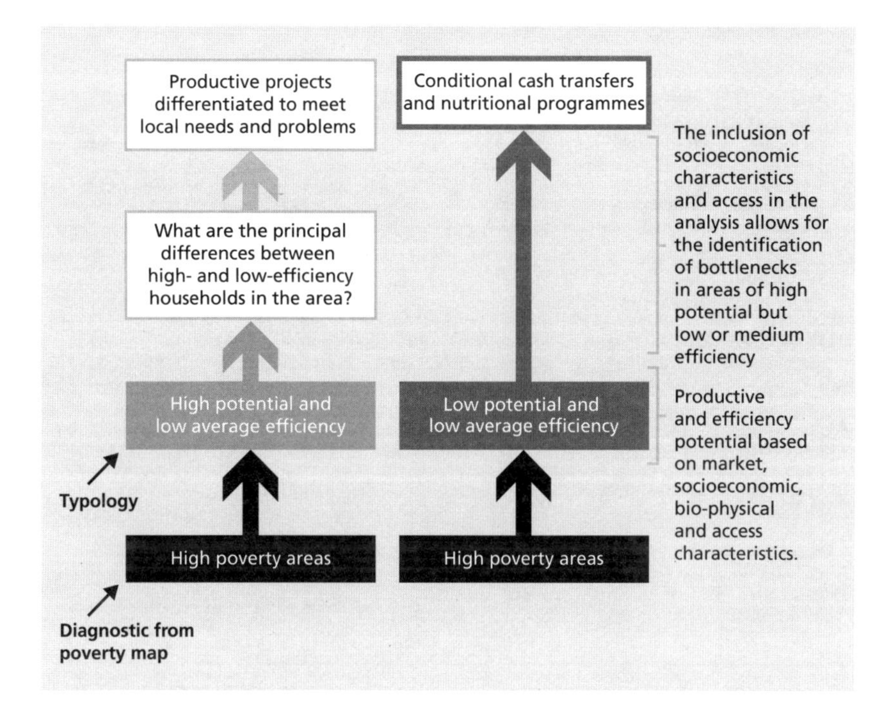

Figure 6.1 Advantages of a typology of micro-regions

Guided by this typology, policymakers can target geographical areas in which there is both profit potential and inefficiency, and identify the appropriate policies to reduce these inefficiencies. Similarly, the typology could identify areas in which the only alternative, given the existing low potential of the land, is to reduce poverty through rural labour programmes or safety-net programmes. Figure 6.2 summarizes some example of these alternatives.

3. Implementing the proposed typology

Among the variety of typologies used to categorize territories, poverty maps are arguably the most widely developed because they allow policymakers to design spatially targeted poverty-alleviation programmes (see Elbers et al., 2004). By imputing consumption and income values from survey data

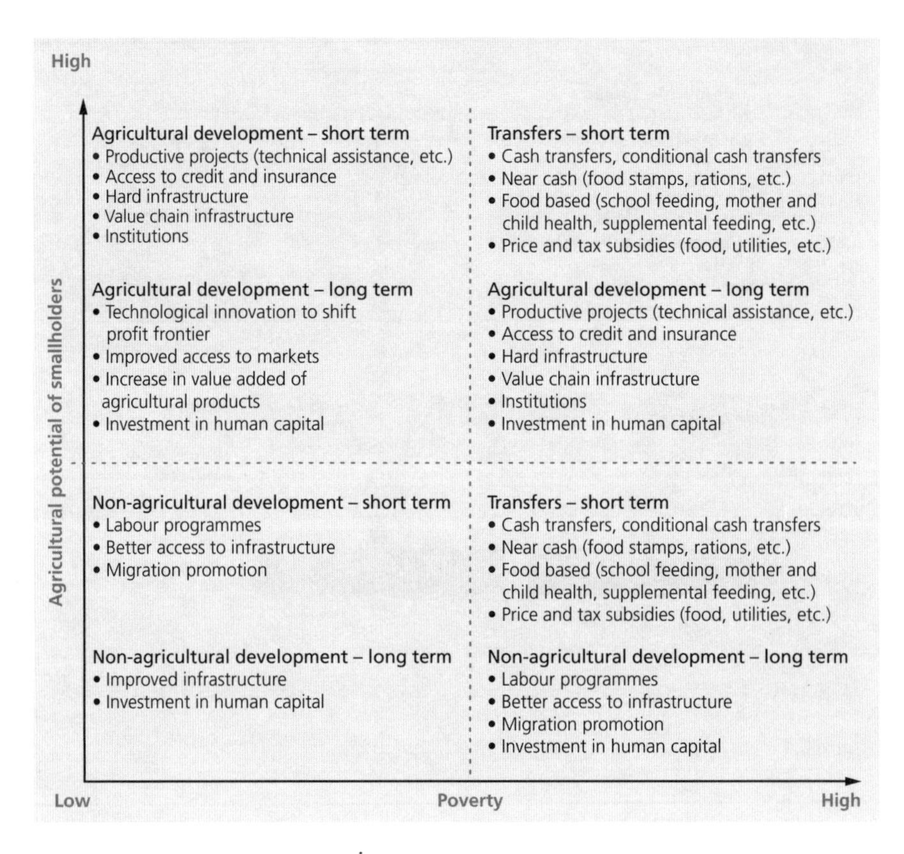

Figure 6.2 Examples of prioritization of interventions based on agricultural potential and malnutrition

estimations and extrapolating them to census data (see Elbers et al., 2003), poverty maps give reasonably static diagnostics of welfare. This is extremely useful in ranking areas by level of poverty when designing a transfer programme. It is much less useful, however, in deciding how to invest resources when designing poverty-alleviation programmes.

Another common tool used to construct typologies is cluster analysis. Cluster analysis methods have become popular because they are data-driven and can be used without the need for rigorous models to define the factors that determine the welfare measure to be analysed. When used to construct a typology to characterize communities' welfare or economic performance, however, the resulting index offers very little information regarding what policies should be implemented to improve current conditions, because the groups are not constructed ordering all variables monotonically (ascending or descending). This can generate confusion about the interpretation of the results. Hence, cluster analysis works well only when differences are determined over a small and relatively homogenous group of variables.

In this chapter, we attempt to construct a typology that takes into account the heterogeneity of small farmers while at the same time relying on a strong economic foundation. By recognizing that farmers are productive units optimizing an objective function subject to a set of constraints, the stochastic profit frontier analysis addresses many of these issues. The efficiency indicator is a continuous measure, similar to a score, and its interpretation is direct and simple. The functional form used for its estimation, which is explained in detail in Annex A, is flexible and imposes a limited structure on the analysis. Moreover, it is possible (with more and better data than is available at this point) to calculate the stochastic profit frontier using non-parametric estimation, in which case no parametric form is imposed for the analysis. Finally, the theory behind stochastic profit frontier estimation methods is standard theory, involving a constrained optimization process and allowing for random shocks. This set-up is suitable for modelling a farmer's decision-making process and analysing the opportunities and challenges each farmer faces.

Profit frontiers have long been used to estimate farm efficiency levels in developing countries. Using data for Basmati rice producers in Pakistan, Ali and Flinn (1989) find a mean level of profit inefficiency of 28 per cent associated with the household's education, non-agricultural employment, credit constraints, water constraints, and late application of fertilizer. Also in Pakistan, Ali et al. (1994) find an average farm profit inefficiency of 24 per cent; they also find that the size of holding, fragmentation of land, subsistence needs, and higher age of farmers contribute positively to inefficiency. Rahman (2003) finds a mean level of profit inefficiency of 23 per cent among Bangladeshi rice farmers, explained largely by infrastructure, soil fertility, experience, extension services, tenancy, and share of non-agricultural income. Using data for Chinese farm households, Wang et al., 1996 find a 39 per cent mean profit inefficiency level, influenced by farmers' resource endowment, education, family size, per capita net income, and family ties with village leaders. All of these studies, however, treat farms as single-output firms. For the purposes of our typology, it is essential to work with a multiple-output profit frontier approach, as this is a more realistic depiction of the farmers' decision-making process. In that respect, we are unaware of any existing studies using the stochastic profit frontier approach in a multiple-output farm setting.

3.1 THE MODEL

Let x denote a $(1 \times m)$ vector of variable and quasi-fixed inputs and y denote a $(1 \times q)$ vector of multiple outputs involved in the farm production process. Let z denote a $(1 \times r)$ vector of environmental variables that, while not directly determining the farmer's profits, could affect the farm's performance. Later in this section, we will discuss our criteria for placing specific variables as elements of x or z.[3]

Let $P \subset \mathbb{R}_+^{m+q}$ be the set of feasible production plans of the farm. We define a measure of output technical inefficiency δ (Farrell, 1957) for some production plans $(x_0, y_0) \in \mathbb{R}_+^{m+q}$ such that:

$$\delta 0 = \delta(\mathrm{x0}, \mathrm{y0}|\mathrm{P}) \equiv \sup\{\delta|(\mathrm{x0}, \mathrm{y0}) \in \mathrm{P}, \delta > 0\} \tag{1}$$

$$\text{For } (\mathrm{x}_0, \mathrm{y}_0) \in \mathrm{P}, \delta(\mathrm{x}_0, \mathrm{y}_0|\mathrm{P}) \geq 1.$$

We now define the restricted profit function $\pi(\mathrm{p}, \mathrm{w}, \delta)$ as the maximum profit attainable by a farm with characteristics z, facing output prices $\mathrm{p} \in \mathrm{P}\ (\mathrm{z})$ and input prices $\mathrm{w} \in \mathrm{W}\ (\mathrm{z})$:

$$\Pi(p, \omega, \delta) \equiv \sup_{\mathrm{x,y}}\{p'y - \omega'x : \delta(x, y)\} \leq \delta \tag{2}$$

Let πi be the observed profits for farmer i. The analyst is confronted with a set of observations (πi,pi,wi,zi) for i = 1,. . .,n, which are realizations of identically and independently distributed random variables with probability density function $f(\pi, p, w, z)$. This function has support over $P \times \mathbb{R}^r$.

We assume that z is not independent from (π, p, w), i.e. $f(\pi, p, w \mid z) \neq f(\pi, p, w)$. This means that the constraints on farmers' choices of prices p and w and on observed profits π, due to the environmental variables z that the farms face, operate through the dependence of (π, p, w) on z in $f(\pi, p, w, z)$. There exist several ways to formulate the model such that the production set is dependent on z (Coelli et al., 1998); however, we consider it to be more appropriate, given the empirical set-up we are analysing, to assume that the environmental variables z influence the mean and variance of the inefficiency process but not the boundary of its support. Hence, in our formulation, the conditioning in $f(\delta i \mid zi)$ operates through the following mechanism:

$$\delta_i = exp\ (z_i\beta + \varepsilon_i) \tag{3}$$

[3] Deprins and Simar (1989) and Kumbhakar and Lovell (2000) discuss the rationale for placing certain variables as elements of x or z, admitting that this issue is frequently a judgment call. In many cases, it is not obvious whether an exogenous variable is a characteristic of production technology or a determinant of productive efficiency.

where β is a vector of parameters and ϵ_i is a continuous i.i.d. random variable, independent of zi.[4] We assume the term ϵ_i is distributed $N(0, \sigma_\epsilon^2)$ with left truncation at $-zi\beta$ for each i.

3.2 ESTIMATION

Because the effect of covariates z operates through the dependence between π and z induced by equation 3, these assumptions provide a rationale for second-stage regressions. Kumbhakar and Lovell (2000) and Kumbhakar (1996) provide the typical set-up in these cases, defining the stochastic profit frontier function as:

$$\pi_i = g(p_i, \omega_i)\exp(v_i - \xi_i) \quad (4)$$

where ν_i is the stochastic noise error and ξ_i is a non-negative random variable associated with inefficiencies in production. Then the profit efficiency of farm i can be defined as:

$$EFF_i = E[exp\ (-u_i)|v_i - u_i| = E[exp\ (-\delta_0 - \sum\nolimits_{\delta=1}^{D} \delta_d X_{di})\xi_i \quad (5)$$

where X_{di} are exogenous (to the production process) variables that characterize the environment in which production occurs and that can be associated with the farm's inefficiencies.

As noted by Simar and Wilson (2007), regressing efficiency estimates obtained from maximum likelihood estimation of a parametric model for $\Pi(p, w, \delta)$ will very likely result in problems for statistical consistency, because the covariates in the second-stage regression (z) are correlated with the one-sided error terms from the first stage (in order for there to be a motivation for a second stage).[5] Consequently, the likelihood that is maximized is not the correct one, unless one takes into account the correlation structure. In order to do so, in the first stage modelling we estimate (4) heteroskedasticity in the one-sided error term ξ as a linear function of a set of covariates. The variance of the technical inefficiency component is then modelled as

$$\sigma_\xi^2 = exp\ (z\theta) \quad (6)$$

We use maximum likelihood estimation and a translogarithmic profit function correcting for heteroskedasticity as shown in (6) and then proceed to the

[4] See Simar and Wilson (2007) for estimation in a semi-parametric set-up.

[5] The errors and the covariates in the first stage will not be independent if the covariates in the second stage are correlated with the covariates in the first stage, which occurs in most empirical applications.

second-stage estimation of the technical efficiency term ξ on the environmental variables z.

3.3 THE DISTINCTION BETWEEN PRODUCTION INPUTS AND ENVIRONMENTAL FACTORS

In this section, we define our criteria for distinguishing quasi-fixed production inputs in x (which also includes variable inputs) from environmental variables z, because in some cases that distinction might seem arbitrary. An input is included in x when it has a clear market and also when its market is active and prices can be identified.[6]

In some cases, prices for certain inputs may not exist (or are not available to the analyst), particularly when studying rural poor populations in developing countries. Active markets and monetary transactions for land- or weather-based insurance, for instance, are rare in these settings, so it is extremely difficult to find a reliable price for land (of varying qualities) and weather (and climate-risk) preferences. Under these conditions, we believe elements like land size and climatic and biophysical conditions should be included in x in order to capture their direct impact on production as fixed or quasi-fixed inputs, even though the argument can be made that these variables capture failures in the land and risk-coping markets, and thus justify their inclusion in z.[7]

3.4 MOVING FROM HOUSEHOLD-LEVEL ESTIMATIONS TO SPATIAL ANALYSIS

The procedure described in the previous section provides profit efficiency estimates at the farm level. A remaining task is to scale up these results to the regional level where they can then be used for the purposes of this typology. According to our model, differences in profits are given by differences in crop choices, local prices, biophysical conditions, and farm efficiency (and, therefore, the exogenous factors affecting it). Hence, the econometric

[6] If there exists any evidence that these prices might not reflect actual market conditions for all the production units in the sample (due to accessibility problems or spatially incomplete markets), then the farm's input levels or stocks can be included in z in order to capture these market failures through their impact on farm efficiency. The idea behind this is that the input price is among the determinants of the production frontier; thus the market failures for that particular input influences the efficiency with which producers approach that frontier.

[7] Forms of land ownership or non-market mechanisms to smooth consumption, however, should be included in z in order to capture their impact on productive efficiency if it is suspected that these markets do not work properly.

estimation of the model described in the first two sections makes it possible to recover technological parameters for the 'representative' agriculture.

As explained earlier, a primary objective in the construction of the typology is to estimate the profit frontier and efficiency for a given region (e.g. community, district, province, or department). If the appropriate information (prices, biophysical factors, farm characteristics, and factors influencing efficiency) at that area's level is available, it can be plugged into the estimated profit function to determine regional efficiency.

Price data comes from household surveys. Ideally, a multi-output frontier model would be estimated, including every single output and input utilized by the farm in the production process. However, data limitations and computational feasibility make this impossible. Therefore, it is necessary to group outputs and inputs into the categories mentioned in the previous section, although such grouping generates other problems. To assign a single price to such broad groups as 'Fruits' or 'Vegetables', the median price per kilogram of all products in that group for a given region is used. How precise this grouping procedure is will depend on how many products in a given group are grown in the region, as well as how different the prices of these products are. For example, if green and red apples are the only fruits grown in a region and their prices are very similar, the median price of all the fruits produced in the region will be an adequate summary statistic. It follows that the choice of a region's size matters as well. If a region is too large, the risk that the median price is a poorer summary statistic is higher because the probability that more products are included in each group and that there is higher price variability increases. However, if a region is too small, the small amount of observations used to calculate any reliable measure of central tendency will also be a problem. Given the data, household information is aggregated at the district level; for those districts with too few observations, the province level is used.

Other farm- or household-specific variables used in the estimation procedures are calculated in a similar way and are subject to the same limitations. The biophysical data and the market accessibility data, on the other hand, have been specifically generated to map out in great detail the entire geography for each of the countries so that no aggregation or grouping issues occur. As mentioned before, with price and farm data aggregated at an adequate level and with perfectly mapped biophysical and accessibility datasets available, all that is left to do is to plug this information back into the model to predict profit frontiers and efficiency levels at the regional level.

The methodology employed to calculate empirically the productive potential and efficiency of the micro-regions is similar to the one used by the World Bank to estimate poverty maps, in which Living Standard Measurement Surveys (LSMS) surveys and census data are combined to take advantage of the richness of information provided by the first source and the representativeness of information provided by the second source. The methodology is

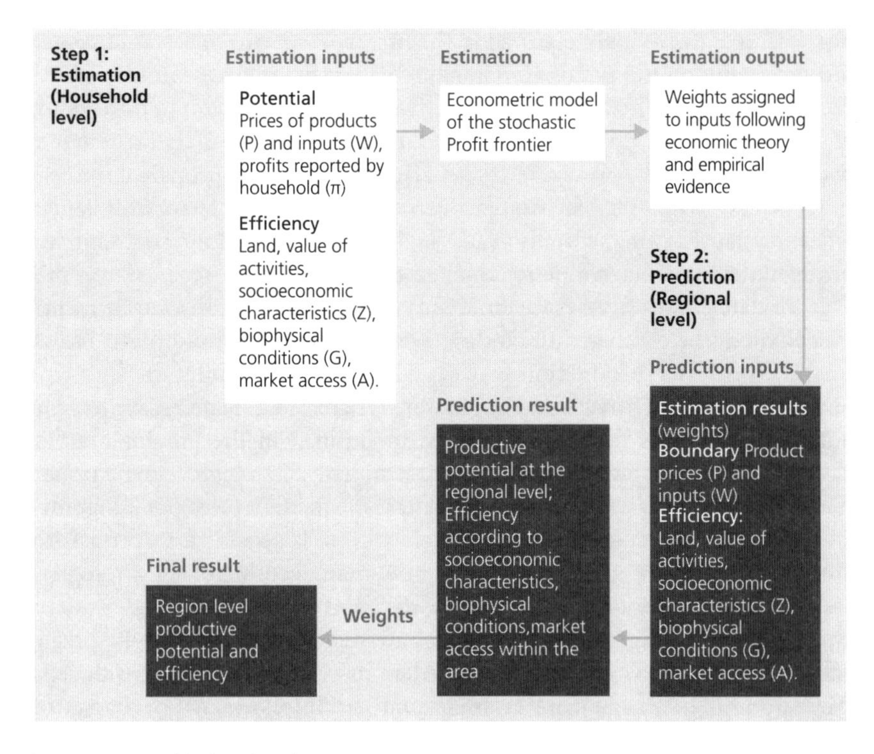

Figure 6.3 Empirical estimation steps

composed of two steps, described and summarized in Figure 6.3 and detailed in Annex A.

4. Implementing the typology in three developing countries: Guatemala, Armenia, and Mozambique

The results of the profit function, which depend on average crop prices and wages (and their interactions) to account for potential complementarities between products and inputs, are presented in Tables C6.2, C6.3, and C6.4 in Annex C for Mozambique, Armenia, and Honduras, respectively. Tables C6.5, C6.6, and C6.7 present the results of the second stage regression of estimated technical inefficiency on environmental variables z for the three countries (farmers' socioeconomic characteristics, biophysical conditions, and land use and market access conditions). These environmental variables affect the observed profits through the technical efficiency component. The association of these covariates with inefficiency is also related to the level of frontier profits, so the model is fully interacted with the predicted level of potential profits. The coefficients for these interactions are shown in the second column of the respective tables. All data sources used for these estimations are presented in Table C6.1.

The correlation between frontier profits and technical inefficiency is negative and significant for Honduras and Armenia, indicating that farmers with higher potential are more likely to have lower efficiency levels in our sample. This presents opportunities for interventions that could improve these farmers' efficiency. However, this is not the case for Mozambique, where the correlation between frontier profits and technical efficiency is positive (0.448) and statistically significant, suggesting that farmers with higher potential are also more likely to have higher efficiency levels in our sample.

When analysing the determinants of inefficiency in Tables C6.5 to C6.7, we found that in all cases, household size is negatively correlated with inefficiency—i.e. the higher the number of household members, the lower the inefficiency (although this association decreases with farm size). In contrast to Honduras and Armenia, in Mozambique ownership of productive assets such as tractors, machinery, and equipment is significantly associated with lower levels of efficiency, but this association basically holds only for farmers with smaller plots. A possible explanation for this result is that on small plots, the ownership of productive assets is not sufficient enough to contribute to production efficiency (and can even be counterproductive) and even more the land area does not seem to be sizable enough to generate the revenue to amortize the capital costs or to meet the running costs of such productive assets.

For farms with high potential, access to formal credit is significantly associated with lower levels of inefficiency. In Mozambique and Honduras, land ownership is correlated with higher levels of efficiency (for all farmers, but particularly for farmers with high potential), which falls in line with common notions of land ownership opening access to credit and encouraging productive investments. In the case of Armenia, the opposite effect is obtained. Unfortunately, to arrive at any conclusion regarding this aspect, we would need to have better controls for land quality, as there might be an endogenous component on which land is owned and which is rented, particularly for smallholders.

Market access costs (cost of transportation) and inefficiency are positively and significantly correlated. The association fades for farms with higher potential, which indicates that accessibility is a more important bottleneck for smallholders. Adequate policies to reduce transportation and transaction costs could be progressive if carried out jointly with assistance programmes that increase the competitiveness of rural areas compared to urban areas.

It is important to note that throughout this study, we use the term 'inefficiency' in ways that are consistent with the literature on stochastic frontiers. However, the model and data cannot capture all the complexities of the farm production process, so the econometric calculation may identify as 'inefficient' decisions that are perfectly rational but difficult to explain due to the analyst's incomplete information. For instance, a farmer facing extremely variable climate conditions may opt for more resilient but less profitable crops in order to reduce the risk of losing all his harvest. If the analyst cannot observe this high variability, he will regard the farmer's decision as sub-optimal. Another example is the over-utilization of available resources such as land or water. In the short run, practices that over-exploit productive assets can result in high profits; however, in the long run, this can cause a premature exhaustion of the productive capacity of the farm. A farmer with better foresight could appear to be 'inefficient' for missing a short-term high-profit opportunity, when in fact he is maximizing his long-term profits.

Unfortunately, until household surveys or other auxiliary sources collect more (and better) information regarding risk preferences, climate variability over time, price variability of inputs and outputs, conservation practices of productive assets, etc., it will be impossible to differentiate the risk-averse or non-depredatory producers from the inefficient producers who are not optimizing long-term profits. If the appropriate data is collected, then we could add covariates that capture preferences for risk, weather variability, etc. in z as environmental variables and test if they can capture a significant fraction of the variance of the technical inefficiency term.

4.1 CONSTRUCTING THE TYPOLOGY FOR GUATEMALA, ARMENIA, AND MOZAMBIQUE

The stochastic profit frontier estimation results in the previous section and the scaling-up steps described in the section on the move to special analysis allow us to construct a typology of micro-regions for these three countries using the notions of market access, profit potential (frontier), profit efficiency (technical efficiency), and priority (poverty rates).

Figures 6.4–6.6 present the results of the accessibility measurement constructed following the methodology explained in Annex B; this measures the number of hours it takes to access the closest market of more than 20,000 inhabitants. We then assess the maximum profit that farmers in a given area can generate, given their average characteristics and assuming an efficient allocation of resources and skills through the scaling-up of frontier profits as estimated in the regression analysis. The results are shown in Figures 6.7–6.9.

Once the profit potential of an area is established, it is necessary to know how far each region is from that frontier. This distance is given by the inefficiency component. The third set of maps in Figures 6.10–6.12 shows the efficiency levels for rural households in the three countries. More efficient areas, or areas closer to their potential (frontier), are depicted in red, while less efficient areas are depicted in green. Some interesting patterns start to appear in these maps. Many areas classified as having a high potential are nevertheless highly inefficient, which might explain some of the poverty rates observed in the areas.

To complete our typology, we need a measure of the optimal prioritization of investments in order to have a greater impact on the overall welfare levels of the rural population. For this purpose, a reasonable criterion is poverty rates. Combining Figures 6.7–6.9, 6.10–6.12, and the poverty maps for each of the countries, we develop the typology for Mozambique, Armenia, and Honduras. With only these dimensions and poverty rates, we can observe the extreme heterogeneity that exists among small farmers in the rural areas of these countries.

For the purposes of exposition, we collapse the types in our typology to seven groups that capture some key characteristics important for policy-making as shown in Table 6.1. These groups are described here and shown in Figures 6.13–6.15.

- Critical areas (high poverty and low potential) [Dark Red]
- High-priority areas (high poverty and medium/high potential) [Dark green]
- Medium-priority areas without opportunities for agriculture (medium poverty and low potential) [Orange]
- Medium-priority areas with opportunities for agriculture (medium poverty, high/medium potential, and low/medium efficiency [Green]
- Low-priority areas with opportunities for agriculture (low poverty, high/medium potential, and low/medium efficiency [Light green]
- High-performance areas (low poverty, high/medium potential, and high efficiency [very light green]
- Low-priority areas (low poverty and low potential) [Light orange]

Figures 6.16–6.18 show the specific areas where different types of interventions can be implemented in Mozambique, Armenia, and Honduras based on the typology developed for the three countries. In all the figures we have also include the transportation costs dimension as a way to incorporate the importance of infrastructure, specifically roads. In each figure, we present four maps: map (a) shows critical areas, i.e. areas of extreme high poverty and very low potential for agricultural development given their low efficiency; map (b) shows high-priority areas, i.e. areas with high poverty and low efficiency but high agricultural potential for agricultural development; map (c) shows high performance areas, i.e. areas with low poverty, high efficiency, and high potential (these are the areas from which policymakers can learn and try to replicate what is happening there but being careful in taking into account the difference in agro-ecological potential), and map (d) shows medium-priority areas with no opportunities for agriculture (medium poverty, low potential, and low efficiency).

Clearly, in all four categories and as detailed in Figure 6.2, investment in infrastructure is essential but, as shown in the maps on the specific case of market access, it is vital to target clearly where the investment is needed and how major complementarities can be obtained. The core debate is more centred on complementarities among different public investments, such as main roads, feed roads, telecommunications, irrigation, water and sanitation, storage, etc., in order effectively and efficiently to reduce rural poverty rates and inequality (Calderón and Servén, 2004). Overall, as public infrastructure contributes to an increase in local and regional productivity in the rural sector, it is expected to have effects on labour productivity and labour incomes. However, in the case of rural roads and telecommunications, a direct effect on the labour markets can be expected, as these contribute to a reduction in job-matching costs. Also, reduced transportation costs allow for greater mobility of the labour supply across different regions and facilitate temporary migration. Temporary migration can prove to be very efficient in the overall employment of labour services in rural areas, as agricultural activities are subject to seasonality.

In the case of the critical areas without agricultural potential, policies should provide direct social assistance in the short term to minimize the risks, especially for children and pregnant women, and major effort should be made properly to target the benefits of such programmes. For example, it will be important to implement existing best practices for such social-protection programmes as conditional cash transfers, direct cash transfers, food for work, or school feeding. Specifically, in the case of Armenia, this will imply more efficient targeting of the country's biggest social assistance programme, i.e. the family benefit system (hereafter, FBS). The proper targeting of this programme will be an important factor in poverty reduction. Similar instruments could be applied to Honduras and Mozambique. In addition, policies to improve human capital to facilitate labour mobility could be important.

The high-priority areas are clearly areas where public investment can play a crucial role by identifying key bottlenecks that explain why producers in these areas show such low levels of efficiency, despite having land with significant potential for agriculture. The policies in these areas should therefore target specific bottlenecks and market failures that prevent producers from reaching their high production potential. For example, Shirak and Aragatzotn have substantial potential, and any improvement in efficiency will create significant positive effects on the income of households in these areas; thus, it will be crucial to focus on identifying the key bottlenecks in these areas. This is clearly not the case for Lori, where the agricultural potential is smaller.

Market failures not only limit smallholders' access to other factors of production and to modern technologies; they also force smallholders to utilize their productive assets for non-productive purposes—for instance, as an insurance mechanism to smooth consumption through the sale of these assets. Therefore, the key idea is to put markets to work in rural areas so that productive resources are entirely devoted in an efficient way to productive purposes. This in turn will lead to higher labour productivity and labour incomes. Four markets deserve special attention: credit, insurance, land, and services. Specific policies must be identified to eliminate market failures and, in some cases, to create these markets, namely markets for agricultural services: extension, legal, accounting, marketing, management, etc. Horizontal and vertical integration arrangements also play a crucial role in rural areas where market failures are widespread by allowing smallholders to combine productive resources through non-market relationships. In this way, labour resources can be combined with other productive resources, modern inputs capital, and services that otherwise would not be possible, given the high incidence of market failures. In the case of vertical integration schemes among smallholders, such as contract farming, we foresee three potential benefits: (a) they allow smallholders to resolve market failures or bottlenecks through the access provided by the contracting party; (b) they allow farmers to specialize in production activities in which they have comparative advantage; and (c) they allow for scales of operation.

In the medium-priority areas with no potential for agriculture, the focus needs to be on non-farm activities or activities that will increase their human capital so that they migrate as a mechanism to move households out of poverty. Public investment in hard infrastructure could help to create these opportunities as investment in rural roads, electrification, or information and communication technologies. Finally, the high-performance areas are where lessons can be learned regarding how these regions were able to reach their maximum production potential, despite them being regions with significant differences in agro-ecological conditions; from there, and also from lessons learned on successful projects in the different types of regions, policymakers can extrapolate the best practices learned to other regions of the country, taking into account the differences.

5. Conclusions and policy issues

Smallholder cultivation and the high intensity and density of poverty are major characteristics of rural areas in sub-Saharan Africa, Latin America, and South Asia. Most smallholders either practise subsistence farming or operate largely in local markets due to a lack of connectivity to more lucrative markets at provincial, national, or global levels. As a result, incentives remain weak and investments remain low, as does the level of technology adoption and productivity, resulting in a low-level equilibrium poverty trap.

Two instruments appear critical to break this deadlock: (i) physical infrastructure, such as roads, electricity, potable water and drainage, water for irrigation, and telecommunications, that connects smallholders to markets; and (ii) the role of accompanying institutions, such as land titling, credit markets, contract farming, vertically integrated schemes, market information systems, commercial rules and laws, commodity exchanges, warehouse receipt systems, and producer and trader associations, that can reduce the marketing risks and transaction costs associated with the process of exchange between producers and consumers.

This paper has attempted to develop a framework that can capture the heterogeneity of smallholders and therefore identify and strengthen the institutional and infrastructural base necessary to support and improve smallholders' links to markets. The chapter brings together three dimensions in its analysis:

- the difference between profit potential and efficiency (technical and allocative);
- the heterogeneity of small farmers and therefore their specific bottlenecks in connecting to markets; and
- the level of market accessibility (measured through transportation costs).

In addition, it applies a common multi-pronged approach to developing countries which aims at the production of an international public good which can be applied in more than one country or region.

REFERENCES

Ali, M. and J. C. Flinn (1989). 'Profit efficiency among Basmati rice producers in Pakistan Punjab', *American Journal of Agricultural Economics*, 71(2), pp. 303–10.

Ali, F., A. Parikh, and M. K. Shah (1994). 'Measurement of profit efficiency using behavioral and stochastic frontier approaches', *Applied Economics*, 26(2), pp. 181–8.

Calderón, C. and L. Servén (2004). 'The Effects of Infrastructure Development on Growth and Income Distribution', Working Paper no. 270, Central Bank of Chile, September.

Coelli, T., D. S. P. Rao, and G. E. Battese (1998). *An introduction to efficiency and productivity analysis*. Boston: Kluwer Academic Publishers, Inc.

Deprins, D. and L. Simar (1989). 'Estimating technical inefficiencies with correction for environmental conditions with an application to railway companies', *Annals of Public and Cooperative Economics*, 60(1), pp. 81–102.

Elbers, C., T. Fujii, P. Lanjouw, B. Ozler, and W. Yin (2004). 'Poverty alleviation through geographic targeting: How much does disaggregation help?' World Bank Policy Research Working Paper No. 3419, World Bank, Washington, DC.

Elbers, C., J. O. Lanjouw, and P. F. Lanjouw (2003). 'Micro-level estimation of poverty and inequality', *Econometrica*, 71(1), pp. 355–64.

Farrell, M. J. (1957). 'The Measurement of productive efficiency', *Journal of the Royal Statistical Society*, Series A (General), Vol. 120, No. 3, pp. 253–90.

Kumbhakar, S. C. (1996). 'Efficiency measurement with multiple outputs and multiple inputs', *The Journal of Productivity Analysis*, 7, pp. 225–55.

Kumbhakar, S. C. and C. A. K. Lovell (2000). *Stochastic frontier analysis*. Cambridge: Cambridge University Press.

Orden, D., M. Torero, and A. Gulati (2004). 'Agricultural Markets and the Rural Poor'. Background paper for workshop of the Poverty Reduction Network, 5 March 2004, International Food Policy Research Institute, Washington, DC.

Rahman, S. (2003). 'Profit efficiency among Bangladeshi rice farmers', *Food Policy*, 28(5–6), pp. 487–503.

Ramaswami, B., P. S. Birthal, and P. K. Joshi (2006). 'Efficiency and Distribution in Contract Farming: The Case of Indian Poultry Growers', MTID Discussion Paper No. 91, International Food Policy Research Institute, Washington, DC.

Simar, L. and P. W. Wilson (2007). 'Estimation and inference in two-stage, semi-parametric models of production processes', *Journal of Econometrics*, 136, pp. 31–64.

Tobler, W. (1993). *Three presentations on geographical analysis and modelling*, vol. 93-1, Technical Report, University of California at Santa Barbara.

Torero, M., E. Maruyama and M. Elias (2009). 'Tipología de micro-regiones de las áreas rurales de Ecuador: Aplicaciones de fronteras estocásticas de utilidades agrícolas'. Finanzas Públicas, Volumen 1, No. 2, Segundo Semestre del 2009.

Wang, J., E. J. Wailes, and G. L. Cramer (1996). 'A shadow-price frontier measurement of profit efficiency in Chinese agriculture', *American Journal of Agricultural Economics*, 78(1), pp. 146–56.

ANNEX A EMPIRICAL ESTIMATION OF STOCHASTIC PROFIT FRONTIER

Step 1: Estimation of the stochastic profit frontier function. Farmer level

The stochastic profit frontier function is defined as:

$$\pi_{ij} = f(P_{ij}, W_{ij})exp\left(v_{ij} - u_{ij}(Z)\right) \tag{1}$$

where π_{ij} is the utility of the farmer i on the area j; P_{ij} and W_{ij} are the vector of median prices of products and inputs at the regional level faced by the farmer; v_{ij} is a two-tailed error or stochastic noise iid distributed with $N(0, \sigma_v^2)$. and independent of u_{ij}, which is a non-negative random variable associated with the production inefficiency distributed independently with a semi-normal distribution $N^+(0, \sigma_{uij}^2)$; Z is a vector of environmental variables that includes: socioeconomic characteristics (z_i,) including the farm's fixed factors (land and capital); biophysical conditions (A_j) and market access costs that the farmer faces (G_j—see Annex B for information on how market access is estimated). All the variables are at farmer level and are obtained from LSMS surveys with the exception of biophysical and market access A_j and G_j, which come from secondary data.[8]

Step 2: Prediction of potential and efficiency. Regional level

After obtaining the parameters in step 1, the values of productive potential and efficiency representatives for each region are predicted. In order to obtain significant results, Living Standard Measurement Survey (LSMS) data is replaced by census data.

The productive potential is estimated in a linear way, using a vector of median prices at the regional level, following the equation presented here:

$$\pi_{predicted,j} = \gamma_1 P_j + \gamma_2 W_j \tag{2}$$

where γ_1 and γ_2 are the parameters obtained in step 1. Given that the vector of prices is at the regional level, there will be one prediction for each region.

The efficiency is estimated in a non-linear way according to the following formula:

$$Efficiency_{jg} = \left\{\frac{1 - \Phi(\sigma* - \mu*_{jg}/\sigma*)}{1 - \Phi(-\mu*_{jg}/\sigma*)}\right\} exp\left(-\mu*_{jg} + \frac{1}{2}\sigma_*^2\right) \tag{3}$$

[8] A_j is based on an accessibility model that includes information on diesel prices, road infrastructure, distance, etc. G_j comes from a biophysical database (see Annex B for details).

Given that the efficiency depends on biophysical conditions and that each region can contain more than one biophysical condition, the prediction of efficiency will be for each region j and for each biophysical condition g within each region, where Φ is a cumulative normal function and:

$$\mu *_{jg} = -\frac{\epsilon_j \sigma^2_{u_{jg}}}{\sigma^2_{u_{jg}} + \sigma^2_v} \quad y \quad \sigma * = \frac{\sigma_{u_{jg}} \sigma_v}{\sigma^2_{u_{jg}} + \sigma^2_v}$$

The variance of the inefficiency, $\sigma^2_{u_{jg}}$, depends on the biophsical and access conditions (i.e. is heteroskedastic). The variance of the random component, is constant and was estimated in step 1. σ^2_v, ϵ_j is the prediction error of the productive potential: $\epsilon_j = \pi_{observado,j} - \pi_{predicted,j}$

To obtain unique values of efficiency per region, a weighted average of the efficiencies in each region is calculated. The weights will be assigned according to the total geographical extension that each biophysical condition occupies in each region.

ANNEX B

Market access estimation

GIS data has recently made it possible to investigate the market access question in a more sophisticated way. With this data, one can calculate the shortest time or distance from any village to a regional or local market using the distance travelled on different road surfaces combined with an impedance measure (for example, rivers or geographical faults), which reflects the speed one can travel on roads of different qualities and on the slope of the terrain through which the road passes. The resulting market access measure can be expressed as a weighted average of the distance travelled on each type of road, where the weights are proportional to the impedance factor.

There are two problems with these measures of access. The first is that they do not incorporate transportation costs, which may well vary with distance and type of road surface in a different way than they do with time. When this is the case, the time-based measure will be misleading because it could imply that for a particular village, one market is closer than another because it takes less time to get there, even though it may cost more. By the same token, it could imply that one village is closer to a market than another as measured by time but not as measured by cost. However, presumably what the farmer wants to know is not how far it is to his market, but rather how much he can sell his produce for in that market or, equivalently, what his farm gate price is, net of transportation cost. In this chapter, we use a measure that incorporates both aspects and report our measure of the merged market distance data for each village in the Republic of Armenia, with a matrix of transportation costs by truck on two different classes of roads and on rivers or by animal on trails where there are no roads. This gives us a measure of accessibility in terms of costs.

The second problem with the typical market access indicator is that it considers only the local market. However, the level of prices in local markets may well vary according to how far they are from the country's largest market. Take the example of Yerevan. It could well be that a farmer would get a higher price for his products by shipping them to a market which, while further from his village, is closer to Yerevan or, equivalently, in which the price of his product is higher. Therefore, we estimate the costs to access simultaneously the local market and the Yerevan market, with the variable reported being the one that minimizes the cost to access both markets simultaneously.

The accessibility model

To 'connect' every household with the closest market, we constructed a series of accessibility indicators. The notion behind this is that accessibility is not a discrete variable (i.e. to have or not have access), but a continuum that reflects the difficulties each household faces when trying to access different types of infrastructure. This accessibility analysis was applied to the entire land surface.

Accessibility is defined as how feasible it is to reach a location from another location, considering factors like distance, moving costs, type of transportation, and time. The base of this analysis assumes that people are likely to move through highways, major roads, or paths when those exist, but otherwise would walk to the nearest market. The final objective is calculating the time a person invests in reaching the nearest market through the fastest route.

The moving time on the land surface depends on different factors, the most important being the distance, but there are other important factors such as the existing road network and its specific characteristics, the slope, and the presence of obstacles such as rivers (except for those cases in which rivers are used as a means of transport).

The accessibility analysis was developed on a raster format, which means that the entire area of analysis was converted into a grid of cells measuring 92.6 by 92.6 metres. Each cell was assigned a 'friction' value based on characteristics of slope, roads, and barriers, which allowed each cell to be allotted a value for the time required to reach the nearest facility (Figures B6.1 and B6.2). Having created the friction grid, the cost-weighted distance algorithm runs over the raster surface, calculating the accumulated time departing from each market available, replacing overlapping values with the least time-consuming route.

The first variable is the slope, which has been used to calculate a walking travel speed that depends indirectly on it. Tobler's (1993) walking velocity has three variations: one corresponding to a footpath; another to a horseback; and finally, one for off-path. The horseback walking velocity has been assigned to the dirt-road tracks and the footpath velocity to walking trails; where no paths are available, the off-path walking velocity has been assigned (see Figure B6.3). Figure B6.4 presents the results for each of the road classifications:

The second variable used in this analysis was transportation infrastructure, of which Armenia has two major kinds: paved roads and unpaved roads. In addition, there are some rivers by which navigation occurs. Each type of road was assigned an average travel speed and the corresponding cell given a crossing time in seconds (see Figure B6.5).

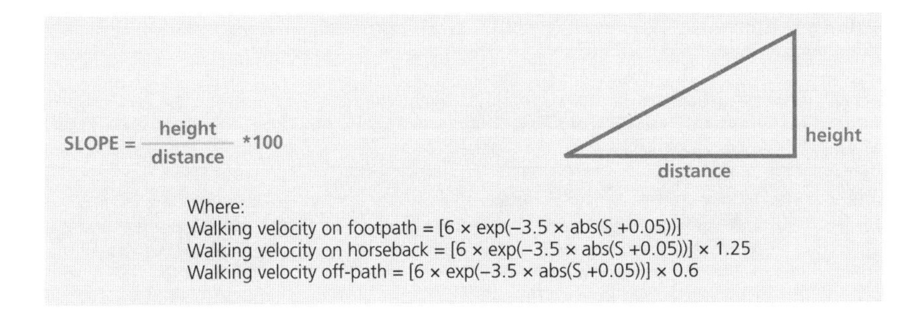

Figure B6.3 Calculation of slope

The third and final variable used in this model corresponds to the presence of natural barriers such as rivers, which prevent people from travelling in a straight line if there is no bridge. Cells corresponding to areas with a river and no bridge are assigned a travel time ten times their value, so that the crossing would only be considered where a bridge is available.

Once the friction model is built and each cell has been allocated a travel time value, cost-weighted distance algorithms are run over the raster surface, calculating the accumulated time required to travel a particular route (choosing the one that is least time-consuming, as in Figure B6.6). This information is then used to simulate the impacts of improvements of road segments. For example, if a road is improved from a walking trail to a dirt road, then the new average speed of the upgraded category is assigned and re-estimates all the accessibility measures.

ANNEX C

Table C6.1 Data sources for estimation of typologies

Data/ Country	Mozambique	Armenia	Honduras
Household Survey	In particular, π_{ij}, P_{ij}, W_{ij}, Z_i are obtained from the 2008 rural household survey (*Trabalho de Inquérito Agrícola*—TIA). The profits (π_{ij}) represent the producer income net of costs (wages, transport, storage, etc.).	Integrated Living Condition Survey 2010: National Statistical Service of the Republic of Armenia <http://www.armstat.am>	LSMS I (2004)
Agro-ecological data	The biophysical condition (G_j) considered is land use obtained from CENACARTA; there can be various land types within a posto: woodlands, secondary woodlands, pastures, croplands, and areas not suitable for agriculture.	<http://world-gazetteer.com/>Environmental, Research and Management Center. American University of Armenia.Spot Satellite images, 2000.	Life zones, Programa Nacional de Desarrollo Urbano (PRONADERS); (1998)

(*continued*)

Table C6.1 Continued

Data/ Country	Mozambique	Armenia	Honduras
Roads, rivers, lakes	*Centro Nacional de Cartografia e Teledetecção*—CENACARTA, *Instituto Nacional de Estatistica*—INE, *Administracion Nacional de Carreteras*—ENA. The model approximates the ease with which a major town (20,000 inhabitants or more) can be reached from each particular location considering factors such as distance, cost of travel, type of transportation, and time.	DIVA—GIS	Roads: Fuente: Secrataria de Obras Publicas de Transporte y Vivienda (SOPTRAVI), elaborated by: Sistema Nacional de Información Territorial (SINIT). 1999Rivers: Instituto Geográfico Nacional. Sistema Nacional de Información Territorial (SINIT).
Populated centres	<http://world-gazetteer.com/>	<http://world-gazetteer.com/>	National Institute of Statistics
Poverty data	Poverty Map, World Bank 2007	2009 Poverty map: National Statistical Service of the Republic of Armenia <http://www.armstat.am>	'Estimación de Indicadores de Pobreza y Desigualdad a Nivel Municipal en Honduras' (2001)—BID-INE

Table C6.2 Mozambique: profit frontier estimation results

Dependent variable: log of profit			
log(price1)	0.346 (0.567)	log(price3)*log(price6)	0.531 (0.519)
log(price3)	−3.101** (1.455)	log(price4)* log(price4)	−0.200 (0.235)
log(price4)	0.487 (0.772)	log(price4)* log(price5)	−0.191 (0.382)
log(price5)	1.073 (0.706)	log(price4)* log(price6)	−0.360 (0.340)
log(price6)	0.250 (0.817)	log(price5)* log(price5)	0.763** (0.301)
log(wages)	0.844 (0.687)	log(price5)* log(price6)	0.443 (0.320)
log(price1)* log(price1)	0.0172 (0.098)	log(price6)* log(price6)	−0.098 (0.193)
log(price1)* log(price3)	0.434 (0.513)	log(price1)* log(wage)	−0.166 (0.197)
log(price1)* log(price4)	−0.030 (0.258)	log(price3)* log(wage)	1.220*** (0.468)
log(price1)* log(price5)	0.034 (0.370)	log(price4)* log(wage)	−0.037 (0.272)
log(price1)* log(price6)	−0.121 (0.223)	log(price5)* log(wage)	−0.489** (0.241)

log(price3)* log(price3)	−1.836***	log(price6)* log(wage)	−0.096
	(0.559)		(0.260)
log(price3)* log(price4)	1.602***	log(wage)* log(wage)	−0.071
	(0.568)		(0.123)
log(price3)* log(price5)	−1.390**	Constant	4.85***
	(0.646)		(1.138)
		log(sigma_v2)	0.448***
			(0.066)
Observations			2533
Log likelihood			−4609.0

Note 1: Standard errors reported in parenthesis.
*** p<0.01,
** p<0.05,
* p<0.1.
Note 2: Prices: (1) Fruits & vegetables, (2) Industrial crops, (3) Maize, (4) Cereals, (5) Beans, (6) Tubers.
Note 3: sigma_v2 is the variance of the stochastic component.

Table C6.3 Armenia: profit frontier estimation results

Dependent variable: log of profit	
log(price1)	13.34***
	(1.532)
log(price2)	−2.750***
	(0.780)
log(price3)	109.8***
	(19.19)
log(price4)	20.07***
	(2.054)
log(price6)	112.5***
	(15.25)
log(price1)* log(price1)	−12.53***
	(1.897)
log(price2)* log(price2)	14.42***
	(1.881)
log(price3)* log(price3)	−28.23***
	(5.207)
log(price4)* log(price4)	48.62***
	(5.637)
log(price6)* log(price6)	43.06***
	(5.692)
Constant	−28.51***
	(9.353)
$\log(\sigma_v^2)$	−0.00100
	(0.0457)
Observations	1633
Log likelihood	−2462

Note 1: Standard Errors are in parenthesis.
*** p<0.01,
** p<0.05,
* p<0.01
Note 2: Prices: (1) 'Fruits', (2) 'Industrial crops', (3) 'Legumes', (4) 'Potatoes', (5) 'Vegetables', (6) 'Grass'
Note 3: σ_v^2 is the variance of the stochastic component.

Table C6.4 Honduras: profit frontier estimation results

Frontier Estimation (dependent variable: lnprofit)

lnprice1	2.787***	lnprice2_lnprice4	−0.195	lnprice1_infertilizer	0.299	lnwage_lnwage	−0.366***
	(0.872)		(0.167)		(0.341)		(0.107)
lnprice2	−0.571	lnprice2_lnprice5	0.122	lnprice2_lnwage	0.0713	lnwage_lnmanure	0.180
	(0.612)		(0.0806)		(0.202)		(0.123)
lnprice4	1.699	lnprice2_lnprice6	0.247*	lnprice2_lnmanure	−0.117***	lnwage_lnfertilizer	1.074*
	(1.675)		(0.150)		(0.0507)		(0.633)
lnprice5	−0.537	lnprice2_lnprice7	−0.0157	lnprice2_lnfertilizer	0.0975	lnmanure_lnmaure	0.00727
	(0.660)		(0.111)		(0.219)		(0.0320)
lnprice6	−3.243***	lnprice4_lnprice4	1.346***	lnprice4_lnwage	−0.113	lnmanure_lnfertilizer	0.377***
	(1.255)		(0.364)		(0.512)		(0.141)
lnprice7	1.371	lnprice4_lnprice5	−0.333*	lnprice4_lnmanure	−0.0364	lnfertilizer_lnfertilizer	−0.484***
	(1.282)		(0.182)		(0.106)		(0.177)
lnwage	0.977	lnprice4_lnprice6	−0.746**	lnprice4_lnfertilizer	−0.0334	Profit Constant	8.763***
	(0.763)		(0.352)		(0.326)		(1.738)
lnmanure	−0.703	lnprice4_lnprice7	−0.843***	lnprice5_lnwage	0.0997	lnsig2v	−0.315***
	(0.439)		(0.309)		(0.194)		(0.103)
lnfertilizer	−0.676	lnprice5_lnprice5	0.183***	lnprice5_lnmanure	0.103**		
	(2.040)		(0.0444)		(0.0486)		
lnprice1_lnprice1	0.0883	lnprice5_lnprice6	−0.111	lnprice5_lnfertilizer	−0.496**		
	(0.112)		(0.142)		(0.212)		
lnprice1_lnprice2	−0.152	lnprice5_lnprice7	0.416***	lnprice6_lnwage	1.090***		
	(0.0935)		(0.141)		(0.383)		
lnprice1_lnprice4	0.695**	lnprice6_lnprice6	0.0170	lnprice6_lnmanure	0.0250		
	(0.287)		(0.0360)		(0.0833)		
lnprice1_lnprice5	−0.275*	lnprice6_lnprice7	0.546***	lnprice6_lnfertilizer	−1.797***		
	(0.141)		(0.219)		(0.396)		
lnprice1_lnprice6	0.281	lnprice7_lnprice7	0.104***	lnprice7_lnwage	−0.680*		
	(0.208)		(0.0362)		(0.378)		
lnprice1_lnprice7	0.143	lnprice1_lnwage	−0.892***	lnprice7_lnmanure	−0.220**		
	(0.244)		(0.273)		(0.0897)		
lnprice2_lnprice2	0.0331	lnprice1_lnmanure	−0.104	lnprice7_lnfertilizer	0.288		
	(0.0441)		(0.0719)		(0.359)		

Note 1: Standard Errors in Parenthesis.

*** p<0.01,

** p<0.05,

* p<0.1

Note 2: Prices: (1) Fruits, (2) Industrial Crops, (3) Corn, (4) Cereals, (5) Vegetables, (6) Beans, (7) Tubers, (lnwage) Wages, (lnmanure) Manure, (lnfertilizer) Fertilizer

Table C6.5 Mozambique inefficiency estimation results

Dependent variable: log(sigma_u2)		
		x land extension
Land extension (hectares)	−0.687***	−0.027
	(0.166)	(0.041)
If owns productive assets	0.559*	−0.498**
	(0.320)	(0.211)
Maximum years of education of the household	0.018	−0.007
	(0.033)	(0.022)
Number of members who work on the farm	−0.178***	0.055***
	(0.041)	(0.017)
If member of farmers association	0.250	−0.225
	(0.492)	(0.291)
Access to technical assistance	0.266	−0.474
	(0.507)	(0.407)
Access to credit	0.571	−0.474
	(0.927)	(0.642)
Log of cost of access to nearest city of 20,000 inhabitants	−0.007	0.013
	(0.032)	(0.019)
Land use: Woodland	−0.027***	
	(0.009)	
Land use: Secondary woodlands	−0.021**	
	(0.009)	
Land use: Pastures	−0.022**	
	(0.010)	
Land use: Croplands	−0.013	
	(0.010)	
sigma_v	1.251	
	(0.041)	
Constant	4.343***	
	(0.910)	
Observations		2533
Log likelihood		−4609.0

Note 1: Standard errors reported in parenthesis.
*** p<0.01,
** p<0.05,
* p<0.1.
Note 2: sigma_u2 is the variance of the inefficiency component.
Note 3: sigma_v is the standard deviation of the stochastic component.

Table C6.6 Armenia inefficiency estimation results

Dependent variable: log of σ_u^2		
		x Land extension
Land extension	0.0956*	−0.000571
	(0.0532)	(0.000360)
Possession of agricultural equipment	−0.0561	−0.0525
	(3.190)	(0.212)
Maximum education level of the household	0.0528	−0.00475
	(0.183)	(0.0110)
Credit access	1.415	−0.185*
	(1.171)	(0.0960)
Time of access to nearest city of 20,000 inhabitants	0.424*	−0.0662***
	(0.243)	(0.0196)
Household size	−0.0906	0.00326
	(0.0577)	(0.00283)
Land use 0: Non-suitable for agriculture	0.0433	
	(0.0513)	
Land use 1: Forest	−0.00997	
	(0.0432)	
Land use 2: Secondary forest	0.0727***	
	(0.0234)	
Land use 3: Grasslands	0.220**	
	(0.105)	
Constant	−2.900	
	(1.922)	
σ_v	0.9995	
Observations	1633	
Log likelihood	−2462	

Note 1: Standard Errors are in parenthesis.
*** $p<0.01$,
** $p<0.05$,
* $p<0.01$
Note 2: σ_u^2 is the variance of the inefficiency component
Note 3: σ_v is the standard deviation of the stochastic component

Table C6.7 Honduras inefficiency estimation results

Inefficiency Determinants (lnsig2u)	
Land	−0.515***
	(0.133)
Land*Land	0.0183***
	(0.00467)
Capital	−0.139***
	(0.0261)
Land Title	−0.187
	(0.164)
Credit Access	−1.206*
	(0.695)
Market Access	−0.252***
	(0.0579)
Education	−0.0234
	(0.0250)
Household Size	−0.105***
	(0.0273)
Technical Assistance	−0.297
	(0.432)
Agroec1: Lower montane moist forest	0.00862
	(0.337)
Agroec2: Subtropical rainforest	−0.500***
	(0.173)
Agroec3: Tropical rainforest	−0.483
	(0.303)
Agroec4: Lower montane wet forest	−0.440
	(0.545)
Agroec5: Subtropical wet forest	−0.706**
	(0.284)
Agroec6: Subtropical dry forest	−0.289
	(0.425)
Efficiency Constant	1.065***
	(0.394)
Observations	1396

Note 1: Standard errors in parenthesis.
*** p<0.01,
** p<0.05,
* p<0.1
Note 2: Some categories were not incorporated to avoid multicollinearity

7 Promoting small farmer market access in Asia

Issues, experiences, and mechanisms

SUKHPAL SINGH[1]

1. Introduction

Nearly 450 million farms in the world are smaller than 2 hectares (ha) each (the average size of farms in Asia and Africa is just 1.5 ha)[2]. They are the largest type of farm in many countries (Bangladesh 95 per cent, China 98 per cent, and Ethiopia 60 per cent).[3] With an average family size of five, they provide work to about 2.5 billion people (about a third of the world population). Further, 800 million people live off farms much smaller than 2 hectares (Polak, 2008); 87 per cent of them are located in Asia and the Pacific region (Thapa and Gaiha, 2011).

Evidence from India, Japan, Taiwan, Philippines, Mexico, Brazil, and Columbia suggests that small farmers are more efficient producers (Chand et al., 2011; Gaurav and Mishra, 2011; Singh and Krishna, 1994; Hazell, 2011). But prices smallholders receive for their output are lower than those obtained by larger farmers due to their weak bargaining power and storage capacity (Agrawal, 2000). In wheat, marginal holders had the highest yield per hectare compared with all other categories in India, but they realized the lowest prices per quintal (Gandhi and Koshy, 2006).

Small farms offer multiple advantages for development in poor countries with large disguised unemployment; they usually maximize labour use and added value, and have higher yields per unit of land. When free of insecurity of tenure or incentive-incompatible systems such as share cropping, they can greatly expand output. Their development leads to a more even distribution

[1] The author thanks the anonymous referee of the chapter and the editors of this volume for their useful comments.

[2] In India, small farmers with less than 2 hectares accounted for 86 per cent of all operational holdings in 2002/03, and 42 per cent of the total cultivated area (Sharma, 2007). Of the total, 64 per cent are marginal (less than 1 hectare each).

[3] They account for 69 per cent of cultivated area in Bangladesh and 87 per cent in Ethiopia.

of income, thus spurring industrial growth through increases in purchasing power and effective demand for local production. Rapid agricultural growth in Korea, China, Japan, and Taiwan, and even in West Bengal in India, have often been attributed to the growth of smallholder farming (Singh and Krishna, 1994; Morris, 2007). All these benefits are in addition to the contribution they make in meeting the basic food needs of a country.

The advantage of smallholder farming is not only higher value added per unit of land; they also have competitive cost advantage due to greater flexibility in their working capability and access to traditional knowledge. The main threats they face are standardization of products in global and national markets, high transaction costs, and the large volume requirements of modern markets (Harper, 2009; Markelova et al., 2009). But even then, there are opportunities in organic and fairtrade markets which are particularly suited for small producers and offer high prices (Harper, 2009). In Kerala's Wayanad district in India, for example, the fairtrade system provides fair doorstep prices to small black pepper farmers who are members of the Fairtrade Labelling Organisation (FLO) certified registered society. This society, The Fair Trade Alliance-Kerala (FTA-K), has about three thousand registered farmers; its trading arm handles dealings with overseas buyers and has collection centres in six districts of the state.[4] Private agencies also stand to gain from small producer links by demonstrating corporate social responsibility and thus gaining political and social legitimacy.

A typical complaint by small farmers is lack of markets for their produce or of access to input markets; on the other hand, exporters or supermarket retailers complain of inadequate supplies of quality produce at the right time and at reasonable cost. This marketing paradox stems from buyers often failing to explore new suppliers, while the farmers lack an understanding of markets and the ability to identify new markets. Thus, farmers cannot take advantage of opportunities with value-added activities such as grading, cleaning, sorting, packaging, and primary processing (Shepherd, 2007).

The second section of the chapter examines the challenges and opportunities faced by smallholders globally and in Asia, and outlines the need for marketing access for such holders. It focuses on crop and allied production sectors, especially high-value produce markets in Asia which are increasingly seen as an opportunity for smallholders. It looks at contract farming mechanisms and introduces the conceptual framework around which the entire issue is discussed and examined. The third section focuses on various

[4] Its turnover in 2007–8 was Rs 200 million, 70 per cent of which came from exports; net profit margin stood at 4–5 per cent. Liberation (a UK-based company that buys these fairtrade goods) is unique, as producer groups from Latin America, Africa, and Asia, including India, own 42 per cent of the company. The farmer members commit their products, and its yearly profits are shared with them (Karunakaran, 2008).

types of market linkages and their performance, including those involving state, NGOs, farmer groups, producer/farmer companies, public–private partnerships, and pure commercial links. In the final section, some of the policy and practice mechanisms for promoting smallholder market access in a globalized world are examined.

2. Constraints and challenges

Small farmers face constraints at all points in the agricultural value chain. Most own tiny plots of land, often in fragments and unirrigated, and many are entirely landless even though agriculture is their main source of livelihood. They lack the financial resources for land improvement or crop insurance, and have limited access to formal credit. They are often ignored by extension agencies and seldom receive information on new technologies or training in skill-intensive agricultural practices. Most importantly, on an individual basis, they lack the bargaining power to deal effectively with government institutions and/or markets (be they markets for land, inputs, or farm produce). For example, major problems of small and marginal farmers in India include unreliable (both in quantity and quality) input supply, inadequate and costly institutional credit, lack of irrigation water and costly access to it, lack of extension services for commercial crops, exploitation in marketing of their produce, high health expenditures, and lack of alternative (non-farm) sources of income (Dev, 2005). Given these constraints, they are also unable to increase their crop output, move to higher-value products, or make profitable market links.

An important issue in globally oriented value chains is whether small producers can participate in and benefit from these chains and markets—which is crucial for their survival as traditional marketing channels weaken or disappear (Pingali and Khwaja, 2004). Small farmers have some advantages for integrating with the supply chains, as they can supply better quality produce. It is also suggested that smallholders are better at managing market risk by diversifying their market and using social networks which may not be available to large growers or corporates undertaking production of raw products such as pineapple in Ghana (Suzuki et al., 2011). But, they cannot benefit from economies of scale in processing and marketing. The net effect of integrated markets on small farmers depends on the nature of the commodity and its market, as well as the ability of small farmers to coordinate marketing activities (Barghouti et al., 2004).

Small producers face production and marketing risks which make them vulnerable to poverty. Commercial farming has risks in addition to the natural phenomena that are intrinsic to farming everywhere (Tomek and Peterson, 2001). Whereas production risks can come from weather, pest and disease attack, low-yielding seeds, or increased costs, commercial/market risks involve market demand, price volatility, seasonality, and quality standards. Risks in agriculture can also be institutional, legal, or financial (availability of credit, sudden rise in interest rate, unexpected demand to repay debt, or an inadequate amount of credit).

It is also important to recognize that small farmers are not a homogenous group. There are small farmers who are fully commercialized, and buy and sell in markets. There are others who participate in the market in a limited manner to buy inputs and sell some part of their output. There are still others who are subsistence farmers consuming most of their farm production while selling labour in the market or buying food grains from the market to meet their consumption needs, thus becoming net buyers of food. For example, in East and South Africa, only 2 per cent of farmers are large commercial growers, 15 per cent regularly sell in markets, another 30 per cent only occasionally and also buy food from markets, while 50 per cent are subsistence growers who depend on off-farm work to buy food from the market (Sahan and Fischer-Mackey, 2011). From a market perspective, some export, and others sell in national or local markets.

Finally, smallholder farmers in many developing countries, especially dry land and resource-poor regions, also draw a large part of their income from non-farm or off-farm sources which include migration for a part of the year for farm or non-farm work. In sub-Saharan Africa, almost 50 per cent of rural incomes was from non-farm sources (incomes from urban sector or from abroad, being mainly remittances and pension payments (Ellis, undated)). In Bihar in India, in 2009, of all the out-migrants, 14 per cent of households were marginal, and small cultivators and 41 per cent had some land (marginal holders) (Bhaskaran, 2011). This shows that such non-farm income can help agricultural development by providing investment capital and new technology, but can also hamper it by withdrawal of labour. This chapter restricts itself to examining market access for farm and allied producers.

2.1 HIGH-VALUE CROPS, MARKET LINKAGE, AND SMALL PRODUCERS

Given the increasing emphasis on the production of high-value crops (which are costly to produce and risky to manage besides being more market-dependent and highly perishable), it is in this sector that effective market access matters more. These crops are more labour intensive, require frequent care, are harvested more frequently, and sold/marketed more often. All this requires knowledge of input markets, output markets, the labour network, and immediate decisions on production and marketing. Similarly, organic production can also be treated as high value as it requires new processes, different bio-inputs and record keeping, and certification of farms and enterprises. All this is about market linkage, as it involves new products, processes, markets, institutions, networks, and information.

For example, in high-value cut-flower production for the Delhi market by growers in an Uttar Pradesh village, the net returns from flower cultivation were many times higher than those from traditional crops of sugarcane or wheat. Further, flower-cultivating households had much higher gross and net returns than the others. However, small farmers received lower prices and incurred higher costs due to smaller volumes. The variability in prices received by small farmers was also greater than those for other categories of farmers. As for the risk, the proportion of households making losses in flower cultivation was very high (12–38) and almost negligible (1–2) in wheat and sugar cane, mainly because of yield fluctuations and, to a lesser extent, price fluctuations. This risk makes it difficult for small and resource-constrained farmers to diversify into high-value crops like flowers (Sen and Raju, 2006). Thus, greater entrepreneurial skill, resources, and risk-taking capability are required for high-value crops (Glover, 1987). A study of farmer entrepreneurship in the suicide-prone cotton district of Andhra Pradesh (Prakasham) found that the risk-taking ability of farmers in general was low, but that of smallholders was much lower than that of medium or large farmers (Singh and Krishna, 1994).

Contract farming is also recommended as the only way to make small-scale farming competitive, as the services provided by contracting agencies can't be provided by any other agency (Eaton and Shepherd, 2001). Contract farming also lowers transaction costs for the farmers, as many of them are internalized by the procuring firm (IFPRI, 2005). In India, expansion of contract farming is expected to be propelled by domestic players (food supermarket chain growth), as well as by the recently permitted Foreign Direct Investment (FDI) in retail, and international trade, especially sanitary and Phyto-sanitary measures, organic, and ethical trade. Contract farming is also promoted by central and state agencies, besides the banking and input industry. The farming crisis and reverse tenancy, coupled with the failure of traditional cooperatives, will help spread contract farming. Even the new Intellectual Property Regime (IPR), which encourages protection and exploitation of proprietary genetics, is likely to accelerate contract farming practice (Wolf et al., 2001). Further, the new agricultural policy sees public–private partnership as the main way to transform agriculture, and the state is providing incentives for corporates to enter the agribusiness sector, partly through contract farming.

Given the predominance of small producers in developing Asian economies, any exclusion of small producers is an important issue. Small producers find it difficult to meet exact standards due to unfamiliarity with supermarket chain systems, the costs involved, or the cultural incompatibilities of the new agricultural practices. Among factors behind the exclusion of small producers are enforcement of contracts, high transaction costs, quality standards, business attitudes and ethics such as non/delayed/reduced payment, a high rate of product rejection, and weak bargaining power (Kirsten and Sartorius, 2002). The organizers of small producers also find it costly to work with them due to their scattered location and smaller volumes, though there are advantages: higher commitment levels, the risk of default is spread, corporates are given a social face, and production costs can be lower. A few cases of small producer groups acting as a collective to deal with modern markets are exceptional (Boselie et al., 2003).

Contracting agencies impose conditions such as minimum land size, no non-contract crop production in neighbouring fields, and certification, especially in seed production (Simmons et al., 2005). The criteria for participation in contract farming projects/schemes such as irrigated, suitable land, proximity to a main road, and the literacy level of the farmer are discriminatory and limit the number of eligible growers. Contracting private agribusinesses everywhere have less interest and ability to deal with small-scale farmers on an individual basis (Hazell, 2005); they prefer large farmers because of their capacity to produce better quality crops due to efficient farming methods and because large volumes reduce the cost of collection. Produce is easier to trace due to absence of pooling, and large producers can bear the risk of crop failure. Services provided by large producers such as transport and storage reduce contractor costs (Key and Runsten, 1999; Simmons et al., 2005).

Even in modern retail-driven arrangements with farmers like that of Bayer Crop Science's Food Chain Partnership programme in India, there are several preconditions for inclusion, including: minimum land holding size of 1 acre (0.4 hectare), irrigation facilities, literacy, and a certain 'business sense', as well as access to a mobile phone. Although there are differences between and within states and regions, such criteria are generally hard to fulfill. This, and the focus on certain core areas of vegetable production, results in the exclusion of many farmers (Trebbin and Franz, 2010). It is also recognized generally that modern markets exclude smallholders and poor regions (Vorley and Proctor, 2008).

2.2 CONCEPTUAL FRAMEWORK FOR MARKET ACCESS

Globally, and more so in the developing world, including India, the various links of primary producers with markets include:

- Farmer to local trader
- Farmer to chain retailer through intermediary (trader or lead farmer)
- Farmer to chain retailer through NGO

- Farmer to chain retailer through farmer co-operative or association
- Farmer to chain retailer with formal contract
- Farmer to chain retailer with informal contract
- Farmer to chain retailer without contract (only 'contact')
- Farmer to processor with formal contract
- Farmer to processor without formal contract
- Farmer to processor through intermediary (trader or lead farmer)
- Farmer to market through co-operative or group
- Farmer to exporter (direct)
- Farmer to exporter through intermediary
- Farmer to dedicated wholesaler
- Farmer to consumer

In these numerous types of arrangements, success depends on the market and the efficiency of operations. Some offer higher prices for growers, while others lower the cost of marketing, thus benefiting either way. But, most of these arrangements, especially indirect ones, do not ensure that small growers are part of them. Many market linkage arrangements just provide another alternative to the primary sellers without any commitment to buy or generate additional value or surplus, as is the case with most of the fresh fruit and vegetable (F&V) retail chains in India which procure only A-grade produce without any contract, leaving the producer to sell the rest of the produce in other channels. Most of these channels also deal with individual growers (Singh and Singla, 2011).

In this context, it is important to understand the issue in a conceptual manner: smallholders deal with input and output markets through various channels; these are often imperfect or inefficient and unfavourable to smallholders for reasons of scale, transaction costs, and bargaining power due to resource constraints. In theory many channels are available, but in practice, smallholders' choice is limited. It is therefore important to expand choice and make market access effective. This scenario is aggravated by policies that are neither neutral nor pro-smallholder (Vorley and Proctor, 2008). Critical evaluation of the alternative markets for small farmers and better market linkages are vital in determining the overall profitability of small farms. This is summarized in Figure 7.1.

3. Market access experience in Asia

3.1 ROLE OF THE STATE

New crops are risky and big investments for smallholders. A supportive role by other stakeholders such as the state, farmers' bodies, and development

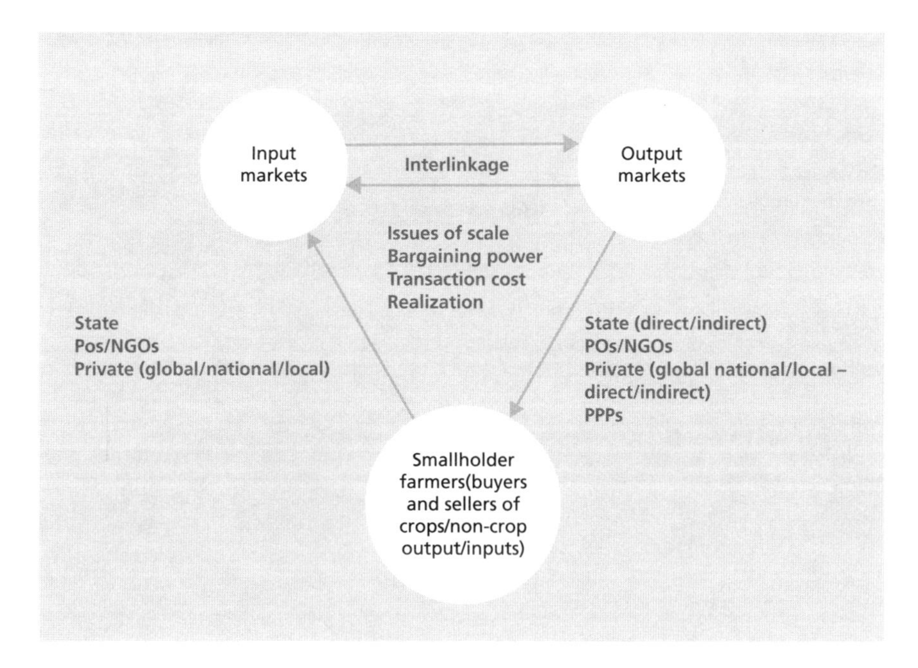

Figure 7.1 A conceptual framework for understanding market access issues

agencies can help lower production and market risk, and the cost of growing new crops. In Thailand and Taiwan, there were two contrasting cases of farmer adoption of new crops. In the former, farmers were reluctant to grow off-season crops (tomato and tomato seed, potato, watermelon seed, and cantaloupe seed) despite all resources being made available, especially irrigation, and a contract package which even covered production risk to some extent. In Taiwan, farmers rapidly adopted new crops on a large scale once the market was established, though it was limited to some areas and could not be replicated elsewhere in the country. Mushroom and asparagus canners gave the responsibility of collection and grading to local farmers' associations which also bargained collectively for contract prices with the mediation of government representatives. A planned production and marketing (PPM) system was put in place to avoid glut and a price crash. The two cases also highlight the creative role of governments in helping the project deliver the outcomes (Benziger, 1996).

But, in Thailand too, in many other contract farming situations where new crops were involved, the state not only provided coordination and support of local authorities such as agricultural extension agents, local administration officers (Figure 7.2), and the Bank of Agriculture and Agricultural Co-operatives (BAAC), it helped train farmers in contract farming

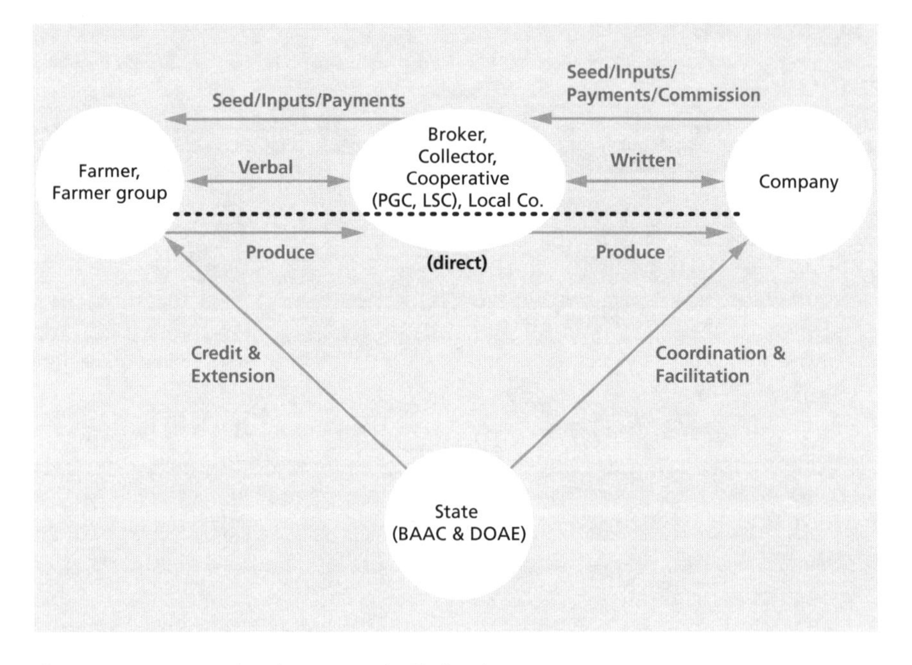

Figure 7.2 Contract farming system in Thailand

Source: Singh (2005).

(Singh, 2005) and also reallocated 250 million Baht deposit in BAAC. The interest compensation for the farmer participants in the programme (3.5 p.a.) was made available to encourage more farmer participation and to reduce production costs. But, later, farmers could obtain only a low interest rate loan (5 per cent p.a.) instead of getting compensation for interest charges.

In the Sarawak region of Malaysia, a government agency, Sarawak Economic Development Corporation (SEDC), ran a contract farming scheme in poultry production aimed at raising incomes and creating agricultural entrepreneurship among small farmers of ethnic minorities (*bhumiputras*). Mainstream production was by a vertically coordinated private poultry chain where the average farm of a chicken contract grower had about 50,000 birds and where private contract growers supplied 85 per cent of the total production in the state to six major players. Since chicken is a predominant part of the local diet, the federal government set ceiling and floor prices and protected the domestic market from imports. The SEDC contract farming scheme helped trainees to set up poultry farms from which its subsidiary bought the chickens and supplied them after processing to the SEDC, which from 1988 supplied state-controlled outlets such as schools, hospitals, and police and army. Trainees with at least 0.8 ha of land were selected for community farming to test the efficiency and quality of a limited number of production cycles. SEDC extended credit for sheds and equipment, and subsidized input costs. The poultry business of the entrepreneur farmers generated 43 per cent of the total income of the household, with some of them depending heavily on poultry income, and most of them were happy with the arrangement. None of the trained farmers left this scheme and took up contract farming with the private sector, perhaps due to their own resource constraints, lack of knowledge and education, and risk averseness. However, the selection criterion adopted (e.g. farms having at least 0.8 hectares of land) was likely to exclude really poor farmers (Morrison et al., 2006).

In India, the state has played a variety of roles to facilitate market access and promote farmers' markets. These have helped participating farmers become aware of the products required by the markets and helped them improve the quality and diversity of their products, but they have not made a major impact because farmers could not increase production. Consequently small farmers form only a small proportion of the sellers in these markets. The more significant government initiatives in terms of scale include the Horticultural Producers' Co-operative Marketing and Processing Society Limited (HOP-COMS—a cooperative) in Karnataka, South India and *SAFAL* (meaning successful) F&V (fruit and vegetable) project of the National Dairy Development Board (NDDB) in north India.

Hopcoms

Karnataka, which is the third largest producer of fruits and fifth largest producer of vegetables in India, set up HOPCOMS in 1959 to help F&V growers sell their produce profitably and free them from the clutches of middlemen, besides ensuring quality supply at reasonable prices. In its present form since 1987, it is almost totally funded and managed by the state government (88 per cent of share capital) and operates across six districts and two cities with 15,000 members and 358 stores (including 318 in Bangalore and 40 in Mysore), which account for almost 10 per cent of the horticultural trade in Bangalore. It has four procurement centres in Bangalore and one in Mysore. Cooperative suppliers included more small and marginal farmers than in southern Karnataka. An average member cultivated only 1.82 ha and a non-member 1.94 ha compared with the average size of holding of 1.74 ha and 73 per cent of the holders being small or marginal in the state (Kolady et al., 2007). Farmers received 10–15 per cent higher prices than the market price of the previous day and did not have to incur any marketing charges. Payment for produce was made in cash on the same day, and there was fair weighing of produce. But, the cooperative bought only quality produce and the rest of the produce was sold in the *mandi* (F&V market) by the farmers. Of the total procurement, 85–90 per cent was directly from farmers (Premchander, 2002), and 72 per cent of the suppliers were satisfied with the operations of the cooperative (Kolady et al., 2007).

Safal

Mother Dairy (an enterprise of the NDDB, an autonomous body of the Government of India for promoting dairy development since 1965) was set up to distribute liquid milk in major cities of India. Its F&V project based in Delhi runs 380 F&V Safal retail outlets. This project, now a separate legal entity as a company (a subsidiary of NDDB), provides remunerative prices to producers and affordable prices to consumers by handling the entire range of operations, from supply management to marketing of fresh F&Vs in and around Delhi and marketing of frozen and processed F&Vs throughout India. This project has a state-of-the-art Central Distribution Facility (CDF) to handle 100,000 tonnes of fresh produce. Retail outlets (totaling about 1,400) remain open ten hours a day, seven days a week and, on an average, about one hundred thousand customers visit the retail outlets every day. Consumers purchase around 95 per cent of the F&V quantities supplied on the same day.

Safal handles approximately 2 million kg of fresh produce every day. It procures fresh produce directly from 110 Growers' Associations in 14 states of India, with membership ranging from 21 growers to as many as 100 in some cases. Total number of growers associated with Safal can be about five thousand or so. Growers assemble their produce for sorting and grading at a Collection Centre (CC) established at a central location within a cluster of villages in the production belts. The CC stores, sorts/grades, repacks if required, weighs the produce in specially designed plastic crates and dispatches the material to Safal in Delhi. Safal provides growers with: professional advice and supply of good quality high-yielding seeds; advice on Integrated Pest Management and good agricultural practices; supply of bio-inputs and agri-implements; and market and environmental information. Innovations include: introduction of chiseling and wide-bed technology with central planting in tomato and potato, and introduction of hybrids in various vegetables; a high-temperature potato storage facility which leads to non-sweet stored potatoes with lower refrigeration cost; introduction of a pneumatic direct seeder; and development and introduction of new agri-implements for mechanization and increased productivity.

Turmeric Cooperative (FAPRO)

FAPRO, India is the largest producer of turmeric, followed by Sri Lanka, Pakistan, and China. Major growing states are Andhra Pradesh, Maharashtra,

Orissa, Tamil Nadu, Karnataka, Kerala, West Bengal, and Assam. India consumes more than 90 per cent of the production and exports the rest. In India, 85 per cent of the turmeric is sold as bulbs and only 15 per cent as powder. It is a high-value crop and product, and has wide consumer and industrial demand.

Since 2001 the Farmers' Produce Promotion Society (FAPRO) in Hoshiarpur (in Punjab, India), promoted by state agriculture and horticulture departments, supported by the Planning Commission of India, and registered under the Societies Act 1860, has been procuring and processing turmeric for member farmers and retailing it in nearby villages and towns. It also produces and sells honey and jaggery. The cooperative procures the turmeric crop, processes it at its own plant, and retails the turmeric powder through its own retailers. A farmer can make Rs 0.5 million per hectare from this nine-month crop after meeting all expenses. About 400 ha were under turmeric crop, and in 2010–11 growing season, 2,645 tonnes of turmeric powder was produced.

FAPRO has a membership of 308 of whom 97 per cent are from within the district of Hoshiarpur and the rest from the adjoining districts. Seventy per cent of the members are smallholders, 20 per cent medium, 5 per cent large, and 5 per cent landless lessee farmers. The farmer members' turmeric acreage ranges from 0.4–4 ha. The choice of crop was based on higher yield and profitability, ease of cultivation, eco-friendliness, and suitability for small and landless farmers. Turmeric was a new crop for local farmers. It started with 71 ha of turmeric crop. An acre (0.4 hectares) of turmeric gives output worth Rs 50,000, which if dried and polished becomes worth Rs 80,000, and if powdered worth Rs 160,000, so there is high added value in the crop.

The seeds to members are supplied on cash payment on a 'no profit, no loss' basis. The cooperative provides technical support to members to grow quality produce. The farmers are paid 50 per cent of the value of the produce on delivery and the rest after one month. Between 2001 and 2011, about 400 ha were placed under turmeric, which led to production of 2,645 tonnes of turmeric powder.

The major costs are seed and organic manure, which accounts for 40 per cent of the total cost. The average yield of the crop is 37 tonnes per ha, and after accounting for a lease rate of Rs 59,000 per ha and costs of cultivation of Rs 40,000/ha, it gives a net return in the order of Rs 124,000 per ha. The crop is also less labour intensive than potatoes, as a ratoon crop[5] is possible for one or two years (author's primary field survey).

[5] A crop where a second crop can be taken from the roots of previous crop left in the field after harvest.

3.2 ROLE OF INTERMEDIARIES IN MARKET ACCESS

NGOs as intermediaries in market access

In Bangladesh, NGOs such as Bangladesh Rural Advancement Committee (BRAC) and Proshika have proactively started introducing new crops and systems to small growers. For example, BRAC started with contract farming of vegetables with 61 farmers in 1997–8 and reached 965 farmers by the end of 2002. Some entrepreneur farmers supplied seeds and other inputs to growers and bought back the produce, which was better quality due to lower use of chemical inputs, at prices much higher than the wholesale market. This was possible because of the supermarket linkage of BRAC, which paid higher prices to growers because it sold at higher prices in their stores than the traditional retailers (Bayes and Ahmed, 2003).

In India, there are many cases of NGOs promoting traditional cooperatives for market access to small producers. *Dhruva* (an NGO) promoted Vasundhara co-operative in 1985 in the Vansda taluka (sub-district) of Navsari district in Gujarat in western India, which covers 39 villages in the taluka. The objective was to look after the marketing and other developmental needs, and to add value to the produce of farmers and marketing it through various channels. Presently, 11 autonomous cooperatives have a total of 2,569 members.

Mangoes are procured from various sources such as traders, other farmers, member farmers, and the markets of Dharampur and Vansda. Of the processed food products, mango accounts for 25 per cent and cashew 75 per cent. Indian Tobacco Company (ITC) buys 40 per cent of total mangoes procured by the cooperative and pays 15–20 per cent higher than market price. The cooperative had an agreement with ITC to organize farmers to produce and then procure, grade, sort, aggregate, and supply certified organic mangoes at pre-agreed price and service charges. The transport cost was borne by ITC.

In 2006–7, cooperatives contributed 40 per cent, traders 27 per cent, and individual farmers 33 per cent of the total procurement. The mango purchase at Dharampur grew from Rs 1.84 lakh in 2004–5 to Rs 7.9 lakh in 2005–6 and further to Rs 11.55 lakh in 2006–7 (Hiremath et al., 2007). *Dhruva* has also set up a centralized mango marketing stall for all cooperatives at Dharampur, which is managed by one of the cooperatives located close to the market (Mandva). A day has been allotted to each cooperative for mango supply to avoid glut or scarcity. *Vasundhara* has provided crates to all primary cooperatives for handling mangoes, which helps fetch higher market prices. The Mandva cooperative sends its transport vehicle to each cooperative on the scheduled day on rotation during the season.

Private sector as intermediaries

In Bangladesh, twenty urban and peri-urban poor households organized into three groups to produce and supply mushrooms to a trader under a contract farming system. The trader was also into mushroom production, supply of spore bags, and buying mushrooms to sell to institutions and retail markets. One of the producers entered into marketing of mushrooms and provides competition to the trader. In this case, the traders bear the risk which producers and buyers are generally unwilling to bear and this sustains the value chain function (Zamil and Cadilhon, 2009).

In China, green onion packers worked with really small contract growers whose holdings were not very different from those of non-contract growers, and provided them with seeds and pesticides. They offered a guaranteed price before planting or market price plus premium (formula pricing) and provided technical assistance to their contract growers. The contract growers received a high price and achieved similar yield, although they had higher input costs. Total household income, net farm, and net crop income were almost 2.5 times higher than that of non-contract growers. These packers were able to export more than 90 per cent of their produce and also sold to supermarkets in China (Miyata et al., 2009).

Lead farmer or contract farming organizer models used by exporters, processors, and retailers require lower external support but much higher investment from farmers themselves. These models attract those farmers who have already shown the ability to meet the quality and quantity demands of modern buyers. Modern buyers (supermarkets, exporters, processors) encourage lead farmers to organize and support neighbouring farmers with little investment beyond the incentives provided by market opportunities. Lead farmers provide various services, which may include production planning, technical assistance, access to inputs, market intelligence, sorting and packing, transportation, and financial administration. Though lead farmers provide these services, farmers pay more for such services than they would to a producer organization (Hellin et al., 2009). In India, there are several examples of enterprising lead farmers, business graduates, and local traders being roped in by processors and exporters who not only coordinate farmers, but also carry out grading and processing where needed. There are many such models in India in the value chains of grapes, baby corn, and banana.

Aggregator model

Over 60 per cent of the 75,000 ha of ponds in Bihar in which thrives the makhana (Euryale ferox), or fox nut (used as an additive, snack, or for medicine) is owned by the government, which leases them to fishermen cooperatives. These cooperatives, in turn, lease the ponds to individuals, i.e.

the impoverished farmers of the *mallah* community, who are skilled in underwater harvesting. The seeds are then popped manually and sold through agents and middlemen across the state. Shakti Sudha Industries[6] has built a robust back-end supply chain. In 2008, it procured over 3,000 tonnes of makhana. For several months, the promoter of the company toured Bihar to study the highly unorganized business and realized it would be impossible to make headway without the participation of panchayats (elected village-level bodies)—even for identifying farmers and hand-holding them. His farm-to-market project later entered into a tripartite agreement involving the panchayat, farmer, and his company in which the roles of all three were defined, from documentation, training in agronomic practices, to purchase guarantees. In 2008, 15,000 families belonging to marginalized communities were associated with the enterprise.

But, no bank would fund farmers, as the pond lease period was for 11 months, a single season. A sense of ownership was missing, and no investment went into pond development. The company had to lobby government, which finally agreed to a nine-year lease period. Next, he persuaded a public-sector bank and coaxed member-farmers to open zero-balance accounts who had no bank accounts previously. Makhana payments were credited directly to the farmers' accounts and the bank promised Kisan Credit Cards to those whose transactions fitted the norm. To earn the trust of the villagers in the 26 panchayats in the first phase, Shakti Sudha functioned like an NGO, providing a literacy mission, subsidies, and assistance. In 2007–8, 42 farmers received Rs 15,000 each.

Transportation was the weak link in the supply chain. The company decided to turn unemployed rural youth into truck owners. Credit linkages and training were set up, and a band of truckers emerged with assured business from the company, which set up an office/collection centre-godown in each of the panchayats staffed by an employee. This functioned like a one-stop window for all makhana issues.

The village agent-trader, who had been rendered jobless, was involved and asked to head his village centre with a 1 per cent commission. They were reluctant with the low rate on offer, but when Singh calculated and showed them that they stood to earn much more than before because of the increased volumes, they turned around. The biggest lesson in supply-chain management was to include all significant players in the old pattern as stakeholders in the new system, in some form or the other. Makhana popping machines, for instance, are being developed and tweaked with the Central Food Technology Research Institute (CFTRI), Mysore; instead of ejecting those depending on this occupation for their livelihoods, Singh wanted to distribute these

[6] A Patna-based company.

machines to farmers' collectives. Shakti Sudha's turnover was nudging Rs 50 crore in 2008–9. Singh hoped to reach 40,000 farmers and a turnover of Rs 100 crore by 2012 (Karunakaran, 2008).

Corporate linkage for smallholders

In India, Namdhari Fresh (NF), a retail venture of Namdhari Seeds Private Limited, a leading vegetable seed seller and exporter, was set up in 2000 in Bangalore to export fresh vegetables. It is a certified organic exporter and had a packing house with British Retail Consortium (BRC) and Hazard Analysis and Critical Control Practices (HACCP) standards. It is also a Primary Marketing Organisation (PMO) under GlobalGAP (Global Good Agricultural Practices) group certification, wherein producers must be members of a PMO to obtain certification. A PMO is supposed to take legal responsibility for the whole operation of a GAP scheme, where each individual producer has to sign a legally binding contract agreeing to meet all the specifications of the GlobalGAP protocol. Importantly, detected non-compliance of one member in the group may result in de-certification of the entire group (Amekawa, 2009). Under PMO, a group of farmers agrees to register under an organization responsible for marketing the certified produce. The commercial terms are agreed between the organization and the group of farmers. The PMO is required to provide technical help to growers and facilitate GlobalGAP certification of the farm. Also the business processes of the organization need to be certified. If a grower wants to sell produce to another buyer, he needs to take a No-objection Certificate (NOC) from the PMO. This system also helps to reduce the cost of certification per grower. The NF fruit and vegetable retail operations are coordinated by two companies: Namdhari Farm Fresh deals with back-end operations and Namdhari Agro Fresh with front-end operations. The chain has a corporate farm of 142 ha. NF has its own chain of 18 retail outlets to sell F&Vs, some with salad bars, and its own cold chain facility. NF practised both captive farming (by leasing-in land) and contract farming in the ratio of 20:80 on about 600 ha (Figure 7.3).

The turnover of the company was Rs 30 crore during 2007–8. It deals with 40 different fruits and vegetables. Over one thousand two hundred contract farmers contributed 110 tonnes (70 per cent) of total production (160 tonnes per week) in 2008. It procured 20 tonnes of baby corn from 700 growers, and 0.5 tonnes of bhindi (lady finger) from 50 growers daily. NF practised planned production to match the daily demand of F&Vs and avoid overproduction. Farmers were given a production plan, and sowing and harvesting dates in advance. The chain gave fixed prices for baby corn, chilli, and bhindi and market-based prices for other F&Vs. If a farmer defaulted on delivery, the chain stopped working with them the next year. The chain also arranged drip irrigation for the selected loyal and regular supply farmers by paying 50 per

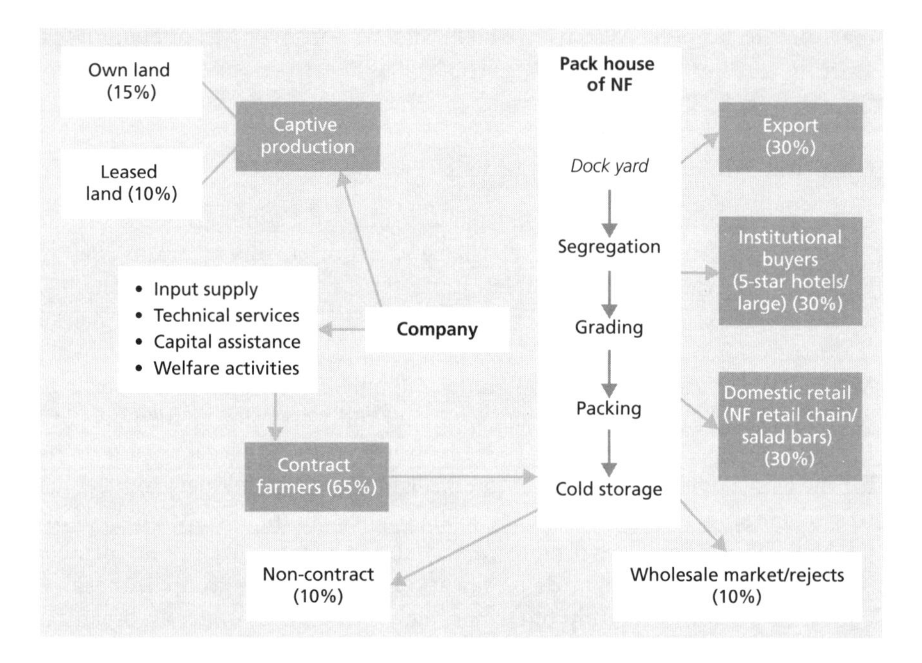

Figure 7.3 Non-farm value chain

Source: Dhananjaya and Rao (2009).

cent of the cost, which was later deducted in instalments from produce payments. The other 50 per cent was paid by the government.

About 75 per cent of the farmers working with NF were small or marginal. NF farmers had higher average yields (97 qtls/ha) in baby corn than non-NF farmers (79 qtls/ha). Though the average price received by NF farmers turned out to be lower than the wholesale market price, since average costs of production and marketing were lower for NF farmers than non-NF farmers, the net income of NF farmers was higher. On average, NF farmers had 41 per cent higher income than non-NF farmers (Singh and Singla, 2011).

Public–private partnership for market access

In India, a NGO in livelihood promotion, Basix entered into a collaboration with Frito Lay India Pvt. Ltd, a subsidiary of Pepsico, for potato contract farming whereby it facilitated the production of chip-grade potatoes by farmers in Jharkhand state in East India. Pepsico agreed prices and specifications while Basix coordinated contracts and provided credit for 424 small contract farmers with 36 ha in 2005–6, which increased to 1,442 contract growers with 237 ha in 2007–8. The partnership led to higher yields, better prices, and higher net returns for the growers compared with those from conventional potatoes, though the cost of production was somewhat higher. Basix made a surplus in 2005–6 and a small loss in 2006–7 (Mishra, 2009). The arrangement subsequently collapsed due to conflict between farmers and the buying company over issues of price and quality.

An outstandingly successful case of linking perishable produce farmers with markets is the creation of Mahagrapes by the Maharashtra State Agricultural Marketing Board (MSAMB), Department of Co-operation, Government of Maharashtra, National Horticulture Board (NHB), National Co-operative Development Corporation (NCDC), Agricultural and Processed Food Products Export Development Authority (APEDA), and the grape growers for the benefit of grape growers. Mahagrapes was set up in 1991 as a marketing arm of the grape growers' cooperatives in Maharashtra by MSAMB, and supported financially by NCDC and APEDA to promote marketing of grapes globally. It has features of both a cooperative and a company in terms of its organizational structure and functioning, which was born as a result of the special provision of the (amended) cooperative law at the provincial level in 1984, which allowed cooperatives to associate with other agencies, including marketing partners. Thus Mahagrapes was registered as a partner to the producer cooperatives under the amended cooperative Act. The two executive partners head the organization, which has an executive council comprising seven elected cooperative heads, followed by a board of directors composed of the heads of sixteen grape growers' cooperatives.

Mahagrapes is a for-profit organization and funding comes from membership equity. Mahagrapes now manages and facilitates the entire value chain of grape production, including production extension and market information to member growers, besides negotiating prices for growers with national and global buyers. Mahagrapes buys inputs for the entire group and manufacturers its own bio-fertilizers, which reduces production costs. It also provides extension and certification services to its members as additional services (Narrod et al., 2009). It only charges a facilitation fee from growers for its services and does not retain the profits it earns. It is totally owned and governed by farmers and their cooperatives. It has been able to deliver better net returns to its member growers than those earned by non-member grape growers. The role of the state agencies is noteworthy: for three years MSAMB paid the salaries of the governing officers of Mahagrapes who seconded from other state government departments. NCDC provided loans to grape cooperatives to help them add value by creating value preservation facilities like pack houses, cold storages, reefer trucks, and containers (Roy and Thorat, 2008).

The tomato market in Uttarakhand highlights the role of an NGO—Himalayan Action Research Centre (HARC) and farmers' organizations—in helping farmers overcome perishable produce marketing problems by linking small-scale farmers producing tomatoes with Mother Dairy's F&V project in Delhi. In 2001 farmers were organized into six federations across 80 villages, which supplied off-season vegetables (mainly tomatoes) to Mother Dairy F&V project through a purchase agreement with the farmers' federations. Tomatoes were brought by farmers to designated collection centres, which were managed by an employee/volunteer of a federation. The farmers graded tomatoes at the collection centre according to the quality parameters provided by Mother Dairy and the process was monitored by a representative of the federation. After grading, tomatoes were packed in plastic crates provided by Mother Diary, which reduced losses during transportation and the cost of packaging by 70 per cent. Although the farmers' federations and Mother Dairy had a legal relationship, the individual farmers were not contractually obliged to sell to Mother Dairy. The farmers sold only about 30 per cent of their tomatoes to Mother Dairy in 2006. Similarly, Mother Dairy was not obliged to buy a fixed quantity of tomatoes from the farmers. The quantity purchased was determined every year through negotiations between the federations and Mother Dairy.

Farmers had no difficulty in forming farmers' federations, but did not comply with the grading standards. Many farmers tried to cheat the system by including poor quality tomatoes, which sometimes led to the rejection of a whole truckload. The federations provided the following services to their member farmers: preparation of an annual production plan and negotiation of supply targets with Mother Dairy; organization of the procurement at Collection Centres (CCs); monitoring the grading of produce at CCs; mediating between farmers and Mother Dairy; renting packaging crates; selling agri-inputs to members; paying farmers for produce delivered; training farmers and selling their surplus production to private dealers during the peak season. The federations received income from the following activities: a one-time membership fee of Rs 250; 1.75 per cent transaction fee from Mother Dairy; service fee (5 per cent of transaction) charged to members for bulk purchases of seeds, fertilizers, pesticides, and other inputs; and renting of plastic crates at Re1/crate per day to members and at Rs2 to non-members.

Mother Dairy farmers had larger landholdings (1.3 ha) than the non-Mother Dairy farmers (1.13 ha), had higher yields (2.5 times), higher cost of cultivation (due to higher use of pesticides [almost double] to meet quality standards), but incurred much lower marketing costs, resulting in higher net returns per kg and per ha than those outside the federations (Alam and Verma, 2007).

Another enterprise in smallholder livelihood improvement is the KNIDS Green Pvt Ltd (KGPL) set up by an Indian Institute of Management, Ahmedabad (IIMA) graduate in agribusiness management, Kaushalendra Kumar who set up the Samriddhi Trust in 2008 with the help of Agricultural Technology Management Agency (ATMA), an extension agency of the government, as a public/private partnership and a loan of Rs 5 lakh from Friends of Women's World Banking (FWWB). It has also secured a loan of US$1 lakh from Punjab

National Bank (PNB). KGPL encourages farmers to produce vegetables and monitors the grading, sorting, and packaging of the products before supplying them to its partnered traditional vegetable retailers for distribution to residential, commercial, and market places. The farmer group gets a commission for managing the collection centre. The farmer procurement prices are announced in the evening for morning supply, and farmers can also forego the option of selling to the cooperative. By 2009 it involved 600 supplying farmers in 11 self-help groups, and 300 traditional vegetable retailers. KGPL supplies seeds to growers at 25 per cent lower than market price. By cutting out the middleman's commission, farmers receive 35 per cent higher price and consumers 15 per cent lower price. It has achieved a turnover of Rs 4 crore within two years and is attempting production and export of exotic vegetables to Dubai (Talukdar, 2010). It sells all the vegetables procured in the morning by the evening. Margins are 20–30 per cent, and farmer incomes are reported to be up by 0.25 to 1.5 times.

Group market access

Even marginal and small farmers sell some of their produce. But they need organizational strength to manage post-harvest storage and obtain the best prices. A way forward is setting up member-based Producer Organizations (POs). POs, unlike individual farmers, can leverage bargaining power to access financial and non-financial inputs and services and appropriate technologies, reduce transaction costs, tap high-value markets, and enter into partnerships with private entities on more equitable terms. They could also help the members move further up the value chain, entering into post-harvest management, direct retailing, value-added services, storage and processing, and engaging in contract production of primary agricultural produce.

Aggregation produces economies of scale and reduced transaction costs for firms and small-scale growers (Key and Runsten, 1999). In contract arrangements with small producers in West African countries in the 1970s, the cotton companies started transferring some of the operational or functional responsibilities, such as distribution of inputs, equipment orders, and credit repayment management, to the village associations and helped them develop the necessary management skills. The companies relied on traditional village authority structures for organizing the associations, but limited the associations to one per village to simplify procedures. This arrangement accounted for a significant part of each cotton company's success (Bingen et al., 2003).

Producers' organizations amplify the political voice of smallholder producers, reduce the costs of input purchase and output marketing, and provide a forum for members to share information, coordinate activities, and make collective decisions. Producers' organizations create opportunities for producers to get more involved in value-adding activities such as input supply, credit, processing, marketing, and distribution. On the other hand, they also lower the transaction costs for the processing/marketing agencies working with growers under contract. Collective action through cooperatives or associations is important not only to be able to buy and sell at a better price, but also to help small farmers adapt to new patterns and much greater levels of competition (Farina, 2002). Vigorous bargaining cooperatives or other agricultural producer organizations are needed to negotiate equitable contracts (Goldsmith, 1985). These types of organizations have been able to standardize contracts and manage their relationships with companies through co-operatives (Ornberg, 2003).

But it is important to remember that producer organizations may not be able to do all that private or state actors do. Further, they are influenced and governed by the local social context, and that means they may not be able to provide an ideal forum for addressing all development issues such as inclusion of various kinds, whether gender or marginalized, and weaker sections of society. Small producers may find it difficult to access producer organizations because they lack the necessary volumes and the capacity to take risks or invest in the organization, which may be needed initially (Penrose-Buckley, 2007). In this context, it is important to discuss a relatively new form of organization which is relevant for agribusiness, has been tried in some countries, and is now legal in India—the producer company.

Producer/farmer companies

There are only a few examples of organizing small farmers under a more business-like entity called farmer or producer companies in Asia (e.g. in Sri Lanka since the 1990s and in India since the mid-2000s). Farmer companies in Sri Lanka and one in the Philippines are investor-owned companies established under the Companies Act as people's companies registered with the Registrar of Companies. They follow rules and regulations like a private company and are registered with a minimum of 50 members to safeguard against possible private ownership by imposing restrictions on membership and share trading. Only farmers and other stakeholders involved in agriculture living within a particular geographical region can become shareholders, and shares cannot be traded except among farmers eligible for membership. In addition, the maximum number of shares a farmer can own is limited to 10 per cent of shares issued at a given time according to the relevant provision of the Act. These companies were organized by different government agencies, and their membership ranged from 200–2,200. They were involved in different stages of the agricultural chains, such as input supply, procurement, selling, packaging, etc. (Hussain and Perera, 2004). A similar law now exists in India under which hundreds of producer companies are being set up by different stakeholders (Singh, 2008).

In Sri Lanka, small farmer companies were established to accelerate commercialization in non-plantation agriculture, as recommended by the

National Development Council (NDC) of Sri Lanka in 1995, which also suggested a national-level farmer company with 50 satellite farmer companies at local levels. It failed to achieve the expected objectives due to such factors as: politicization of farmer companies; lack of managerial and entrepreneurial skills due to poor recruitment of management staff; lack of sound plans and poor management by incompetent boards of directors without professional advice; lack of proper mechanisms to monitor and evaluate; mistrust between farmer company management and farmers; farmer perception of the farmer company as a service provider; an awareness gap between shareholders and the farmer company; and restriction on share capital ownership which inhibited the producer company from expanding its commercial activities (Esham and Usami, 2007).

In the case of a relatively successful farmer company in an irrigation scheme in Sri Lanka, the nominal value of a share was Rs 10; to become a shareholder, a minimum of 10 shares was needed, and to become involved in the activities of the farmer company, the minimum was 25 to 40 shares. Shareholders increased from 430 in 1998 to 1,898 in 2011 (71 per cent of all farmers in the irrigation scheme) and to 2,234 in 2004, which was 80 per cent of the total of 2,796 farming families living within the scheme. Accordingly, the share capital of the farmer company steadily increased from Rs 183,450 in 1999 to Rs 839,303 in 2004. Farmers living within the scheme were literate and, on average, a farm household head had 8.3 years of schooling and owned 1.37 ha (0.55 ha) of land, of which 0.78 ha (0.3 ha) was irrigated. Agricultural input sales, paddy (rice) seed production, a group loan programme, and basmati rice production were the major revenue-generating commercial activities undertaken by the company. The company was either directly involved in or facilitated links between farmers and agribusiness. In broiler and maize production the company acted as a facilitator between farmers and agribusiness firms (Table 7.1). The farmer company selected suitable farmers from its shareholders and entered into a contract with the agribusiness firm on behalf of the farmers. The company distributed inputs, provided extension service, and monitored and assisted agribusiness firms in collection of the produce. In basmati rice production, seed paddy production, and vegetable seed production programmes, the company had a contract with a supermarket chain to market basmati rice and, with the agrarian service centres, to market seed paddy and vegetable seeds under the company's own brand name. In these operations, the company provided all inputs on credit and free extension services to the farmers as shareholders. The company was able to fulfil input requirements for about 1,000 ha and extension needs for one-third of the irrigation scheme. The farmers were able to buy agricultural inputs at a lower price compared to the open market, as the company retained only a low profit margin to cover its operational costs (Esham and Usami, 2007).

The government provided seed money of Rs 10 million and met the operational costs for the first three years, which was Rs 7 million, besides

Table 7.1 Major commercial activities of the farmer company in 2003–4

Activity	No. of Farmers	Amount/qty distributed/ procured	Revenue (Rs. 000)	Linkage partner
Group loan	1,035	Rs 8.6 million	1,364*	–
Broiler production	83	115,000 birds/month	947	Ceylon Agro-Industries
Seed paddy	83	Rs 0.7 million 144 MT/season	5,068	Agrarian service centres
Basmati (rice)	35	39 MT/season	1,284	Cargills supermarket chain
Vegetable seeds	25	–	514	–
Maize	30	–	–	Ceylon Agro-Industries
Dairy/livestock	20	20 cows	–	–
Agricultural inputs sales	–	–	14,783	Many input suppliers

Source: Esham and Usami (2007).

provision of government buildings and storage space. The board of directors was entirely chosen from among farmer members, including tenant farmers. The company undertook more than twenty-five activities, ranging from seed, crop, dairy, and poultry production to supply of inputs and group loans, as well as domestic and export market linkages, almost half of which failed by 2004 (Hussain and Perera, 2004). Despite the group's joint liability to repay loans, the loan recovery of 80 per cent was low, attributed to the farmers' perception of the farmer company as a service provider.

The company reported operating losses for the first three financial years, but the performance improved, with operating profits increasing in the last three financial years (2001–4). The initial losses of the company were covered by government grants, which were made available until 2001/2002 when the company became financially viable. All the farmers were aware of the group loan programme and inputs sales by the farmer company, but awareness of the other activities such as seed paddy production and broiler production was low among both active and passive shareholders. There was a contrasting level of satisfaction with irrigation management (74 per cent) and the commercialization of agriculture in terms of input supply and marketing of produce (29 per cent). Since the company started making profits in the last three financial years, it has been able to reward the farmer shareholders by issuing bonus shares. Fifty-five per cent (17 out of the 31 farmers) were active shareholders involved in at least one commercial activity organized by the farmer company. Many farmers just own the minimum number of shares to have access to the services provided by the company. Furthermore, a majority of farmers believed that the company should neither charge a fee for coordinating links between farmers and agribusinesses nor charge a market-based interest rate for credit. This situation could have far-reaching implications on the capital base as well as the business orientation of the company. The farmer company was overly dependent on the group loan programme to generate profits, as more than 33 per cent of the total profit came from this programme (Esham and Usami, 2007).

A producers' company (NorminCrop) in the Philippines has been successful in interfacing between large buyers and its small vegetable farmers by working on cooperative lines. It plans production at cluster level and provides marketing facilitation for a fee. The farmer and buyer agree quality standards, delivery arrangements, and conditions (Vorley et al., 2009).

In India too, small producers' companies provided for by Indian company law are being set up to provide market access. A producer company can be registered under the provisions of the Companies Act 1956. The objective of the company can be production, harvesting, procurement, grading, pooling, handling, marketing, selling, export of members' primary produce, or import of goods and services for their benefit. Its members can be individuals (ten or more) or producer institutions or a combination of both. Under this law, dozens of companies already exist in many states (for selling spices, seeds, fruit, vegetables, organic inputs, etc). There are 131 producer companies in India across states, promoting agencies, crops, and products. Of these, the largest number (44) is in the west, followed by the east (34) and the south (30). Most of them are into input supply, seed production, and farm produce-related business, and one-third are more than two years old, but most of them are facing problems. For example, banks often refuse to lend without a government guarantee (which is given to cooperatives). They also face difficulties in getting Agricultural Produce Marketing Committee (APMC) licences and mobilizing capital from the market (Singh, 2008).

4. Conclusions and ways forward for effective market access

Though there are concerns about the ability of small farms to survive in the changing environment of agribusiness, there are still opportunities for them to exploit, such as organic production and other niche markets. But the major route has to be through other factors such as economies of scale by networking or clustering and such alliances as contract farming (Kirsten and Sartorius, 2002). Linking small farmers with global or national markets in processing and marketing is a must (Lipton, 2002) whether by private enterprise, a state or para-statal organization, or a cooperative, farmer association, or producer company.

The above analysis of various market access experiences shows that there is no single model for such links. The appropriate model depends on the context and nature of the problem and local agro-ecology. But the examples also show that it is important to have a producer agency to orient and involve producers as stakeholders. The state cannot deliver it from above, as seen in the case of HOPCOMS and Safal in India and farmer companies in Sri Lanka. It should facilitate from outside, as was the case in Thailand's contract farming programme.

Market access for small producers depends on: understanding the markets, organization of the firm or operations; communication and transport links; and an appropriate policy environment (Page and Slater, 2003). Understanding markets involves understanding value chains and networks and their dynamics from a small producer perspective. Innovations in smallholder market links are needed in terms of partnerships, use of Information and Communication Technologies (ICTs), leveraging networks, value chain financing, smallholder policy, and even in contracts which can promote efficiency and inclusiveness of the linkage in terms of roping in smallholders (Mendoza and Thelen, 2008). Effective market participation is at the heart of success for any market links for primary producers. Markets should be more profitable than the existing arrangement and with only a little more risk for the primary producers. These markets could be niche export markets, niche domestic markets, organic produce markets, fairtrade and ethical markets, fresh and cut perishable produce export markets, or local markets for fresh and value-added products (Shepherd, 2007).

In the case of smallholders, market access boils down to education and training, business networks for capital and markets, resources, and infrastructure and institutional support. Small farmers also require professional training in marketing and in technical aspects of production, which can come through small farmer organizations and can help increase productivity, provide help in improving the quality of produce, and encourage them to participate more actively in the marketing of their produce (Schwentesius and Gomez, 2002). Other requirements are marketing extension, which includes better product planning at farmer/group level, provision of market information, securing markets and alternative markets for farmers, and improving marketing practice at farm level in terms of grading, sorting, packaging, and primary processing. There is also a need to make producers understand the various benefits of the links, not just the price benefits. It could be lower cost of production, lower transaction cost, or better quality of produce, or other benefits such as resource conservation or brand building in the market.

The first step is bringing the right agencies together and creating social capital, as seen in the Vasundhara cooperative case, while the need for new knowledge comes later and can be provided by different agencies. This is followed by private enterprise helping to scale up production (Reddy, 2010) to the required standards.

Government has to play an enabling role through legal provisions and institutional mechanisms to help farmer cooperatives and groups, and to facilitate such links as the contract farming system. It need not intervene in contract farming directly, as seen in the case of the Punjab where the experiment failed (Kumar, 2006). On the other hand, in Thailand the state's external provision of credit and extension was successful (Singh, 2005).

Fairtrade and alternative trade networks provide scope for participation of small and marginal producers, as seen in the case of black-pepper farmers in Kerala. There is, therefore, a need to bring organic and fairtrade movements into the mainstream to ensure the participation of a large number of producers in these markets, without involving them with the ills of conventional chains (Raynolds, 2004). These movements require policy input and support for a sustainable expansion of production.

Marketing is all about identifying and serving relevant markets. Further, selecting the right market and development strategy needs to avoid a 'race to the bottom' by innovative products and business models (GTZ, 2007). These business models need to provide not just for value creation and conservation, but also value sharing with smaller partners like smallholders.

Innovative ways to provide finance for small producers and their collectives have to be found, as well as favourable credit policies, crop insurance, and even institution-building subsidies. In Benin in Africa, for example, the yield of organic cotton per hectare was lower, but this was compensated for by waiver of input credit loans and a premium of 20 per cent over the local conventional produce price. The producer price was set at the start of the season and purchase of the entire organic cotton crop was guaranteed. This helped the producers of organic cotton to organize themselves (Verhagen, 2004). The state can also step in to provide initial support to small producer organizations as in Sri Lanka or India's Madhya Pradesh state, which provided Rs 25 lakh to each producer company for three years.

Furthermore, it is important to address the entire value chain of a product, and not just specific bottlenecks, to produce tangible benefits up to the farmer level. The surplus should be shared among chain partners, including primary producers. The essence of such sharing is the contract, which has to be flexible so that primary producers are not tempted to defect when there is a small price increase in open markets or alternative marketing channels. Farmers or their representing agency need to be involved in contract design and negotiation.

The private sector can deliver market access, but it may not be willing to bear the costs, with few exceptions like Namdhari Fresh in India. Public–private partnerships are an alternative, but for them to work, NGOs, their staff, and farmers need to be more market-oriented and capable of negotiating fair contracts with private agencies. This requires training of NGO personnel and farmers in modern markets and their dynamics, which includes contract negotiation, business management, market research, supply or value chain analysis, basic business documentation, crop and farm plans, and budgets. Farmers also need to be made aware of the need to respect contracts and specific terms and conditions, including prices, rejections, and penalties for default. Private-sector agencies also need to invest in building links. For example, a supermarket in South Africa provided interest-free loans to 27 growers (Shepherd, 2007). Similarly, many contracting agencies provide credit to their growers in India, as costs of production and transaction for high-value crops are generally higher.

Finally, it can be inferred from case study analysis that group and intermediate market links are faster routes to expand market access for smallholders, but they are not necessarily the most appropriate for them. Group and institutional links require many value-chain and process innovations and, of course, policy backup. Producer companies in India are a case in point, where an innovative farmer organizational model is undeveloped because there has been no policy support, not even as much as a cooperative receives.

REFERENCES

Agrawal, R. C. (2000). 'Perspectives for Small Farmers in Developing Countries: Do They have a Future?', *Forum zur Gartenkonferenz 2000.* Retrieved from: <http://userpage.fu.berlin.de/garten/Buch/Agrawal(englisch).htm> (6 April 2005).

Alam, G. and D. Verma (2007). *Connecting small-scale farmers with dynamic markets: A case study of a successful supply chain in Uttarakhand.* Dehradun, India: Centre for Sustainable Development.

Amekawa, Y. (2009). 'Reflections on the growing influence of good agricultural practices in the Global South', *Journal of Agricultural and Environmental Ethics*, 22(6), pp. 531–57.

Barghouti, S., S. Kane, K. Sorby, and M. Ali (2004). 'Agricultural diversification for the rural poor—guidelines for practitioners', ARD Discussion Paper No. 1, World Bank, Washington, DC.

Bayes, A. and M. S. Ahmed (2003). 'Agricultural diversification and self-help group initiatives in Bangladesh', paper presented at the IFPRI-FICCI Workshop on Vertical Integration in Agriculture in South Asia, 3 November, New Delhi.

Benziger, V. (1996). 'Small fields, big money: two successful programs in helping small farmers make the transition to high value-added crops', *World Development*, 24(11), pp. 1681–93.

Bhaskaran, R. P. (2011). 'From the village to the city: the changing dynamics of migration in contemporary India', *Indian Journal of Labour Economics*, 54(3), pp. 579–95.

Bingen, J., A. Serrano, and J. Howard (2003). 'Linking farmers to markets: Different approaches to human capital development', *Food Policy*, 28, pp. 405–19.

Boselie, D., S. Henson, and D. Weatherspoon (2003). 'Supermarket procurement practices in developing countries: redefining the roles of the public and private sectors', *American Journal of Agricultural Economics*, 85(5), pp. 1155–61.

Chand, R., P. A. L. Prasanna, and A. Singh (2011). 'Farm size and productivity: Understanding the strengths of smallholders and improving their livelihoods', *Economic and Political Weekly*, 46 (26/27), 25 June, pp. 5–11.

Dev, M. (2005). 'Agriculture and rural employment in the budget', *Economic and Political Weekly*, 40(14), 2 April, pp. 1410–13.

Dhananjaya B. N. and A. U. Rao (2009). Namdhari Fresh Limites (Case Study 1) in M Harper: Inclusive Value Chains in India—linking the smallest producers to modern markets, *World Scientific*, Singapore, 26–41.

Eaton, C. and A. W. Shepherd (2001). *Contract Farming: Partnerships for Growth.* Rome: FAO.

Ellis, F. (n.d.). 'Small farms, livelihood diversification, and rural-urban transitions: strategic issues in sub-Saharan Africa'. Retrieved from: <http://www.uea.ac.uk/ . . . fs/1.53421!2005% 20future%20small%20farms.pdf> (20 December 2011).

Esham, M. and K. Usami (2007). 'Evaluating the Performance of Farmer Companies in Sri Lanka: A case study of Ridi Bendi Ela Farmer Company', *The Journal of Agricultural Sciences*, 3(2), pp. 86–100.

Farina, E. M. M. Q. (2002). 'Consolidation, multinationalisation, and competition in Brazil: Impacts on horticulture and dairy products systems', *Development Policy Review*, 20(4), pp. 441–57.

Gandhi, V. P. and A. Koshy (2006). 'What marketing and its efficiency in India', Working Paper No. 2006-09-03, IIM, Ahmedabad, September.

Gaurav S. and S. Mishra (2011). 'Size, class and returns to cultivation in India—a cold case reopened', IGIDR working paper No. 27, IGIDR, Mumbai.

Gibbon, P. (2003). 'Value chain governance, public regulation and entry barriers in the global fresh fruit and vegetable chain into the EU', *Development Policy Review*, 21(5–6), pp. 615–25.

Glover, D. (1987). 'Increasing the benefits to smallholders from contract farming: Problems for farmers' Organisations and Policy Makers', *World Development*, 15(4), pp. 441–8.

Goldsmith, A. (1985). 'The private sector and rural development: Can agribusiness help the small farmer?', *World Development*, 13(11/12), 1125–38.

GTZ (2007). 'International Conference: Value chains for broad based development', 30 May–1 June, German Technical Co-operation (GTZ) Conference Report.

Harper, M. (2009). 'Development, value chains and exclusion', in Harper, M., 'Inclusive Value chains in India—Linking the Smallest Producers to modern markets', *World Scientific*, Singapore, Chapter 1, pp. 1–10.

Hazell, P. B. R. (2005). 'Is there a future for small farms?', in Colman, D. and N. Vink (eds), *Reshaping Agriculture's Contributions to Society*, Proceedings of the 25th International Conference of Agricultural Economists (ICAE), 16–22 August, Blackwell, USA, pp. 93–101.

Hazell, P. (2011). 'Five Big Questions about Five Hundred Million Small Farms', paper presented at the IFAD Conference on New Directions for Smallholder Agriculture, IFAD, Rome, 24–25 January.

Hellin J., M. Lundy, and M. Meijer (2009). 'Farmer organisation, collective action and market access in Meso-America', *Food Policy*, 34, pp. 16–22.

Hiremath, B. N., H. K. Misra, and J. Talati (2007). 'Adivasi Development Programme: Valsad and Dang Districts, Gujarat-Evaluation III', IRMA, Anand, January.

Hussain, I. and L. R. Perera (2004). 'Improving agricultural productivity for poverty alleviation through integrated service provision with public–private sector partnerships: Examples and issues', International Water Management Institute (IWMI) Working paper No. 66, IWMI, Colombo, Sri Lanka.

IFPRI (2005). 'High value agriculture and vertical co-ordination in India—will the smallholders participate?', a draft research report, IFPRI, Washington.

Karunakaran, N. (2008). 'Makhana's gold', *Business Outlook*, 3(21), pp. 64–5 (special issue on Farming 2.0).

Key, N. and D. Runsten (1999). 'Contract farming, smallholders, and rural Development in Latin America: The organisation of agro-processing firms and scale of outgrower production', *World Development*, 27(2), pp. 381–401.

Kirsten, J. and K. Sartorius (2002). 'Linking agribusiness and small-scale farmers in developing countries: is there a new role for contract farming?', *Development Southern Africa*, 19(4), pp. 503–29.

Kolady, D., S. Krishnamoorthy, and S. Narayanan (2007). 'An "Other" Revolution?—Marketing Co-operatives in a New Retail context', a case study of HOPCOMS, The Regoverning Markets project report, August.

Kumar, P. (2006). 'Contract farming through agribusiness firms and state corporation: A case study in Punjab', *Economic and Political Weekly*, 52(30), 30 December, A5367–75.

Lipton, M. (2002). 'Access to assets and land in the context of poverty reduction and economic development in Asia', paper presented at the Global Distance Learning Course 'Reaching the Rural Poor: Strategies for Rural Development for the East Asia Region', Bangkok, 3–6 June.

Markelova, H., R. Meinzen-Dick, J. Hellin, and S. Dohrn (2009). 'Collective action for smallholder market access', *Food Policy*, 34, pp. 1–7.

Mendoza, R. U. and N. Thelen (2008). 'Innovations to make markets more inclusive for the poor', *Development Policy Review*, 26(4), pp. 427–58.

Mishra, B. S. (2009). 'Contract Farming of Potato: An attempt to include poor farmers into value chain (A learning from BASIX intervention)', in Harper, M., 'Inclusive value chains in India—Linking the smallest producers to modern markets', *World Scientific*, Singapore, Case Study 5, Chapter 4, pp. 92–109.

Miyata, S., N. Minot, and D. Hu (2009). 'Impact of contract farming on income: Linking small farmers, packers and supermarkets in China', *World Development*, 37(11), pp. 1781–90.

Morris, S. (2007). 'Agriculture: A perspective from history, the metrics of comparative advantage and limitations of the market to understand the role of state in globalising world', Working paper No. 2007-02-02, February, IIM, Ahmedabad.

Morrison, P. S., W. E. Murray, and D. Ngidang (2006). 'Promoting indigenous entrepreneurship through small-scale contract farming: the poultry sector in Sarawak, Malaysia', *Singapore Journal of Tropical Geography*, 27(2), pp. 191–206.

Narrod C., D. Roy, J. Okello, B. Avendano, K. Rich, and A. Thorat (2009). 'Public–private partnerships and collective action in high value fruit and vegetable supply chains', *Food Policy*, 34, pp. 8–15.

Ornberg, L. (2003). 'Farmers' Choice: Contract farming, agricultural change and modernisation in Northern Thailand', paper presented at the 3rd International Convention of Asia Scholars 3 (ICAS3), Singapore, 19–22 August.

Page, S. and R. Slater (2003). 'Small producer participation in global food systems: Policy opportunities and constraints', *Development Policy Review*, 21(5–6), pp. 641–54.

Penrose-Buckley, C. (2007). *Producer Organisations—A Guide to Developing Collective Rural Enterprises*. Oxford: Oxfam GB.

Pingali, P. and Y. Khwaja (2004). 'Globalization of Indian diets and the transformation of food supply systems', *Indian Journal of Agricultural Marketing*, Vol. 18, No. 1, pp. 26–49.

Polak, P. (2008), *Out of poverty—what works when traditional approaches fail*. New Delhi: Tata Mcgraw Hill.

Premchander, S. (2002). 'Co-operative for sale of fruits and vegetables—a success story of urban horti marketing (HOPCOMS)', *Sampark*, Bangalore, August.

Raynolds, L. T. (2004). 'The Globalisation of Organic Agro-Food Networks', *World Development*, 32(5), pp. 725–43.

Reddy, T. S. V. (2010). 'Exploring mechanisms for putting agri value chain oriented research into use: empirical cases from research into use program', paper presented at the *Globelics 2010 Conference, Making innovation work for society: Linking, Leveraging and Learning*, Kuala Lumpur.

Roy, D. and A. Thorat (2008). 'Success in high value horticultural export markets for the small farmers: The case of mahagrapes in India', *World Development*, 36(10), pp. 1874–90.

Sahn, E. and J. Fischer-Mackey (2011). 'Making markets empower the poor; programme perspectives on using markets to empower women and men living in poverty', Oxfam Discussion Paper, Oxfam GB, Oxford, 7 November.

Schwentesius, R. and M. A. Gomez (2002). 'Supermarkets in Mexico: Impacts on horticulture systems', *Development Policy Review*, 20(4), pp. 487–502.

Sen, S. and S. Raju (2006). 'Globalization and Expanding Markets for Cut-Flowers: Who Benefits?', *Economic and Political Weekly*, 30 June, pp. 2725–31.

Sharma, V. P. (2007). 'India's agrarian crisis and smallholder producers' participation in new farm supply chain initiatives: A case study of contract farming', IIMA Working Paper No. 2007-08-01, August, Indian Institute of Management, Ahmedabad (IIMA), Ahmedabad.

Shepherd, A. W. (2007). 'Approaches to linking producers to markets—a review of experiences to date', AMMF Occasional paper 13, FAO, Rome.

Simmons, P., P. Winters, and I. Patrick (2005). 'An Analysis of contract farming in East Java, Bali, and Lombok, Indonesia', *Agricultural Economics*, 33, pp. 513–25.

Singh, S. (2005). 'Contract farming system in Thailand', *Economic and Political Weekly*, 40(53), 31 December, pp. 5578–86.

Singh, S. (2008), 'Producer Companies as New Generation Co-operatives', *Economic and Political Weekly*, 34(20), 17 May, pp. 22–4.

Singh, K. A. and K. V. S. M. Krishna (1994). 'Agricultural entrepreneurship: The concept and evidence', *The Journal of Entrepreneurship*, 3(1), pp. 97–111.

Singh, S. and N. Singla (2011). *Fresh Food Retail Chains in India—Organisation and Impacts.* New Delhi: Allied.

Suzuki, A., L. S. Jarvis, and R. J. Sexton (2011). 'Partial vertical integration, risk shifting, and product rejection in the high value export supply chain: The case of Ghana pineapple sector', *World Development*, 39(9), pp. 1611–23.

Talukdar, T. (2010). 'Vegetable Soup for the Soul', *The Economic Times*, Ahmedabad, 22 August, p. 9.

Thapa G. and R. Gaiha (2011). 'Smallholder farming in Asia and the pacific: challenges and opportunities', presented at the IFAD conference on new directions for smallholder agriculture, 24–25 January, IFAD, Rome.

Tomek, W. G. and H. H. Peterson (2001). 'Risk management in agricultural markets: A review', *Journal of Futures Markets*, 21(10), pp. 953–85.

Trebbin A. and M. Franz (2010). 'Exclusivity of private governance structures in agrofood networks: Bayer and the food retailing and processing sector in India', *Environment and Planning* A, 42, pp. 2043–57.

Verhagen, H. (2004). 'International sustainable chain management—Lessons from the Netherlands, Benin, Bhutan and Costa Rica', *Royal Tropical Institute (RTI) Bulletin 2/360*, KIT publishers, Amsterdam.

Vorley, B. and F. Proctor (2008). 'Inclusive business in agrifood markets: evidence and action', a report based on proceedings of an international conference held in Beijing, 5–6 March.

Vorley, B., M. Lundy, and J. MacGregor (2009). 'Business Models that are Inclusive of small farmers', in da Silva, C. A., D. Baker, A. W. Shepherd, L. Jenane, and S. Miranda-da-Cruiz (eds), *Agro-industries for Development.* Oxfordshire: FAO, UNIDO and CABI, pp. 186–222.

Wolf, S., B. Hueth, and E. Ligon (2001). 'Policing mechanisms in agricultural contracts', *Rural Sociology*, 66(3), pp. 359–81.

Zamil, M. F. and J. J. Cadilhon (2009). 'Developing small production and marketing enterprises: mushroom contract farming in Bangladesh', *Development in Practice*, 19(7), pp. 923–32.

8 Financing smallholder farmers in developing countries

ATIQUR RAHMAN AND JENNIFER SMOLAK[1]

1. Introduction

Smallholder farmers need appropriate financial support to expand their opportunities and meet their potential. The recent food price crisis and lasting higher food prices have added greatly to the importance of improving the productivity of small-scale farms in developing countries (Dethier and Effenberger, 2011). The persistence of about 450 million small farms speaks of their vast impact and their prominence in the rural landscape.[2] Advancing smallholder agriculture can build on efficiency of production in comparison to large farms (Eastwood et al., 2010), stimulate pro-poor economic growth, and contribute to a vibrant non-farm rural economy (Hazell, 2011). With appropriate investments and technologies, there is significant opportunity for raising productivity[3] to meet their own food needs, to commercialize and produce higher value crops for local and export markets.

Despite the opportunities, the conditions needed for smallholders to meet their potential are often lacking. Access to modern input supplies is problematic, as are connections to markets and an ability to engage with value chains (Poulton et al., 2010). Rapidly transforming food retailing systems and the need to interact with new players pose further challenges. Uncertainty about climate change and lack of adaptation measures adds to the vulnerability of smallholders, whose livelihoods are severely affected if the risks they face materialize (IPCC, 2007). Without their own resources to invest, and

[1] The authors wish to thank the referee comments of this chapter; errors and omissions still remaining are however the authors' own responsibility.

[2] An estimated 450 million small farms (concentrated in Asia and Africa) are home to over 2 billion people, produce 80 per cent of the food supply in developing countries, and use and manage more than 80 per cent of farmland—and similar proportions of other natural resources—in Asia and Africa (Collette et al., 2011).

[3] Cereal yields increased by 29 per cent in sub-Saharan Africa between 1961–3 and 2003–5, and 177 per cent in developing Asia (Staatz and Dembele, 2007). India realized 2.6 times higher yields with 90 per cent on farmland of less than 10 ha; China doubled yields between 1991 and 2001 based on small farms averaging 0.4 ha (Byerlee and de Janvry, 2009).

lacking access to financial markets, exposure to shocks and risk traps them in a cycle of poverty (Carter, 2008).

This chapter explores how rural financial markets are evolving in developing countries and how far they are helping smallholders to seize emerging market opportunities. With such developments in mind, this chapter takes a broad sweep of the context, challenges, and prospects. Claims of innovative forms of financing are examined to see whether they can reach smallholder farmers and contribute to attaining development goals. The second section presents the current context of formal and informal finance for smallholders; the third examines value-chain finance and weather-index insurance as innovative financial products with potential for reducing barriers to finance; the fourth looks at the role of the public sector and donors in smallholder agriculture and access to financial services; conclusions are set out in the final section.

2. The context of finance for smallholder farmers

Providing financial services to smallholder farmers in developing countries has long been a challenge, adding to the constraints of agricultural development. Financial systems in rural areas have struggled due to geographically dispersed clientele, high transaction costs, and high information costs (Carter, 2008; Poulton et al., 2010). Serving the agriculture sector brings further challenges: high covariate risk among clients affected by the same climate in a specified region; and irregular income flows linked to crop cycles which do not correspond to traditional financial products. Food price variability affects income from agricultural production and adds to the stress of repaying loans. Smallholder farmers with low net worth and lacking collateral are among the most excluded from formal financial services.

Developing-country policies and donor efforts in the 1960s through to the 1980s were specifically designed to expand the outreach of formal financial markets and overcome the above challenges in rural areas. However, few of these objectives were achieved, and dissatisfaction triggered a major shift in paradigm (AZMJ, 2011; Meyer, 2011). Rural finance moved from a supply driven directed credit approach to a market-oriented approach. It came to be viewed more broadly, reversing the policies of subsidized credit and focusing on institutional viability and sustainability. Even so, access to formal financial institutions for smallholders remains low.[4]

[4] Less than 1 per cent of farmers in Zambia and 2 per cent of the rural population in Nigeria have access to formal credit; 45 per cent of smallholders in India do not have a formal savings account and

2.1 THE PERSISTENCE OF INFORMAL FINANCE

Informal markets have been and remain a major source of finance for most smallholder farmers (Collins et al., 2009; Meyer, 2011). The very nature of informal finance[5] makes it hard to quantify, but evidence paints a picture of widespread reliance. Ghate and Das-Gupta (1992) compile evidence among Asian countries; informal finance accounts for about one-third to two-thirds of total rural credit in Bangladesh[6] and China, about two-fifths in India, Sri Lanka, and Thailand and two-thirds to three-quarters in Malaysia, Nepal, Pakistan, and the Philippines. Evidence from El Salvador shows that only 41 per cent of loans are formal, yet this accounts for up to 76 per cent of the amount borrowed (Wenner and Proenza, 2000). In Mexico, 54 per cent of rural poor were found to access supplier or commercial credit, 10 per cent used moneylenders, 14 per cent borrowed from friends and relatives, and 8 per cent engaged in forward sales (Wenner and Proenza, 2000). Figure 8.1 shows informal sources of financing to be significant in East and southern Africa, especially in Uganda and Kenya.

Informal finance was generally viewed unfavourably, even exploitative, in early development literature due to high imputed interest rates (cases of up to 10 per cent per day (Adams, 1992) and transactions occurring outside the range of regulatory and financial authorities). Classic cases of moneylender–landlords illustrate the argument; farmers in India were discouraged from adopting income-enhancing technologies in order to perpetuate subordinate relationships, and small farmers in the Philippines received loans from larger neighbours who hoped for default and acquisition of land (Armendariz and Morduch, 2005). Among vast numbers of non-exploitative informal transactions, clientele tend to be the poorest of the poor and the most vulnerable (small farmers, the landless, hawkers, women, etc.). This fuels an argument that informal finance is used to sustain an income level rather than increase it, and is not a substitute for formal finance in stimulating economic growth

69 per cent lack a formal credit account; nearly 40 per cent of farmers in Honduras, Nicaragua, and Peru are credit-constrained (AgriFin, 2010). Regarding overall coverage, 65 per cent of adults in Latin America are unserved, 80 per cent in sub-Saharan Africa, 58 per cent in each of South and East Asia, and 8 per cent in OECD countries (Chaia et al., 2009).

[5] Informal finance refers to providers who are neither formal nor semi-formal financial institutions, and who themselves fall along a continuum from strictly governed transactions to casual relationships. These include traditional moneylenders, cooperatives, communal savings clubs, rotating savings and credit associations (ROSCAs), input suppliers or other traders, moneyguards, friends, relatives, etc.

[6] Rahman (1992) estimated that about two-thirds of the financial market in Bangladesh was accounted for by informal transaction.

(Ayyagari et al., 2008).[7] The directed-credit approach of the past was intended, in part, to crowd out informal operators (Armendariz and Morduch, 2005; Meyer, 2011).

A distinct comparative advantage to informal finance is in monitoring and enforcing contracts, enabling it to persist despite efforts to suppress it. Due to its highly localized scope, informal lenders can closely monitor borrowers to ensure that loans are being used for agricultural production as intended, thus reducing information asymmetries (Ayyagari et al., 2008; Conning and Udry, 2005). Lenders can enforce the loan contract with a credible threat of sanction if farmers default; they are unlikely to have any other source of credit.

The persistence of informal financial activity to this day is also due to it being highly interlinked with the formal financial sector, a linkage which has been viewed in two ways. On the one hand, informal finance is seen to compete with formal institutions, with demand spilling over from the formal sector to the informal (Bell, 1990; Varghese, 2005). This is illustrated by the rapid expansion of Bank of Agricultural Cooperative (BAAC) in Thailand; an alternative to informal lenders led the informal interest rates to decline from 4.5 to 2.8 per cent per month, with a similar decline in the informal share of total loans (Meyer and Nagarajan, 2000).

On the other hand, informal and formal finance are complementary, with finance flowing from formal institutions through informal lenders to clients (Rahman, 1992; Conning and Udry, 2005). In this sense, the infrastructure, systems capacity, and funds of formal institutions are combined with the information and enforcement advantages of informal institutions. The whole arrangement benefits from the close monitoring of clients, possible with informal lenders, and greater flow of funds to smallholders unserved by formal institutions. Indeed, a successful linkage depends on the effective monitoring of every entity in the channel (Conning and Udry, 2005).

The potential for complementary linkages has gained traction with donors (Ritchie, 2010): either as a means for capital to flow directly to an informal intermediary who lends on and expands their lending portfolio; or, as a means to facilitate access by engaging an informal intermediary to act on behalf of a formal institution (Pagura and Kirsten, 2006). Self-help groups (SHGs) of 20 to 40 members are a prime example of intermediaries formed with the clear purpose of mobilizing savings and linking to formal institutions. From a policy perspective, SHG-bank linkages have provided a viable way of expanding the outreach of India's already existing commercial bank

[7] Ayyagari et al. (2008) estimate the effects of informal and formal financing on growth in a sample of firms in China, where informal finance is given credit for the country's vast economic growth despite weak financial systems. The authors failed, however, to find evidence that informal borrowing leads to faster firm growth and assert that informal finance does not effectively substitute for formal financial institutions.

network (Meyer and Nagarajan, 2005). In India, 69,000 federations of SHGs were reported as of 2007 (Ritchie, 2010), with the National Bank for Agriculture and Rural Development (NBARD) alone reaching over 67 million clients through its network of SHGs (Reed, 2011). Pagura and Kirsten (2006) find the vast majority of linkages to be direct provision of credit, and recommend focusing on development of facilitating linkages to expand the range of financial services available to the poor.

2.2 THE EVOLUTION OF INFORMAL FINANCE; ITS POTENTIAL ROLE IN SUPPORTING SMALL FARMERS

The persistence of informal finance in rural areas is also attributed to its innovativeness and flexibility in meeting the demands of smallholder farmers and others. As paradigms shifted and markets liberalized, informal finance itself evolved in a number of directions: (a) shifting to trade and market-linked credit to facilitate and capture a share of trade and production profits; (b) institutionally integrating social capital as a means to enforce compliance; and (c) mobilizing savings and achieving financial intermediation. Additionally, innovations in mobile technology and its usage affect how informal transactions occur and information flows. Informal finance has evolved to semi-formal finance in the process, with increasing controls and regulations, a decline in spontaneity, and increasing transaction costs.

Informal lenders have adapted to change brought about by the new rural finance paradigm by moving away from the exploitative moneylender–landlord system and taking on a facilitating role between markets and smallholders (Rahman, 1992). Trade credit often involved advancing funds to smallholders (i.e. dadan system[8] in South Asia), sourced from the trader's own equity, from financial institutions or processors and wholesalers to whom they sell (Miller et al., 2011). In this way, traders use their specialized knowledge of markets and local agriculture to enhance profit potential. Due to its central role in connecting farmers to markets, trade credit remains a prevalent form of value-chain finance.

A key development is the group structures in which members pool contributions and distribute funds in an agreed-upon manner, and rely on social capital to enforce the arrangement. Since group members are also owners, managers, and users, incentives are aligned and monitoring is embedded into the group structure (Conning and Udry, 2005). Members thus have a safe method to accumulate savings and access to loan funds. Groups often receive outside technical assistance, and the transactions retain the informal finance

[8] Money provided to agricultural producers (for meeting production costs) as advanced purchase of crops was and still is widely practised in many parts of South Asia.

characteristics of small amounts over short time spans. Functional forms include rotating savings and credit associations (ROSCAs), an original form of member-based organization which mobilizes savings from its members and extends loans on a revolving basis (Armendáriz and Morduch, 2005). Forms of Savings and Credit Co-Operatives (SACCOs), the SHG described earlier, credit unions or Community-managed Village Savings and Credit organizations are also viable, especially if well governed and linked to outside sources of funding. In many countries, these latter forms come under the purview of government legislation, developed to regulate and encourage them (Conning and Udry, 2005). In many others the issue of regulation, governance, and supervision remains unresolved (Meyer and Nagarajan, 2005).

Strengthening the capacities of membership-based organizations is a useful way forward for expanding the outreach of rural finance. Not only have they proved effective to mobilize resources from within the group, but they are seen as viable intermediaries through which commercial banks can reach new clients. Smallholder producer organizations can also facilitate the process, as described among Ethiopian coffee producers in Box 8.1. UNDP and IFAD have been supporting 'Sanadiq' organizations in Syria to intermediate financial resources and link to financial institutions, while still meeting Islamic financing principles. An IFAD-supported project currently counts 76 sanadiq and 13,500 members, with almost 12,000 loans distributed totalling US$17 million (Mahieux et al., 2011).

BOX 8.1 FARMER COOPERATIVES IN ETHIOPIA: ACCESSING FINANCE AND ADDING VALUE

Coffee farmer cooperatives in Ethiopia can access working capital finance from Nib International Bank S.C. (NIB) through The Coffee Initiative, its partnership with the IFC, The Bill and Melinda Gates Foundation, and TechnoServe. Underlying the financing is the bank's need to mitigate risk; this comes in the form of a US$10 million IFC guarantee facility for credit losses up to 75 per cent, and technical assistance for crop processing provided by Technoserve. Cooperatives purchase fresh coffee cherries from its farmers who hold, on average, 3/4 ha of land. A wet milling process is used to yield a high-quality washed coffee, which earns up to 50 per cent premium at market. Farmers earn a competitive price for the cherries and a further dividend from the cooperative's sale of washed coffee, the sum compensation to farmers amounting to two-thirds of the cooperatives' gross profit. Loans have been extended to 62 cooperatives reaching 45,000 farmers. Cooperative revenues have increased by US$1.5 million, with potential for growth in the high-quality coffee sector. NIB holds 29 per cent of the market share for agriculture sector loans which represents 6 per cent of its own lending portfolio. Physical access is a strength for NIB, as 40 per cent of its branches are located outside of the capital city.

Source: IFC (2011).

The same group dynamic underlies the microfinance model, pioneered famously by Grameen Bank in Bangladesh. Microfinance Institutions (MFIs) have flourished with group lending models, where social capital replaces collateral requirements in order to make financing more accessible to the poor. As such, MFI clients are generally poorer than those reached by financial institutions, and predominantly women (Armendariz and Morduch, 2005). Positioned to bridge a gap between informal and formal financial sectors, MFIs are neither completely informal nor are they regulated financial institutions. Nonetheless, MFIs have become vastly more commercialized and have leveraged massive donor investment; global commitments from public and private funders were US$24 billion in 2010, up 13 per cent from the previous year, of which 86 per cent is lent on to retail clients and 14 per cent dedicated to capacity building (CGAP, 2012). In terms of MFI outreach, the 2011 Microcredit Summit Campaign Report counts 3,589 institutions reaching 190.1 million clients, of whom 140.1 million are women and 128.2 million were reported as living under the poverty line (Reed, 2011).

However, MFIs have yet to reach smallholder farmers in any great number (Morduch, 1999; Meyer, 2011). In Bangladesh, the birthplace of microfinance in its group lending form, data suggests a rather limited reach of microcredit to the agricultural sector. Alamgir (2009) describes seasonal loan products for farmers as a 'new frontier'; mainstream NGOs have yet to be engaged and the sector currently offers limited savings and credit services. Of 25–30 million total microcredit borrowers in 2008, 6–7 million were agricultural producers, yet only 1.5 million of them accessed seasonal loans specifically designed for agriculture (Alamgir, 2009).

The limited outreach of microfinance to the agriculture sector is partly explained by a lack of innovation in MFI products. Short-term loans with frequent repayment periods, often beginning soon after the loan is disbursed, do not match the cash-flow realities of agricultural production (Morduch, 1999). Additionally, joint lending models which rely on a group's shared liability for repaying loans is ill-suited to agriculture where borrowers share common weather risks and a long gestation period (World Bank, 2007). Lending portfolios cannot sustain the covariate risk. As for savings and deposits, the lack of trust and knowledge of banking options remains a barrier to take-up. There is a tendency for distrust because MFIs fall outside regulated banking, and savings are not protected by deposit insurance (Dupas et al., 2011).

A further drawback for microfinance is its questionable impact on poverty reduction. Empirical evidence remains ambiguous, and the quality of past evidence is under scrutiny.[9] A recent impact evaluation found that

[9] Morduch and Roodman (2009) re-estimate a seminal work establishing the positive impact of microfinance in Bangladesh. The empirical techniques are questioned, and with updating, the results

households with small enterprises use credit for investing in their business, while households without enterprises (and unlikely to begin one) spent more on consumption goods (Banerjee et al., 2010). Critics argue that the latter perpetuates poverty; since borrowing will not increase productivity or income, households are borrowing against their future. However, loans for consumption purposes play the important role of preventing households from selling off assets during times of distress. Events in Andhra Pradesh, India, illustrate concerns over impact; enormously profitable MFIs (including a very successful initial public offering by SKS Microfinance Limited) are juxtaposed with incidences of farmer suicide possibly attributed to coercive pressure by microcredit loan collectors. Underlying the situation is a lack of credit infrastructure and an inability for MFIs to gauge the creditworthiness of borrowers, a problem not unique to India. This contributes to persistently high interest rates (Conning and Udry, 2005; Gine et al., 2010) and over-indebtedness of borrowers (Chakrabarti and Ravi, 2011).

Yet the sheer volume and outreach of MFIs make them an integral part of the financial landscape. Without MFIs, access to finance can be greatly reduced, as seen in Andhra Pradesh. The government's 2010 ordinance following the incidents of farmer suicide had the effect of halting the industry, with repayment rates dropping from 90 to 30 per cent and overall availability of credit greatly reduced (Chakrabarti and Ravi, 2011). It is the same for access to savings deposit facilities: Grameen Bank alone added over 2 million new savers between 2005 and 2008, with deposits totalling 147 per cent of the outstanding portfolio (Glisovic et al., 2010). Lessons can be drawn from successful MFIs to adapt and improve the outreach of rural MFIs and agricultural-sector lending. Christen and Pearce (2005) identify ten such lessons, including basing loans on household income rather than income from agriculture, acknowledging the cyclical nature of smallholders' income, using existing infrastructure or technology, and diversifying loan portfolios. They also stress the need for savings mechanisms (see Box 8.2) to be a feature of microfinance for agriculture, as should insurance schemes and links to member-based organizations. The extent to which these lessons have been successfully institutionalized remains to be seen, and, if they have not, an understanding of the reasons.

Mobile technology provides an outlet for MFIs to better reach smallholder farmers. A first generation of mobile money services has reached maturity—such as with remittance facilities, money transfers, and bill payments—but

do not hold. The original authors have responded (see <http://www.brown.edu/research/projects/pitt/sites/brown.edu.research.projects.pitt/files/uploads/reply_0.pdf>) with updated results that do hold, without counter-response from Morduch and Roodman.

MFIs have yet to play a large role in this domain (Payne and Kumar, 2010). There needs to be an evolution from mobile money to mobile banking to assist smallholders in becoming market-oriented. M-Pesa mobile savings in Kenya is an example of exceptionally rapid growth (8.5 million users) and sheer volume of movement (US$3.7 billion) between 2007 and 2009. However, a recent evaluation estimates a transaction 'velocity' of every 2–3 days, implying that M-Pesa has yet to become a means of storing savings (Mbiti and Weil, 2011). Pilot projects are looking at the potential for linking MFIs with mobile network operators (MNO) in order to offer mobile savings, credit, and insurance products. Ventura et al. (2011) (see Box 8.3) find significantly lower operating and transaction costs with mobile banking, greatly reducing barriers to providing rural financial services. Lessons show the need to improve financial literacy, build trust in financial service providers (Payne and Kumar, 2010; Ventura et al., 2011), and for MFIs to develop financial products that match the seasonal needs and income flows of smallholders.

BOX 8.2 SAVINGS AND SMALLHOLDER FARMERS

Poor households are willing and able to save money (Collins et al., 2009) and an expanding literature is examining safe savings methods. These are important for farmers whose income and expenses are 'lumpy' and intermittent, yet who need to smooth consumption through crop cycles. Savings also provide a risk transfer mechanism in the event of shock or extreme weather; otherwise households sell assets, impeding future productivity and income, or reduce consumption with long-term effects on health and development (Skees, 2008). For capital investment, mobilizing savings underlies the ability for banks to extend long-term credit. Informal means of savings are widespread—as seen with ROSCAs and self-reliant village savings and credit banks, and, for example, Susu money collectors in West Africa—and formal and semi-formal institutions (MFIs) are placing more emphasis on savings.

A randomized control trial among farmers in rural Malawi found a significant impact of 'commitment' accounts, where deposits are locked in for a determinate period of time (Brune et al., 2011). The latter led to 9.8 per cent increase in cultivated land, 26.2 per cent more input use, 22 per cent greater crop output in the next harvest, and 17.4 per cent increase in household expenditure following that harvest. The results point to the benefit of formal savings, but in a way that protects a household from the social pressures of sharing accumulated savings. They also point to suboptimal savings behaviour and the usefulness of product design that induces farmers to commit to saving. Likewise, Kendall (2010) lists privacy as the first of four design requirements of savings products for the poor. Risk needs to be minimal, calling for a sound regulatory and supervisory context. Reliability is also key, making a case against informal savings groups which fail, for example, when farmer-members are affected by a shared weather shock. Finally, the cost needs to be minimized, both in terms of economic costs (i.e. minimum balances or transaction fees) and opportunity costs (i.e. time and travel costs).

BOX 8.3 MOBILE BANKING IN SOUTH-EAST ASIA AND WEST AFRICA

Evidence from a pair of pilot studies shows potential for MFI/MNO collaborations to significantly expand financial services to unbanked rural poor. A start-up branchless bank was paired with an established MNO in South-East Asia (SEA), with potential for reaching a subscriber base up to ten times the size of a comparable MFI. Likewise, a local MFI and leading MNO in West Africa (WA) was found to be able to increase its subscriber base by 60 per cent. In the project 80–90 per cent of subscribers had never opened a bank account. For clients in SEA, using mobile banking reduced transaction costs by 80 per cent, due to lower travel and opportunity costs. WA clients also saw a cost reduction of about 50 per cent. Uptake and usage of simple, stand-alone savings accounts was low; the benefit was unclear and not considered practical for households with highly variable income. However, bundling savings with another product and using a pre-determined schedule increased savings rates by up to three times. Low financial literacy and inadequate regulation are challenges to the introduction of mobile banking, as is lack of expertise among banks and MNOs operating in a new territory. In this sense, a progression from first generation (mobile transfer mechanisms) to second generation (mobile financial services) can help ready the market and ease introduction of MFI products.

Source: Ventura et al. (2011).

Developments in rural finance (informal and semi-formal) are significant in that they allow for greater mobilization of savings, expansion of trade and business, reduced transactions costs, and expanded reach and efficiency. In many ways, rural finance has come to be seen as a combination of formal, semi-formal, and informal institutions, providing financial services to a heterogeneous mix of farm and non-farm clients of various income levels. By definition (or design), informal finance is limited in its infrastructure and resources, and to its geographical area. It also cannot provide the full range of services needed by rural clients; for example, money transfer mechanisms and long-term loans for the purpose of capital investment (Meyer and Nagarajan, 2005). The latter represents a specific barrier for the potential of informal finance to contribute to productivity enhancement and improved livelihoods. Nonetheless, lessons can be drawn from the persistence of informal finance in reaching the poorest, most vulnerable, and hardest to reach. Adams (1992) lists six areas: variety in types of services; flexibility to adapt products to client needs; discipline among providers and users; capacity for large savings mobilization; expectations of reciprocity in relationship-building; and innovatively keeping transaction costs low.

Continuing to promote open and competitive financial market development is important to provide smallholders with options, limit rent extraction, enhance bargaining power (Miller and Jones, 2010), and expand access to finance (Beck et al., 2009). A well-functioning financial system will reach as many users as possible with savings, payments, and risk management products, and will allocate financing for investment in promising growth opportunities (Beck et al., 2009). The risk-averse smallholder farmer would be more willing to invest in technology and inputs if he can smooth consumption through shocks, periods of variable weather, and price fluctuations. This is true no matter where a farm household falls along the continuum between subsistence agriculture and fully commercialized production,[10] although greater market-orientation invariably requires more sophisticated financing. Informal financial mechanisms have primarily served this function for smallholders, and improving rural financial systems should capture the interaction of informal and formal finance to build on the strengths of each.

3. Overcoming development challenges: innovative approaches to financing small farmers

Smallholder farmers are being challenged in at least two important ways, with implications for further evolution of rural financial markets: (a) integrating with expanding value-chain processes; and (b) overcoming the risks and uncertainties emerging from environmental hazards and changing climates. The first draws on informal–formal financial linkages and advances commercialization of small-scale agriculture. The second relates to developments in financial markets to support risk-averse smallholders in overcoming constraints and investing their limited resources in productivity-enhancing technology and capital assets.

3.1 FINANCING VALUE CHAINS

The interest in value-chain finance (VCF) lies in the dual potential for expanding access to finance and advancing commercialization of small-scale agriculture (Miller and Jones, 2010). Agricultural value chains have expanded in recent years, driven both by growing local demand and increased exports of high-value commodities (Swinnen and Maertens, 2010). These changes in local and global markets come amidst rising incomes, faster urbanization, liberalized trade, greater foreign investment, and advances in technology (World Bank, 2007). Linking to emerging value chains provides vast opportunity for pro-poor growth for smallholder farmers (World Bank, 2007) as well as addressing rural development challenges in a coordinated manner. However, smallholders face significant barriers: buyers demand specific quality standards, timely

[10] As defined in Hazell (2011).

BOX 8.4 AGRICULTURAL VALUE CHAIN FINANCE—FINANCIAL INSTRUMENTS

Type of Product	Financial Instrument
Based on product	Trade credit Supplier (inputs) Broker/wholesaler Lead firm/contract farming
Based on accounts receivable	Sale of receivables Factoring Forfaiting
Based on physical assets	Credit warrant Repos agreement Financial leasing
Risk reduction	Insurance Futures contracts Futures and hedging
Improve the quality of the credit supply	Securitization Guarantees Joint Ventures

Source: Miller et al. (2011).

deliveries, and scale economies that exceed sourcing from many small farmers (World Bank, 2007). Growth opportunities are by no means assured; smallholders need to upgrade production in order to link to value chains, which necessitates investment and financial services (Downing and Harper, 2008).

The growth of VCF, which evolved as a sophisticated form of trader credit, has coincided with the new quality requirements of expanding value chains (Swinnen and Maertens, 2010). Its emergence is also in response to the gap left by the privatization and liberalization of state-controlled supply chains, where formal financial institutions have failed to expand in rural areas (Swinnen and Maertens, 2010). Defined simply, VCF is the flow of funds through the various transactions within a value chain or made available to borrowers linked to a chain (Miller, 2008). In practice, VCF can be as simple as a trader providing a cash advance and accepting payment in kind at harvest time. Or, it can be a highly sophisticated configuration of farmers, traders, and agribusinesses, it can leverage formal financial flows, and it can involve the variety of financial instruments listed in Box 8.4.

Forming the basis of analysis are theories of interlinked transactions, said to exist when prices are determined simultaneously based on exchange goods and services (Basu, 1995). VCF can be internal to the chain, when financing is extended from one actor to another, and is possible when a transaction is

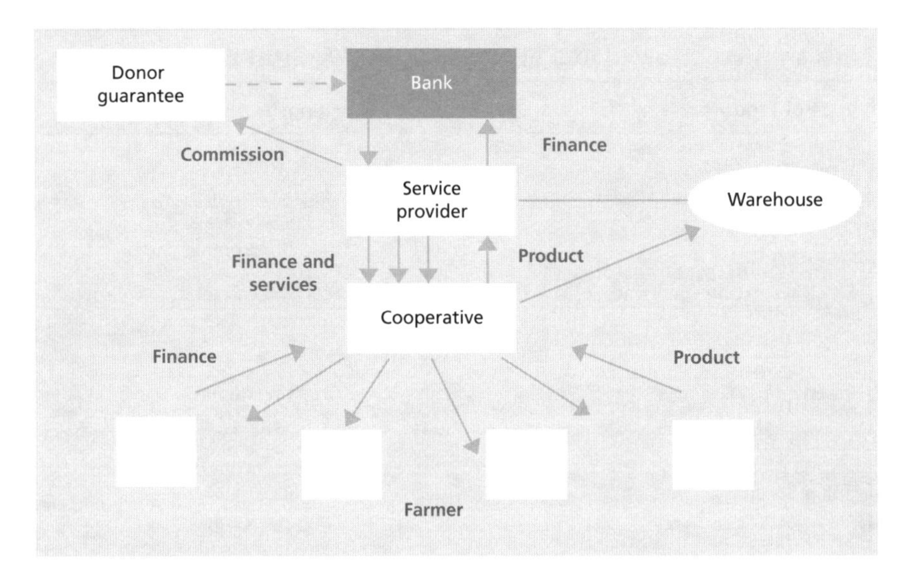

Figure 8.2 New model for value-chain finance

Source: Shwedel (2011).

secured and production, marketing, and price risks are reduced.[11] When based on the underlying crop product, financing is generally limited to credit with short-term maturities, and prices and interest rates become less transparent (Gonzalez-Vega et al., 2006). The other perspective holds that linkages within value chains are a means of improving creditworthiness (Gonzalez-Vega et al, 2006). Smallholders thus leverage their relationship with buyers to acquire external financing. Likewise, actors further along the chain (buyers, processors, retailers, etc.) can obtain outside financing and lend on to farmers. VCF arrangements which formalize the linkage have attracted remarkable attention among large-scale institutional investors, multinational food companies, and financial institutions.[12]

The model in Figure 8.2 illustrates the formal financial linkage of Rabobank Foundation with the local banks and value chains it supports (Shwedel, 2011), financing flows through the chain and ultimately to farmers via producers' cooperative. The model also illustrates the potential for packaging finance

[11] For example, an impact study of PepsiCo's contract farming programme (described in Box 8.7) found contracted farmers to be protected from potato market price risk as compared to non-contracted farmers who incurred losses in the year of study (Pandit et al., 2009).

[12] Notable examples include Rabo Development's strategy in Africa of investing in financial institutions (see Figure 8.2); Alliance for a Green Revolution in Africa's input-supplier-led model of value chain development; PepsiCo's contract farming arrangement with potato farmers in India and chick-pea farmers in Ethiopia.

with inputs and extension services. In this sense, VCF is highly integrated with the value-adding activities of the chain. Yet they are nonetheless distinct elements; the paths for achieving sustainability in marketing channels and financial service provision are separate but complementary (Mensink and de la Rive, 2011). The structure of the value chain in itself provides a framework for comprehensive analysis of financing needs among chain actors (Miller, 2008).

Value chains represent a new market opportunity for smallholder farmers, and VCF itself helps overcome the barriers to connecting with chains. VCF (cash credit or pre-paid inputs) was found to be the primary motivation for entering contracts among cotton farmers in Kazakhstan and vegetable farmers in Mozambique and Senegal (Swinnen and Maertens, 2010). Production contracts with Hortifruti supermarket have helped smallholders in Central America obtain external finance from financial institutions (Gonzalez-Vega et al., 2006). Similarly, the World Food Programme's Purchase for Progress pilot initiative is finding financial institutions to be more willing to lend to farmers who hold forward contracts that guarantee sale at a future date (WFP, 2011). Having a sustained relationship with a buyer signals the creditworthiness of smallholders and helps resolve information asymmetries on behalf of the financial service provider. For instance, half the regulated rural financial institutions surveyed in Latin America require agriculture clients to have formal sales contracts, and 39 per cent request clients be part of a value chain (Wenner et al., 2007). The overall cost of lending is reduced in this way, since screening and monitoring functions are partly absorbed by buyers and intermediaries, who themselves have vested interest in the farmer's success.

The incentive to provide direct finance lies in the assurance that a certain quality and quantity of product will be delivered, and that the loan will be repaid. Value-chain lenders can threaten sanction in case of default (the buyer no longer sources from that farmer), but face the same information asymmetries as financial institutions. However, they can screen, monitor, and enforce contracts more effectively with the information gained from the value-chain relationship, and the costs can easily be absorbed into value-chain interactions (Johnston and Meyer, 2008). For example, 99 per cent of farmers surveyed in Madagascar report that the buying firm knows the exact location of their plot, and 92 per cent report the firm knows the number of plants on the plot (Minten et al., 2009). In this case, a meticulously kept database is used to follow contracted farmers. Similarly, the DrumNet platform in Kenya (see Box 8.5) provides channels of communication, making monitoring functions and interactions more efficient.

Governance structure within the chain will have a bearing on the effectiveness and sustainability of VCF, as it directly relates to monitoring and enforcement of contracts. Directed governance structures are found to be most conducive to direct VCF (Johnston and Meyer, 2008). These involve a

BOX 8.5 LEVERAGING INFORMATION TECHNOLOGY TO FACILITATE VCF: DRUMNET MOBILE PLATFORM

The DrumNet IT platform connects farmers to buyer firms, suppliers, and financial institutions using the Internet, mobile phone networks, and other wireless devices. It plays a networking role and applies a value chain framework to make transactions and interactions more efficient. Once a group of farmers receives a fixed-price contract from a buyer, they can request credit from the bank to finance the purchase of inputs. Funds are transferred directly to the supplier so that the farmer receives the credit in kind. Likewise, the buyer's payment is transferred to the bank, the farmer receiving the excess of any amount owed to the bank.

Beyond its facilitating role, the platform improves access to finance by assuring banks of farmer income, of credit going towards input purchase, and by lowering default risk by channelling buyers' payments. Despite the strength of the IT platform, Drumnet relies on the commitment and compliance of the actors involved. Cases of farmer side-selling, buyers failing to honour contracts, banks charging additional fees, and simply poor crop yields have posed challenges to the arrangement. As Drumnet commercializes its operations in Kenya, it will respond by incorporating crop insurance, soil analysis, improved payment systems, and a method of tracking users' performance.

Source: Campaigne and Rausch (2010).

lead buying firm (acting as a monopsony) which holds significant control over the quality and quantity of output, the price at which goods are traded, and limits the negotiating power of buyers—in other words, conditions characteristic of contract farming. In situations with competing buyers, farmers have more bargaining power and the direct governance structure weakens. For example, Kakira Sugar Works Ltd in Uganda operates an estate (directed governance) model of value chain, including financing, but chose to take advantage of low market prices and purchased sugar cane from non-contracted farmers. Contracted farmers, in need of liquidity and distrustful of the contract, side-sold to competing buyers and defaulted on their loans to Kakira (Johnston and Meyer, 2008). Side-selling has been avoided in other circumstances by setting a contract price sufficiently higher than the market price (Minten et al., 2009).

Market-based structures trading undifferentiated commodities are deemed least conducive to VCF (Fries and Akin, 2004; Johnston and Meyer, 2008) and value chain development in general. Numerous buyers and sellers, and high price variability leave little incentive to draw up contracts. Farmers faced with upside price risk can be tempted to side-sell, as would a buyer faced with falling market prices, and neither is conducive to an ongoing business relationship.

This raises questions of the extent to which VCF can expand access to finance for smallholder farmers. The vast majority of them produce food staples for local markets, yet value chains are concentrated in higher-value

BOX 8.6 WAREHOUSE RECEIPT FINANCING

Warehouse receipt schemes use securely stored goods as collateral for financing. Smallholder farmers, traders, and others can deposit commodities of a particular quantity, quality, and grade in a secure warehouse in exchange for a deposit certificate. The certificate supports a loan request from a financial institution. Warehouses can be classified into three different types, (i) private, (ii) terminal, and (iii) field warehouses. The latter two are conducive to financing.

South Africa's Electronic Silo Certificates system (ESC) handles most of the grains produced in South Africa, up to 5 million tonnes. ESC serves as a basis for exchange delivery systems, for warehouse receipt finance, electronic warehouse receipts trading, and information supply. Certified elevators/warehouses receive deposits and input the information into an electronic database, which interfaces directly with banks, facilitates trading of warehouse receipts, and supplies information to users and regulators. This system is instant, secure, and paper-free, as it relies on the Internet. National Bulk Handling Corporation (NBHC) has become, since mid-2005, India's largest support company for warehouse receipt financing. It operates through 3,500 public and field warehouses and, at the peak season, organizes up to US$2 billion worth of finance for agricultural commodities on behalf of agents for over twenty banks. The efficiency of the system is such that an individual farmer can arrive at a warehouse with a bag of products on the back of his bicycle, have them tested and accepted, fill out loan forms (with assistance), and by the next day have a loan arranged and money available.

Source: Rutten et al. (2011).

products for local and export markets (Doran et al., 2009; Fries and Akin, 2004). As these farmers become more integrated with markets, different approaches are needed to ensure their access to finance. Furthermore, strategies such as Rabo Development focus on the 'missing middle', defined as small cash-crop farmers with low marketable surplus who are too large for microfinance and remain under-served by commercial banks (van Empel, 2010). This strategy notably excludes a layer of farmers defined as 'subsistence', for which support in different forms is needed to lift them out of poverty. Warehouse receipts are suggested as a more viable option for commodity producers (Johnston and Meyer, 2008), and Box 8.6 illustrates the potential for efficient warehouse receipt financing systems. It requires the public sector to devise an appropriate institutional framework, legislation to recognize a receipt as legal tender, licensing and inspection of warehouses, performance oversight, and collaboration with the private sector to establish commodity quality standards (Giovannucci et al., 2000). With tighter regulation and a better business environment, VCF for small commodity producers is more feasible.

The willingness to provide direct finance is ambiguous; lead buying firms are not in the business of offering financial services and prefer to do it as needed (Miller et al., 2011; Nyoro, 2007). They have a limited capacity for carrying debt themselves; financing smallholders reduces their overall liquidity and creates an added cost of doing business (Gonzalez-Vega et al., 2006; Johnston and Meyer, 2008). Illustrating this point is the case of a sunflower oil outgrower scheme led by Mukwano Industries in Uganda. The firm chooses not to provide financing to smallholders under contract, despite having the necessary relationships and governance structure. It claims to have enough producers willing and able to self-finance their inputs, that side-selling is too large a risk, and that the interest rates necessary to recover the costs of lending would be perceived as exploitative and attract negative goodwill (Johnston and Meyer, 2008). In contrast, where value-chain buyers are willing to lend, and since lending is not their core business, they could justify taking on more lending risk than financial institutions providing sufficient profit and return on investment followed (Johnston and Meyer, 2008).

VCF leverages the business relationships within a value chain in order to improve the availability of finance to the various actors. But it is sustainable in situations where governance structures and production characteristics permit. The onus is on the chain actors to ensure the sustainability of VCF, whether the financing flows directly from actors within the chain or is accessed from outside sources. The focus for governments and donors is to encourage a sound business environment, including well-functioning enforcement institutions to assist smallholders with credit contract negotiations and dispute settlements.

3.2 INDEX-BASED WEATHER INSURANCE

Beyond the risks associated with prices and linking up to sustainable value-chain processes, smallholder farmers' production and livelihoods also face considerable risks associated with weather variations. Climate change[13] poses an additional challenge, as smallholders face more frequent and severe extreme weather events, and are limited in their capacity to adapt to gradual ecosystem changes. Smallholders have persisted in using traditional risk management strategies but these often perpetuate poverty[14], and are ineffective when covariate risks affect whole communities and regions.

[13] Africa is most vulnerable with the least adaptive capacity. Multiple stresses can lead to 50 per cent lower rain-fed agriculture yields by 2020. Agricultural land, length of growing season, and yield potential are all expected to decrease. Crop yields are projected to rise by 20 per cent in East and South-east Asia, but fall by 30 per cent in Central and South Asia. With rapid population growth, net impact in Asia will be adverse. Latin America will experience depletion of tropical rainforest, biodiversity loss, changing precipitation patterns and melting glaciers. Adaptation efforts are underway, but suffer from a lack of information and poor quality (IPCC, 2007).

[14] See Collins et al. (2009) for a detailed discussion of ex-ante and ex-post strategies.

Weather index insurance (WII) has emerged as a micro-insurance product for farmers (including smallholders) which, by design, improves on traditional products meant to protect against income losses. Major challenges for insurance providers are the high transaction and information costs of covering geographically dispersed clients, and covariate risk among borrowing smallholders (Carter, 2008). WII improves on traditional crop insurance by basing policies and payouts on data from weather stations; the transaction costs of assessing individual contracts and evaluating claims are thus significantly reduced. Farmers can neither influence that data nor does it matter if they are more prone to loss than others; this relieves moral hazard and adverse selection which plague traditional insurance. Furthermore, covariate risk (i.e. weather affecting all producers in a region) can be eased and diversified to international markets via reinsurance. These advantages prompted great interest among governments, donors, and researchers in developing WII. In terms of improving access to finance for smallholders, there is the potential for greater flow of credit, better household risk management and enhanced willingness to invest in risky inputs and technology, and healthier interaction between smallholders and growing rural financial markets (Hazell et al., 2010).

WII policies employ an index of weather variables that is correlated with crop yields so that indemnities are triggered when actual weather deviates from the index's pre-specified pattern. Given a strong correlation between indexed weather variables and yields, a smallholder farmer can use this product to hedge against actual production (and income) losses incurred by weather distress. Conversely, a mismatch between weather fluctuations projected by the index and actual yields will create basis risk, defined as the risk of losses that are not covered by index-triggered payouts (Hazell et al., 2010). Basis risk manifests itself in three ways (Skees et al., 2007): as differences between physical places, such as with microclimates causing a farm to experience different weather from the local weather station; across time if weather variables occur as anticipated, but earlier or later than is optimal for the insured crops and thus not covered; and finally, as losses that are unrelated to index variables, such as price fluctuations, poor health, damage to assets, crop disease, etc.

WII is categorized along technical lines into 'insurance for disaster relief' and 'insurance for development' (Hazell et al., 2010). Both are intended to reach smallholder farmers at the micro-level by protecting assets and smoothing consumption, yet disaster relief insurance is a macro-level product for governments and donors to provide safety nets more effectively. Insurance for development is a micro-level product targeted to smallholders and, by improving household risk management, can contribute to expanding access to finance.

WII is still in its initial phase and limited in geographical coverage, and empirical evidence has yet to show conclusively the impact on smallholder agriculture development, production, and livelihoods.[15] Nonetheless, a decade's worth of experience and lessons can be drawn from over 36 projects, at least 30 of which are development-oriented schemes.[16] These lessons shape the following discussion.

There needs to be effective demand for insurance markets to develop, yet take-up among smallholder farmers is persistently low (Hazell et al., 2010). The main cause tends to be the prices of insurance contracts, which are deemed too high, but the problem runs deeper to liquidity constraints. Demands for a household's limited cash resources are highest at the pre-crop-cycle moment when insurance should be purchased. To illustrate, a randomized experiment in India transferred cash to smallholders and found take-up increased by 150 per cent of the baseline rate, an effect several times larger as compared to reducing the price of the policy (Cole et al., 2010). Similarly, studies have found take-up to be highest among the wealthiest and most educated, albeit generally at low rates, in Ethiopia, Malawi, and India (Vargas Hill et al., 2011; Clarke and Dercon, 2009). As a rule of thumb, if index-triggering weather is expected to occur more than once every seven years, the pure risk will be too costly for smallholders to insure (Hazell et al., 2010). Basis risk can drastically affect demand; the value proposition of WII is compromised if farmers experience (or anticipate) loss not covered by the policy. Also reducing demand for WII is a lack of trust in insurance providers and lack of awareness about the product (Cole et al., 2010; Vargas Hill et al., 2011; Gine and Yang, 2009).

Transaction costs notwithstanding, flexibility among insurers (or intermediaries) to tailor products can stimulate demand for insurance among smallholder farmers. Given the choice, clients in India opted for less coverage, as it incurred a smaller premium (Skees, 2008). Likewise in Mongolia, herders were offered between 30 and 100 per cent coverage and the majority opted for the lower amount.[17] Lessons from input technology adoption in Kenya suggest that the timing of marketing could affect insurance take-up, the best time being the period after harvest when household has the most liquidity (Duflo et al., 2010). Finally, an Oxfam–WFP–SwissRe–USAID partnership is

[15] A recent study of the Agroasemex programme in Mexico finds that insuring weather risk leads to a 6 per cent increase in maize yields, and an 8 per cent increase in household income. Greater investment led to higher productivity, and spillover effects included less land dedicated to the insured crop (maize) and freed-up land diversified into cash crops. Productivity for non-insured crops improved, suggesting better access to finance. The results can be interpreted insofar as Agroasemex is 'disaster relief' insurance where farmers are automatically enrolled and the government pays the premium (Fuchs and Wolff, 2011a).

[16] Table 3 in Hazell et al. (2010) lists pilot projects by location, weather variable, and salient characteristics.

[17] The paper by Adesina et al. (Chapter 9 of this book) describes 'small packs' of fertilizers and seeds as a marketing innovation that is leading to increased take-up of inputs among cash-constrained smallholders in Kenya.

scaling up a programme in Ethiopia and Senegal which offers WII to farmers in exchange for labour (WFP, 2011a). Insurance-for-work reduces the liquidity constraint to demand and, by encouraging take-up, helps build awareness of the product and insurers.

Improving access to credit by insuring away weather risk could encourage farmers to take up WII, and lead formal lenders to increase their share of portfolio lending to agriculture. Thus far, the scale and maturity of pilot projects have yet to show the impact of WII on credit flows to smallholders, and inference from case studies is indeterminate. A World Bank-led livestock insurance scheme in Mongolia did not bundle credit with policies, yet herders were found to have greater access to credit at a lower interest rate by virtue of decreased exposure to livestock mortality risk (Skees et al., 2007). PepsiCo in India has succeeded (at scale) in bundling WII with credit, extension, and inputs in its contract farming arrangement, thus improving productivity among potato farmers (Hazell et al., 2010) (see Box 8.7). In contrast is a World Bank initiative in Malawi, established for the specific purpose of

BOX 8.7 PEPSICO CONTRACT FARMING IN INDIA: BUNDLING WEATHER INDEX INSURANCE WITH INPUTS

PepsiCo uses contract farming to secure a supply of potatoes for its FritoLay potato chip business. As part of the arrangement, contracted farmers receive a direct link to markets and a comprehensive package of support: high-quality seed, fertilizers and pesticides, technical support, credit for input purchase, and weather index insurance. WII is intended to manage risk to the supply chain and help foster long-term relationships with farmers. PepsiCo sets a base price at the beginning of the season with price incentives for quality, use of fertilizer and pesticides, and index insurance purchase. By 2008, 10,000 farmers were contracted, and the aim for 2009 was to increase it to 12,000–15,000. The volume of potatoes sourced grew from 2,920 tonnes in 2002 to 57,000 tonnes in 2007.

Weather Risk Management Services (WRMS) manages weather stations and brokers the WII product to PepsiCo's contracted farmers, in coordination with ICICI Lombard General Insurance Company. The WII protects against losses from late blight disease in potato crops, and is triggered by consecutive days of relative humidity over 90 per cent, average temperature between 10–20°C, and a frost index when temperatures fall below 1–2°C. The premium costs 3–5 per cent of the sum insured, and the index is constructed to pay out when losses exceed 40 per cent of yield. Farmers have incentive to undertake better farming practices since they are exposed to minor losses. Take-up rates of 95 per cent are attributed to trust in the insurer, the price incentive with insurance purchase, and timely pay-outs in prior periods. Farmers in the programme can obtain loans through an agreement between PepsiCo and the State Bank of India (SBI), in which farmers pay 7 per cent interest rate for loans covering most of their costs per acre. PepsiCo directly finances seeds by supplying them to farmers on credit. The potato contract farming programme is being expanded to other regions in India and, as of 2011, PepsiCo has partnered with USAID and WFP to extend its model to Ethiopia for chick-pea cultivation.

Source: Hazell et al. (2010) and PepsiCo Inc. (2011).

improving provision of credit to groundnut and maize farmers (Skees et al., 2007). While access to credit was improved in the first year, the project met design flaws; gaps in delivery channels led farmers to sell their output outside the contractual arrangement in order to avoid repaying loans (Hazell et al., 2010). Gine and Yang (2009) found farmers less willing to take on credit if bundled with a WII policy; the inability of the formal lender to enforce a loan contract implicitly limits the liability of the smallholder to repay, so WII became an added cost of borrowing.

The way insurance is delivered and marketed to smallholders will have a bearing on the costs and take-up rates. Experience suggests that using established institutional channels to deliver WII to smallholders is more cost-effective and efficient than investing in a new channel (Hazell et al., 2010). Channels are typically producers' groups or organizations that aggregate farmers, or trusted financial institutions with established rural networks. Using a 'meso-level' intermediary, rather than marketing directly to smallholders, can achieve lower administration costs and greater reach, and eventually bring WII to scale (Skees et al., 2007; Hazell et al., 2010a). It also enhances the possibilities for bundling insurance with credit, inputs, and extension, thus improving the value proposition for smallholders reluctant to take-up WII (Hazell et al., 2010). For example, Nyala Insurance Co.'s (Ethiopia) partnership with the Lumme–Adama Farmer's Cooperative Union as an intermediary succeeded not only to market WII, but to provide consistent educational support and awareness-building to farmers (Araya, 2011). BASIX in India has benefited from its extensive rural network in very poor areas, the reputation and trust it has garnered, and its efficient business model (Skees et al., 2007). It can be noted, however, that the BASIX scheme has encountered challenges, since it does not bundle insurance with credit (Hazell et al., 2010).

There are limits and potential costs to delivering insurance through 'aggregators', and a need for legal and regulatory support. For instance, the Nyala–Lumme partnership found that purchasing a policy on behalf of its member farmers was problematic if the producers' organization has limited financial capacity and farmers cannot pay their portion up front (Araya, 2011). Furthermore, aggregating organizations may suffer a collective action problem; claims paid out to members of a producers' organization may be used inefficiently, or a portion may be captured before redistribution to farmers (de Janvry et al., 2011). An alternative is to target WII to MFIs as a means to defray their portfolio exposure to weather risk. In this way, greater and more efficient rural financial services can emerge (from institutions with less risk exposure) and the next step of reaching out to smallholders will be easier (Skees et al., 2007). Conversely, not reaching smallholders directly would prevent farm households from using WII as a risk management tool (Hazell et al., 2010). Careful assessment of the context is needed to determine for whom WII is most suited.

WII is an important consideration for climate change, but is not an adaptation tool because it does not address the underlying impacts on agriculture yield. Investments are possible that would directly improve the resilience of agriculture to weather factors: developing improved seed varieties that are resistant to drought or temperature; improving irrigation infrastructure and natural resource management; extension services to educate farmers in new farm systems; and even developing an exit strategy in regions where climate stresses overwhelm agricultural production (Collier et al., 2009). Complicating the matter is the possibility that a financial solution to climate change removes incentive to invest in adaptation technologies. As an example, the AGROASEMEX insurance scheme in Mexico insures only rain-fed agriculture, which may diminish the incentive to invest in irrigation (Fuchs and Wolff, 2011b). Nonetheless, WII can be a tool to protect livelihoods against disruptions caused by climate change, but requires continually adapting rates and coverage to reflect changing weather histories and indices. Collier et al. (2009) point out that the price of insurance (if competitive and actuarially fair) signals the magnitude of the risk, which, in turn, informs smallholders' production decisions and livelihood strategies. Likewise, the incremental cost of insurance adjusted for changing weather data can signal the cost of climate change to smallholders. Subsidizing the increment may be a method for governments to support climate change adaptation efforts specifically (Hazell and Hess, 2010).

There remains potential for WII to live up to its claims, but the case evidence suggests that it is only a viable product in the right context. In a lesson drawn from WFP experience in Ethiopia, it is difficult to justify introducing a weather-index insurance product in regions where non-weather-related risks are dominant (Hazell et al., 2010). Thorough regional-level risk assessments are necessary to determine if and where there is a business case for WII (Collier et al., 2009). Much work is being done in the field of micro-insurance more generally, and, while this chapter considers WII, an important way forward is in area-based insurance. In essence, the index can comprise any regional agriculture-production related variables such as average yield or temperature. For smallholder farmers, the importance of insurance lies in its ability to unlock access to inputs and credit (Hazell and Hess, 2010).

Governments and donors will play an important role if WII is taken on, by carrying out assessments, bearing the public good costs of weather station construction and maintenance, adapting regulation and insurance laws, providing reinsurance, and building awareness and capacity (Hazell and Hess, 2010). Providing this support is an alternative to subsidizing the price of insurance, which may be necessary for the poorest smallholders to be reached, but is approached with caution based on past experience (Hazell et al., 2010; Collier et al., 2009). The private sector should be involved early and implicated deeply in WII schemes, since preserving the market-orientation underlies the capacity for viable financial market development, access to finance for smallholders, and hence improvement in investment, productivity, and livelihoods.

4. Public sector financing of small farmers

A market-oriented approach to financial-sector development necessarily involves the private sector, to collaborate and even lead. However, since smallholder farmers have been traditionally marginalized and starved of resources, and remain under-served by formal finance, they are at a disadvantage; they are often not deemed a profitable proposition for the private sector. There is thus a distinct need for continued public-sector support. After decades of being pushed to the wayside, agriculture is once again at the forefront of the development agenda. Official Development Assistance (ODA) to agriculture is seeing a slight rebound,[18] and dialogue among multilateral and bilateral donors is creating new partnerships and commitments. Governments are seeing the value of committing greater proportions of their budget spending to agriculture.[19] Increased awareness and new investments need to follow carefully designed policies which support small-scale agriculture, and provide the well-functioning social safety nets needed in a sustainable market-based food security system (Christiaensen, 2009).

There is also concern that a market-oriented approach to finance, and smallholder agricultural development more broadly, will only reach smallholders to the extent that there is a business case. Farmers at the subsistence end of the commercialization continuum, and likely out of the scope of the private sector, will continue needing assistance from governments to raise their ability to connect with value chains, access resources, and be protected from the shocks and uncertainty that can devastate their livelihoods.

Having better access to financial services is not a goal in itself; rather, the accumulation of capital and adoption of technology can raise productivity and improve livelihoods over the long run. A long literature describes the risk-averse farmer unwilling to adopt new inputs or technologies because of uncertainty regarding the return on investment (Fafchamps, 2010). In this

[18] Total ODA to agriculture amounted to US$7.2 billion in 2007–8. Aid to agriculture has decreased by 43 per cent since the mid-1980s, but recent data indicate a slowdown in the decline and an upward trend. The share of agriculture in total bilateral ODA had also fallen since the late 1980s from 17 per cent and has stabilized at 6 per cent from 2003–4 to the most recent data (OECD, 2010).

[19] For example, CAADP in Africa aims to raise spending on agriculture to 10 per cent of GDP, as compared to countries in Asia where 8.5–11 per cent is spent, with upwards of 15 per cent during the Green Revolution (Fan et al., 2008).

sense, developing rural financial markets is key so that smallholders can insure away risk, accumulate savings and access credit to invest and smooth consumption (Carter, 2008; Fafchamps, 2010). Markets for agricultural output also need to be strengthened; without them, incentive to invest is weak, adoption of technology is low, and poverty persistent. Fafchamps (2010) argues that farmers are not risk-averse but rather loss-averse. This suggests that transactions/products that protect farmers from downside risk (i.e. weather shock, falling output prices) but preserve upside benefits (i.e. higher output prices) can create incentives for smallholders to invest. Intensification of agriculture for both food security and poverty reduction goals depends on this investment, from both private and public sources. FAO estimates that US $83 billion of net capital investment will be needed annually simply to meet food demand to 2050 (Schmidhuber et al., 2009). The Global Harvest Initiative (2011) puts this amount at US$90 billion. Long-term capital needed for investment is a notable gap in the financial products available to smallholder farmers.

The question remains how governments can provide support to smallholder farmers in achieving their potential. The strategy for rural finance put forward by the World Bank (2006) is based on three pillars: i) government policies and a legal, regulatory, and supervisory framework; ii) financial-sector and real-sector infrastructure; and iii) financial institutions. This echoes the approach of organizations including FAO, GTZ, and IFAD in advocating a financial systems approach of well-managed and sustainable financial intermediaries to ensure a long-term supply of financial services to rural areas (Meyer, 2011).

Yet, less than one in five African households has an account in a financial institution (Beck et al., 2009), and evidence shows MFI outreach to the 'poorest half of the poor' to be low and diminishing as they grow and mature as institutions (ADB, 2010). The shortcomings of a market-oriented approach in engaging the poorest end of the continuum beg the question, can subsidies[20] be used in a more strategic way to bridge this gap? For example, subsidization of WII risks distorting emerging insurance markets, but it is becoming apparent that take-up will remain low without subsidies in at least the initial stages (Skees et al., 2007; Hazell et al., 2010).

The flaws and benefits of subsidies lie in their design. A rule of thumb is that subsidies should be time-bound, capped, and decreasing over time (World Bank, 2006). These are the characteristics of 'smart subsidies',

[20] The directed-credit approach depended heavily on subsidies and credit lines, which distorted market forces and stunted rural development. Eliminating subsidies became an objective of the new paradigm for rural finance. Consultative Group to Assist the Poor's key principles of microfinance warn against subsidies as scarce and in uncertain supply (CGAP, 2006). The World Bank's policies stipulate that subsidies used within the scope of its work cannot affect the interest rates paid by borrowers (Meyer, 2011).

intended to increase food security over the short run and achieve social and market development goals over the long run. While the concept of smart subsidies is well defined, how to administer them is less known (Meyer, 2011). A voucher-based fertilizer subsidy programme in Malawi used a public-service exchange to target less-well-off farmers and increase uptake of fertilizer (World Bank, 2008). Vouchers were redeemable from private vendors and entitled the farmer to pay a third of the retail price. The scale of the programme demanded significant budgetary costs, highlighting the trade-off between subsidies and other forms of spending. However, an evaluation of a similar programme in Ghana found the scheme suffered from poor initial planning and may not have been accessible to the poorest farmers (Brookings, 2010).

Donors can leverage their funds to stimulate private funding, in effect complementing subsidization with commercial capital (Morduch, 2006). AGRA and its partners use guarantee funds to create loan facilities with partner banks in Africa; for example, an agreement with Equity Bank in Kenya, with support from IFAD and the Kenyan Ministry of Agriculture, backed a US$50 million loan facility with a US$5 million cash guarantee fund. In this way, US$160 million of loanable funds have been made available in five African countries (AGRA, 2011). The guarantee scheme serves to address the underlying risk perceptions of lending to agriculture and stimulates further penetration of financial institutions into the agriculture sector. A private–public partnership between IFAD and the Government of Yemen, the Economic Opportunities Fund (EOF), serves not only to manage IFAD-financed activities, but provides a platform to attract additional financiers to development programmes in Yemen. Finally, donors can subsidize the start-up capital needed for establishment of rural financial institutions, both to encourage entrants into emerging financial sectors and to keep costs/rates at the market level without passing on start-up costs to clients (Morduch, 2006).

Recognizing the capital and time required for building institutions, there is renewed attention in reforming agricultural development banks and leveraging their already established network of rural banking services. There is a vast literature on the deficiencies of agriculture development banks in either promoting financial markets or substituting for them, particularly in Africa (see Chapter 9) and Latin America. While many agricultural development banks have been unwound or downsized, those that remain have seen a resurgence in recent years. Advocates cite already-established networks of retail outlets,[21] the void left in rural finance when banks unwind, and the

[21] Tapping existing networks is also an argument for expanding financing through postal networks, for example in an IFAD-Universal Postal Union initiative to streamline remittance flows. Lemon Bank in Brazil (now named Banco Bracce S.A.) rapidly expanded due to its innovative network of service points in gas stations, convenience stores, etc.

capacity for reform based on learning from past errors (Trivelli and Venero, 2007). Examples of successful reforms include The Agricultural Bank of Mongolia (later called XAAH), BAAC in Thailand, and BRI in Indonesia.[22]

A common thread across successful reforms of development banks is changes in ownership and governance structure. As part of their reform, agricultural development banks have sought funds from capital markets in order to limit the political influence that contributed to past failures. The success of XAAH led to its complete privatization in 2003 (Dyer et al., 2004). Tanzania's National Microfinance Bank was born of the former National Bank of Commerce to serve rural populations using micro-finance lending techniques, and its success also led to privatization (Meyer, 2011). The government-owned Bandesa in Guatemala was reformed to create Banrural S.A., a mixed-capital company with 70 per cent private and 30 per cent public ownership (Meyer, 2011). Initial success greatly improved outreach, but with average loan amounts rising from about US$900 in 1998 to US$3,400 in 2005, concern over outreach to the poorest smallholder farmers has been raised (Trivelli and Venero, 2007). The case of Banrural S.A. highlights the juxtaposition of institutional reform and ability for the public sector to support the rural financial sector and improve the access to finance for smallholder farmers.

Public spending is best directed to institutions and financial infrastructure, capacity building, and efforts to improve the efficiency and competitiveness of emerging financial sectors. Care must be taken not to intervene in areas already served by the private sector. Investment in public goods is greatly needed and, since they benefit the entire financial sector, can generate higher returns than subsidization of institutions (Meyer, 2011). For example, evidence from Malawi finds that a system of borrower identification (fingerprinting in this case[23]) increases loan repayment rates and contributes to expanding availability of credit, since lenders can use credit history information when evaluating loan requests (Gine et al., 2010). Financial-sector reform in Columbia had the government shift from intervener to facilitator, implementing 'Know-Thy-Banker' reforms, interest rate ceiling reforms, a regulatory framework for branchless banking, and administrative efficiency gains to facilitate establishing small accounts (AZMJ, 2011). Similar to the warehouse receipts systems described earlier, the Banking Correspondents system is made possible by the oversight and regulatory responsibility of the Reserve Bank of India (see Box 8.8).

[22] See case studies in Dyer et al. (2004) for XAAH and Meyer and Nagarajan (2001) for BAAC and BRI.

[23] Fingerprinting has been instrumental in relieving illiteracy constraints and fraud risk for savings account holders with PRODEM FFP in Bolivia, and underlies the Banking Correspondents system in India (see Box 8.8).

BOX 8.8 EXPANDING THE REGULATORY FRAMEWORK OF FINANCIAL SYSTEMS—BANKING CORRESPONDENTS IN INDIA

Bank Correspondents (BCs) are outsourced operators hired by banks in India to provide financial services to unbanked populations, particularly in rural areas where bank coverage has not reached. The State Bank of India (SBI) is one public sector bank which uses this system to offer smart-card-based, no frills savings accounts. BCs operating on behalf of SBI travel with a laptop computer, a wireless modem, and a fingerprinting terminal to open savings accounts, handle deposits, and issue withdrawals. As of 2009, SBI engages 33 BCs to reach over 27 million clients.

It is innovative not only in the use of technology, but also for its ability to overcome physical barriers to accessing finance, as well the barriers of illiteracy, financial illiteracy by capacity building, and economic access by reducing transaction costs for users. Nonetheless, viability of the system will depend on the business proposition for BCs, since in many cases their costs exceed their revenues. Other challenges include having non-bank employees handle cash, funding capacity building activities and outreach to SHGs, and managing fraud and misappropriation. Firm support for the scheme is provided by the Reserve Bank of India, and rapid scaling-up is testament to its potential for outreach.

Source: Access Development Services, 2009.

The Comprehensive Africa Agriculture Development Programme (CAADP) provides a guiding framework for financial-sector investment, for the countries which have signed on. Of the four programme Pillars,[24] Pillar 2 is dedicated to market access, including a key strategic focus on value-chain development and financial access. The programme sets out policy and regulatory actions for an enabling business environment that is sustainable and inclusive of smallholder farmers (see Box 8.9) (CAADP, 2009). However, government commitment to the programme is slow, and only eight countries have met the 10 per cent of public spending target set out by the Maputo Declaration and adopted by CAADP. Another nine are within five percentage points of the target; however twenty-eight countries are not within range (Omilola et al., 2010).

Food security infrastructure at the international level has reached maturity. As such, rather than creating new organizations to address the issues, new partnerships have arisen to raise funds and channel them to food security. Overall, there is a growing consensus for the need to leverage private-sector (and philanthropic) resources in order to achieve goals of poverty reduction and economic development. For example, refinancing as described in Box 8.10 illustrates not only the potential for public–private partnership, but a mechanism in which public investment stimulates private investment.

[24] See <http://www.nepad-caadp.net> for documents and description of the program and its pillars.

BOX 8.9 CAADP'S FRAMEWORK FOR VALUE-CHAIN DEVELOPMENT AND FINANCIAL SERVICES

1. **Levers for agricultural enterprise growth and value-chain development**
a. Targeted actions on country- and sector-specific determinants of agribusiness enterprise growth;
b. Priority interventions that raise the value accruing to actors at each level of the chain, focusing on:
i. Unit cost reduction;
ii Volume expansion;
iii. Product value addition.

2. **Financial services sector development for value-chain growth**
a. Development of adapted credit information systems;
b. Creation of new and strengthening of existing regulatory mechanisms for collateral enforcement;
c. Provision of training and other capacity-building mechanisms to improve risk perception and thus reduce the barriers to as well as cost of borrowing;
d. Mobilization of non-bank financial institutions (insurance, pension funds, microcredit systems) to ease access to long-term investment funds and expand financial services in rural areas;
e. Enhancement of agribusiness enterprises' capacity to improve business reporting and provide satisfactory documentation in support of loan applications;
f. Enhancement of banks' capacity to provide affordable, flexible, and innovative financial products that meet the varied needs of AREs.

3. **Sector policy and governance**
a. Stable and well-balanced exchange rate, monetary, and fiscal policy regimes;
b. Conducive bank regulations, business and trade taxation, regulation of business start-up and closing, legal compliance of business operations, employment laws, and contract enforcement.

Source: CAADP (2009).

The G20 Global Agriculture and Food Security Program (GASFP[25]), under its Private Sector Window, has recently launched a call for proposals from financial institutions and firms. This is a first effort of GASFP to address the difficulties smallholders face in accessing commercial finance, and currently involves CAD$48 million available to finance private investments and CAD$2 million for advisory services (GASFP, 2011). Among numerous private commitments to rural finance, of note is the 2010 announcement

[25] GAFSP is a multilateral mechanism assisting the implementation of US$22 billion in pledges made by the G8 and G20 in 2009 to strengthen food security in low-income countries (AFSI, 2009). Donor funding is channelled to public and private initiatives developed by client countries to improve productivity, competitiveness, and governance in their agribusiness sectors. The Private Sector Window is initially funded by Canada and administered by the International Finance Corporation.

BOX 8.10 PUBLIC–PRIVATE PARTNERSHIPS FOR FINANCE: LEVERAGING PUBLIC FUNDS TO STIMULATE PRIVATE SECTOR

Refinancing has been used in IFAD-supported projects to support financial sector development in the transition economies of Moldova, Macedonia, and Armenia. A semi-independent refinancing unit is established in a relevant government ministry and funded by the government who receives a highly concessional IFAD loan. The third actors in the arrangement are the participating financial institutions, from whom rural clients apply for loans. Once a bank receives and approves a loan application, it makes a formal request to the refinancing facility. If accepted, the facility transfers funds to the bank to finance the rural client's loan. In contrast to lines of credit and on-lending to rural clients, refinancing capital is triggered once a loan has been applied for. Released capital is used specifically for that loan. Loans range from US$2,000 to US$150,000 and cater to all actors in supply chains. Financial institutions retain the credit risk, and pay back the refinancing loan in full. They have incentive to expand their loanable funds through refinancing, but incentive to learn how to do business with farmers and rural businesses. This initiative has stimulated lending from banks out of their own funds, and has also encouraged banks to explore 'softer' collateral requirements such as unregistered land, household assets, and business assets. Since inception, cumulative refinancing has amounted to US$22.6m in Moldova, €11.5m in Macedonia, and US$7.1 in Armenia, with recovery rates of 98.5 per cent, 96 per cent, and 98 per cent, respectively. The initiative has recently attracted co-financing: Macedonia has received loans of €21.2m from the World Bank, €20m from the European Investment Bank, and US$0.85m from SIDA; to Armenia the World Bank allocated US$6m and Millennium Challenge Corporation US$8.5m.

Source: IFAD (2008).

from the Bill and Melinda Gates Foundation of US$500 million in grants to inclusive finance projects, with a particular focus on developing savings products, in addition to US$530 million already committed as part of its 'Financial Services for the Poor' programme (Bill and Melinda Gates Foundation, 2010).

5. **Conclusion**

Advances in rural finance are promising and are yielding more and better access to finance for smallholder farmers. However, much work still remains to build comprehensive financial systems. In its myriad forms, informal finance has best overcome the information asymmetries and the transaction costs associated with serving smallholders in rural areas. Lessons can be drawn from this experience, and the way forward could build upon the linkage of informal, semi-formal, and formal finance. In any case, competitive rural financial markets are needed to intermediate and channel funds to the most productive investments. While focus has been predominantly on credit, savings and insurance products designed to meet the needs of smallholder farmers are recognized as equally important. Innovations in technology and product design are, in their own way, enhancing how financial services are being delivered to smallholder farmers.

Progress has lagged in expanding the range of formal financial services tailored to the needs of the agriculture sector, and smallholder farmers in particular. This chapter has focused on two innovative forms of financing with potential to expand access to finance sustainably. By design, weather index insurance has adapted insurance products for financial institutions serving smallholders and for smallholders themselves. However, scaling-up is limited to regions with particular climate characteristics and risk environments, and low take-up rates signal marketing difficulties. As for value-chain finance, underlying any sophisticated VCF model is a close analysis of the value-chain actors and their financing needs, with the side-effect of supporting the commercialization of small farms. Achieving a highly tailored product is, in this case, resource-intensive, not easily replicated, and ill-suited to the low value-added food commodities that the majority of small farms produce. Improving the regulatory and business environment will make VCF more feasible and more inclusive of the poorest.

The new paradigm has seen the public sector significantly roll back its financial-sector intervention, and this has left a void that markets arguably will not fill. Smallholder farmers at different points along the notional continuum from subsistence to commercialization have different needs and use financial services in different ways. The challenge for governments is to combine an often fragmented set of services into an efficient, private sector-led financial sector. Equally important is fostering the conditions where financial services are demanded, for example, enabling rural market development, and advancing agricultural technologies. Smallholders need to see the business prospect in farming; this will stimulate the risk-taking necessary for entrepreneurship, the willingness to engage on a commercial level with a rapidly changing food system, and the incentive to take on irreversible investments. Financing in itself will not provide an exit out of poverty, but rather it is a mechanism for enabling productivity-enhancing investment in order to generate higher income.

REFERENCES

Access Development Services (2009). 'Business correspondents and facilitators: Pathway to financial inclusion?' Proceedings of the retreat, College of Agricultural Banking, Access Development Services and CGAP, New Delhi.

Adams, D. W. (1992). 'Taking a fresh look at informal finance', in Adams, D. W. and D. A. Fitchett (eds), *Informal Finance in Low-Income Countries*. Boulder, CO: Westview Press.

ADB (2010). 'Making microfinance work: Evidence from evaluations', ECG Paper 2, Evaluation Cooperation Group, Independent Evaluation Department, Asian Development Bank, Manila.

AFSI (2009). 'L'Aquila', 'Joint Statement on Global Food Security'. L'Aquila: L'Aquila Food Security Initiative.

AGRA (2011). 'AGRA in 2011, Investing in Sustainable Growth; A Five-Year Status Report', AGRA, Nairobi.

AgriFin (2010). *Program Strategy. Agriculture Finance Support Facility.* Washington, DC: The World Bank.

Alamgir, D. (2009). *Microfinance in SAARC Region: Review of Microfinance Sector of Bangladesh.* Dhaka: Institute of Microfinance.

Araya, N. S. A. (2011). 'Weather insurance for farmers: Experience from Ethiopia', in *IFAD Conference on New Directions for Smallholder Agriculture: Proceedings of the Conference.* Rome: IFAD.

Armendariz-de Aghion, B. and J. Morduch (2005). *The Economics of Microfinance.* Cambridge, Ma.: MIT Press.

Ayyagari, M., A. Demirguc-Kunt, and V. Maksimovic (2008). 'Formal versus informal finance: evidence from China', Policy Research Working Paper 4465, Development Research Group, Finance and Private Sector Team. Washington, DC: The World Bank.

AZMJ (2011). *Cracking the Nut: Overcoming Obstacles to Rural & Agricultural Finance.* Falls Church, VA: AZMJ.

Banerjee, A., E. Duflo, R. Glennerster, and C. Kinnan (2010). 'The miracle of microfinance? Evidence from a randomized evaluation', BREAD Working Paper No. 278, Bureau for Research and Economic Analysis of Development, Washington DC.

Basu, K. (1995). 'Rural credit and interlinkage: Implications for rural poverty, agrarian efficiency, and public policy', in Quibria, M. G. (ed.), *Critical Issues in Asian Development. Theories, Experiences and Policies.* Oxford: Oxford University Press.

Beck, T., A. Demirguc-Kunt, and P. Honohan (2009). 'Access to financial services: Measurement, impact, and policies', *The World Bank Research Observer*, 24(1).

Bell, C. (1990). 'Interactions between institutional and informal credit agencies in Rural India', *World Bank Economic Review*, 4(3), pp. 297–327.

Bill and Melinda Gates Foundation. 2010. 'Melinda Gates challenges global leaders: Create savings accounts and bring financial security to the world's poorest', Press release, November 16. Retrieved from: <http://www.gatesfoundation.org/press-releases/Pages/melinda-gates-at-global-savings-forum-101116.aspx>.

Brookings Institution (2010). '"Market-Smart" subsidies? Analyzing the agricultural input dealer sector in Ghana', presentation at the Africa Growth Initiative, The Brookings Institution, Washington, DC.

Brune, L., X. Giné, J. Goldberg, and D. Yang (2011). 'Commitments to save: A field experiment in rural Malawi', Policy Research Working Paper 5748, Development Research Group, Finance and Private Sector Development Team, The World Bank, Washington, DC.

Byerlee, D. and A. de Janvry (2009). 'Smallholders unite', letters to the editor, *Foreign Affairs*, April.

CAADP (2009). *Framework for the Improvement of Rural Infrastructure and Trade-Related Capacities for Market Access (FIMA)*, Strategic area C: Value-Chain Development and Access to Financial Services, the African Union and NEPAD.

Campaigne, J. and T. Rausch (2010). 'Bundling development services with agricultural finance: The experience of DrumNet', Focus 18, Brief 14, 2020 Vision Research Briefs, International Food Policy Research Institute and The World Bank, Washington, DC.

Carter, M. (2008). 'Inducing innovation: Risk instruments for solving the conundrum of rural finance', Keynote Paper Prepared for the 6th Annual Conference of the Agence Française de Développement and The European Development Network, 12 November, Paris.

CGAP (2006). *Good practice Guidelines for funders of Microfinance: Microfinance Consensus Guidelines.* Washington, DC: Consultative Group to Assist the Poor.

CGAP (2012). *Current trends in cross border funding.* Retrieved from: <http://www.cgap.org/publications/current-trends-cross-border-funding-microfinance> (21 July 2013).

Chaia, A., A. Dalal, T. Goland, Maria Jose Gonzalez, Jonathan Morduch, and Robert Schiff (2009). 'Half the world is unbanked', Framing Note, Financial Access Initiative and McKinsey & Company, New York.

Chakrabarti, R. and S. Ravi (2011). 'At the crossroads: Microfinance in India', *Money & Finance*, forthcoming.

Christen, R. P. and D. Pearce (2005). 'Managing risks and designing products for agricultural microfinance: Features of an emerging model', Occasional Paper No. 11, CGAP.

Christiaensen, L. (2009). 'Revisiting the Global Food Architecture: Lessons from the 2008 Food Crisis'. Discussion Paper No. 2009/04, UNU-Wider, Helsinki.

Clarke, D. and S. Dercon (2009). 'Insurance, Credit and Safety Nets for the Poor in a World of Risk', DESA Working Paper No. 81, United Nations Department of Economic and Social Affairs, New York, NY.

Cole, S., Xavier Gine, J. Tobacman, P. Topalova, and R. Townsend (2010). 'Barriers to household risk management', Policy Research Working Paper 5504, Development Research Group, Finance and Private Sector Development Team, The World Bank, Washington, DC.

Collette, L., T. Hodgkin, A. Kassam, P. Kenmore, Leslie Lipper, Christian Nolte, Kostas Stamoulis, and Pasquale Steduto (2011). *Save and Grow: A policy makers guide to the sustainable intensification of smallholder crop production.* Rome: FAO.

Collier, B., J. Skees, and B. Barnett (2009). 'Weather Index Insurance and Climate Change: Opportunities and Challenges in Lower Income Countries', *The Geneva Papers*, 34, pp. 401–24.

Collins, D., J. Morduch, S. Rutherford, and O. Ruthven (2009). *Portfolios of the Poor.* Princeton, NJ: Princeton University Press.

Conning, J. and C. Udry (2005). 'Rural Financial Markets in Developing Countries', Chapter 56 in Evenson, R. E. and P. Pingali (eds), *The Handbook of Agricultural Economics, Volume 3, Agricultural Development: Farmers, Farm Production and Farm Markets.* Amsterdam: North-Holland.

Dethier, J.-J. and A. Effenberger (2011). 'Agriculture and development: A brief review of the literature', Policy Research Working Paper 5552, Development Economics Research Support Unit, The World Bank, Washington, DC.

Doran, A., N. McFayden, and R. C. Vogel (2009). *The Missing Middle in Agricultural Finance: Relieving the capital constraint on smallhold groups and other agricultural SMEs.* Oxford, UK: Oxfam International.

Downing, J. and M. Harper (2008). 'Crossfire: Microfinance has upstaged enterprise development, and finance is now in danger of doing the same to value chain interventions', *Enterprise Development & Microfinance*, 19(4).

Duflo, E., M. Kremer, and J. Robinson (2010). 'Nudging farmers to use fertilizer: Theory and experimental evidence from Kenya', *American Economic Review*, 101(6), pp. 2350–90.

Dupas, P., S. Green, A. Keats, and J. Robinson (2011). 'Challenges in banking the rural poor: Evidence from Kenya's Western Province', unpublished paper.

Dyer, J., P. Morrow, and R. Young (2004). 'The Agricultural Bank of Mongolia', Chapter 3, in *Scaling up Poverty Reduction: Case Studies in Microfinance*. Washington, DC: CGAP and World Bank.

Eastwood, R., M. Lipton, and A. Newell (2010). 'Farm Size', Chapter 65, in Evenson, R. E. and P. Pingali (eds), *The Handbook of Agricultural Economics, Volume 4*, pp. 3323–97. Amsterdam: North-Holland.

Fafchamps, M. (2010). 'Vulnerability, risk management, and agricultural development', *African Journal of Agricultural Economics*, 5(1), pp. 243–60.

Fan, S., M. Johnson, A. Saurkar, and T. Makombe (2008). 'Investing in African agriculture to halve poverty by 2015', IFPRI Discussion Paper 00751, Development Strategy and Governance Division, International Food Policy Research Institute, Washington, DC.

Fries, R. and B. Akin (2004). 'Value chains and their significance for addressing the rural finance challenge', microREPORT #20, USAID.

Fuchs, A. and H. Wolff (2011a). 'Concept and unintended consequences of Weather Index Insurance: The Case of Mexico', *American Journal of Agricultural Economics*, 93(2), pp. 505–11.

Fuchs, A. and H. Wolff (2011b). 'Drought and retribution: Evidence from a large scale Rainfall Index Insurance in Mexico', Unpublished Working Paper, version May 15.

GASFP (2011). *Call for Proposals*. Washington, DC: Global Agriculture and Food Security Program, Private Sector Window Secretariat.

Ghate, P. B. and A. Das-Gupta (1992). *Informal finance: some findings from Asia*, Published for the Asian Development Bank by Oxford University Press, 1992.

Giné, X., J. Goldberg, and D. Yang (2010). 'Identification strategy: A field experiment on dynamic incentives in rural credit markets', Policy Research Working Paper 5438, Development Research Group, Finance and Private Sector Development Team, The World Bank, Washington, DC.

Giné, X., and D. Yang (2009). 'Insurance, credit, and technology adoption: Field experimental evidence from Malawi', *Journal of Development Economics*, 89(1), pp. 1–11.

Giovannucci, D., P. Varangis, and D. F. Larson (2000). 'Warehouse Receipts: Facilitating credit and commodity markets', in *Guide to Developing Agricultural Markets and Agroenterprises*. Washington, DC: The World Bank.

Glisovic, J., M. El-Zoghbi, and S. Foster (2010). 'Advancing savings services: Resource guide for funders', (Draft for Public Review), Technical Guide, Consultative Group to Assist the Poor, Washington, DC.

Global Harvest Initiative (2011). 'Enhancing private sector involvement in agricultural and rural infrastructure development', Policy Issue Brief, June, Washington, DC.

Gonzalez-Vega, C., G. Chalmers, R. Quiros, and J. Rodriguez-Meza (2006). 'Hortifruit in Central America: A case study about the influence of supermarkets on the development and evolution of creditworthiness among small and medium agricultural producers', microREPORT #57, USAID.

Hazell, P. (2011). 'Five big questions about five hundred million small farms', in *IFAD Conference on New Directions for Smallholder Agriculture: Proceedings of the Conference*. Rome: IFAD.

Hazell, P. and U. Hess (2010). 'Drought insurance for agricultural development and food security in dryland areas', *Food Security*, 2, pp. 395–405.

Hazell, P., J. Anderson, N. Balzer, A. H. Clemmensen, U. Hess, and F. Rispoli (2010a). *Potential for scale and sustainability in weather index insurance for agriculture and rural livelihoods*. Rome: International Fund for Agricultural Development and World Food Programme.

IFAD (2008). *Refinancing facilities: IFAD introduces an innovation in rural finance development.* Rome: International Fund for Agriculture Development.

IFC (2011). 'Case study: Nib International Bank', in *Inclusive Business Models—Guide to the Inclusive Business Models in IFC's Portfolio.* Washington, DC: International Finance Corporation.

IPCC (2007). 'Summary for Policymakers', in Parry, M. L., O. F. Canziani, J. P. Palutikof, P. J. van der Linden, and C. E. Hanson (eds), *Climate Change 2007: Impacts, Adaptation and Vulnerability. Contribution of Working Group II to the Fourth Assessment Report of the Intergovernmental Panel on Climate Change.* Cambridge, UK: Cambridge University Press, pp. 7–22.

de Janvry, A., V. Dequiedt, and E. Sadoulet (2011). 'Group contracts for index-based insurance', Policy Brief 30, Fondation pour les Etudes et recherches sur le Développement International, Clermont-Ferrand, France.

Johnston, C. and R. L. Meyer (2008). 'Value chain governance and access to finance: Maize, sugar cane and sunflower oil in Uganda', *Enterprise Development & Microfinance,* 19(4), pp. 281–300.

Kendall, J. (2010). 'A penny saved: How do savings accounts help the poor?' *Focus Note,* Financial Access Initiative, New York.

Mahieux, T., O. Zafar, and M. Kherallah (2011). 'Financing smallholder farmers and rural entrepreneurs through IFAD-Supported operations in the Near East and North Africa', in *Conference on New Directions for Smallholder Agriculture: Proceedings of the Conference.* Rome: IFAD.

Mbiti, I. and D. N. Weil (2011). 'Mobile Banking: The impact of M-Pesa in Kenya', Working Paper 17129, NBER Working Paper Series, National Bureau for Economic Research, Cambridge, MA.

Mensink, M. and J. de la Rive (2011). 'How to support value chain finance in a smart way?' Policy Statement of the European Microfinance Platform, Rural Outreach & Innovation Action Group, e-MFP, Luxembourg.

Meyer, R. (2011). 'Subsidies as an instrument in agriculture finance: A review', Joint Discussion Paper, German Federal Ministry of Economic Cooperation and Development (Bmz), Food and Agriculture Organization of the United Nations (FAO), German Agency for International Cooperation (Giz), International Fund for Agriculture Development (IFAD), The World Bank, and United Nations Capital Development Fund (UNCDF).

Meyer, R. and G. Nagarajan (2000). *Rural Financial Markets in Asia: Policies, Paradigms, and Performance.* Oxford: Oxford University Press.

Meyer, R. and G. Nagarajan (2001). 'Development of rural financial markets in Asia', in *Agricultural Credit in Asia and the Pacific.* Tokyo: Asian Productivity Organization, 2001, pp. 75–89.

Meyer, R. and G. Nagarajan (2005). 'Rural finance: Recent advances and emerging lessons, debates, opportunities', Working Paper AEDE-WP-0041-05, Department of Agricultural, Environmental and Development Economics, The Ohio State University, Columbus.

Miller, C. (2008). 'A baker's dozen lessons of value chain financing in agriculture', *Enterprise Development and Microfinance,* 19(4), pp. 273–80.

Miller, C. and L. Jones (2010). *Agricultural value chain financing: Tools and Lessons.* Rome: FAO, and Rugby, UK: Practical Action Publishing.

Miller, C., A. Campion, M. D. Wenner, and A. Nair (2011). 'Lessons learned in agricultural value chain finance', Chapter 3, in *Summary of the conference, 'Agricultural Value Chain Finance'.* Rome: FAO.

Minten, B., L. Randrianarison, and J. F. M. Swinnen (2009). 'Global retail chains and poor farmers: Evidence from Madagascar', *World Development*, 37(11), pp. 1728–41.

Morduch, J. (1999). 'The microfinance promise', *Journal of Economic Literature*, 37, pp. 1569–614.

Morduch, J. (2006). 'Smart subsidy', *ESR review*, 8(1), pp. 10–17.

Morduch, J. and D. Roodman (2009). 'The Impact of microcredit on the poor in Bangladesh: Revisiting the Evidence', NYU Wagner Research Paper No. 2010–09, New York University, New York.

Nyoro, J. (2007). 'Financing agriculture: Historical perspective', presentation at the AFRACA Agribanks Forum.

OECD (2010). *Measuring Aid to Agriculture.* Paris: OECD-DAC.

Omilola, B., M. Yade, J. Karugia, and P. Chilonda (2010). 'Monitoring and assessing targets of the Comprehensive Africa Agriculture Development Programme (CAADP) and the First Millennium Development Goal (MDG) in Africa', Working Paper 31, Regional Strategic Analysis And Knowledge Support System (ReSAKSS), International Food Policy Research Institute, Washington, DC.

Pagura, M. and M. Kirsten (2006). 'Formal-informal financial linkages: Lessons from developing countries', *Small Enterprise Development*, Volume X, March.

Pandit, A., N. K. Pandey, R. K. Rana, and B. Lal (2009). 'An empirical study of gains from potato contract farming', *India Journal of Agricultural Economics*, 64(3).

Payne, J. and K. Kumar (2010). 'Using mobile money, mobile banking to enhance agriculture in Africa', Briefing Paper, United States Agency for International Development.

PepsiCo Inc. (2011). 'PepsiCo, World Food Programme and USAID partner to increase food production and address malnutrition in Ethiopia', Press Release, 21 September, New York.

Poulton, C., A. Dorward, and J. Kydd (2010). 'The future of small farms: New directions for services, Institutions, and Intermediation', *World Development*, 38(10), pp. 1413–28.

Rahman, A. (1992). The informal financial sector in Bangladesh: An appraisal of its role in development, *Development and Change*, Volume 23, Issue 1, pp. 147–68, January.

Reed, L. R. (2011). *State of the Microcredit Summit Campaign Report 2011.* Washington, DC: Microcredit Summit Campaign.

Ritchie, A. (2010). 'Community-based financial organizations: Access to finance for the poorest', Focus 18, Brief 3, 2020, Vision Research Briefs, International Food Policy Research Institute and The World Bank, Washington, DC.

Rutten, L., R. Galindo, E. Vargas, R. L. Alba, M. Osorio, F. Chilavert, and B. Obara (2011). 'Models of agricultural value-chain financing. Perspectives of chain members', Chapter 4, in *Summary of the conference, 'Agricultural Value Chain Finance'.* Rome: FAO.

Schmidhuber, J., J. Bruinsma, and G. Boedeker (2009). 'Capital requirements for agriculture in developing countries to 2050', paper presented at the Expert Meeting on How to Feed the World in 2050, Food and Agricultural Organization of the United Nations, Rome.

Shwedel, K. (2011). 'Agriculture value chain finance: Four years on', Chapter 2, in R. Quirós (ed) 2011. Agricultural Value Chain Finance, Procedings of the Conference, FAO and Academia de Centroamérica, 2011 (English edition), Jan Jose, Costa Rica.

Skees, J. R. (2008). 'Challenges for use of index-based weather insurance in lower income countries', *Agricultural Finance Review*, 68, pp. 197–217.

Skees, J., A. Murphy, B. Collier, M. J. McCord, and J. Roth (2007). 'Scaling up index insurance: What is needed for the next big step forward', paper prepared for Kreditanstalt für Wiederaufbau (KfW), Microfinance Centre, LLC and GlobalAgRisk, Inc.

Staatz, J. M. and N. Nango Dembele (2007). 'Agriculture for development in Sub-Saharan Africa', background paper for World Development Report, 2008.

Swinnen, J. F. M. and M. Maertens (2010). 'Finance through food and commodity value chains in a globalized economy', paper prepared for KfW Financial Sector Development Symposium 2010, *Finance for Food—Towards new Agricultural and Rural Finance*, Berlin.

Trivelli, C., and H. Venero (2007). 'Agricultural development banking: Lessons from Latin America', Instituto de Estudios Peruanos, Lima, Peru.

van Empel, G. (2010). 'Rural banking in Africa: The Rabobank Approach', Focus 18, Brief 4, 2020 Vision Research Briefs, International Food Policy Research Institute and The World Bank, Washington, DC.

Vargas Hill, R., J. Hoddinott, and N. Kumar (2011). 'Adoption of weather index insurance: Learning from willingness to pay among a panel of households in rural Ethiopia', IFRPI Discussion Paper 01088, International Food Policy Research Institute, Washington, DC.

Varghese, A. (2005). 'Bank–moneylender linkage as an alternative to bank competition in rural credit markets', *Oxford Economic Papers*, 57, pp. 315–55.

Ventura, A., G. Rung, T. Seck, and P. Singh (2011). *Beyond Payments: Next Generation Mobile Banking for the Masses.* PlaNet Finance and Oliver Wyman. Study commissioned by Bill and Melinda Gates Foundation. Retrieved from: <http://www.gsma.com/mobilefordevelopment/wp-content/uploads/2012/06/planetfinance54.pdf> (21 July 2013).

Wenner, M. and F. Proenza (2000). 'Rural Finance in Latin America and the Caribbean: Challenges and opportunities', Working Paper, Department of Sustainable Development, Inter-American Development Bank, Washington, DC.

Wenner, Mark, S. Navajas, C. Trivelli, and A. Tarazona (2007). *Managing credit risk in rural financial institutions in Latin America.* Washington, DC: Inter-American Development Bank.

WFP (2011). 'Purchase for Progress: Achievements', Rome: World Food Programme.

WFP (2011a). 'Scaling up innovative climate change adaptation and insurance solutions in Senegal', Rome: World Food Programme.

World Bank (2006). 'Meeting development challenges: Renewed approaches to rural finance', Agriculture and Rural Development Department, World Bank, Washington, DC.

World Bank (2007). *Agriculture for Development: World Development Report, 2008.* New York: Oxford University Press.

World Bank (2008). 'New approaches to input subsidies: Agriculture for development policy brief', in *Agriculture for Development: World Development Report, 2008.* New York: Oxford University Press.

9 Improving farmers' access to agricultural inputs and finance: approaches and lessons from sub-Saharan Africa

AKINWUMI A. ADESINA, AUGUSTINE LANGYINTUO, NIXON BUGO, KEHINDE MAKINDE, GEORGE BIGIRWA, AND JOHN WAKIUMU

1. Introduction

For over three decades, per capita food production has been declining in Africa, even though aggregate production increased through expansion of cultivated area. Agricultural productivity remains low (at 1 tonne per hectare, it is about a fourth of the global average) due to slow adoption of agricultural technologies (Alston et al., 2000). This is in sharp contrast to the experience of Asia, where food production increases came from rapid uptake of improved agricultural technologies such as high yielding wheat and rice varieties, the use of fertilizers, and irrigation. Public subsidies drove down the unit cost for production inputs, and raised land and labour productivities (Hazell and Ramasamy, 1991).[1]

Several factors contribute to the low adoption of agricultural technologies, including poorly functioning and ineffective extension systems, high cost of inputs, poorly developed input and output markets, and weak infrastructure, which raise transaction costs. With market liberalization reforms, state marketing agencies were privatized, disbanded, or reformed with much reduced roles in markets (Dorward et al., 1998). Although these reforms had positive impacts such as increased share of international

[1] The Asian Green Revolution boosted food availability, reduced malnutrition, raised rural wages, increased employment, drove down the price of food which benefited poor rural and urban consumers, lowered inflation, stimulated forward and backward linkages in the agricultural sector, and helped fuel industrial and economic growth.

markets than before, they had negative consequences as well. Input supply systems remain underdeveloped and fragmented, and farmers have had difficulties in accessing inputs in rural areas (Smale and Heisey, 2001; Phiri et al., 2004). Kherallah et al. (2002), in a comprehensive review of reforms and effects on agricultural input and output markets, noted that 'evidence shows . . . that economic growth in Africa has stagnated, or declined, especially in agriculture'. The general consensus is that reforms have failed to spur agricultural growth as Africa lags behind the rest of the world (Commander, 1989; Mosley and Weeks, 1993; Spencer and Badiane, 1995; Eicher, 1999; World Bank, 2006).

Lack of access to finance for farmers, traders, seed companies, fertilizer producers, fertilizer importers and wholesalers, and agro-processors has further limited the capacity of farmers to pay for new agricultural technologies and hindered the capacity of the private sector. The limited success of prior efforts to improve access to finance for farmers through government credit programmes, agricultural development banks, limited lending by commercial banks, and inability of microfinance institutions to meet the specific needs of the agricultural sector (although this is changing gradually), have contributed to undercapitalization of the agricultural sector.

To accelerate food production in Africa there is a need for measures to improve access of farmers to improved technologies, finance, and markets. The chapter reviews challenges facing farmers in accessing agricultural inputs and finance, and provides recommendations to address these in order to spur agricultural productivity and income growth. For the purpose of this chapter, we will focus on improved seeds, fertilizers, and seasonal credit.

The rest of the chapter is divided into six sections. Section 2 reviews sector-specific challenges and approaches for expanding demand for improved seed. Section 3 discusses demand and supply constraints to expanded use of fertilizers. Section 4 discusses approaches for improving access to agricultural inputs through the development of rural markets, with special focus on the development of agro-dealers. Section 5 discusses approaches for improving affordability, with focus on subsidies, vouchers, and small-size packaging of agricultural inputs. Section 6 discusses approaches that have been used to improve access to finance, including agricultural development banks, microfinance, savings, and credit and credit guarantees. Section 7 ends with conclusions and recommendations.

2. Improved seed supply and demand challenges

2.1 FARM-LEVEL SEED DEMAND CHALLENGES

Despite decades of research to develop improved crop varieties, the adoption of improved seeds continues to lag behind in Africa compared to other developing and developed regions (Tripp, 1998).[2]

The cross-sectional data presented in Table 9.1 is typical of the level of use of improved seed of maize, the predominant food crop in sub-Saharan Africa, and many other staple food crops by farmers in Africa. Over 72 per cent of the maize area in eastern, western, and southern Africa (or 12 million hectares) is planted to traditional, unimproved varieties from the informal seed sector (mainly farm saved seeds), with average yields of less than 1 tonne/ha compared to 3–4 tonnes/ha for improved seed.

Recent empirical evidence suggests that increased use of improved seed in sub-Saharan Africa is constrained by (i) level of awareness by farmers of the availability and value of improved seeds, (ii) the price of seed relative to output prices, (iii) farmers' unwillingness to change, (iv) ecological adaptation of the seeds sold, and (v) availability of credit to farmers to purchase the improved seed and/or complimentary inputs such as fertilizers (Langyintuo et al., 2010). At the institutional level, strategies that commonly enhance farm-level seed usage include (a) approaches involving direct seed distribution; (b) market-based approaches, and (c) seed fairs. Each has its advantages and disadvantages.

Direct seed distribution: Direct procurement, transportation, and distribution by governments and development agents when farmers face disasters is considered a short-term measure and based on the assumption that farmers cannot find seed (Rohrbach et al., 2005; Remington et al., 2002). It is therefore believed that distributing seeds freely would allow farmers to recover from the disaster and subsequently participate in the seed market, and at the same time gives the agent the satisfaction that their monies are being spent visibly.

A number of disadvantages have been observed with the direct distribution approach. First, much of the seeds distributed are often low-quality recycled grains. Bramel et al. (2004) observed that in several countries (e.g. Ethiopia, Eritrea, South Sudan, and Burundi), implementers of direct seed distribution had to fall back on low-quality seed procured from seed/grain markets supplying low-quality seeds as there are no strong commercial seed sectors. Suppliers also try to maximize profits by supplying cleaned-up grains at

[2] It is difficult to assess whether it is due to lack of demand for new seeds. Estimating seed demand poses an empirical challenge due to the wide scale of distribution of free seeds under various government and non-governmental emergency programmes (Sperling et al., 2008). A useful surrogate, therefore, is seed usage, which is rather easy to estimate based on cropped area.

Table 9.1 Adoption rate of improved maize varieties in selected countries in Africa

Country	Area (million ha)	Seed requirement (1,000 t)[1]	Adoption rate (per cent of area)
Ethiopia	1.7	42	19
Kenya	1.6	39	72
Tanzania	2.6	64	18
Uganda	0.7	17	35
Angola	0.8	19	5
Malawi	1.4	35	22
Mozambique	1.2	30	11
Zambia	0.6	14	73
Zimbabwe	1.4	34	80
Benin	0.7	16	Na
Ghana	0.7	19	1
Mali	0.3	8	0.3
Nigeria	3.6	89	5
Total	17.3	427	28

Note: [1] Estimate based on area and planting rate of 25 kg/ha.
Source: Langyintuo et al. (2010).

certified seed prices (Makokha et al., 2004; Rohrbach et al., 2004). Second, the direct distribution can lead to the distribution of incorrect varieties (Anon, 1997; Sperling, 2000), because the seeds are often procured from outside the region by the suppliers that win the tenders (Anon, 1997; Sperling, 2000). Third, where the distribution is repetitive, it tends to distort farmers' own seed procurement strategies (Sperling, 2002; Phiri et al., 2004) and undermines local seed/grain market functioning, particularly in retail sales (Rohrbach et al., 2004; Walsh et al., 2004). Finally, the approach could compromise the development of long-term commercial seed supply systems (Tripp and Rohrbach, 2001; Bramel et al., 2004; Rohrbach et al., 2005).

***Market-based approaches*:** The market-based approach commonly involves the use of seed vouchers. Farmers are given vouchers that they can take to local, private input suppliers to acquire seed. The cost of the seed for the farmer is reduced by the value of the voucher (Minot and Benson, 2009). The supplier then takes the voucher to a designated bank or similar redemption centre/agency to be reimbursed for its value, plus any handling fees.

It is a form of subsidy and is believed to be effective in improving the purchasing power of farmers to acquire seeds. The retailers, on the other hand, are also assured of a guaranteed market for their seeds. A similar system could also be used to expand fertilizer use among smallholder farmers.

Due to the cumbersome and lengthy process to redeem vouchers, however, some retailers are often reluctant to accept them. Many researchers question the sustainability of voucher schemes owing to their burden on national budgets. Moreover, their administration can be a source of abuse among technocrats associated with the implementation. In addition, staff capacity and logistics to scale up the voucher programme over many and large agro-ecological zones can be daunting.

***Seed fairs*:** Sometimes voucher schemes are complemented by seed fairs, which provide an ad hoc market place to facilitate access to seeds, or specific crops and varieties, from other farmers, traders, and the formal sector (Longley, 2006). Combining seed vouchers and fairs not only facilitates farmers' access to seeds in a timely manner, it also allows farmers to maintain seed diversity within their communities at the same time (Sperling and Loevinsohn, 1993; Jarvis et al., 2000; Nathaniels and Mwijage, 2000). The approach seems broadly empowering—allowing, *inter alia*, poor households, including women, to sell—and at the same time provides economic support to local seed system entrepreneurs and private seed companies willing to compete on a level playing field.

Seed fairs also provide an opportunity for exchange of knowledge among farmers and traders (Remington et al., 2002; Makokha et al., 2004; Walsh et al., 2004; Bramel et al., 2004). However, the quality of the seed on the fairs is a major concern for implementers, most of whom would consider themselves accountable for the product they deliver and may shy away from promoting the use of farmer-produced seed (Sperling et al., 2004; West and Bengtsson, 2005).

2.2 BOTTLENECKS AFFECTING SEED SUPPLY IN AFRICA

A combination of non-conducive policy environment and technical problems has continued to hinder the supply of improved seeds. As far back as the 1970s, most African governments and development agencies recognized the critical role of improved seeds in agricultural transformation, but limited their investments in the seed sector to parastatals (Maredia and Howard, 1998). These agencies were often bureaucratic, inefficient, and subject to volatile government budget restrictions (Bay, 1998).

In the past two decades most governments deregulated the seed sectors, leading to an increased participation of private, local, regional, and multinational seed companies. Nevertheless, a number of institutional and policy bottlenecks still hinder the expansion and smooth functioning of the seed sector, as documented by Langyintuo et al. (2010).

First, procedures for release of seed varieties in some countries tend to take several years (up to six years in some cases). This has financial and economic implications for both the seed companies and the farmers. In addition, there is a lack of coordinated and harmonized variety release systems, which would allow for simultaneous release of varieties across countries to enhance regional spillover of benefits of new varieties. Several countries have signed on to the harmonized regional seed laws and regulations within the economic communities, but these are rarely implemented at the country level for unknown reasons.

Second, although private seed companies have been encouraged to participate in the seed industry, in some countries they have been deprived of access to foundation seeds of publicly developed germplasm which they need to produce certified seeds for farmers. It is common for governments to monopolize the production and distribution of the foundation seeds through parastatals. Consequently, in countries where public breeding dominates, such a policy can cause private seed companies to operate below capacity and render them unable to effectively meet farmers' demands for certified seed.

Third, most seed companies lack capital to finance seed production and processing equipment. Financial institutions are often reluctant to lend to seed companies just as they are for agricultural entrepreneurs, primarily due to risks associated with agricultural lending, as will be seen later.

Finally, many countries do not have plant variety protection laws that would assure seed companies of their control of intellectual property rights over released varieties, especially for the open pollinated crops and hybrids (Langyintuo, et al., 2010; Tripp and Rohrbach, 2001).

As a consequence of these constraints, even when farmers have the means to purchase the seeds, supply may be the limiting factor. For example, in Ethiopia where seed requirements are aggregated by the government based on seeding rate and cultivated area, recent data suggest that only 3–6 per cent of the required improved seed is produced (Alemu and Spielman, 2006). There are a number of strategies, which if effectively implemented could reduce or eliminate the supply side constraints as discussed in the next section.

2.3 EXPANDING THE SUPPLY OF SEED

In a bid to support governments to address seed supply problems, the Alliance for a Green Revolution in Africa (AGRA) has been investing in agro-ecology based breeding by private seed companies and public breeding institutes. The number of crop varieties released by seed companies supported by AGRA in the 13 target countries has increased from 29 in 2007 to 154 in 2010. The number of commercialized varieties has also expanded, rising from 28 in 2007 to 97 in 2010. Most of the varieties released are the predominant staple crops, including maize (44 per cent), cassava (38 per cent), groundnuts (25 per cent), and pigeon peas (17 per cent). The total volume of seed produced has expanded significantly, from 2,346 metric tonnes (MT) in 2007 to 25,844 MT in 2010. However, there is still much more to be done. While the quantity of seed required of all major staple cereal crops is estimated at 549,471 MT per annum across the 13 countries where AGRA focuses, seed production falls way below the demand projections.

In addition to strengthening internal seed laws, facilitating access to foundation seed and supporting the implementation of harmonized regional seed laws, and providing support to seed companies to procure seed production and processing equipment, long-term investment capital and affordable credit to finance seed production operations would potentially enhance seed production on the continent. Recently, AGRA initiated two special loan facilities targeting working capital and investments in capital equipment for small and emerging seed companies[3] that focus on serving the smallholder farmers: the African Seed Investment Fund (ASIF) and the West Africa Agricultural Investment Fund (WAAIF). ASIF and WAAIF are dedicated to fostering the growth of seed companies in eastern/southern Africa, and West Africa, respectively. Current investments total about US$5.9 million spread across eight African seed companies in Kenya, Uganda, Tanzania, Malawi, Mali, and Nigeria. AGRA has brokered the creation of a new working capital (short-term) loan facility operated by Root Capital, Inc. in 2010, which has invested US$150,000 in Tanzania and an additional US$200,000 in Kenya.

The provision of credit should be complemented with the enhancement of access for seed companies to appropriate and adapt germplasm. This could be facilitated through contractual arrangements between the public and the private sector. To allow seed companies to ramp up seed volumes from their current levels so that more is produced to reach a larger number of farmers, AGRA has supported selected seed companies with one-time grants. Companies could also use part of the grant to educate farmers and create awareness about the value of using improved seed through field days, demonstrations, radio programmes, agricultural shows, and print material in local languages.

Private seed companies should also be facilitated to access foundation seed of public materials. As noted, policies in some countries such as Ethiopia, Ghana, Nigeria, and Tanzania slow down development of the seed sector by controlling foundation seed production. AGRA, together with some development partners, is working with the respective governments to reform such

[3] The small emerging seed companies have comparative advantage in serving smallholder farmers because (i) they are often located in close proximity to the small scale farmers and hence transportation costs are minimized, (ii) the diverse agro-ecological and social conditions in African countries minimize any economies of scale that favour large companies in Europe, Asia, and the Americas, instead favouring highly fragmented market niches, (iii) because of relatively low overheads, small and emerging seed companies are able to produce and sell seed at a lower per-unit cost than large companies, (iv) small and emerging seed companies handle a range of crops which are commonly grown by smallholder farmers, unlike the multinationals which focus primarily on maize, and (v) the macroeconomic environment in Africa is highly dynamic and changes drastically across countries, which favours smaller companies with local orientation that can adapt quickly.

policies. Such efforts have resulted in the review of the Ghana seed law in 2010 to completely liberalize the seed sector. In Nigeria and Tanzania, governments have liberalized foundation seed production and are in the process of reviewing the laws to provide the legal basis for the policy change. There is progress towards liberalization of the sector in Ethiopia, though the change is slow.

For rapid spillovers of varieties released in one country to similar agro-ecologies in different countries, the implementation of regional seed laws and regulations should be expedited. One of the reasons for the slow pace of implementation of the harmonized seed law is the laxity in establishing institutional frameworks for implementation. This is evident in the Economic Community of West African states (ECOWAS) where implementation has been constrained by the slow pace of setting up the regional varietal release committee and the establishment of the minimum requirements for release. Similar problems exist in the different regional trading blocks (SADC, EAC, etc.). AGRA is working with the regional economic communities to speed up the processes.

3. Demand and supply for fertilizers among farmers

Rising population, lessening of fallow periods, and extensive agricultural production practices have led to extensive nutrient mining in many parts of Africa. At the Abuja Fertilizer Summit of African Heads of State in 2006, it was agreed that measures should be taken to raise the level of use of fertilizers to an average of 50 kg/ha to accelerate food production. To date, fertilizer use in Africa averages less than 10 kg per hectare, the lowest in the world. Both supply and demand side factors limit the use of fertilizers (Crawford et al., 2006; Morris et al., 2007), as discussed later.

3.1 DEMAND SIDE CONSTRAINTS TO THE USE OF FERTILIZERS

We now review some of the factors that constrain the demand for fertilizers by farmers:

***High fertilizer prices and low value-to-cost ratio*:** The price of fertilizers has been one of the major factors limiting the use of fertilizers for all farmers, especially for smallholders in agriculture. In terms of profitability, and incentive to use fertilizers, traditional cash crops (cotton, coffee) and other high-value non-traditional crops (horticulture products) are better placed than low-value crops such as cereals and other staple crops.

In addition to prices, value-to-cost ratio (VCR) (defined as total returns divided by the total cost of fertilizers used in production) is also critical for creating incentives to adopt technologies. It is generally accepted that a VCR of two and above will be needed to stimulate fertilizer adoption, but this has been declining in many African countries due to rapid increases in the price of fertilizers compared to that of grains (Kelly, 2006).

Poorly functioning output markets: Access to markets is necessary for farmers to adopt and sustain the use of fertilizers, which is high in cash crop-based farming systems. In West Africa, the average level of use of fertilizers is higher in the cotton-based systems than in the non-cotton based systems. Cotton companies sell fertilizers to farmers on credit and provide assured markets for their crop. Such tied input–output markets reduce risks for farmers and encourage the use of expensive farm inputs. As a result, the productivity, profitability, and sustainability of the production systems are much higher in cotton-based zones, although some researchers have questioned both the profitability (Briand et al., 2006) and environmental sustainability (Moseley and Gray, 2008) of these cotton-based systems.

Output prices are important for ensuring incentives for farmers to use fertilizer. In Asia and Latin America, guaranteed prices for grains (Cummings et al., 2006) were important for high usage of fertilizer. Governments established strategic grain reserves and announced guaranteed minimum prices for grains ahead of the planting season. In many parts of Africa, strategic grain reserves are no longer in operation, while the private sector has not invested optimally in storage facilities. As a result, it is common for prices for food crops to collapse following bumper harvests.[4] Development of improved and efficient output markets including local grain banks, warehouse receipt systems, market information systems, strategic grain reserves, and guaranteed minimum prices are needed to encourage farmers' investments in expensive cash inputs such as fertilizers.

Limited knowledge: Because fertilizers are expensive, it is important that farmers know how to use them efficiently. The importance of such knowledge has been demonstrated through the Sasakawa/Global 2000 (SG-2000) programmes in the 1990s. These programmes supported half-hectare demonstration plots, typically in productive areas, where farmers were supplied with credit, inputs, and extension advice. They successfully raised yields, but sustainability was compromised by input and output market constraints that reduced access to inputs and profitability, or increased risks (Crawford et al., 2006).

[4] To illustrate the point, in 2001 and 2002, farmers in Ethiopia got high yields for their cereal crops and consequently bumper harvests as a result of a major effort by the government to subsidize use of fertilizers and improved seeds in 2001. The two consecutive bumper harvests led to surpluses and the price for grains collapsed. The collapse was primarily due to lack of market information systems, limited access to financing by grain traders, lack of grain storage systems (at farm level and/or poor operation of the national strategic grain reserves), and restrictions on the movement of grains across borders (Gebreselassie, 2006).

When farmers lack adequate knowledge as in much of smallholder agriculture in Africa, fertilizer use is inefficient (Crawford et al., 2006). This lack of adequate knowledge can be blamed partly on the lack of extension agents. This is especially the case in the public sector following reforms that accompanied structural adjustment programmes, leading to massive reduction in their expenditure. As observed by Aina (2006), the ratio of extension agents to farmers is low and can be as low as 1:30,000 in some locations, reducing the reach of farmers, frequency of contact, and effectiveness of extension messages.

Risk: The profitability and risk of the use of fertilizers varies by agro-ecology depending on type of soil, farming systems, and rainfall levels. In semi-arid West Africa, where rains are low and highly variable, the risks are high, and higher in the case of late rains.

Inappropriate fertilizer recommendations: Fertilizer recommendations in many parts of sub-Saharan Africa are out-dated. Many are based on fertilizer responses for cash crops and are often based on what is technically optimal, and not what is economically optimal, invariably resulting in higher levels of use than what is necessary. They, also, do not take into consideration the diversity of the farming systems, rainfall, risks, soil types, farmers' resource constraints, and objectives for production, which should form the basis for new recommendations.

3.2 SUPPLY SIDE CONSTRAINTS TO FERTILIZER USE

Lack of access to finance by importers: Importers face major challenges in accessing foreign exchange and financing for import of fertilizers. High interest rates charged by banks and demand for collaterals make it difficult for small importers to secure commercial lines of credit. High rates of interest on financing are eventually transferred to farmers through highly priced fertilizers (Morris et al., 2007). Lack of timely access to foreign exchange often leads to delays in procurements, thereby increasing the probability that farmers will apply the wrong fertilizers at the wrong time, with negative consequences on crop yields. Even in countries where governments use tenders for import, delays in placement of orders have been found to negatively affect crop production and food security.

Small volumes of imports: Fertilizers are bulky and costly to transport; therefore, the higher the volumes imported, the lower the cost of transport. In addition, fertilizer manufacturing companies offer price discounts for high-volume importers. The need to achieve economies of scale in procurement and transport therefore requires consolidation of orders (Morris et al., 2007; Kelly et al., 2003). But African countries purchase fertilizers individually and in small quantities. Over twenty-five countries import less than 10,000 metric tonnes, too little to achieve economies of scale in transport. The lack of ability of importers to predict local fertilizer demand does not allow them to import larger quantities. It is not uncommon for oversupply to occur, leading to high inventory costs and risks. Government interventions in markets, especially in directly procuring fertilizers, and subsidizing them, create uncertainties for the private sector. Private-sector traders therefore order small volumes to lower their risks. When there are delays in government's import processes, and the private sector has not ordered enough to reduce their risks, farmers end up not getting fertilizers on time.

High transport costs: High inland transport cost accounts for a significant share of the high farm gate prices for fertilizers. Poor road and rail systems lead to high costs and risks. Transportation by rail can lead to significant reduction in farm gate prices. Railway systems are poorly developed in Africa and do not reach many rural towns. Among the many problems for the railway system are: not enough railcars; lack of enough carriages; many of the carriages being open, leading to high risks of transit losses and reduction in the quality of fertilizers from exposure to sunlight and rainfall; limited frequency of operations; and other operational inefficiencies (Morris et al., 2007). As a result, traders rely on small trucks to transport fertilizers, which raise transportation costs, as several trips may need to be made to cities to buy small volumes.

Inland transport costs are high compared to the cost of shipping fertilizers into Africa. For example, it costs US$50 to ship a tonne of fertilizers from the US to the port of Mombasa in Kenya, for a distance of 11,000 kilometres. Transporting the same tonne of fertilizer from Mombasa to neighbouring Uganda, a distance of less than 1,000 kilometres, costs US$80–90 (Morris et al., 2007).

Handling at ports: In addition to high transport costs, importers face several problems at the port when fertilizer imports arrive. Inefficiency in port handling can add up to 10–15 per cent additional costs. This comes from taxes, high demurrages, low rate of discharge, inefficiencies in dockside bagging and stevedoring operations, and storage costs.

Limited local manufacturing and blending: Africa has about 70 per cent of the world's deposits of rock phosphates. However, this is not being exploited optimally. While many countries will need to continue to rely on commercial imports for the foreseeable future, efforts are needed for promoting local fertilizer manufacturing, especially blending. Due to wide variability in soils, standard fertilizers (N, P, and K) do not address micro-nutrient deficiencies that limit crop growth and efficiency of uptake of nutrients. Local blending of fertilizers will allow the development of right kinds targeted to match soil niches, reduction in dependence on imports, and efficient (and safer) uses through labelling in local languages. When the volume of domestic fertilizer use is below 25,000 tonnes it is better to rely on commercial imports. As the fertilizer market reaches between 25,000 and 50,000 metric tonnes, it becomes economical to shift into local blending using imported local materials. Once market size reaches 500,000 metric tonnes and above, local manufacturing becomes economical.

***Lack of fertilizer market and trade information systems*:** Because there are few large manufacturers and suppliers of fertilizers, the international market is dominated by cartels. This leads to non-competitive pricing behaviour and high margins. Lack of market and trade information systems also make it difficult for traders to know about alternative sources of supply. It also limits their ability to take advantage of lower cost routes for transportation.

4. **Improving affordability of farm inputs**

Building sustainable rural input markets to accelerate the delivery of productivity enhancing technologies for poor farmers is critical for achieving the African Green Revolution. To achieve this, the following three elements are especially important: (a) increasing the volume, range, and quality of agricultural inputs in rural areas using commercial channels (accessibility); (b) reducing input costs to farmers so they can profitably use new agricultural technologies (affordability); and (c) improving the functioning of agricultural output markets so that farmers can achieve higher prices and incomes from the sale of their farm produce (incentives).

4.1 EXPANDING 'SMART INPUT SUBSIDIES'

Improving affordability of agricultural inputs can be achieved through improving the purchasing power of farmers, especially those who are poor. Additionally, inputs can also be packaged in appropriate smaller sizes to make them more affordable, with 'smart subsidies' to improve affordability and with the use of small input packs for farmers.

African farmers face large climate-induced risks, much more than what Asian farmers faced during their Green Revolution. Measures to mitigate them include better weather predictions; training farmers on the optimum time to apply fertilizers depending on the state of nature; use of water harvesting techniques to retain water and improve soil moisture; using small doses of fertilizers ('micro-dosing'); development of weather-indexed crop insurance schemes; targeting fertilizers for areas in crop fields with higher chances of achieving crop responses; and better methods of application to reduce risk to crops. Sub-Saharan Africa relies predominantly on rain-fed agricultural systems, much more than in Asia.[5] Therefore, the risk of using fertilizers is much higher than in Asia. Where it is economically feasible, investment in small-scale irrigation can help to reduce the risk of fertilizer use, and improve the efficiency and crop responses to the added soil nutrients. As noted by You (2008), countries such as Nigeria stand out for having particularly high potential for both large and small-scale schemes, while Niger stands out as a particularly lucrative site for irrigation investments of all sizes. Kenya, Tanzania, and Zambia show significant potential for large-scale schemes, while Burkina Faso, Chad, Cameroon, and Senegal have potential for small-scale schemes.

Fan et al. (2008), in a revisit of the experience of subsidies in the Green Revolution in Asia, concluded that investments in other public goods such as education, road investments, irrigation, and agricultural research and development have a higher rate of returns on agricultural GDP than subsidies. They also have higher impacts on reduction in poverty. These estimates are being used to assert that Africa does not need to focus on subsidies. While it is important to ensure that investments in subsidies are compared to alternatives to improve cost effectiveness (Crawford et al., 2006), using the Asian experience to argue against use of subsidies in Africa is misplaced.[6]

Factors leading to the rethinking of the role of subsidies in Africa include the stagnation of agricultural productivity, rising poverty and food insecurity, concerns about soil fertility depletion, failure of the market liberalization policies to stimulate socially and environmentally optimal level of use of fertilizers, and concerns on negative externalities from continued deforestation in Africa with negative consequences on climate change (Dorward, 2009). The global food crisis of 2008 led to a push for raising agricultural productivity and food security. National governments, the World Bank, the IFAD, the African Development Bank, and other development agencies called for use of subsidies for seeds and fertilizers to address the short-term negative effects of the high food prices on poor smallholder farmers and consumers.

[5] The irrigated area, extending over 6 million hectares, makes up just 5 per cent of the total cultivated area (two-thirds in Madagascar, South Africa, and Sudan), compared to 37 per cent in Asia and 14 per cent in Latin America (You, 2008).

[6] Subsidies played a major role in the Asian Green Revolution (Djurfeldt et al., 2005; Fan et al., 2008). These included subsidies for seeds, fertilizers, irrigation, electricity, and credit. These were complimented with other public goods such as research and development, and investment in transport infrastructure such as rail and roads.

Without subsidies, farmers in Asia would not have achieved the Green Revolution from high-yielding crop varieties, fertilizers, and irrigation, as the costs would have been too high for farmers. Subsidies were used to jump-start the rapid adoption of agricultural technologies. Dorward (2009) noted that 'later ineffectiveness and inefficiencies of input subsidies should not obscure their initial contribution in driving growth forward' (p. 8), arguing that state-led support for subsidies is needed for Africa to achieve its Green Revolution.

BOX 9.1 USE OF SUBSIDIES FOR ACCELERATING FOOD PRODUCTION IN AFRICA: PAST PROGRAMMES

The government of Malawi initiated the Malawi Agricultural Input Subsidy Programme in 1998/99 and 1999/2000 as a nationwide subsidy programme targeting smallholder farmers. The programme initiated 'starter packs' that provided farmers with 15 kg of fertilizers, and 2 kg of seed, enough to plant 0.4 ha. The programme reached over 1.5 million households. Due to the high cost of the programme it was subsequently scaled down to provisioning of subsidized inputs for 25 per cent of the farming households. The government scaled up the programme in 2005/6, providing subsidies of 100 kg of fertilizers, complemented with small quantities of seed and grain legumes. Food production has expanded significantly, with Malawi exceeding its national food requirements every year for the past four years (Denning et al., 2009). It has moved from being a food deficit country to a food exporting country, exporting maize to Zimbabwe and Kenya, while also providing food aid to its neighbours. The success of the Malawi subsidy programme has encouraged other African countries to launch subsidy programmes for their farmers, including Zambia, Nigeria, Rwanda, Ghana, and Tanzania. The use of subsidies for farmers in Rwanda has led to rapid productivity growth and achievement of food security within two years. Tanzanian government's subsidy programme has led to significant production of maize in the southern highlands, where 700,000 farmers produced 5 million metric tonnes of maize which allowed the country to supply food to its deficit areas experiencing drought.

Prior to the food crisis, a number of governments had been providing subsidies to accelerate food production and the food crisis simply reinforced these policy directions (see Box 9.1 for examples of the use of subsidies in Africa).

There is wide diversity in the approaches used for subsidies. However, a number of trends are notable. There has been a general move away from universal fertilizer and seed subsidies to targeted subsidies, where beneficiaries, selected based on poverty level, farm size, or geographic locations, are targeted for support. Vouchers are being used to improve the targeting of subsidies to households. The exclusivity of the public sector in import and distribution of fertilizers has waned, with increased participation of the private sector.[7]

Xu et al. (2009) evaluated the impact of the Zambia governments' Fertilizer Support Program on private-sector commercial sales. Under the programme, the government encouraged the private sector to distribute fertilizers on behalf of the government, especially in rural areas, as a way of reaching the poor while 'crowding in' the private sector by lowering their fixed costs for the

[7] Public-sector fertilizer exclusivity has been known to erode incentives for participation by the private sector and for commercial sales. Most governments combine public sector subsidy programme and market development or promotion to expand private-sector fertilizer and seed markets, especially in remote rural areas.

development of markets in rural areas. Their results show that where the focus is on poor farmers as beneficiaries, the programme was effective in expanding fertilizer use by farmers, crowding in the private sector, than when the beneficiaries were wealthier farmers. They note that 'when programs are appropriately targeted at poorer areas in which the private sector is relatively inactive, subsidies may indeed encourage greater commercial demand for fertiliser and hence the development of commercial input distribution system' (p. 92). Where the subsidies were distributed in areas where commercial sales were normally high, and leakages existed to benefit farmers that would normally have purchased fertilizers, evidence of displacement of commercial sales was found.

A number of challenges have to be resolved before making subsidies useful for smallholders. These include controlling costs, designing exit strategies and targeting systems, possible overuse of external inputs, leakages to larger farmers, and potential market distortions when public institutions distribute subsidized inputs and displace commercial sales (Dorward, 2009; Xu et al., 2009). Morris et al. (2007) call for the use of 'smart subsidies' which combine public support with market development. Measures are needed to ensure that subsidies do not displace commercial interests from developing input market systems where the private sector is engaged in distribution of inputs. It is also important to avoid uncertainties that arise from unpredictability of timing and duration of government intervention programmes, which may leave the private sector with unsold inventories or unable to invest in long-term market development, as they are unsure whether they will be able to recoup profits.

Careful design of effective subsidy programmes requires public–private partnerships to build trust, transparency, mutual accountability, and commitment to long-term development of markets. The cost effectiveness of subsidies will depend on how well they can be targeted to avoid leakages and/or development of secondary markets for vouchers which may arise because recipients sell their inputs to other non-target farmers because of a need for cash, or to traders who resell at non-subsidized prices. The administrative costs of using vouchers could also be high. Electronic smart cards, mobile phones, and scratch cards could be used to lower transaction costs of disbursing subsidized inputs.

4.2 PACKAGING INPUTS IN SMALL SIZES TO IMPROVE AFFORDABILITY

Since risk-averse poor smallholders in agriculture suffer from cash constraints and are probably more keen on experimenting with incremental inputs depending on weather conditions, packaging of improved seeds and fertilizer into small packs holds promise in the context of Africa. In Kenya, the percentage of farmers using fertilizers in low potential areas (western lowlands) could be as low as 8 per cent compared to high potential areas (central highlands), where it is as high as 90 per cent (Ariga and Jayne, 2006). In low potential areas, small packs of inputs (starter packs) can be effective to gradually build confidence of poorer farmers in using improved seed and fertilizer technologies (Kelly et al., 2003).

As a marketing innovation, this is rapidly achieving success in expanding the access of poor farmers to improved technologies. In Kenya, the Sustainable Community-Oriented Development Program (SCODP) initiated a programme to package seeds and fertilizers in small quantities to meet the cash needs of poor farmers in Nyanza province of western Kenya. This Province is characterized by low crop yields, high levels of poverty, and soil fertility depletion. Taking advantage of a new law which removed restrictions on breaking down 50 kg bags of fertilizer and re-packaging into smaller sizes (Kelly et al., 2003), the SCODP initiated re-packaging of fertilizers in sizes from 1 to 2 kg. The small-size packs have been a hit with farmers. Most of the farmers started with small sizes for experimentation, and over time increased their expenditure and moved on to larger sizes.

The Farm Inputs Promotion Service (FIPS) in Kenya has continued this work, with complimentary focus on improving the agronomic practices and knowledge of farmers, coupled with extension services. Recognizing that farmers face soil micronutrient deficiencies, FIPS worked with a local fertilizer manufacturer (Arthi River Mining) to develop new blends that address these micronutrient deficiencies. The new fertilizer blends are packaged in small sizes that are affordable. The production and commercialization of the new fertilizer has expanded substantially. Seed companies have also begun to package their commercial seeds in small sizes.

4.3 DEVELOPING RURAL INPUT MARKETS

Private-sector service providers, replacing the public-sector service providers in the agricultural sector, generally are concentrated in areas with relatively good infrastructure, high population densities, and presence of large numbers of commercial farmers. Poorer farmers in areas with low population density and weak infrastructure hardly benefit from the market development. This has been the motivation for developing rural agro-dealers to fill this gap. This is not new; in Asia rural agri-input dealers play a critical role in expanding the access of farmers in Asia to fertilizers and improved seeds. This contributed significantly to the agricultural and rural economic growth witnessed in several parts of Asia.

Table 9.2 Services offered by stockists to farmers, Uganda

Services provided	Number of stockists	%
Credit	1,086	48
Advice	1,921	85
Drug Application	994	44
Chemicals Application	1,058	47
Other	97	4

Source: AT Uganda (2005a).

Expanding agro-dealers networks

Agro-dealers can offer a whole range of support services for poor farmers, aside from selling agricultural inputs. In a national survey of 2,264 stockists across Uganda, AT Uganda (2005a) found that the range of services provided included: extension advice (85 per cent), credit (48 per cent), advice on drug application, and advice on chemical application, among others (Table 9.2). Studies in other countries have reported similar findings, especially in Nigeria, Ghana, Malawi, Kenya, Tanzania, Zambia, and Zimbabwe (Krausova and Banful, 2010; Kibaara et al., 2009).

Where traditional extension services are weak, new forms of extension, based on private sector company collaborations with the rural stockists are rapidly replacing traditional forms of extension. In Malawi, Monsanto conducted national demonstrations of reduced tillage technologies for farmers using agro-dealers. In Kenya, over 10,000 farmers were exposed to new technologies in joint demonstrations conducted by agro-dealers and several seed, fertiliser and agrochemical supply companies (CNFA, 2005). Creating such knowledge engines and networks in rural areas is necessary for accelerated learning and adoption of agricultural technologies among smallholders in African agriculture.

Notwithstanding the critical role of agro-dealers, their emergence in Africa has been hampered by government control and monopoly over input marketing. Where they exist, they also tend to be concentrated in high potential areas compared to low potential areas. This has implications for limited access to inputs in the latter areas where poverty and food insecurity are higher. For example, in Kenya, where use of improved seeds and fertilisers has expanded rapidly in the past decades, the density of stockists is uneven. The average density of stockists per farming population in high potential areas varies from 1:4,333 farmers in the Mount Elgon area to 1:5,252 in Bungoma district. For low potential areas the density per farming population is as low as 1:24,000 in the district of Teso.

A number of factors may explain the apparent lack of incentives even when government control of the sector is absent. Firstly, most of the local importers have limited human and financial resources to implement large scale training programs needed to create rural input dealers. Secondly, entrepreneurs consider many rural areas to be risky to do business especially those where the rural poor reside and are unable to spend much on agricultural inputs. Thirdly, the transaction costs of reaching poor farmers are considered too high as compared to concentrating their supply chains towards fewer and richer farmers who form a more stable market due to higher incomes and purchasing power. Fourthly, the business of agricultural inputs is highly seasonal and hence supply companies are reluctant to open depots and employ full time staff. Finally, resources spent for developing such a network of rural input traders cannot be captured exclusively by the firm alone. This is a classic case of public goods or the 'free riders' problem. Thus public resources are needed to create a cadre of input dealers in rural areas. Improving the access of farmers in those areas to inputs will require public policies to subsidise farmers and/or stockists, given the higher risks in those regions.

Challenges faced by agro-dealers

Lack of access to finance: Access to credit for agro-dealers will allow them to expand range of inputs stocked and sold and the timeliness of supply of inputs. Omamo and Mose (2000) found that access to credit was very significant in determining volume of fertilisers sold in rural markets. Yet agro-dealers face significant constraints in accessing finance to stock up a range of improved seeds, fertilisers and other inputs needed by farmers. In Kenya, no more than 30 per cent of the stockists had used credit at one time or the other, with the majority relying on informal credit or own savings to purchase stocks. Where agro-dealers have access to commercial lending, the size is limited. In Kenya, Omamo and Mose (2000) found that loans ranged from 100,000–1 million Kenya Shillings (US$1,400–US$14,000) for a period of 1–12 months at the interest rate of 28 per cent per annum. In Uganda, only 19 per cent of stockists have access to capital and the probability of accessing credit seems to positively correlate with size of business. While only 14 per cent of stockists with capital base of less than 500,000 Uganda Shillings (US$312.5) reported that they had access to credit, 31 per cent of those with capital base of over 2 million Uganda Shillings (US$1,250) reported having access to credit (Table 9.3). Table 9.4 suggests that females have a significantly lower credit limit than their male counterparts.

Limited knowledge of inputs: Many stockists have very little experience in selling agricultural inputs and hence lack technical knowledge about safe storage, handling and use of agrochemicals. A majority of them have been

Table 9.3 Financing challenges facing agro-dealers in Uganda by size of business (Uganda Shillings)

Size of business	Number	%	Capital a problem (%)	Have access to credit (%)	Credit a problem (%)
No answer	159	7	51	11	22
< 500,000	776	34	82	14	45
500,000 to 1 million	501	22	71	19	44
1 million to 2 million	396	17	68	21	40
> 2 million	432	19	60	31	36
Total	2264	100	71	19	40

Source: AT Uganda (2005a); Exchange rate: USD 1.0 = UGS 1,600.

Table 9.4 Credit limits available for rural stockists in Uganda

Category of recipient	Available credit limit (Amount in Uganda Shillings)	
	Mean	Median
Women	1,447,413	400,000
Men	2,595,621	350,000

Source: AT Uganda (2005b).

in input retail business for less than 10 years. A national census of rural stockists in Uganda (AT Uganda, 2005a) found that about two-thirds of stockists (or 66 per cent) who are men have been in the business for between two and five years, compared with 63 per cent of women. In Malawi, Phiri (2004) found that over 60 per cent of stockists have been in business for over three years.

High costs of certification: Many stockists sell inputs without licence or certification by regulatory agencies, apparently due to high cost of registration. Evidence suggests that the probability of registration is correlated to the size of business. Over 73 per cent of stockists in Uganda have business of between 500,000 and 2 million Uganda Shillings, which means they have too little funds to meet the high registration and licensing costs. In Uganda, a recent national census of stockists found that over one thousand five hundred stockists (or 66 per cent of all stockists) are currently selling inputs without requisite training, certification, and licencing (AT Uganda, 2005). Capital constraint seems to be the major problem.

In the absence of stockist registration it becomes difficult to trace the supply of poor quality or fake inputs within the supply chain, which can cause low yields, health hazards, and negative externalities on the environment.

Weak business management skills: Weak business, financial, inventory, and credit management skills have undermined the viability of agri-businesses and the ability of entrepreneurs to secure loans from commercial financial institutions. Business development services have been shown to have positive and significant impacts on the growth of their businesses. AGRA's investments have led to training and certification of close to ten thousand agro-dealers in about thirteen countries. An external evaluation of the agro-dealer programme in Malawi by Phiri (2000) found that the range of services provided by the Citizens' Network for Foreign Affairs (CNFA) to agro-dealers included product knowledge, business management, record keeping, trade credit management, technology demonstrations, etc. A majority of the trained agro-dealers indicated that they are benefiting from the training sessions. For example, 79 per cent of agro-dealers indicated that they are now selling more inputs than before they were trained. When asked what contributed to this, the responses ranged from technical and business training received (38 per cent) to greater volume of stocks due to the credit guarantee (31 per cent).

Impacts of agro-dealers on input supply in rural areas

Expansion in the number of agro-dealers has helped to reduce the distance travelled by farmers and their search costs in securing farm inputs. In western Kenya, rural penetration of agro-dealers has led to a decline in average distance travelled by farmers from 10 kilometres to 4 kilometres (Kibaara et al., 2009). In Ghana, Krausova and Banful (2010) found that the average distance travelled by farmers has also declined as a result of greater penetration of rural areas by agro-dealers.

The volumes of agricultural inputs supplied to rural markets have increased significantly (Table 9.5). Over 373,000 metric tonnes of seeds and 768,000 metric tonnes of fertilizers have been sold by agro-dealers over the past three years. In addition, agro-dealers have accessed loans of over US$40 million from financial institutions based on credit guarantees used to reduce risk.

Table 9.5 Selected indicators of agro-dealers development across Africa

Indicator	Achievements
Number of agro-dealers trained in business management	9.34
Number of agro-dealers trained in technical knowledge	14.32
Number of agro-dealers certified	6.56
Number of input demonstrations held	3.11
Number of field days held	1.00
Quantity of seeds sold (MT)	373.28
Quantity of fertilizer sold (MT)	768.71
Value of loans issued to agro-dealers (000 US$)	44,501.62

Source: AGRA programme for African Seed Systems (PASS)—Personal communication.

5. **Farm credit constraints**

Agriculture is the predominant activity in African economies, yet less than 3 per cent of total commercial bank lending goes into the agricultural sector (Figure 9.1). Financial institutions have not been inclined to lend to the sector for a variety of reasons: (a) high transaction costs for service providers due to the remoteness of the clients and heterogeneity among communities and farms, dispersed demand for financial services, small size of farms, and individual transaction; (b) the lag between investment needs and expected revenues; (c) lack of usable collateral; (d) lack of irrigation, and pests and diseases, contributing to high covariant risks due to variable rainfall; (e) underdeveloped communication and transportation infrastructure; and (f) weather and price risks.

While the literature on access to financial services is extensive, we focus only on four approaches that have been used to stimulate greater lending to agriculture: (a) agricultural development banks; (b) microfinance, savings, and credit; (c) interlocked markets for input, output, and credit, and value-chain financing; and (d) credit guarantees to banks.

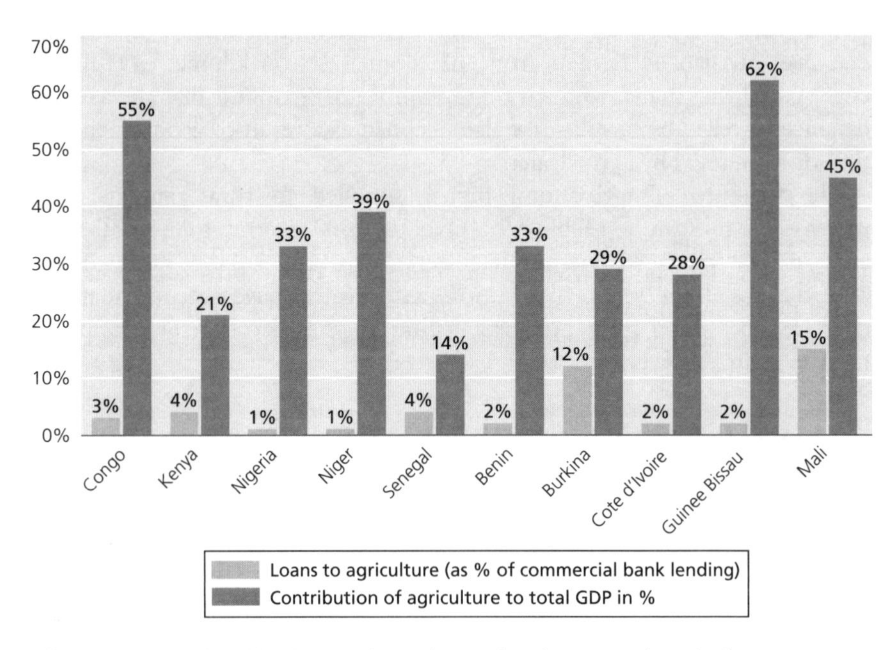

Figure 9.1 Agricultural lending as share of agricultural GDP in selected African countries

Source: CIA, World Factbook.

5.1 AGRICULTURAL DEVELOPMENT BANKS

Specialized state-led agricultural development banks were created by Governments in Africa to supply longer-term agricultural credit that the commercial banks were not prepared to grant or to provide loans 'needed' by specific (risky) clientele such as medium and small farmers (who lacked access to the financial services of the traditional banking sector, but who were considered by the governments to be of priority). Even though they received the largest share of their funds from international donor agencies, governments, and/or central banks at subsidized interest rates, their costs were high and they found it difficult to remain financially viable. They had to rely on subsidized recapitalization by governments to continue their operations. Governments eventually withdrew their support, partly as a result of the fiscal difficulties and macroeconomic stabilization programmes of the 1980s.

A number of factors contributed to the gradual decay of state-supported banks. These include (a) high monitoring/transaction costs (due to a large number of scattered holdings, and market imperfections), (b) over-emphasis on borrowers than depositors (which undermined supply of funds, and hence longer-term sustainability), (c) inadequate assessment of borrower's capacity to repay, etc., (d) repressive agricultural policies, which dampened loan demand, reduced creditworthiness of potential farm borrowers, and weakened their ability to save and repay (Adams, 1994), and (e) pursuit of 'development objectives' by such financial institutions (through cheap loans, and priority to lending operations rather than on resource mobilization[8]), which undermined their profitability and financial viability.

The state ownership of banks also had several negative consequences. It made these banks vulnerable to political intrusion, in the sense that decisions about whom to lend to, what to lend for, and on what terms and conditions have not been autonomously taken by the financial intermediary, but have been imposed from outside by presumed government 'owners' and external sources of funds.

As a result of this, the primary goal of improving rural access to formal credit was poorly met. A majority of the rural poor lack access to a broad range of reasonably priced financial services. A reduction in the cost of borrowing was achieved only for large borrowers, while small producers

[8] Given the public-sector nature of agricultural banks, the capacity to mobilize domestic savings represents a genuine advantage of state-owned agricultural development banks over other non-bank intermediaries. But due to mis-directed regulations, they were not allowed to do so, nor did the banks develop any capacity to do so over time. The banks were more concerned with making sure that interest rate ceilings and loan targeting directives were complied with (repressive supervision). Thus, policymakers took advantage of state ownership to levy on the development banks even more repressive directives about prices and quantities in financial transactions than in other cases, in a futile attempt to influence the allocation of resources.

were saddled with high transaction costs. Credit portfolios have been concentrated in a few hands, with regressive consequences on distribution.

5.2 MICROFINANCE, SAVING, AND CREDIT ASSOCIATIONS

The majority of the population of sub-Saharan Africa does not have access to banks or other formal financial services. On an average, less than 20 per cent of households have access to formal financial services. The unsatisfied demand for financial services is met by a variety of Microfinance Institutions (MFIs) which have emerged over time in Africa. Some of these institutions concentrate only on urban areas and provide credit, while others are engaged in providing both deposit and credit facilities, and others are involved only in deposit collection.

However, microfinance has not been able to significantly expand financing for agriculture for several reasons. Microfinance institutions are often located in urban areas where agricultural activities are not important, or in rural growth centres with strong non-farm sectors that are dominated by small-scale traders and micro-entrepreneurs (Dorward et al., 2001). Besides a high interest rate, the size of loans given by microfinance institutions is often too small to meet the financial needs of farmers. The repayment schedules for microfinance loans are not synchronized with the seasonality of agriculture and the timing of cash flows of farmers. The frequency of the savings and repayments tends to be more appropriate for consumption loans and traders and micro-entrepreneurs who generate incomes on weekly or monthly basis than for farmers whose incomes come several months after loans have been received. MFIs also have limited capital bases to support agricultural lending. Linking them to formal financial institutions could help them to raise their capital base (Poulton et al., 2006), but these could also raise additional problems, as experienced recently in India with MFIs, where investors demand high rates of returns that create incentives for pushing loans to the poor, poor risk assessment for loans, and high interest rates on borrowers. Excessive influence of external financiers and overly rapid expansion of loans could also weaken management of MFIs and put savings and deposits at greater risk (Poulton et al., 2006; Goldstein and Barro, 1999).

Microfinance institutions face additional challenges in lending to the agricultural sector: (1) their outreach remains very small compared to the demand; (2) they have inadequate capacity to properly conduct credit analysis and loan appraisals; (3) their risk management and control systems are generally inadequate; (4) their shares of total deposits and loans in the country remain small compared to commercial banks; (5) public policy, the regulatory framework and non-performing collateral laws limit the effectiveness of the MFIs; (6) most are inadequately capitalized which limits the level

Table 9.6 Agricultural financing in selected African countries by MFI and rural savings and credit associations

	No. of institutions	Average loan ($US)	Ag. Credit as % of total bank loans
Kenya (2002):			
Rural SACCOS	113	180	10
Tanzania (2001):			
Rural SACCOs	400	230	14
Ethiopia (2003):			
Rural microfinance institutions	18	100	2.5
Benin (2003):			
Rural Mutual Credit Association	101	550	13

Source: Sacerdoti (2005).

of lending they can support; and (7) transactions costs are still relatively high, particularly in assessing the possibilities of default by a prospective borrower.

While several informal financial institutions exist such as savings and credit cooperative (SACCOs), Caisse Villageios and Village Banks and Accumulative Savings and Credit Associations, they have not been able to expand lending to agriculture (Table 9.6), even though they mobilize savings, due to weak linkages with banks (Poulton et al., 2006). In Kenya, SACCOs are well developed but they account for only US$30 million or 0.5 per cent of the total asset of the banking system and give loans that average US$180, too low for farmers. In Senegal, where agricultural credit unions are strong, they also do not account for much lending to agriculture, with total lending to agriculture being no more than 5 per cent. Estimated figures in 2004 for Côte d'Ivoire, Mali, and Burkina Faso were 2.2 per cent, 5.4 per cent, and 6.4 per cent respectively (Sacerdoti, 2005).

5.3 INTERLOCKED MARKETS FOR CREDIT AND VALUE-CHAIN FINANCING

Interlinked markets for inputs, outputs, and credit can help to overcome missing markets in the supply of credit to farmers for purchasing improved seeds and fertilizers (Poulton et al., 1998).[9]

Several studies have reviewed the contributions of such interlocking transactions (Poulton et al., 1998; Govereh et al., 1999). Farmers growing cash crops have been known to receive fertilizers for these crops which they also

[9] Marketing boards for cash crops have traditionally provided seeds and fertilizers to farmers on credit and purchased back the produce from farmers at a guaranteed price after deducting the cost of inputs. Under this arrangement, outputs, inputs, and credit are all interlocked in one transaction (Jayne et al., 2002; Poulton et al., 1998).

use on their food crops. Such spillover effects have helped to raise and sustain cotton and maize crop production in the Sahel (Dione, 1991). The introduction of cotton in Mali, Burkina Faso, Niger, and Benin led to rapid uptake of fertilizers primarily for cotton production. The expanded demand for cotton attracted input suppliers to develop markets within the cotton zones, which also helped to improve the accessibility to farm inputs for farmers growing food crops. Farmers often participate in contract farming arrangements because they provide them access to agricultural inputs and credit. In Kenya, Jayne et al. (2002) found that the use of fertilizers on non-interlinked staple food crops, grown within tea and coffee production systems, rose significantly due to participation of farmers in the interlocking credit markets for coffee and tea.

There are many advantages to interlocked credit market arrangements. The risk of default is low since the cost of inputs is deducted before the farmers receive payments for their delivered produce. Monitoring and supervision costs are also reduced since the input loan is delivered in kind to farmers to be applied on their crops. However, where farmers engage in side-selling of their produce at harvest, monitoring costs and defaults may rise, unless there are strict contract enforcement rules. The cost of inputs may also be reduced under interlocked systems since farmers' demand for inputs are aggregated and ordered in bulk.

Market liberalization of the cash crop sector has not always augured well for farmers' access to input and credit. Credit and fertilizer use declined among cotton farmers in West Africa and coffee in Tanzania, following the market reforms that disbanded or reduced the influence of the commodity boards that relied on interlinked markets (Kelly et al., 2003). In Benin, following the dismantlement of the marketing board in 2000, the three largest cotton producer groups organized themselves into a strong farmers' association, the Centrale de Securitisation des Paiements et du Recouvrement (CSPR). The CSPR is linked to the ginning companies under an interlocked credit system. At the beginning of the season, the ginning company pays 40 per cent of agreed quota for delivery which is used to pay for the amounts due to input dealers or commercial banks, with the balance of 60 per cent paid at the end of the season when the cotton is delivered by the farmer groups to the ginneries (Sacerdoti, 2005).

Value-chain financing has traditionally been done through government controlled parastatal agencies (Poulton et al., 1998). For many farmers, these agencies often represented the major source of credit (IFAD, 2003). The market liberalization reforms led to reduction in availability of credit and input and eroded support for subsidized inputs by the state-led value chains (Kherallah et al., 2002; Rozelle and Swinnen, 2004). New forms of value-chain financing emerged, led by the private sector and dominated by agro-processors, agribusinesses, traders, and food-processing companies, either local or international, operating under various forms of contract farming, vertically integrated operations, or out-grower schemes (Swinnen and Maertens, 2010).

Table 9.7 Motivations of small vegetable farmers for participating in high-value commodity chains

	Madagascar	Senegal	
	Reasons for contracting	Reasons for contracting	Most important reason
Stable income	66	30	
Stable prices	19	45	15
Higher income	17	15	
Higher prices	—	11	10
Guaranteed sales	—	66	32
Access to inputs and credit	60	63	44
Access to new technologies	55	17	
Income during the lean period	72	37	

Source: Swinnen and Maertens (2010).

These value-chain financing arrangements are increasingly becoming important for farmers. Studies of commodity value chains have reported that the primary reasons driving farmers to participate in contract farming arrangements are related to having access to finance and agricultural inputs (Table 9.7).

This is especially so for export-oriented value chains, such as horticulture in Kenya (Neven, 2009), and Senegal (Maertens and Swinnen, 2009; Minten et al., 2009). Value-chain financing has been found to lead to significant reduction in poverty for participating farmers (see Maertens et al., 2006 for Senegal and Minten et al., 2009 for Madagascar).

Value-chain financing offers several advantages, including assured markets, guaranteed prices, reduction in marketing risks, sharing of risks in lending and input supply by all participants in the interlocked arrangements, positive spillover effects on other crops, and increased and stable cash flows for farmers (Swinnen and Maertens, 2010).

For value-chain financing to be more effective in improving access of farmers to finance and agricultural inputs, several policy areas of intervention should be prioritized (Swinnen and Maertens, 2010). First, there is need for enforcement of contracts. Side-selling by participating farmers erodes confidence of agribusinesses and could lead to reduction in capital outlays for credit and inputs. Farmers also need to be trained on how to negotiate contracts. Second, a stable macroeconomic environment and lower cost of business will attract greater foreign direct investments by agribusinesses, which will expand access to financing for farmers. Third, greater competition is also needed to expand options for farmers, to reduce opportunities for excessive rent seeking by companies and to create greater equity in sharing of benefits within the value chains. This can be achieved through lowering entry barriers and enhancing competition policies. Fourth, to avoid farmers being locked into monopsony arrangements that offer limited flexibility in terms of financing, alternative approaches for accessing credit should be promoted. This will further enhance the bargaining power on terms of credit they negotiate. Fifth, government regulatory framework should be improved to support inclusive innovations in value-chain financing, especially for food crops grown by poorer farmers.

5.4 CREDIT GUARANTEE SCHEMES

A principal reason why banks don't lend to agriculture is the perception of high risk. To encourage banks to lend to agriculture, credit guarantees have been used. By covering part of the default risk, a lender's risk is lowered—this guarantees secure repayment of all or part of the loan in case of default (Levitsky, 1997). Credit guarantees are useful in addressing challenges faced by farmers, especially those linked to insufficient collateral and poor credit history. Therefore, credit guarantee schemes can improve loan terms and facilitate access to formal credit. Additionally, by allowing loans to be made to borrowers that otherwise would have been excluded from the lending market, farmers and SMEs would be able to establish a repayment reputation in future (De Gobbi, 2002). Although measuring 'additionality' (i.e. the credit that would have occurred in the absence of the guarantee) is difficult (Levitsky and Prasad, 1987), evidence does suggest that credit guarantees can help to leverage more credit for SMEs and farmers, lower their transaction costs, and help to raise productivity (Ruiz Navajas, 2001; Green 2003).

There is a strong role for the use of credit guarantees to expand access to finance for farmers (World Bank, 2008), but their use must be guided by best practices, some of which are discussed here.

Risk sharing: An improperly designed guarantee scheme can increase moral hazards among borrowers by reducing the default risk they otherwise would face. This could lead to 'strategic defaults' (Gudger, 1998). However, a properly designed guarantee scheme can limit moral hazard. For this to occur, it is important that the loan risk is shared amongst the lender, the borrower, and the guarantors.

The extent to which each party should share in the risk shapes the effectiveness of guarantee schemes. The guarantor should accept enough risk to be able to persuade banks to participate in the scheme, while reducing scope for moral hazards or adverse selection. A World Bank study revealed that among the 76 schemes in 46 developed and developing countries, 40 per cent offered 100 per cent guarantees (World Bank, 2008). This is ill-advisable. Coverage rates below 50 per cent reduce the potential for moral hazard and encourage the adequate assessment and monitoring of loans. On the other hand, a coverage rate below 50 per cent reduces banks' incentives to participate in the guarantee programme, especially because loan administration costs can be quite high. The level of risk sharing will also depend on which part of the agricultural value chain the scheme intends to focus on. For lending towards the upper part of the agricultural value chain, agro-processors, agro-dealers, fertiliser and seed companies, etc., a direct risk sharing at 50:50 would be sufficient, as the risk of lending is lower. However, for lending to the lower part of the agricultural value chain, especially to poor smallholder farmers, higher levels of risk sharing will be required. This includes, where necessary, the use of first loss arrangements, especially when considering how to leverage large commercial corporate banks into the sector.

Fees: The fees charged for use of credit guarantees impact not only the incentives lenders and borrowers have in participating in the programme, but also determine the financial sustainability of the fund. Fees must be high enough to cover administrative costs, but low enough to ensure adequate lender and borrower participation. While it is unrealistic to expect credit guarantees to cover its full costs through fees, they need to at least cover the administrative costs of running the scheme. Among the 76 countries studied by the World Bank, 56 per cent of fees were paid by borrowers and 21 per cent were paid by the financial institution receiving the guarantee. Only 15 per cent of schemes impose a membership fee, while 30 per cent impose an annual fee and 48 per cent of the 76 schemes charge a per-loan fee. Fifty-seven per cent of the schemes base the fee on the amount of the guarantee, while 26 per cent base it on the loan amount (World Bank, 2008).

Types of loans: Another important element is whether the guarantee scheme should focus on individual or portfolio loans. A loan-level or individual model applies when applications are approved by the guarantor. In this case, there is a direct link between the borrowers and the lenders since the application assessment is done on case-by-case basis. This allows for a more careful risk management and likely reduces the probability of moral hazard. Such a scenario probably results in a higher-quality loan portfolio. However, this method can also be more costly for the fund to manage. According to the World Bank, 72 per cent of credit guarantee schemes use this selective or individual loan approach (World Bank, 2008).

If the objective of the scheme is to increase guarantee and credit volume, the portfolio model might be a better approach. Under this approach, the guarantor negotiates the criteria of the portfolio. For example, a fund can specify that loans made with its guarantees are targeted at the SMEs sector, a particular location, or a specific loan size.

Defaults: The default rate is an important indication of a scheme's sustainability. When applications are appropriately assessed and monitored, an adequate default rate is possible. Research has shown that a sustainable scheme should aim to have a default rate between 2 and 3 per cent. Newly established schemes in developing countries might consider a higher default rate (i.e. over 5 per cent) in their early years of operation. However, prolonged high default rates should be avoided. A scheme's credibility is also based on how defaults are handled. Guarantee payouts should only be used as a last resort. Before it comes to this, guarantors (or lenders) should negotiate rescheduled payments.

Risk management: In order to reduce the exposure of schemes to default and diversify risk, funds might use risk management mechanisms such as reinsurance, loan sales, or portfolio securitizations. However, these mechanisms require relatively well-developed local capital and financial markets. Nevertheless, the World Bank study revealed that 76 per cent of the schemes studied use risk management tools, about 20 per cent purchased some form of loan insurance, 10 per cent securitized the loans portfolio, and 5 per cent used risk management strategies (World Bank, 2008).

5.5 SUCCESSFUL EXPERIENCES IN EXPANDING ACCESS TO FINANCE FOR FARMERS

As noted earlier, agricultural lending is limited in the continent due to varied reasons. This has prompted the implementation of different models to address the problem. This section reviews the experiences with selected financial models to promote agricultural lending. These include those involving state-led development banks, microfinance, savings and credit schemes, interlocked input, output and credit markets, and credit guarantees. It concludes with lessons from emerging experiences with the use of innovative risk-sharing facilities currently being used in Africa to leverage commercial banks to lend to farmers and agricultural value chains that support them.

AGRA has negotiated and put in place a number of lending programmes with African commercial banks. These lending programmes, which are based on some of the best practices of use of credit guarantees discussed earlier, were set up on the premise that the perceived risk of lending to agriculture is much higher than the real risk. AGRA and its partners have provided risk-sharing facilities to the banks to make it commercially viable to lend as well as reduce the interest rate paid by farmers. The banks are all making significant progress and some achieving impressive results of lending to agriculture. These include the following:

Equity Bank

A risk-sharing facility of $5 million was provided by AGRA ($2.5 million) and IFAD ($2.5 million) to Equity Bank of Kenya in 2008. The facility is expected to leverage $50 million in loans to the agricultural sector in Kenya.

As a result of the risk-sharing facility, the interest rate has been lowered from 18 per cent to 10 per cent. The bank has been strategic and aggressive in promoting the facility. To manage the loan scheme, the bank recruited over one hundred new agricultural graduates. It also established a general manager to take over the operation. The bank has rapidly expanded lending to the agricultural sector, recruited a large number of staff to manage the facility, developed a portfolio of loans (small, medium, and large farmers, and across different districts) that allows it to reduce its risk of lending, and is spearheading the development of new innovative platforms for lending, including the M-PESA (which is now the largest money transfer platform for the country) and the recently launched new initiative with SAFARICOM for mobile banking. These new products will help to reduce the transaction costs of lending and provide for wider reach of its lending products in rural areas.

As at 30 November 2010, 1.6 billion Kenya Shillings (US$20 million) in loans have been provided to the agricultural sector, representing 40 per cent of targeted loan portfolio. Loans to small-scale farmers amounted to Kshs.1.04 billion (US$13.0 million), representing 66.6 per cent of total borrowing. Large-scale farmers have borrowed Kshs.410.6 million (US$5.1 million), representing 26.3 per cent of total lending. Agribusinesses have borrowed Kshs.111.0 million (US$1.4 million) or 7.1 per cent of total borrowing. Loans were extended to 37,645 small-scale farmers, 979 large-scale farmers, and 313 agribusinesses. As a result of expanded lending from the bank, it has been reported that several agro-dealers no longer needed working capital loans from the bank, since their stocks of seeds and fertilizers are moving very rapidly as farmers can now afford to pay for the inputs in cash. This is a clear indication that the model is helping to expand demand for agricultural inputs in rural areas. There has not been any call to the guarantee funds.

The bank has rapidly gained an international reputation for its innovation in expanding lending into agriculture. The bank has won the African best bank of the year, while its CEO also won the 2010 Africa's best banker of the year citation. The approach of the bank in accelerating lending to smallholder farmers has become a model for other banks in Africa, many of whom are sending representatives to the bank to learn from its approach. The risk-sharing facility of AGRA and IFAD with Equity Bank was featured as one of the successes in lending in African agriculture at the United Nations Private Sector Forum during the 2010 UN Summit on the Millennium Development Goals held in September in New York.

National Microfinance Bank

A loan guarantee facility was provided by AGRA (US$1.1 million) in partnership with the Financial Sector Deepening Trust (US$1 million) to the National Microfinance Bank in Tanzania. The facility leveraged US$10 million in loans to agro-dealers in Tanzania.

The facility resulted in the lowering of interest rate on loans from 28 per cent to 15 per cent and a reduced demand for collateral by the bank. As at 30

September 2010, Tshs.9.05 billion (US$7 million) in lending has been provided to 709 agro-dealers to stock and sell improved seeds, fertilizers, and agrochemicals, representing 87 per cent of targeted lending. The program started in five districts in 2008 and has since expanded to 38 districts nationwide. So far there have been delays in payment for Tshs. 527 million, which represents about 6 per cent of total lending. However, the bank expects to fully recover these amounts using its normal loan recovery process and does not expect to revert to the guarantee fund.

As a result of the lending, access to agricultural inputs has significantly improved, with expansion in the range, volume, and timeliness of supply of farm inputs. The success of the bank in achieving this convinced the government of Tanzania to use the bank to manage its subsidized farm inputs programme nationwide. Under the agreement, the government issues input vouchers to farmers, which they redeem from the agro-dealers. The agro-dealers redeem the vouchers from the bank. This has led to a dramatic increase in agricultural production in the southern highlands especially, where the success of the bank in lending to and supporting voucher reimbursement for agro-dealers have supported some seven hundred thousand farmers to produce 5 million metric tonnes of maize. To further build the pipeline of eligible agro-dealers, CNFA has trained and certified 6,200 agro-dealers, all of whom will be eligible to access credit under this scheme.

Standard Bank

In 2009 AGRA and its strategic partners (Millennium Challenge Corporation, Millennium Development Authority—MIDA, Ghana, Millennium Challenge Account, MCA-Mozambique and Kilimo Trust) put in place a US$10 million risk-sharing facility to encourage the Standard Bank (Africa's largest bank) to get into agricultural lending, especially for smallholder farmers and agricultural value chains that support them. The facility was able to leverage $100 million in lending commitments to farmers and agricultural value chains in four countries—Tanzania, Uganda, Ghana, and Mozambique, respectively. Good progress is being made on the facility, despite a slower take-off than anticipated. Standard Bank provides loans carrying interest rates of base plus 3 per cent for smallholder farmers and up to base plus 5 per cent for agri-businesses. As of 4 November 2010 loan deals approved has targeted thousands of farmers: in Uganda US$3.8 million towards 53,000 farmers, in Tanzania US$2.3 million towards 21,000 farmers, in Mozambique US$190,000 towards 2,000 farmers, and in Ghana US$4.9 million towards 13,000 farmers.

6. Conclusions

Improving farmers' access to agricultural technologies is critical for efforts to achieve greater agricultural productivity and food security in Africa. Despite the availability of adaptable agricultural technologies, including a wide range of improved crop varieties, fertilizers and other complimentary technologies, farmers' use of technologies has fallen way below expected levels. The majority of the technologies are on the shelf. Some of the major reasons for this are poorly developed agricultural input and output markets, as well as missing financial markets for accessing needed finance to invest in new technologies.

This chapter has reviewed the alternative approaches that can be used to improve access of farmers to agricultural inputs and finance. A number of areas deserve attention for policymakers. To accelerate the access of farmers to agricultural inputs, priority should be accorded to the rapid expansion of agro-dealers across rural Africa. In areas where they exist at higher densities, the distances travelled by farmers to find farm inputs have declined substantially. Challenges remain in the low potential areas where the density of agro-dealers is much lower; risks to the use of external purchased inputs are higher and demand for inputs are too low to sustain private-sector markets. To reduce inequality in the development of agro-dealers in low potential areas, policy measures should be developed to subsidize the training of agro-dealers and improve their access to affordable financing. Agro-dealers can be used to complement traditional extension systems to improve the demonstration of agricultural technologies in rural areas. Measures to improve affordability of farm inputs include the use of 'smart subsidies' which targets poorer farmers, uses vouchers, and relies on the private sector to deliver subsidized inputs. Great care should be taken to avoid displacement of commercial sales, especially in areas where there exist wealthier commercial farmers that would normally have purchased farm inputs. The packaging of seeds and fertilizers in small quantities will help poor farmers to experiment with new technologies at lower cost and risks, while making inputs more affordable for a greater majority of the rural poor, especially those in low potential areas.

Access to finance is a more difficult challenge for farmers due to the highly covariate nature of risks, lack of collateral from farmers, absence of credit bureaus to determine risk profiles for loans, and limited coverage of formal financial institutions in many rural areas. Agricultural lending programmes through agricultural development banks have been tried in the past. These have generally failed in Africa due to their reliance on subsidized interest rates, pervasive defaults arising from a culture of writing off loans, high cost of delivering loans, political interference, and poor managerial capacities. Microfinance, savings, and credit institutions can also play a role in agricultural lending, but they suffer from the miniature nature of their loans, non-appropriateness for agriculture, and low levels of liquidity to support agricultural lending unless they are linked to formal financial institutions to access larger levels of capital. These institutions charge very high interest rates to cover their high costs of lending. As the recent experience with microfinance in India shows, microfinance institutions tend to transfer their high costs of operations to borrowers, which may raise the risk of default since their clients are generally poor farmers or rural households.

Market liberalization has led to a reduction in the importance of agricultural parastatals that engage in interlocked markets that provide inputs, credit, and purchases output at guaranteed prices. The growing importance of high-value crops and global food chains has opened up new opportunities for farmers to secure credit through interlocked markets operated by agribusinesses, traders, and processors. Farmers participate in these arrangements mainly to secure credit and agricultural inputs. Value-chain financing should be promoted, under contract farming or out-grower schemes, to improve access of farmers to finance and markets. However, attention should be given to issues of equity in the value chains, regulations to avoid exploitation of farmers, training of farmers in contract negotiations, enhanced competition by agribusinesses through reduction in barriers to entry and encouraging the development of alternative financial institutions for credit to provide farmers with wider options.

Commercial banks hold the greatest scope for expanding lending to agriculture, for seasonal loans and term financing. African banks have excess liquidity which should be tapped to expand lending into agriculture. This will require the design of appropriate risk-sharing instruments to reduce their risks of lending and leverage them into the agricultural sector. The successful experiences of AGRA, IFAD, and their partners in leveraging commercial banks to lend to agriculture should be carefully evaluated and scaled up across African countries. To further expand access to finance, there is need for insurance products that are appropriate to farmers. Weather-indexed crop insurance can help reduce the risks faced by commercial banks in lending to farmers.

REFERENCES

Adams, D. W. (1994). 'Altruistic or production finance? A donors' dilemma', *Economics and Sociology*, Occasional Paper No. 2150, Columbus, Ohio: The Ohio State University.

Aina, L. O. (2006). 'Information provision to farmers in Africa: The library-extension service linkage', paper presented at the *World Library and Information Congress: 72nd IFLA General Conference and Council*, 20–24 August, Seoul, Korea.

Alemu, D. and D. J. Spielman (2006). 'The Ethiopian seed system: Regulations, institutions and stakeholders', a paper presented at *ESSP Policy Conference 2006 Bridging, Balancing, and Scaling Up: Advancing the Rural Growth Agenda in Ethiopia*, 6–8 June, Strategy Support Program and International Food Policy Research Institute.

Alston, J. M., C. Chan-Kang, M. C. Marra, P. G. Pardey, and T. J. Wyatt (2000). 'A meta-analysis of rates of returns to agricultural R & D. Ex Pede Herculem?', Research Report 113, International Food Policy Research Institute, Washington DC.

Anon (1997). *Monitoring of Relief Seed Distribution.* Nairobi: Ministry of Agriculture.

Ariga, J. and T. S. Jayne (2006). 'Can the market deliver? Lessons from Kenya's rising use of fertiliser following liberalization', Policy Brief No. 7, Tegemeo Institute for Agricultura and Policy, July.

AT Uganda (2005a). 'The input distribution sector at a glance: Summary results of the 2004 National Agro-input Dealers Census', report submitted to Rockefeller Foundation.

AT Uganda (2005b). 'Facilitating Agricultural Input Distribution Linkages in Northern and Eastern Uganda', report submitted to The Rockefeller Foundation.

Bay, A. P. M. (1998). 'The seed sector in sub-Saharan Africa: Alternative strategies', *Proceedings of the Regional Technical Meeting on Seed Policy and Programmes for Sub-Saharan Africa*, FAO, 23–27 November, Abidjan.

Bramel, P., T. Remington, and M. McNeill (eds) (2004). 'CRS Seed Vouchers & Fairs: Using Markets in Disaster Response', 21–26 September, Symposium, Lake Baringo Kenya (Nairobi: CRS).

Briand, V., N. Diop, and Q. Wodon (2006). 'Challenges in Mali's cotton sector: Profitability and competitiveness', chapter 2 of Draft Working Paper 'Cotton and Poverty in Mali', World Bank, Washington DC.

CNFA (Citizens' Network for Foreign Affairs) (2005). 'Distribution and characteristics of stockists in agricultural inputs in Western Kenya', a survey report prepared by CNFA in collaboration with Agricultural Market Development Trust (AGMARK), Kenya, April 2005.

Commander, S. (1989). *Structural adjustment and agriculture: Theory and practice in Africa and Latin America*, Overseas Development Institute.

Crawford, E. W., T. S. Jayne, and V. A. Kelly (2006). 'Alternative approaches for promoting fertilizer use in Africa', Agriculture and Rural Development Discussion Paper 22, The International Bank for Reconstruction and Development, The World Bank, Washington DC, USA.

Cummings, R., S. Rashid, and A. Gulati (2006). 'Grain price stabilization experience in Asia: What have we learned?', *Food Policy*, 31, pp. 302–12.

De Gobbi, M. (2002). 'Making Social Capital Work: Mutual Guarantee Associations for Artisans', Social Finance Programme, Employment Sector, International Labour Organisation, September.

Denning G., P. Kabambe, P. Sanchez, A. Malik, R. Flor, et al. (2009). 'Input subsidies to improve smallholder maize productivity in Malawi: Toward an African Green Revolution', *PLoS Biol*, 7(1), e1000023, doi:10.1371/journal.pbio.1000023.

Dione, J. (1991). 'Food security and policy reform in Mali and the Sahel', in Dasgupta, P. (ed), *Issues in Contemporary Economics: Proceedings of the Ninth World Congress of the International Economic Association*, Volume 3. London: Macmillan.

Djurfeldt, G., H. Holmen, M. Jirstrom, and R. Larsson (eds) (2005). *The African Food Crisis: Lessons from the Asian Green Revolution.* Wallington: CABI Publishing

Dorward, A. (2009). 'Rethinking agricultural input subsidy programmes in a changing world', Paper prepared for the Trade and Markets Division of the Food and Agriculture Organization of the United Nations, April 2009.

Dorward, A., J. Kydd, and C. Poulton (eds) (1998). *Smallholder Cash Production under Market Liberalization: A New Institutional Economics Perspective.* Wallingford: CAB International.

Dorward, A., C. Poulton, and J. Kydd (2001). 'Rural and farmer finance: an international perspective (with particular reference to Sub-Saharan Africa)', Drakensburgs, South Africa, paper presented at the Agricultural Economics Society of South Africa Conference.

Eicher, C. (1999). 'Institutions and the African farmer', Third Distinguished Economist Lecture, CIMMYT Economics Program. CIMMYT, Mexico.

Fan, S., A. Gulati, and S. Thorat (2008). 'Investments, subsidies and pro-poor growth in rural India', *Agricultural Economics*, 39(2), pp. 163–70.

Gebreselassie, S. (2006). 'Intensification of Smallholder Agriculture in Ethiopia: Options and Scenarios', paper prepared for the Future Agricultures Consortium Meeting at the Institute of Development Studies, 20–22 March, Future agricultures.

Goldstein, G. and I. Barro (1999). 'The role and impact of savings mobilization in West Africa: a study of the informal and intermediary financial sectors (Benin, Burkina Faso, Ghana, Guinea, Mali and Togo)', MicroSave-Africa/West Africa.

Govereh, J., J. Nyoro, and T. Jayne (1999). 'Smallholder commercialization, interlinked markets and food crop productivity: cross-country evidence in Eastern and Southern Africa', East Lansing, MI: Michigan State.

Green, A. (2003). 'Credit guarantee schemes for small enterprises: An effective instrument to promote private sector-led growth?', the United Nations Industrial Development Organization (UNIDO) Working Paper No. 10, August.

Gudger, M. (1998). 'Credit guarantees: An assessment of the state of knowledge and new avenues of research', FAO Agricultural Services Bulletin 129.

Hazell, P. and C. Ramasamy (1991). *Green revolution reconsidered: The impact of High Yielding Rice Varieties in South India.* Baltimore: USA, Johns Hopkins University Press.

IFAD (2003). 'Agricultural marketing companies as sources of smallholder credit in Eastern and Southern Africa. Experiences, insights and potential donor role', Rome, December 2003.

Jarvis, D., L. Myer, H. Klemick, L. Guarino, M. Smale, A. H. D. Brown, M. Sadiki, B. Sthapit, and T. Hodgkin (2000). 'A Training Guide for in situ Conservation On-farm', Rome: International Plant Genetic Resources Institute.

Jayne, T. S., T. Yamao, and J. Nyoro (2002), 'Interlinked credit and farm intensification: evidence from Kenya', Contributed paper submission for the 25th International Conference of Agricultural Economists, 16–22 August 2002, Durban, South Africa.

Kelly, V., A. A. Adesina, and A. Gordon (2003). 'Expanding access to agricultural inputs in Africa: a review of recent market development experience', *Food Policy*, 28, pp. 379–404.

Kelly, V. A. (2006). *Factors affecting demand for fertilizer in sub-Saharan Africa.* Washington, DC, USA: The International Bank for Reconstruction and Development/World Bank.

Kherallah, M., C. Delgado, E. Gabre-Madhin, N. Minot, and M. Johnson (2002). *Reforming Agricultural Markets in Africa.* Baltimore: IFPRI and Johns Hopkins University Press.

Kibaara, B., J. Ariga, J. Olwande, and T. S. Jayne (2009). *Trends in Kenyan Agricultural Productivity: 1997–2007.* Nairobi, Kenya: Tegemeo Institute of Agricultural Policy and Development. WPS 31/2008.

Krausova, M. and A. B. Banful (2010). 'Overview of the agricultural input sector in Ghana. International Food Policy Research Institute', Discussion Paper 01024, September.

Langyintuo, A. S., W. Mwangi, A. O. Diallo, J. MacRobert, J. Dixon, and M. Bänziger (2010). 'Challenges of the maize seed industry in eastern and southern Africa: A compelling case for private–public intervention to promote growth', *Food Policy*, 35, pp. 323–31.

Levitsky, J. (1997). 'Best Practice in Credit Guarantee Schemes', *The Financier*, Vol. 4, No. 1 and 2, February, May.

Levitsky, J. and R. N. Prasad (1987). 'Credit guarantee schemes for small and medium enterprises', World Bank technical paper, no. 58. Industry and finance series, v. 19. World Bank, Washington DC.

Longley, C. (2006). 'Agricultural input vouchers in emergency programming: lessons from Ethiopia and Mozambique', HPG background paper, Overseas Development Institute, London.

Maertens, M., L. Dries, F. A. Dedehouanou, and J. F. M. Swinnen (2006). 'High-value global supply chains, EU Food Safety Policy and smallholders in developing countries. A case-study from the green bean sector in Senegal', in Swinnen, J. F. M. (ed.), *Global Supply Chains, Standards and the Poor.* Cambridge, MA: CABI Publishing.

Maertens, M. and J. F. M. Swinnen (2009). 'Trade, standards, and poverty: Evidence from Senegal', *World Development*, 37(1), pp. 161–78.

Makokha, M., P. Omanga, A. Onyango, J. Otado, and T. Remington (2004). 'Comparison of seed voucher & fairs and direct seed distribution: Lessons learned in eastern Kenya and critical next steps', in Sperling, L., T. Remington, J. M. Haugen, and S. Nagoda (eds), *Addressing Seed Security in Disaster Response: Linking Relief with Development.* Cali, Colombia: International Center for Tropical Agriculture, pp. 45–68.

Maredia, M. and J. A. Howard (1998). 'Facilitating seed sector transformation in Africa: Key findings from the literature', Policy Synthesis No. 33, United States Agency for International Development (USAID) Bureau for Africa, Washington, DC.

Minot, N. and T. Benson (2009). 'Fertilizer subsidies in Africa: Are vouchers the answer?', International Food Policy Research Institute, Washington DC. IFPRI Issue Brief 60, July.

Minten, B., L. Randrianarison, and J. F. M. Swinnen (2009). 'Global retail chains and poor farmers: Evidence from Madagascar', *World Development*, forthcoming.

Morris, M., V. Kelly, R. J. Kopicki, and D. Byerlee (2007). 'Fertilizer use in African agriculture: lessons learned and good practice guidelines', Directions in Development, Agricultural and Rural Development, World Bank.

Moseley, W. G. and L. C. Gray (2008). *Hanging by a Thread: Cotton, Globalization, and Poverty in Africa.* Athens, Ohio: Ohio University Press.

Mosley, P. and J. Weeks (1993). 'Has recovery begun? Africa's structural adjustment in the 1980's revisited', *World Development*, 21, pp. 1583–606.

Nathaniels, N. Q. R. and A. Mwijage (2000). 'Seed fairs and the case of Marambo village, Nachingwea District, Tanzania: implications of local informal seed supply and variety development for research and extension', AgREN network paper no. 101, January 2000, Overseas Development Institute, Agricultural Research and Extension Network, London.

Neven, D., M. M. Odera, T. Reardon, and H. Wang (2009). 'Kenyan supermarkets, emerging middle-class horticultural farmers, and employment impacts on the rural poor', *World Development*, 37(11), 1802–11.

Omamo, L. O. and S. W. Mose (2000). 'Fertilizer trade under market liberalization', preliminary evidence from Kenya, Kenya Agricultural Marketing and Policy Analysis Project, Tegemeo Institute of Agricultural Policy Development/Ergerton University/Kenya Agricultural Research Institute, and Michigan State University.

Phiri, M. A. (2004). 'An evaluation of RUMARK's input distribution strategies for smallholder agricultural incomes in Malawi', Report prepared for CNFA and the Rockefeller Foundation, September 2004.

Phiri, M. A., R. R. Chirwa, and J. M. Haugen (2004). 'A review of seed security strategies in Malawi', in L. Sperling, T. Remington, J. M. Haugen, and S. Nagoda (eds), *Addressing Seed*

Security in Disaster Response: Linking Relief with Development. Cali, Colombia: International Center for Tropical Agriculture, pp. 134–58.

Poulton C., A. Dorward, J. Kydd, N. Poole, and L. Smith (1998). 'A new institutional economics perspective on current policy debates', in Dorward, A., J. Kydd, and C. Poulton (eds), *Smallholder Cash Crop Production under Market*.

Poulton, C., A. Dorward, and J. Kydd (1998). 'The revival of smallholder cash crops in Africa: public and private roles in the provision of finance', *Journal of International Development*, 10 (1), pp. 85–103.

Poulton, C., J. Kydd, and A. Dorward (2006). 'Overcoming market constraints to pro-poor agricultural growth in sub-Saharan Africa', *Development Policy Review*, 24(3), pp. 243–77.

Remington, T., J. Maroko, S. Walsh, P. Omanga, and E. Charles (2002). 'Getting off the seed and tools treadmill with CRS seed vouchers and fairs', *Disasters*, 26(4), pp. 302–15.

Rohrbach, D., R. Charters, and J. Nyagweta (2004). *Guidelines for Agricultural Relief Programs in Zimbabwe*. Bulawayo: International Crops Research Institute for the Semi-Arid Tropics.

Rohrbach, D., A. B. Mashingaidze, and M. Mudhara (2005). *The Distribution of Relief Seed and Fertilizer in Zimbabwe, Lessons Derived from the 2003/04 Season*. Bulawayo, Zimbabwe: The International Crops Research Institute for the Semi-Arid Tropics.

Rohrbach, D., K. Mazvimavi, T. Pedzisa, and T. Musitini (2006). *A Review of Seed Fair Operations and Impacts in Zimbabwe*. Bulawayo, Zimbabwe: The International Center for Research in the Semi-Arid Tropics.

Rozelle, S. and J. Swinnen (2004). 'Success and Failure of Reforms: Insights from Transition Agriculture', *Journal of Economic Literature*, 42(2), pp. 404–56.

Ruiz Navajas, A. (2001). 'Credit Guarantee Schemes: Conceptual frame', Financial System Development Project, GTZ/FONDESIF, November.

Sacerdoti, E. (2005). 'Access to bank credit in sub-Saharan Africa: Key issues and reform strategies', IMF Working Paper, WP/05/166, IMF, Washington DC.

Smale, M. and P. W. Heisey (2001). 'Maize technology and productivity in Malawi', in Byerlee, D. and Eicher, C. (eds), *Africa's emerging maize revolution*. USA: Lynne Reinner.

Spencer, D. S. C. and O. Badiane (1995). 'Agriculture and economic recovery in African countries', in 'Agricultural competitiveness: Market forces and policy choices', ed. Peters, G. H. and D. D. Hedley, *Proceedings of the Twenty-Second International Conference of Agricultural Economists, Harare, Zimbabwe*, 22–29 August 1994. Aldershot, UK: Dartmouth.

Sperling, L. (2000). 'Emergency seed aid in Kenya: a case study of lessons learned', Report submitted to the United States Agency for International Development, Washington, DC.

Sperling, L. (2002). Emergency seed aid in Kenya. 'Some case study insights from lessons learned during the 1990s', *Disasters*, 26(4), pp. 283–7.

Sperling, L., H. D. Cooper, and T. Remington (2008). 'Moving towards more effective seed aid', *Journal of Development Studies*, April, 44(4), pp. 586–612.

Sperling, L. and M. Loevinsohn (1993). 'The dynamics of improved bean varieties among small farmers in Rwanda', *Agricultural Systems*, 41, pp. 441–53.

Sperling, L., T. Osborn, and H. D. Cooper (2004). 'Towards effective and sustainable seed relief activities, report on the workshop on effective and sustainable seed relief activities, 26–28 May 2003', in Tripp, R. (ed.) (1997), *New Seed and Old Laws: Regulatory Reform and the Diversification of National Seed Systems*. London: Intermediary Technology Publications on behalf of the Overseas Development Institute.

Swinnen, J. F. M. and M. Maertens (2010). 'Finance through food and commodity value chains in a globalizede', paper presented at the KfW Financial Sector Development Symposium 2010, *Finance for Food: Towards New Agricultural Rural Finance, 2010*.

Tripp, R. (1998). Regulatory issues: Varietal registration and seed control, in Morris, M. L. (ed.), *Maize Seed Industries in Developing Countries.* Lynne Rienner Publishers, Inc., pp. 159–74.

Tripp, R. and D. Rohrbach (2001). 'Policies for African seed enterprise development', *Food Policy*, 26(2), pp. 147–61.

Walsh, S., J. M. Bihizi, C. Droeven, B. Ngendahayo, B. Ndaoroheye, and L. Sperling (2004). 'Drought, civil strife and seed vouchers and fairs: the role of the trader in the local seed system', in Sperling, L., T. Remington, J. M. Haugen, and S. Nagoda (eds) (2004), *Addressing Seed Security in Disaster Response: Linking Relief with Development.* Cali, Colombia: International Center for Tropical Agriculture, pp. 45–68.

West, J. and F. Bengtsson (2005). 'The wider context of emergency seed vouchers and fairs', Masters thesis, Development Studies, Management of Natural Resources and Sustainable Agriculture, Norwegian University of Life Sciences, As.

World Bank (2006). 'Promoting increased fertiliser use in Africa: lessons learnt and good practice guidelines', Africa Fertilizer Strategy Assessment, Technical Report. Discussion Draft, June 2006.

World Bank (2008). 'The typology of partial credit guarantee funds around the world', Policy Research Working Paper 4771. Development Research Group, The World Bank, Washington DC.

Xu, Z., W. J. Burke, T. S. Jayne, and J. Govereh (2009). 'Do input subsidy programs "crowd out" commercial market development? Modeling fertiliser demand in a two-channel marketing system', *Agricultural Economics*, 40, pp. 79–49.

You, L. Z. (2008). 'Africa infrastructure country diagnostic: Irrigation investment needs in Sub-Saharan Africa', Summary of Background Paper 9, International Food Policy Research Institute (IFPRI), Washington DC, USA.

10 Corporate agribusiness development and small farms

MARTIN EVANS[1]

1. Introduction

The term 'agribusiness' covers a wide range of activities in the agricultural sector. In this chapter it refers to enterprises engaged in: (i) the provision of specialized agricultural inputs, such as seeds, animal feeds, fertilizers, agrochemicals, and mechanical services; (ii) 'farming' itself, including crop cultivation, agro-forestry, livestock rearing, and aquaculture; and/or (iii) the primary procurement, handling (including storage and transportation), processing, and distribution of agricultural raw materials,[2] which are organized, managed, and financed in order to achieve what are fundamentally commercial objectives. Thus the performance of such enterprises is usually judged on the basis of criteria such as turnover, profitability, market share, growth in net worth, and so on.

'Corporate agribusiness' here refers to agribusiness enterprises that are formally constituted as companies owned by shareholders, officially recognized and registered as such where necessary. They fall into two categories: private companies which are typically owned by family members or a small number of partners whose shares are 'unlisted'; and public companies, whose shares are traded on stock markets. Private agribusiness companies come in all sizes and some can be very large (for example, Cargill and Compagnie-Fruitière).[3] It needs to be borne in mind at this point, however, that much of the off-farm value addition in the agricultural economies of many countries is accounted for by the 'non-corporate agribusiness' sector, composed of myriad

[1] The author is greatly indebted to Geoff P. Tyler for his highly valuable comments and contributions on an earlier draft of this chapter. He is also very grateful to Peter Hazell for his help in identifying key references and source material.

[2] Agribusiness, as defined here, excludes the food and beverage sector, which is typically engaged in secondary processing and manufacturing.

[3] Arguably, private agribusiness companies may be better able to focus on business strategies for long-term profitability than their public counterparts, which also need to manage the short-term opinions and expectations of stock markets.

commercial agriculture-based enterprises that are informally organized and often not registered.

The tendency of corporate involvement at all levels of agricultural value chains to expand as economies grow and develop, and the reasons for this, have been known for some time. Like urbanization and industrialization, this process of agricultural 'corporatization' is an inevitable—and probably necessary—feature of development.[4] However, there is currently new interest in the process, for probably two main reasons. First, in the two decades before the global financial crisis of 2008/9, the pace of structural transformation in many developing and emerging market economies, especially in Asia, had been accelerating, with corresponding rapid changes in agricultural systems. Second, starting a few years earlier, but dramatically revealed by the commodity price 'spike' of 2007/8, there has been a recent uplift in most agricultural prices towards levels in real terms not seen since the 1950s and 1960s. This is widely regarded as probably irreversible and has suddenly and sharply focused attention on the ability of the global agricultural system to satisfy the fast-growing demand for food, feed, fibre, and fuel, particularly for food and for higher-value products that use a lot of agricultural resources. It seems that the availability of good quality, safe, diversified, and affordable food, even for relatively affluent urban populations, let alone poor rural communities, can no longer be taken for granted.

The greater part of the world's food supply is still the responsibility of small-scale or smallholder farmers (see Chapter 1 of this book for a definition of 'small farm') and most of those who have 'surpluses' to sell do so into markets still organized on very traditional lines. To some governments, this poses a strategic risk to national food security in that it will take too long and be too costly on public funds to source much of the additional food supplies required from the small farm sector. This is not to deny the developmental case for increasing the productive capacity of small-scale farmers and improving their welfare as essential elements of poverty reduction strategy. Rather, it reflects growing doubt in some quarters that this small farmer-centred route to economic security and prosperity will, by itself, be sufficient. Private investment from the corporate agribusiness sector is increasingly now seen as playing an essential role in unlocking national agricultural potential, both to safeguard domestic food security and to contribute significantly to overall economic growth. At the same time, corporate interest in investing in the agricultural sector has been stimulated by the perception that prospects appear considerably more attractive than they have for some time.

Why should the simultaneous development of the small farm sector and the corporate agribusiness sector, the former for social as well as economic

[4] In the same sense that agro-industrialization in general is necessary, according to Reardon and Barrett (2000).

reasons and the latter for purely economic reasons, be an issue? Basically, it is because the two sectors are very unequal in terms of their respective economic power. Due to their greater commercial presence, companies can gain access to resources, including land and water, and to agricultural input and output markets more easily than can individual small farmers. By the same token, companies can exert greater influence over how markets are structured, managed, and regulated. A growing corporate agribusiness sector represents an increasingly potent force for change in agricultural systems and, as such, can work both for and against the interests of the small farm sector.

On the positive side, corporate agribusiness can offer small farmers and rural communities a number of significant opportunities. Principal among these is access to output markets, including connection to national and even global markets, which small farmers would otherwise have difficulty accessing themselves on any sustained and remunerative basis without a corporate intermediary in the supply chain. Another important opportunity is access to superior agricultural technology, including the necessary material inputs and often finance, to related information and knowledge, and the acquisition of relevant skills. There is frequently a spillover effect that benefits other parts of the farm household economy outside this specific linkage (Barrett et al., 2012). Corporate off-farm activity downstream[5] of agricultural production, especially product processing, also creates opportunities locally for employment, small business development, marketing of food produce, and vocational training usually on a larger scale and in a more organized manner than informal agribusiness enterprise can deliver.

However, there can also be a negative side, actually in some situations and potentially in others, to greater corporate involvement in the agricultural economy. This largely stems directly from the tendency towards industrial concentration in the markets both upstream and downstream from farming.[6] This is particularly marked in agricultural input supply, especially of improved seeds from biotechnology companies, where barriers to entry can be considerable. Industrial concentration in value-chain activities downstream from farming tends to increase the nearer to the final consumer. This is particularly so as countries become more industrialized, with a very small number of food processing and marketing companies being responsible

[5] This chapter follows the general convention, which is to refer to primary producers as being at the upstream end of supply chains and at the bottom of value chains, and to final consumers as being at the downstream end of supply chains and at the top of value chains.

[6] In 2010, it was reported that the combined worldwide annual sales of the three largest agri-producers/traders exceed US$237 billion, with the corresponding figure for agribusiness input suppliers (different companies) being US$96 billion [Business Monitor International Ltd, *Argentina Agribusiness Report*, Q3 2010]. In the Venezuelan coffee industry, there are more than half a million producers and over 5 million consumers, but only four roasters and processors.

for the bulk of procurement from large numbers of farmers.[7] Insufficient competition in the markets for farm product is always a concern. It can be counteracted by regulation,[8] although this is often difficult to make effective in practice. In principle, it can also be counteracted by farmers organizing themselves into collaborative alliances, such as production or marketing associations and cooperatives, and thus increasing their bargaining power collectively. Such groupings can be difficult and costly to establish and manage, however.

Actual experience shows that the interaction between corporate agribusiness and small farms can be either good or bad for the latter or have little effect at all one way or the other, depending on the specific circumstances. Clearly, the development challenge is how to widen and deepen corporate involvement in the agricultural economy where this can bring significant benefits to the small farm sector, while at the same time keeping any detrimental impact to a minimum.

2. **Corporate investment in agricultural value chains**

It is useful to start by considering where corporate agribusiness has been stepping up its investment in the agricultural sector in recent years, since this will provide some clues as to what is currently driving greater corporate involvement.

2.1 THE AGRICULTURAL TRANSFORMATION PROCESS

There is a substantial literature on the changes that the agricultural sector undergoes as countries follow the common development path from predominantly 'agriculture-based', through 'transforming' to 'urbanized' economies, to adopt a recent World Bank typology (World Bank, 2008). A useful characterization of *food* systems is given in McCullough et al. (2010), which distinguishes three main types: 'traditional', 'structured', and 'industrialized'. There is an implicit transition from the first, through the second, to the third as the agricultural sector becomes increasingly commercialized and the overall economy develops.

[7] Livestock farmers in the US protested a few years ago at the 'unfair' economic power wielded by meat processing companies [*The Times* (London), July 2010]. These protests echoed similar concerns voiced by Australian cattle raisers some years earlier.

[8] Vorley et al. (2009) cite the example of the appointment of a supplier ombudsman in Australia with an independent regulatory role to oversee the way in which powerful buyers such as supermarkets engage with their suppliers.

Corporate enterprises and small farms co-exist, and may engage with each other, at all stages of agricultural transformation. Companies growing and processing plantation crops, such as cotton, cocoa, coffee, oil palm, sisal, sugar-cane, tea, and tobacco, have long been established in countries where agriculture is predominantly smallholder-based and the food system is mainly of the *traditional* type. Some of these smallholders also grow plantation crops, sometimes as outgrowers for the plantation companies.[9] In *structured* food systems supplying a more diversified product mix, with a greater degree of processing in the value chain, and the emergence of specialized wholesalers alongside traditional ones and with greater market regulation, small farms become more specialized. The corporate presence in food processing becomes more extensive, with higher value-adding activities such as meat, dairy, fruit, and vegetables becoming more important compared with milling and oil extraction for staple foods. Many of these processors and wholesalers, which include corporate as well as non-corporate small and medium enterprises (SMEs), procure their raw materials from small farms as well as large ones. There is some development of supermarkets and vertical coordination of value chains. The latter tend to be pronounced features of *industrialized* food systems, in which processing plays a very large role and distribution of fresh food is highly organized. Economic relationships between value-chain participants are increasingly governed by individual contract, particularly for fresh produce for which demanding consumer requirements for product quality, safety, traceability, and ethical sourcing must be met, or by highly standardized transactions in the case of linkages between open, competitive markets. Corporate business dominates agricultural activity, both upstream and downstream from farming, and sometimes farming itself as well.

This corporatization of value chains in the agricultural sector as it becomes more complex does not by itself mean that small farms get 'squeezed out'. Much depends on the prevailing agrarian structure. In parts of China, India, and some South-East Asian countries where land tenure laws restrict the size of agricultural holdings over large areas, corporate agribusiness necessarily has to source directly or indirectly through intermediaries mainly from small farms. In contrast, in parts of Africa and Latin America where there are fewer legal restrictions on sizes of holdings, corporate agribusiness has more of a choice about whether or not to source from small farms. This has long been the situation. In recent years, however, there have been some shifts of emphasis in patterns of corporate investment in agribusiness, some of which

[9] It is interesting to note that a review of agribusiness investments made over 50 years by the Commonwealth Development Corporation (CDC) concluded that whereas outgrower-only schemes and corporate farm (estate)-only schemes enjoyed the same rate of success, the combination of nucleus estate with outgrowers was more successful than both these business models (Dixie, 2011).

are 'new' and some of which represent an acceleration of earlier trends, which are changing this picture. These are discussed in the next section.

2.2 SOME RECENT DEVELOPMENTS

Private-sector recognition of agriculture as a 'growth industry'

There is a widespread perception among corporate investors, both agribusiness operators and general fund managers (private, public, sovereign), that profitability in the agricultural sector is now higher and more sustainable—relative to other sectors—than it has been for many decades. Since the commodity price hikes of 2007/8, the terms of trade are seen as having swung in agriculture's favour and are expected to remain so. At the same time, the greater uncertainty that the market volatility of very recent years has induced, has focused agricultural investor interest even more on primary product supply from locations where domestic resource costs are internationally low, and the product can be relatively easily and cheaply brought to market, including the global market. This includes both high-value products, such as Peruvian asparagus and Ethiopian cut flowers destined for Europe, as well as low-value ones such as Cambodian paddy rice exported by Vietnamese and Thai companies to their own countries (FAO, 2010) and Sierra Leone rice produced with Chinese investment.[10] This acceleration of the global integration of agricultural markets is a potential threat to parts of the small-farm sector, in that those farms which are not part of this process may lose out comprehensively, being able neither to compete in their domestic markets nor benefit from export markets. Bulk handling, for example, has greatly reduced the cost of international freight, and it is often cheaper for coastal cities to import grain than to source this internally from remote and scattered small farms at a high transportation cost.

Agribusiness concern over security of raw material supply

The recent price hikes have led to a re-evaluation by agribusiness of the security of its raw material supplies—both their physical availability and their cost. There is greater interest by processors in investing upstream in farming as owners or operators and in cementing strategic partnerships with farming organizations. Some traders and distributors, including retailers, are looking at investing upstream in processing and wholesaling. Again the extent to which small farms benefit or suffer from such developments depends largely on agrarian structures and land availabilities. In Indonesia, a sugar trading company is currently investing in a new sugar factory on Java, already

[10] <http://www.vanguardngr.com/2012/01/chinese-firm-investments>.

overcrowded with sugar factories, in order to secure its supply base. All the sugar-cane for the factory will be grown by small, independent farmers, the plan being to entice them away from their existing relationships with other factories by offering better prices and support. Another example in the same industry is a foreign-owned agribusiness producing the popular food flavour enhancer monosodium glutamate (among other products), which is considering for the first time investing upstream in sugar-cane production.[11]

Increased corporate interest in large-scale farming

This revised view of agricultural prospects and priorities has in very recent years triggered an upsurge of corporate interest in securing primary agricultural resources at relatively low cost, while they are still 'available', to assure future production capacity. Much of this interest has been directed at poor countries in sub-Saharan Africa reliant on smallholder agriculture, still with considerable expanses of land in only low-intensity use (for example, pastoralism and fuel wood gathering) and where rural communities often lack legal title to the land they use, this being governed by customary tenure arrangements or ultimately controlled by the state, such as Angola, DRC, Ethiopia, Mozambique, South Sudan, and Tanzania. The corporate farming ventures involved are often very large, sometimes over 100,000 ha, although the amount of land to date actually developed, as opposed to planned for development, is typically very much less. These highly visible ventures, which have been dubbed as 'land grabs' have been widely publicized as a threat to rural communities dependent on small-scale farming by cutting off their access to local land and water resources that have long been traditionally used and obliging farming households to work instead for the corporate project as waged labourers.[12] Another reason these projects have attracted so much attention is that some of them are partly or wholly financed by foreign corporate interests, and the projects' output, often food product, is primarily intended for shipment back to the investors' home country markets, where demand is strong and the scope for expanding domestic supply capacity is limited. Hence they are seen in some quarters as a threat to food security in the 'host' country. Few of these ventures make provision for small farmer participation on anything other than a token scale. The foreign investors concerned include both private and public companies and

[11] Private communication, Booker Tate Ltd, 2011.

[12] Where the welfare of smallholder farmers and rural communities is threatened by corporate agribusiness, it is probably due in most cases to weak governance by both government and the corporate sector of the processes of land acquisition and local community engagement. Guidance on good practice is available; for example, that drawn up by FAO, IFAD, UNCTAD, and the World Bank concerning responsible agricultural investment (FAO et al., 2010) and the longer-standing Performance Standards of IFC.

government agencies. Various studies[13] have attempted to estimate the total area subject to 'land grabbing', most of which is in sub-Saharan Africa and Latin America. The most recent and detailed study, based on 1,217 transnational land deals, estimates that these involve over 80 million ha, said to be 1.7 per cent of the world's agricultural area (Anseeuw et al., 2012).

Corporate large-scale farming, especially of plantation crops, is not new; it goes back centuries. Nor are large, company-owned agricultural estates an immutable mode of production organization in particular situations. As the history of the Kenya Tea Development Agency (formerly Authority) (KTDA) and the Malawi tobacco sector show, smallholder production can emerge to dominate a sector that was previously largely in corporate hands in the form of large estates. Furthermore, the KTDA in due course extended its activities from growing tea to processing and marketing it as well (Leonard, 1991). Conversely, smallholdings may be consolidated by their owners into larger operational units for cost-efficiency reasons and professionally managed or leased out to agribusiness. Much of the recent FDI in large-scale farming, however, involves non-plantation arable crops including cereals and oilseeds. Until fairly recently, relatively few such projects proved to be commercially viable in sub-Saharan Africa,[14] but farming experience gained over many years in parts of Latin America, particularly Brazil and Argentina, is changing this (World Bank and FAO, 2009). Again until fairly recently, it was thought that the scale economies of farming operations per se were relatively limited, the cost efficiencies of large farms as single operational units deriving mainly from the lower unit cost of transactions at the upstream and downstream interfaces (procurement, sales, compliance, access to supporting services, infrastructure, etc.). However, technological innovations such as precision farming (GPS-control over mechanized operations, including input application), automated remote control and monitoring of systems such as drip fertigation,[15] and minimum tillage (often combined with GM seed and agrochemical packages) have increased the potential economies of scale in farming operations. This has probably improved the commercial prospects for large arable farms even in areas with under-developed infrastructure, communications, and agri-support services.[16] Nevertheless, smallholder labour

[13] For example, by IFPRI, the Mo Ibrahim Foundation, and the World Bank. The most comprehensive recent study is that by Anseeuw et al. (2012).

[14] See, for example, Poulton et al. (2008), which concludes that while large-scale agriculture has proved more competitive in export horticulture, sugar, and flue-cured tobacco, smallholders dominate in cotton, cashew, and food staples.

[15] Drip fertigation supplies both water and nutrients to crops through a single drip line delivery system that goes directly or close to the root zone. Given the high capital installation costs of such systems, the economics of this process work in favour of large-scale farming.

[16] See Deininger and Byerlee (2011) for an analysis of what the future holds for large farms in land-abundant countries.

costs are normally much lower than estate labour costs, due to access to (usually) unpaid family labour. Estate farming must therefore deliver higher productivity and/or quality to have a chance of competing.

Growth of bio-energy production

Driven by renewable energy policies in industrialized countries, the production and international trading of 'first generation' bio-fuels has expanded rapidly in the last few years. In developing countries, large areas of land are planted with sugar-cane, cassava, and maize for bio-ethanol production and with soy bean, oil palm, and Jatropha for bio-diesel production. Some of this represents diversion from food and feed use, but more recently it has come from investment (including some of the FDI referred to earlier) in additional agricultural production capacity. How far the new bio-fuel value chains have opened up opportunities for small farmer participation again depends very much on the pre-existing agrarian structures, rural institutions, and land availabilities. Thus in India, Mission NewEnergy Ltd is implementing a project to source Jatropha from 140,000 small farmers in over 15,000 villages across 5 states.[17] However, where 'unused' land and water do exist in substantial amounts, companies invariably opt to grow the bio-fuel feedstock themselves on a large scale, as in Mozambique, where several energy companies are active.

Growing South–South investment

Recent years have seen acceleration in agricultural investment in developing and transition countries by Asian and Latin American companies, to some extent between these two regions but particularly into sub-Saharan Africa. There are abundant under-exploited land and water resources in many parts of sub-Saharan Africa and some parts of South East Asia that can be made available to investors who have the capital, management, and global market access needed to develop these areas. Current examples include Indonesian food conglomerates investing in Brazil, Ghana, Laos, and Zaire, Malaysian palm oil companies investing in West Africa, and Brazilian companies based in Mato Grosso investing in arable cropping in Mozambique. An additional attraction can be the similarity of the business environment and general operating conditions in the two countries or regions concerned. In other cases it is possession of farming system experience in a similar agro-ecological zone that the investors perceive as giving them a competitive advantage. It is also alleged that in some instances—and this applies to North–South investment flows too—a laxer regulatory environment concerning environmental

[17] See <http://www.missionnewenergy.com>.

and social aspects of agribusiness development than prevails in the investor's home country is also an incentive. Much of this South–South investment is also in large corporate farms, with relatively little provision for, or possibility of, smallholder participation.

Increasing interest in social enterprise

Businesses that are run strictly according to commercial principles but which nevertheless have clearly defined social objectives as well as conventional financial objectives are not new phenomena—the Grameen Bank, whose business model has been much imitated, was founded in 1983—but they have been proliferating rapidly in recent years as experience has accumulated and the effectiveness of traditional models of development aid have been challenged. Venture-capital and private-equity funding of agribusiness has become increasingly important in recent years. With these funds there is a very clear division between the management and the providers of capital. Fund managers negotiate with capital providers a specific set of criteria, objectives, and targets for a fund, which do not have to conform to the expectations of investors in traditional stock markets. Hence substantial funds have been created which specialize in emerging markets and agribusiness; for example, the African Agriculture Fund promoted by Phatisa Fund Managers[18] includes smallholder participation as one of its aims. The Silverlands Fund, promoted by Silver Street Capital LLP,[19] is claiming a highly specific focus on hub-farms with outgrowers in targeted African countries as its core business model. These specialized funds do not have to be fully profit maximizing, and funds can be established aimed at attracting 'patient' capital from wealthy philanthropists and charitable foundations. AgDevCo, for example, explicitly seeks 'patient' capital in order to make its model of green-field agricultural development, with smallholder participation, financially viable.[20] Innovative approaches can thus be stimulated by incentivizing fund managers to promote smallholder participation.

It would seem to be important to establish the circumstances in which corporate engagement with smallholders requires the use of capital, technology, or management resources being available to these companies at less than market cost (including relevant risk premia). For if social enterprises can achieve long-term viability without resources having to be supplied on concessional terms, as with partnerships between philanthropic and commercial organizations, it would certainly beg the question of why corporate engagement with small farms is not more widespread. This point is considered later.

[18] <http://www.Phatisa.com>.
[19] <http://www.silverstreetcapital.com>.
[20] <http://www.agdevco.com>.

3. Corporate engagement with small farms

This section considers some features of the interface between corporate agribusiness and the small-farm sector—as buyer and seller of agricultural material, respectively[21]—in the context of agricultural value-chain organization. According to Miller and Jones (2010) the development of agricultural value chains can be characterized as producer-driven, buyer-driven, facilitator-driven, or integrated.

3.1 BUSINESS MODELS AND VALUE CHAINS

The evidence from around the world is that there is no inherent incompatibility between corporate agribusiness and smallholder agriculture, but the difference in operational scale between the two sectors does pose some problems that generally increase the cost to each party of doing business together. Corporate agricultural processors and distributors want suppliers who are reliable in terms of quality, quantity, and delivery schedules and who are relatively easy to engage with logistically. This applies both to the provision and supervision of technology/standard practice packages as well as to the procurement of farm products. A small number of selected, large commercial farms are more likely to meet these requirements than a myriad of small ones, which may still be fulfilling a partly subsistence role. This is not universal: some forms of smallholder agriculture may actually be better at meeting some of these requirements, especially where intensive management supervision is essential to ensure high quality; for example, in horticulture. In general, though, corporate agribusiness professes to prefer dealing with a few large farms than with several small farms, other things being equal.

The reality, though, is a little more complicated. As Swinnen and Maertens point out, though dealing with larger farmers is indeed the preference expressed by companies, in practice they actually contract with many more small farms than might have been expected (Swinnen and Maertens, 2007, p. 99). This is consistent with Barrett et al.'s conclusion concerning contract-farming arrangements, that there is little empirical evidence of a positive or negative correlation between farm size and value-chain participation by farmers (Barrett et al. 2012, p. 724). However, it is not entirely clear whether this also applies to farms owned or operated by people who belong to some form of farmer organization, such as an agricultural producer association or cooperative, which then becomes the companies' primary point of engagement with the small-farm sector.

[21] The reverse situation, namely small farms as purchasers of inputs from agribusiness, also raises several issues, but these are outside the scope of this chapter.

From the small farmers' perspective, there are both advantages and disadvantages of engaging with corporate buyers in formalized arrangements regulated by contract, as opposed to spot selling into open markets. As already noted, the advantages can include guaranteed access to credit, more advanced agricultural technology with advice and training on how to use it, good quality inputs, and a stable market. On the other hand, meeting the requirements of corporate buyers can mean capital outlays on specialized equipment needed for a specific production process or on particular facilities needed to ensure compliance with value-chain certification, a higher level of indebtedness, and less decision-making autonomy in production and marketing.[22] The advantages of a guaranteed market outlet may be offset by a contractual price that is lower than prices perceived by farmers to be available outside the corporate market.[23] Hence Barrett et al.'s finding that membership of a cooperative or some other farmer organization matters to small farmers. This is partly because it lowers transaction costs and helps attract contract offers from companies, but also because the contract terms available through farmer organizations are usually better than those available to farmers acting individually (Barrett et al., 2012, p. 725).

While some agribusiness ventures would prefer to grow their own crops, as well as process and trade them, they often remain nervous of committing significant capital to high-risk, emerging markets. They may therefore prefer some form of outgrower arrangement,[24] because this may mean the company can more easily access land from government and low-cost capital from development agencies for infrastructure (such as irrigation, roads, power), at least for the outgrower component of the project.

In the majority of cases, therefore, the initiative to engage probably comes from the agribusiness side, looking for sources of raw material supply, rather than from the small farm side, looking to the corporate sector to provide markets and perhaps also agricultural support services. However, recent years have seen a growing emphasis on value-chain interventions aimed at interesting and encouraging, through various inducements, corporate agribusiness to engage with small farmers. These 'promotional' efforts on behalf of small farmers are usually led by aid agencies and NGOs, and sometimes by government.[25]

[22] Indeed, Reardon and Barrett (2000) go so far as to conclude that 'contracts are not the institutional panacea for small farmer involvement in agro-industrialisation' (p. 200).

[23] It is not uncommon, however, in such circumstances for farmers to overlook the fact that high prices in local traditional markets may reflect relatively low traded volumes, and prices could collapse if the bulk of the corporately contracted supply was diverted into these markets.

[24] Some agribusinesses take this 'out-sourcing' principle a stage further by strategically focusing on processing/trading, contracting others to handle raw material procurement.

[25] For example, AusAid has a major value-chain development programme in Cambodia that includes linking small farmers with agribusinesses, and Farm Africa is doing similar work in selected East African countries.

The 'ideal' model from the corporate perspective

What kinds of value chain, in terms of participant organization and governance of their activities, are generally preferred by corporate agribusiness? This question assumes that companies have already decided to source some or all of their agricultural supply requirements from independent farms, either being unable or unwilling to rely on using farms of their own. Fundamentally, companies look for opportunities that will give them:

(i) a high degree of control over total value-chain coordination in order to minimize supply risk (quantity, reliability, quality, traceability)—where relevant, this applies to supply risk on both sides of their position in the chain as with agricultural processors who must source input and market output;
(ii) a sufficiently large exploitable supply base upstream and a sufficiently large potential market downstream (as applicable) to reach the minimum scale of operation required for acceptable cost efficiency;
(iii) controllable, or as a second best, predictable margins, with the ability to pass back upstream price changes coming from downstream closer to the final market—the less competition the better from the company's perspective, as is little or no official regulation of prices or margins; and
(iv) economic space in which to grow, whether due to market demand that is likely to remain unsatisfied, the prospect of accessing niche markets that pay a premium price, or the potential available for reducing unit production costs through efficiency improvements that enhance the company's competitiveness.

The consequences for small farmers' participation in value chains

This business model—'ideal' from the corporate perspective—is not an approach that is especially likely to favour greater participation by small farms in agricultural value chains or provide the most favourable terms on which they can participate, although it does not necessarily prevent either of these things. In attempting to position itself as closely as possible to the 'ideal' model, corporate agribusiness tends to push the growth and evolution of value chains in three directions:[26]

(1) Towards *vertical integration* as a means of achieving greater vertical coordination.[27] Some companies prefer to integrate their processing or

[26] Hitchcock (2008) identifies three similar tendencies in relation to agricultural sectors in the Asia-Pacific region, namely increasing concentration, vertical coordination, and contracting.

[27] For a wide-ranging discussion of the impact of vertical linkages on the roles of participants in agricultural supply chains, see Young and Hobbs (2001).

trading operations completely with their own farming operations. Some companies may be unwilling to envisage any kind of reliance on independent farmers even if sufficient land is not available for their own farm on the scale deemed necessary—such companies will simply look elsewhere.[28] Other companies, however, are prepared to accept that part of their raw material supply will have to come from independent farms if the company is to process/trade at sufficient scale. Nevertheless, they are likely to prefer to have some base level of supply from their own farm, wherever possible, to ensure a minimum rate of utilization of plant capacity. Fixed costs form a high proportion of total processing cost (net of raw material cost) for most agricultural products. However, the possibilities for upstream integration by a corporate processor/trader may be limited by the lack of available suitable land or by the cost of constructing a corporate farm, which may mean building roads, irrigation and drainage systems, workshops, housing, etc. in addition to land development operations. Much of this may already be in place in an area already farmed by others. The company may in any case have little experience of farming and not regard it as a core competency.[29]

The next best 'substitute' for having one's own farm, in terms of corporate control over value chain coordination, is to be in a close relationship with a few (and by implication, relatively large) trusted, independent farms.[30] If these are non-existent, then the third-best option is to deal with well-organized groups of small farms. The least attractive option is to have to rely on large numbers of unorganized small farms to supply the greater proportion of corporate throughput. Transaction cost and supply risk will be high.[31] From the corporate perspective, therefore, the existence of producer associations or cooperatives of small farmers will

[28] It is hard, for example, to interest the private sugar industry in Indonesia, which is mostly outside Java, in sourcing sugar-cane from outgrowers on any significant scale. It prefers instead to search further and further afield for large blocks of land for corporate farm development, despite such areas being very difficult to find. This may reflect the poor performance of the State-owned corporate sugar industry on Java which does depend on small farmers, though this is not the reason for its poor performance.

[29] The possibilities for *downstream* integration by a corporate processor/trader may also be limited by the difficulty of market entry into the relevant segment of the value chain, particularly if the degree of concentration in it is high, as with supermarkets or internationally branded products, or access to the necessary technology is expensive in terms of acquisition or learning costs. The compromise is to enter into long-term off-take agreements and strategic alliances.

[30] 'Trusted' is the key word here. Barrett et al. point out that, generally, while contracting with larger (and better-off) farmers may reduce agribusiness companies' transaction costs, it may also oblige the companies to offer better contract terms and face a higher risk of contractual non-compliance (Barrett et al., 2012, p. 725).

[31] The exception is when the processor/trader has a monopsony over the agricultural product concerned and is likely to be potentially 'over-supplied', for a given handling capacity, by local farms.

generally be an inducement to engage with the small-farm sector, particularly if they are voluntarily formed for just this purpose.

(2) Towards *larger-scale operations* in processing and trading in order to exploit the cost advantages of large size. 'Economies of scale' are very significant in most agricultural systems downstream and upstream from farming itself.[32] The cost advantages to corporate agribusiness of operating on a large scale, and there can sometimes be quality advantages too in certain agricultural processes where size justifies the expense of automation, puts small farmers at a disadvantage unless they can 'horizontally integrate' into single management units for transactional purposes as noted earlier.

(3) Towards *less competitive market structures*, for the obvious reasons that it is easier to preserve margins and maintain market share in these circumstances. This does not of course mean that corporate agribusiness cannot thrive in fully competitive markets—there are plenty of examples of it doing that—rather it means that corporate agribusiness is more likely to be induced into engaging with small farms in a particular location if it faces relatively few competitors there. This reduces the opportunities for 'side-selling', whereby farmers contracted to one company switch deliveries to another at harvest time or sell into local open markets, and takes the pressure off processing/trading margins by allowing the corporate buyer to pay lower prices to farmers than would otherwise be the case. In this context, there is a major practical difference between value chains that involve processing facilities that need to be close to the sources of production (for example, those for horticulture, oil palm, sugar-cane, and tea) and therefore have geographically restricted sourcing options, and those which are normally located close to the point of consumption, such as cereal flour and arable crop oil mills, and which can source much more widely.

However, price is far from being the only criterion by which to judge the quality of the transactional relationship between corporate agribusiness and small farms. A recent study of the cotton sector, for example, found that while higher farm prices for raw cotton were indeed associated with greater competition among ginneries, farmers received better support and services from ginneries when the market conditions they offered farmers were more like those of an oligopoly. Generally, in the real world of imperfect markets and weak states, there is likely to be a trade-off between competition and coordination (Tschirley et al., 2009). Also,

[32] There is abundant evidence for this: from industrial history (tendency for continuous capacity rationalization in industries, resulting in fewer large units over time and the limited use of small-scale processes except in the absence of competition from large scale), from engineering (cost/capacity relationship for plant and equipment), and from econometric analysis (estimated cost functions).

being faced with a sole buyer in the market may be a condition that farmers have to accept for having a market at all: some farming areas may be too remote and costly for corporate agribusiness to access unless farm product can be obtained relatively cheaply. In Cambodia, for example, a successful exporter of organic rice relies to some extent on this principle; nevertheless, an independent study concluded that the farmers involved had benefited (Cai et al., 2008).

3.2 FORMS OF ENGAGEMENT

Nature of the transaction

Corporate agribusiness engages with independent farms, small or large, basically as a seller or as a buyer. In either case, the company concerned may also provide additional services. These are often included as an integral part of the exchange, as with the 'package' of specially selected seeds, agrochemicals, technical information, advice, and training offered by biotechnology companies supplying farm inputs. Agricultural processing and trading companies procuring product from farmers may also require their suppliers to use certain approved agricultural technologies and adopt certain husbandry practices, which the company may or may not provide or train farmers in, respectively. Particularly important to small farmers is the corporate partner's facilitation of the farmers' access to credit, sometimes by acting directly as the lending agency itself (often by providing inputs in kind rather than by advancing cash) or indirectly by standing as 'guarantor' of small farmer loans for local banks (and may also act as the debt collection agency on behalf of the banks, for example by deducting repayments from farmers' sales proceeds received on delivery of product to the company). In the case of tree crops, which may have an economic life of 20–30 years, small farmers may have the opportunity of securing title to land prepared and planted on their behalf by the corporate partner, usually a processor, after paying off a long-term loan to cover the cost of plantation establishment. There are many forms of value-chain financing arrangements, several of which involve multiple agents, and the more frequently used ones are considered in detail by Miller and Jones (2010).

The arrangements for procuring product from small farmers by corporate agribusinesses range from simple agreements to quite complicated ones, depending on the extent of additional services provided, the nature of the undertakings given by each party, and the extent to which third parties are involved, whether as economic actors in 'multi-dimensional' relationships (e.g. farmer/company/bank or farmer/cooperative/company/bank) or as endorsers of eligibility or guarantors of performance. These are not necessarily written agreements: in many parts of the world traders acting on behalf of corporate agribusiness procure farm product on the basis of verbal understandings only, with spot prices at the time of collection. The relationship between the particular traders and farmers involved may be a long-standing one, with mutual trust being the basis on which inputs and credits are advanced to the farmers each season and security of purchase availability is assured for the trader. Contracts may be renegotiated for each crop, or annually, or may be multi-year agreements.

Terms and conditions

These can range from minimal specification in advance, perhaps only involving the verbal striking of a minimum procurement price that may or may not be uplifted at sale time depending on market conditions, to multi-page written agreements setting out in considerable detail the undertakings and obligations of each party.[33] The latter is more likely to be required where compliance with minimum standards or specific operating procedures at the farming stage is an important part of the value of the final product. The terms and conditions involved can sometimes prove too onerous for small farmers, which is why their participation in supply chains for certain high-value products, such as those for export horticulture, is often problematical. An oft-cited example is that of green beans in Kenya, where small-farmer participation in the fresh market value chain dropped significantly after the introduction of international food safety standards (IFSS), with some farmers switching to the canning market where standards are less exacting. On the other hand, many smallholders successfully meet IFSS requirements with the help of their contracting companies, sometimes in public–private partnerships for this purpose (see Narrod et al., 2008). In some agricultural value chains, failure by farmers to meet terms and conditions may result in, among other things, exclusion from the next round of contracting or dispossession of the right to continue using land in cases where this has been developed by the company on the farmer's behalf.

Price determination

There is a wide variety of ways in which the level of the procurement price is determined, the point in the farming cycle at which the price becomes fixed, and the extent and the conditions under which the price may be subsequently adjusted. Prices may include the value of co-products and by-products from processing the raw material supplied by the farmer or the farmer may have the right to take physical possession of all or a proportion of these products.

[33] The website of the Mumias Sugar Company in Kenya (<http://www.mumias-sugar.com>) sets out on its 'Farmer' page what is required of each party. Mumias has a long-established farmer association, the Mumias Outgrower Company (MOCO).

There may be deductions from a basic price for poor quality or late delivery and additions for good quality and timely delivery. Farmers may be paid individually for their deliveries or collectively as part of a group delivery. Farmers may be entitled to a pre-agreed share of the corporate processor/trader's own sales revenue or paid on the basis of a specific unit price for their deliveries to the company. Some agro-industries pay farmers on the basis of specific, publicly announced formulae that link farm procurement prices for the raw material to the prices of the final processed product on domestic or international markets using technical conversion coefficients for either industry average or factory-specific processing outcomes. More sophisticated versions seek to reward the farmer and processor/trader separately for efficiency improvements that are solely within their control and not dependent on the performance of the other party. This formulaic approach to pricing has been particularly developed in the sugar industry, where the capital cost of improving processing efficiency can be very high. The room for manoeuvre that the procuring company has in setting its offer price to farmers will depend on the regulatory regime (if any) for that particular market. In some agro-industries, the rules are formulated and enforced by multi-stakeholder bodies representing the interests, for example, of farmers, processor/traders, consumers, and government. In some market regimes, both procurement (into-factory) and processed product (out-of-factory) prices, and hence processing margins, are fixed by government. In others, only the former is fixed. Attempts are sometimes made to regulate distribution margins too.

Outcomes

Can valid generalizations be made about which forms of engagement between corporate agribusinesses and small farmers are likely to be of greater benefit to the latter than others? A comprehensive attempt was made 30 years ago by Glover and Kusterer (1990), and many of their conclusions still apply today. For example, they observed that contract farming is not a 'zero sum game' in that the distribution of benefits between the company and its growers can affect the magnitude of total benefits available (mainly via the effects of pricing on investment and participation levels). This makes it difficult to determine the distribution that will maximize grower benefits in the longer run. A contemporary reassessment of many of the same issues addressed by Glover and Kusterer is that of Barrett et al. (2012), which reviews results from surveys of over six thousand six hundred households, mostly farm households,[34] in Ghana, India, Madagascar, Mozambique, and Nicaragua. As with most other economic transactions, the answer to the question posed depends largely on the extent to which one party is able—and is prepared to—'exploit'

[34] Additionally, in India alone 42 companies were surveyed.

the other by virtue of the balance of power between the two, in this case market power.[35] Clearly, small farmers will generally be in a stronger position to take advantage of opportunities to link with corporate agribusiness in circumstances where:

(i) there is a strong and sustained demand for the final product of the value chain;
(ii) the corporate sector has no option but to source from small farms all or part of its supply of the raw material needed to manufacture or trade in this product;
(iii) there are multiple corporate buyers competing for the farmers' output; and
(iv) farmers are organized into large groups for the purposes of negotiating, on behalf of all group members, off-take agreements with the corporate sector.

Beyond these fairly obvious conclusions, however, it is more difficult to generalize. The farm products that are technically the least demanding of farmers' skills, knowledge, and management time and as such pose less supply risk to corporate agribusiness if sourced from smallholders, are the plantation crops (with some exceptions). However, these are finally sold into commodity markets where prices have in the past been relatively volatile and marked a declining trend in real terms (although the trend at least may have turned upwards very recently). The production of higher-value items for sale such as horticultural crops, livestock (including poultry), and milk, is technically more demanding, but here this may work to the advantage of small farmers in that they can often provide the more intensive management supervision required at least as cost competitively, if not more so, as large farmers. Thus the out-sourcing of poultry rearing to small farmers through the distribution to them of day-old chicks and the subsequent buy-back of the grown bird is a quite common business model in the poultry industry. However, as the retailers and exporters at the top of agricultural value chains increasingly seek to 'lock' the final products into branded markets that differentiate themselves on the basis of specification, quality, origin-with-traceability, and certification of compliance with ethical production practices, it can become harder and more costly for small farms to participate in these higher-value supply chains.

[35] Sartorius and Kirsten (2007) discuss three case studies and find that smallholder contracting was unsatisfactory where the power balance was very unequal (for timber), and satisfactory in the other cases (for sugar-cane) where power was less skewed.

4. Inducing greater, and more beneficial, corporate involvement with small farms

This section considers the circumstances in which, and the ways in which, agribusiness companies are choosing to engage with small farmers. How does the prospective corporate investor in the agricultural sector currently perceive business opportunities that may or may not involve engagement with small farmers; how is the decision to invest made; and what mainly influences it? It needs to be recognized at the outset that this is not a 'one-way street'; small farmers also exercise discretion about whether or not to engage with companies.[36]

4.1 THE AGRIBUSINESS INVESTMENT DECISION

There is nothing specifically unique about corporate investment behaviour in agribusiness as compared with other sectors, but agribusiness investors do need to be willing to wait a considerable time for the returns on their investment and to accept relatively high levels and multiple sources of risk. Significant business opportunities that are identified will be evaluated within a strategic framework, effectively the company's business model, in order to gauge the extent to which the company is likely to have, or can develop, a competitive advantage in the market. Increasingly this effectively means the international market, because the globalization of supply chains and market liberalization are progressively reducing the degree of both natural and policy protection of domestic markets. Individual investment projects are the vehicles for delivering that advantage in practice. Each project will be considered on its own merits (its benefits, costs, and risks), even though it may be intended to form only part of a larger investment programme in pursuit of the same business opportunity.

For agribusiness companies whose shares are listed on stock markets, the markets' perceptions of these companies and their credibility are very important. Managers need to maintain a high P/E ratio to avoid hostile takeovers. A strategy of heavy exposure to 'commodities', to emerging markets and reliance on smallholder production has not only to be objectively sound, but also to be convincing in the eyes of stock-market analysts. Analysts in turn are influenced by fashion; hence the notorious herd mentality of stock markets.

A fundamental consideration at both the strategic and project levels is whether the business opportunity essentially means doing more of what the

[36] As small farmers do in other economic relationships—see Minten et al. (2011) on how farmers in Uttarakhand self-select long-term relationships with state-regulated brokers in wholesale markets.

company is already doing or whether it means undertaking a very different kind of venture. The latter invariably means higher costs (mainly of learning) and greater uncertainty (due to lack of institutional or in-house experience). A pertinent example would be an opportunity that involves having to establish an agricultural value chain more or less from scratch or to participate in a value chain during the early stages of market development, typical of much recent bio-fuel activity. A review of 179 agribusiness investments made over the course of 50 years by the Commonwealth Development Corporation (CDC) found that new start-ups are significantly more risky than investments into existing agribusinesses (Dixie, 2011).

4.2 BENEFITS, COSTS, AND RISKS

For agribusiness projects in general

Agriculture is a long-term and risky business. Those investing in it in the expectation of making quick and certain returns from farming per se (as opposed to speculative investment in land as a financial asset) are likely to be disappointed.[37] The same applies to agribusiness, because of its dependence on farming, although any one agribusiness company may be able to reduce its supply risk by sourcing from diversified origins. Much of agribusiness is peculiarly reliant on physical infrastructure for long-distance transportation of purchased inputs and manufactured outputs from and to, respectively, distant market centres, because of the need to be close to raw material supplies that are typically of low value, bulky, and deteriorate quickly once harvested, and for energy distribution for processing (except for those agricultural industries such as sugar and palm oil which generate their own power from the waste from processing).

The recent upsurge of interest among general investors, as opposed to sector specialists, in agribusiness is likely to result in an increased failure rate due to inadequate understanding of agricultural risk. This is already apparent with bio-fuels, particularly bio-diesel production from Jatropha oil. Several new enterprises established for this purpose have already gone out of business, fundamentally due to under-capitalization reflecting a lack of appreciation of the time and cost required to develop land and water resources, provide infrastructure and services, and build up to an economic level of commercial operating capacity. Crucially, the companies concerned generally failed to understand the need for, and to invest adequately in, R&D

[37] A quite extraordinary level of promotional 'hype', for example, accompanied the early ventures into Jatropha cultivation for bio-diesel production. Such investments were marketed to the UK public as highly suitable for personal pension portfolios. In one case at least, the legality of the promotion concerned is being officially investigated.

to identify appropriate farming systems and develop agricultural technologies suited to the environment in which they operated. The latter point also highlights the risks for small farmers who may be persuaded to participate in 'new' markets by becoming suppliers to corporate agribusiness that has not yet developed a fully tested, technically and commercially proven operating model. Governments and development agencies thus need to be wary of pushing for smallholder involvement in pioneering production systems until their technical and commercial viability have been proven by the promoter.

For agribusiness projects involving small farmers

The generic supply risks, the transaction costs, and the coordination challenge for corporate agribusiness associated with significant reliance on small farmers for raw material supply have already been discussed. In localities serviced by relatively good infrastructure and services and populated by well-organized and trained farmers, many companies are perfectly prepared to accept these risks and manage them as a matter of routine. Indeed, even where the alternative may be possible, namely sourcing from the company's own farms, the additional cost to the business of developing the latter may be unattractive.

The benefit, cost, and risk calculation changes dramatically, however, when the enterprise is a new venture located in a remote and under-developed area.[38] The start-up investment is likely to be very high, with the company having to provide almost everything it needs in terms of infrastructure and services itself, resulting in large and costly inventory holding to overcome basic logistical problems caused by weak supply lines. It will probably also have to pay premium wages to attract and retain skilled staff. On top of this, it will need to train and support local farmers so that they can become reliable suppliers at an economic cost. Expectations, both among the farmers and their local government organizations, may be high. So may be the scope for misunderstanding what each party has agreed to put into the venture in terms of land, labour, technology, capital, and so on. Several agribusinesses have failed because of a breakdown in trust between them and the local farming community, and the consequent withdrawal of local political support for the venture.[39] All this presents a daunting challenge to corporate agribusiness,

[38] Barrett et al. (2012) point out that the geographic placement effect of corporate decisions to contract with smallholders are commonly overlooked in the literature (p. 723).

[39] In one case in Senegal, for example, the deal was for the company to provide materials and equipment (Jatropha seedlings, drip irrigation, organic fertilizer, termiticide, technical expertise, management, etc.) and for the villagers to provide unpaid labour on the land, made available by agreement with the local community under the auspices of the *Conseil General*. The harvest would belong to the local people and the company would buy it off them. However, after only eighteen

and most companies in such circumstances opt to develop their own farms first (and solve the problems that involves) before engaging with small farmers on a commercial, as opposed to a pilot, scale. This approach carries its own risks in terms of local public relations, although this may be mitigated if the processing/trading side of the business provides significant new local employment.

4.3 THE STRATEGIC CHOICE TO ENGAGE WITH SMALL FARMERS

'Internal' influences

These refer to aspects of the corporate culture and business strategy of individual agribusiness companies that do or do not motivate them to consider engaging with small farmers. Broadly speaking, these factors fall into two categories: (i) those that reflect directly commercial considerations; and (ii) those that reflect broader considerations that are not directly concerned with the commercial performance of the company, even if indirectly they do impact on this.

Fundamental among commercial considerations is the likely unit cost of production or value added ($/ton) for the agribusiness if it sources its supply from small farms, compared with sourcing from its own farms or a few large independent farms. In the case of crop-based agribusiness, yields per hectare (even harvested yields, before accounting for losses between field and factory) are generally a good guide—inversely—to overall production cost, although very significant economies of scale in the processing/trading part of the business may offset poor yields to some extent, as in the Thai sugar industry. In plantation crop agribusinesses that operate their own farm as well as buying in from outgrowers, the latter's yields are typically less than the companies' and certainly more variable, even if the outgrowers receive technical support from the company. There are a number of reasons for this that are beyond the scope of this chapter[40], but it raises transaction costs for the company (and may also reduce profitability for the farmers). It also partly accounts for the quite widespread view in corporate agribusiness that smallholder farms are inherently 'inefficient' compared with large commercial

months or so, local disenchantment with the project started to set in, and people began insisting on being paid for working on the land.

[40] Management can enforce reasonable standards of agricultural operation throughout the corporate estate, while often struggling to motivate workers to do more than the minimum. On the other hand, those outgrowers who are committed (often women) know that every extra tonne in yield is extra cash in their pockets and they can achieve excellent results. Conversely, there will be other outgrowers who will neglect their plots, satisfied with a basic level of income for little effort. In addition, outgrowers show the same general tendency as other smallholders with regard to 'packaged' technical advice, namely to adopt some of the recommendations, but seldom all of them or at least not to the full extent.

farms. In the case of non-plantation crops, the yield differential between small and large farms is less clear cut and may actually reverse. Intensive animal-based production on small farms, as in some poultry and dairy farming systems, can also be cost-competitive with large-farm production from the perspective of the company doing the purchasing. Again, supply risk and quality control issues also need to be factored into the decision, as well as labour costs in the case of very labour-intensive crops, such as tobacco, cotton, tea, coffee, and green beans. Smallholders often have access to unpaid or poorly paid family labour, and they can hire casual labour more cheaply than agribusiness companies, which in principle are obliged to respect labour laws and welfare provisions.

Particularly important among non-commercial considerations are the perceived threat of political interference in the business on the one hand and public image on the other. In some parts of the world, some sections of corporate agribusiness see engagement with small farmers as 'asking for trouble' by creating challenges in terms of unnecessarily complicating local relationships and weakening corporate 'control' over the value chain by unnecessary broadening of the stakeholder base. This is usually compounded with reservations about the supply risks, transaction cost, and coordination problems of dealing with large numbers of small farmers and an overall negative view of the latter as possible economic partners. If 'justification' of this position is needed, it may be offered in the form of a rationalization that members of the local rural community are likely to be just as well off or even better off, working as waged employees on the corporate farm than as independent growers for the company.[41]

However, there are probably more situations where the opposite view is taken and small farm engagement is viewed by corporate agribusiness as a necessary, if not sufficient, condition for local political and social acceptance, even if such engagement is not mandated by the authorities. Indeed, in some circumstances, having a small farmer 'constituency' that is economically dependent on the continuing commercial viability of the agribusiness is seen as providing a degree of protection against potentially damaging local government and even national government policy changes.

From the small farmers' perspective, the critical issue is whether the engagement is essentially tokenism, i.e. just enough engagement on the part of the company to demonstrate it is socially responsible, or actually represents a fundamental commitment by the company to make small farm engagement an integral part of its business model. The latter is likely to mean a better economic deal overall for the farmers. Their public image is important to

[41] This kind of argument can be valid for rural women if they can find employment in the processing or packing plants of agribusinesses, which usually prefer female workers on the grounds of their greater skill at the tasks involved.

most companies and particularly so for agribusinesses that produce food, manage natural resources, handle animals, operate in impoverished parts of the world, and so on. The larger the company, the more visible it becomes, especially with modern ICT that gives the public easy access to corporate information. Listed agribusinesses are extremely conscious of their reputation for social and environmental good behaviour. This is unlikely to affect materially the decision whether or not to engage with small farmers, in situations where the company has other options, but it certainly affects how the company engages once it has decided to do so. At the very least, the company will want to be perceived as treating the farmers fairly, not exploitatively, and at best as actively working to improve their economic circumstances.

External influences

These are factors in the business environment that are perceived by corporate agribusiness as helpful—in the context of its chosen business strategy or business model—in its management of the costs and risks of engaging with small farmers and of increasing the likely corporate benefits from this. Since the external business environment is heavily influenced by third-party policies and programmes, the factors specific to agribusiness are discussed under 'Policy Implications'.

4.4 THE SMALL FARM 'EXCLUSION' ISSUE

There is a range of views about the general desirability of promoting smallholder agriculture. In some quarters, small farms are seen as poverty 'traps', with many subsistence farmers having neither the interest in nor aptitude for commercial farming and opting for waged employment as and when opportunities are available. It is the author's contention, however, that while not all small farmers (for various reasons) have the potential to develop some or all of their agricultural activities into commercially sustainable enterprises, many definitely can—with the right kind of help—and many have already done so. In certain circumstances, engagement by the corporate agribusiness sector with the small-farm sector can be a very effective way of lifting small farmers out of poverty (although it is certainly not the only way).

How, then, to approach this issue? First, legislation aimed at 'pushing' corporate agribusiness into engagements that they would otherwise avoid is unlikely to succeed. The last thing small farmers need is reluctant corporate partners. Second, government needs to accept that there will be some parts of the small-farm sector that will always remain unattractive to companies that are in business solely for commercial reasons. Farming communities in remote places that are costly and difficult to access or are in marginal, low productivity agro-climatic zones are two examples.[42] Here small farmers will inevitably be 'left behind' and miss out on the kind of opportunities to be linked up to markets and services provided by the corporate sector in more favourable locations. The needs of such communities will have to be addressed by agencies in government and civil society that are pursuing social as well as economic objectives. Third, even where corporate engagement exists, government needs to be careful not to relinquish its core responsibility for being the primary provider or enabler of basic infrastructure and services to the small-farm sector. In parts of Kalimantan, Indonesia, for example, it sometimes seems that the basic business of agricultural extension has been left to the mining companies to do as part of their CSR.[43] This will lead to confusion in rural administration (and difficult politics) at the local level and uneven provision.

If development strategy does envisage sharing provision with the corporate agribusiness sector or even transferring the bulk of provision to it, then this should be implemented on the basis of formal public–private partnership arrangements that spell out clearly the roles and contributions of the respective parties. Related to this is a final observation, namely that for very many small farmers, especially those in sub-Saharan Africa, improving existing 'traditional' value chains and marketing systems and small farmers' access to them is far more relevant and a much higher priority than trying to link them up with the more 'structured' systems in which most corporate agribusinesses will be operating (see McCullough et al., 2010 and Vorley et al., 2009).

5. Policy implications

Several agricultural sub-sectors are successfully organized around value chains based on small farmers 'by default', due to historical agro-industrial development processes, land availability and tenure constraints, or traditional agrarian structures and institutions that preclude large-scale farming. Mars Inc., for example, has recently announced a major investment in the Ivory Coast to help farmers triple cocoa yields in seven years, involving the replacement of 150 million trees.[44] DoubleAPaper in Thailand[45] is a world-class, fully integrated pulp and paper business that has been developed on the basis of

[42] Thus Barrett et al. (2012) conclude that companies commonly, but not always, opt not to buy from areas where infrastructure and agro-ecology conspire to make agriculture less profitable Barrett et al. (2012, p. 724).

[43] Personal communication, PT Gramar Samudra Nusantara.

[44] <http://www.businessweek.com/news/>, 10 February 2012.

[45] <http://www.doubleapaper.com>.

1.5 million contracted outgrowers of eucalyptus, mainly planted on the bunds separating paddy fields. This concept was developed in part because the Thai promoters realized that they would not be able to acquire large land areas on an estate basis for political reasons; indeed, early attempts to develop estate plantings caused great local controversy. Corporate investment in such value chains will continue wherever opportunities to capture added value in existing or new markets become apparent. This final section considers what can feasibly be done by governments and other important stakeholders, such as consumers and civil society, to encourage more of the kind of corporate agribusiness involvement that is pro-small farm and to discourage the opposite kind.

5.1 THE PARTIES INVOLVED

At its most basic, an enduring economic linkage between corporate agribusiness and small farms is a partnership. It can be categorized by reference to several qualities such as the degree of formality and legal underpinning, frequency and relative economic importance of transaction, balance of management control, risk-sharing, cost and benefit allocation, etc. In these respects there are a great variety of forms, but in most cases successful linkage between small farmers and corporate agribusiness is a result of collaboration between *three* parties, as Vorley et al. (2009) have observed, namely:

(i) *trained and organised* small farmers;
(ii) a *receptive* business sector; and
(iii) *a facilitating* public sector (this author's emphasis).

Following Berdegué et al. (2008), Vorley et al. (2009) depict this collaboration as a triangular set of relationships emerging from a process of partnership facilitation (Figure 10.1). It may be noted in passing that NGOs are frequently involved in the construction or maintenance of the farmer–agribusiness side of this triangle. While this reflects NGOs' partiality in supporting the small farmer cause, it also testifies to their expertise and experience in mobilizing, organizing, and motivating (and often empowering) small farmers and in helping them improve their agricultural practices and ability to link to markets. For example, Farm Africa is performing such a role with SABMiller, which has recently invested in a brewery in South Sudan and requires high-quality cassava to be locally supplied in order to replace expensive imported sugar.[46] After a long civil war, farmers in South Sudan are in great need of assistance with some of the most basic aspects of agriculture. Fritschel

[46] <http://www.farmafrica.org.uk>.

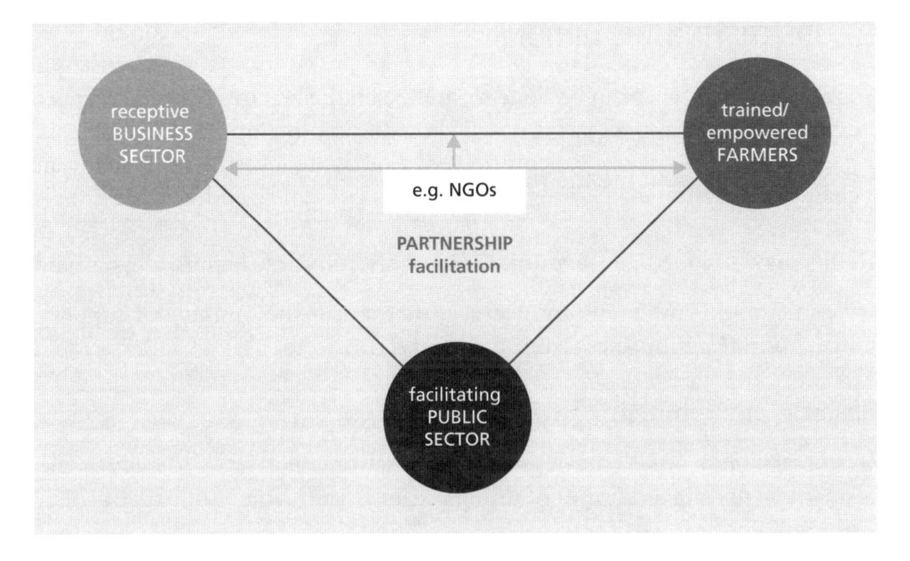

Figure 10.1 Foundations of sustainable market linkages between small-scale producers and agribusiness

Note: Adapted from Vorley et al. (2009).

(2012) provides several other examples of NGOs playing similar roles in agricultural value chains.

Between them, the public sector and the private business sector, partners (ii) and (iii), have to address what is basically market failure, whether it is small farmers poorly served by commercial input suppliers, or corporate agribusinesses that choose either to bypass small-farm suppliers in favour of other suppliers or not engage with small farms at all. If, for some reason, both these particular potential market opportunities are unattractive to private commercial capital, then either they need to be made attractive to this type of capital in some way through specific policies and programmes or the investment needed to develop them will have to come from other kinds of capital.[47] However, most market situations are dynamic, not static. There is always the possibility that some entrepreneurs (and entrepreneurial development agencies) will spot potential, create and develop new market opportunities, and build the critical mass in investments required to sustain change.

[47] Sometimes, it must be accepted that the market is 'correct'; i.e. there simply is not a good business opportunity, because the small farmers concerned are neither a significant market nor competitive suppliers.

5.2 IDENTIFYING THE OPTIONS

The debate about what can be done to strengthen the linkages between corporate agribusiness and small farms usually becomes centred on two themes:

- creating and sustaining the right kind of (enabling) business environment
- joint government—private-sector initiatives (public–private partnerships)

To this may be added another quite commonly considered approach, namely

- specific market intervention measures that favour the sourcing of supplies by corporate agribusiness from the small-farm sector

Business environment

A general policy, legal, and regulatory framework that seeks to reduce the costs, risks, and uncertainties of doing business with large numbers of small farms will need, in particular, to:

(i) provide corporate agribusiness with a clear, unambiguous, consistent, and unchanging message from government about the role required of the small-farm sector in the agricultural economy and the respective roles that the state, civil society, and the public and private sectors are expected to play in furthering this objective;
(ii) provide a technical reference or code of practice to guide the formation and management of contractual relationships between small farmers and corporate agribusinesses, with provisions for independent dispute resolution processes, whether formal (arbitration) or informal (mediation);
(iii) facilitate the voluntary, non-mandatory establishment and operation of associations of small farmers with the legal power, where necessary, to enter into contracts with agribusiness companies on behalf of the whole group, and generally to represent and act on behalf of the group in transactions and discussions of all kinds with these companies and other agencies, including banks and government departments; and
(iv) ensure agricultural marketing is conducted according to 'open market' principles, particularly with regard to transactional transparency, competitiveness, and reference to standardized product/delivery specifications.

With regard to (i), government has a lot of room for manoeuvre in terms of the measures it can take to back up its message. In Indonesia, for example, release of state land for corporate plantation agriculture requires a minimum proportion of the total area to be developed by the company to be assigned specifically for smallholder use. In Mozambique, prospective private investors in agribusiness are required to engage in extensive consultations with local communities in the areas identified by companies as their preferred project locations. Government's role is particularly critical at the early stages of market development, nowhere better demonstrated than with bio-fuels (Hazell and Evans, 2011). Without policy support to create market demand (through mandatory blending with fossil fuels, taxes, and subsidies) and stimulate supply (through minimum prices for feedstock, distribution of inputs, and technical advice), corporate agribusiness will hesitate to step into the proto-value chain and establish the necessary processing and distribution linkages between small farmers growing the feedstock and the end-users of the bio-fuel.

The point of (iii) is largely to make it more attractive for corporate agribusiness to engage with small farmers by institutionally consolidating them into larger entities that effectively offer companies a 'one-stop' transaction service for sales or procurement. This opens up a number of possibilities for greater empowerment of the farmers in the value chain. The routes to this are depicted graphically in Figure 10.2, adapted from CIAT (undated), all of which are likely to prove much more feasible for farmers acting jointly as economic agents in value chains than acting individually.

Public–private partnerships

Poulton et al. (2008) distinguishes between four main types of mechanism for leveraging private-sector involvement and investment in poorly functioning value chains. His typology is adopted here:

1. *Capital development and maintenance* for infrastructure such as rural roads, irrigation and drainage, marketplace facilities, rural ICT. This can in principle mean joint public–private financing of general agriculture infrastructure programmes, but this is relatively rare. More common is (a) project-based allocation of financing responsibility between the public and private parties in relation to specific project components; for example, government financing the infrastructure in outgrower areas supplying a new corporately owned processing plant,[48] and (b) government and the private sector providing facilities intended specifically to make it logistically easier and less costly for small farmers and corporates to transact with each other, such as agro-processing and trading parks/centres, farm infrastructure bases, or wholesale marketplaces.[49] 'Growth corridors', 'poles', 'clusters', etc.[50] are in some respects a large-scale extension of this

[48] Thus the government of NgheAn Province in Vietnam financed the main and feeder roads to connect sugar-cane growers to the state-of the-art sugar factory that Tate & Lyle Ltd built for NgheAn Tate & Lyle J V Co.

[49] For example: incorporated agro-processing parks in India; and AgDevCo/InfraCo's irrigation provision in eastern Africa for smallholder farmers.

[50] Such as the Beira Agricultural Growth Corridor traversing parts of Malawi, Mozambique, Zambia, and Zimbabwe, and the Southern Agricultural Growth Corridor in Tanzania.

approach, in that value-chain participants are encouraged to concentrate their activities in areas selected for relatively intensive infrastructure development and service support.

2. *General delivery of services* to the small-farm sector, with the private sector collaborating with, or actually being contracted by, government and donor agencies to provide agricultural R&D, extension, inputs, credit, and the like. This is different from the provision of such services by corporate agribusiness under contract farming arrangements, which are usually highly product-specific and only tangentially linked to government programmes for servicing farmers. An example is FARM-Africa's Maendeleo Agricultural Enterprise Fund, which provides grants on a competitive basis to consortia of local partners to assist small-scale agricultural producers incubate new enterprises, access new technologies, skills, and inputs, and link to markets.[51] It builds on the success of its predecessor, the Maendeleo Agricultural Technology Fund, which during 2002–10 benefited more than one hundred and fifty thousand rural households.
3. *Targeted delivery of services* aimed at the 'bottom of the pyramid', which in agriculture means poorer, small-scale farmers in remote, low-potential, or marginal areas which are inadequately serviced by both the public and private sectors. Government agriculture budgets get spread too thinly in these areas, and their weak linkage with the cash economy means the level of purchasing power is too low to sustain commercial expansion into them by the corporate sector. By definition, private-sector involvement is initially only possible using 'soft' or 'patient' capital that does not require market rates of return. This might come from the public sector, aid agencies, and civil society donors, such as philanthropic foundations. If investment on this basis is successful in raising the level of economic activity, then such partners can eventually withdraw, leaving the field to wholly commercial, self-sustaining ventures, or shift their own activities on to a more commercial footing. Micro-finance is an example of significant progress made in this respect. Another FARM-Africa example is its wholly owned subsidiary Sidai Africa Ltd, which was established in early 2011.[52] This is a social enterprise aimed at revolutionizing the way that livestock and veterinary services are offered to pastoralists and farmers by building a network of 150 franchises across Kenya, owned and managed by qualified livestock professionals. Sidai, which is based on FARM-Africa's successful earlier experience with establishing village-level businesses providing basic agricultural products and services to poor farmers, is specifically intended to bring livestock care services to underserved communities in more remote locations as well as to more

[51] <http://www.farmafrica.org/downloads/resources/BP%202007%20MATF.pdf>.
[52] <http://www.Sidai.com>.

favourably placed farmers elsewhere. A grant provided the start-up capital for the business, which will in due course become self-financing from sales and franchise licences.

4. *Coordination* in this context refers to deliberate efforts to introduce small farmers and agribusiness companies to each other as prospective clients/customers in value chains and generally foster the emergence of economic relationships between them. Such activities feature extensively in the programmes of aid donors and NGOs, often supported by funds established specifically for this purpose. For example, in the cassava-into-beer project in South Sudan referred to earlier, the NGO (Farm-Africa)'s work with small farmers to enable them to supply the corporate buyer (SABMiller) is financed by the Agricultural Enterprise Challenge Fund.

Market intervention measures

In terms of specific taxes, tariffs, subsidies, price supports, and so on that are intended to reward procurement by corporate agribusiness from small farmers, these appear to be more theoretically appealing than readily applicable in practice. This is probably due to problems of targeting and the risk of benefit misappropriation. More common is subsidized or facilitated credit. In Cambodia, for example, the IFC and World Bank are providing a risk-sharing facility that partially guarantees a portfolio of new commercial bank loans to (mostly SME) agribusiness in an effort to accelerate lending to a sector traditionally perceived as high risk.[53]

Among other market intervention measures is the use of 'seals of approval' to badge good practice in this respect and enable corporate agribusiness to develop specialist markets for discriminating consumers who are prepared to pay a premium for ethically sourced agricultural products. One of the best-known labels is that of the Fair Trade movement which oversees a system whereby (small) farmers receive a 'guaranteed, fair price' in value chains in which the number of intermediaries is reduced as much possible. Bananas and coffee are Fair Trade's biggest-selling products. Others products include cocoa, sugar, and tea. The focus is on export crops, since the final market is predominantly in the industrialized world.[54]

[53] <http://www.worldbank.org/projects/P121809/cambodia-agribusiness-access-finance-project?lang=en&tab=overview>.

[54] Similar ethical supply-chain systems such as the Roundtable on Sustainable Palm Oil, the Round Table on Responsible Soy Association, and the Better Sugarcane Initiative are mainly concerned with managing social and environmental issues and only incidentally involve small growers.

5.3 WHAT WORKS?

There is an enormous array of different projects, programmes, movements, and campaigns throughout the world aimed at encouraging agribusiness companies and small farmers to do business together (mostly with the corporate sector buying from the small-farm sector rather than the reverse).[55] There is a very extensive literature,[56] documenting numerous case studies, drawing inferences from these about the reasons for success and failure, and suggesting principles to guide good practice in the future. Many are very valuable analyses, particularly for initiatives being considered in similar circumstances in terms of type of product, the stage of market development, the underlying agrarian structure, the sophistication of the financial system, etc.

Contract farming represents one of the more economically significant forms of engagement between these corporate entities and small farmers, and Barrett et al.'s recent study analyses the patterns of farmer participation in such contracts (Barrett et al., 2012). A key, though perhaps unsurprising, revelation is that contract-associated transaction costs and the likelihood of contract compliance vary with geography, farmer type, and commodity. An important implication of geographic choice is that they tend to reinforce geographically defined poverty traps and regional inequality. Thus companies commonly, but not always, opt not to buy from areas where infrastructure and agro-ecology combine to make agriculture less profitable. Another suspicion, confirmed by the empirical evidence from this study, is that corporate 'hold ups' in contract implementation, namely specious rejection of farm product or outright failure to be physically available to buy it, appears to increase as the number of smallholders increases. A less expected finding is the relatively high frequency with which both farmers and companies move in and out of contract farming arrangements (participant turnover) as circumstances change, particularly market conditions. A counter-intuitive finding is that there is no correlation between initial landholding (and other welfare-related assets) and value-chain participation. While contracting with larger, better-off farmers may reduce corporate transaction costs, the company may have to offer better terms and the risk of supplier non-compliance may be greater than in the case of smaller, poorer farmers. It was not the intention of this study to decide whether or not contract farming is a 'good thing', and indeed it draws no such conclusion either way. It also reiterates the important

[55] There are also movements and campaigns working in the opposite direction, proclaiming that developing country farmers are exploited by multinationals, that hunger is caused by excessive cash-crop production, that foodstuffs should be grown and consumed locally in order to reduce the number of environmentally damaging 'food miles', and so on.

[56] See, for example, Poulton (2009), Vorley et al. (2009), McCullough et al. (2010), Miller and Jones (2010), and Barrett et al. (2012).

point that the bulk of market sales of all food *sold* worldwide emanates from a small number of relatively well-capitalized producers in the more favourable agro-ecological zones.

Since so much seems to depend on the specifics of the situation, sound project formulation is critical. This basically means identifying clearly the incentives for each party to engage with the other and to maintain the engagement, and ensuring processes can be established to provide those incentives on a sustainable basis. In the author's view, two factors above all could prove to be critical in this respect in the longer term.

Product branding for smallholder supply

There is a growing proportion of consumers who discriminate between agricultural supply chains on the basis of their ethical or social responsibility attributes in addition to the usual determinants of buying choice. Thus many consumers are already willing to pay a premium for Fair Trade bananas, chocolate, coffee, and tea, because they believe that by doing this they are helping poor (and by definition, small) farmers in developing countries. Modern marketing techniques could be used to take this further and create brand value on the basis of smallholder supply per se, emphasizing its 'green' credentials and real or imagined quality differences ('hand-picked') in addition to the poverty reduction impact. This would give a competitive advantage to those supply chains, and hence to the corporate agribusinesses linked into them, that incorporate small farmers.

Corporate commitment to new sourcing from smallholders

There is a growing understanding by corporate agribusiness, particularly among the bigger players in it, that token 'corporate social responsibility' (CSR) is not going to be enough in an increasingly communicative and connected world. Big business in developing countries is increasingly accepting that to maintain its 'social licence to operate' means adapting its business model in ways that contribute fundamentally, not just superficially, to meeting the needs of the poorer segment of the population.[57] In some sectors, including low-cost consumer goods and telephone banking, this has been proved to be a commercially sound business strategy in its own right. However, for much of corporate agribusiness that is not already operating within the more traditional type of managed supply chains involving smallholders such as those for the plantation crops, the motivation to work purposely with small farmers is likely to continue to require some inducement from third

[57] It is interesting that two of the three corporate examples of Fritschel (2012) on agricultural value-chain stimulation are major TNCs: PepsiCo and Walmart; the other company is La Colonia, a leading supermarket chain in Nicaragua.

parties which have social objectives, are resourced with non-commercial capital, and can help reduce the commercial corporate sector's perception of the costs and risks involved.[58]

REFERENCES

Anseeuw, W., M. Boche, T. Breu, M. Giger, J. Lay, P. Messerli, and K. Nolte (2012). *Transnational Land Deals for Agriculture in the Global South. Analytical Report based on the Land Matrix Database*, ILC/CIRAD/GIGA German Institute of Global and Area Studies/GIZ, April.

Barrett, C., M. Bachke, M. Bellemare, H. Michelson, S. Narayanan, and T. Walker (2012). 'Smallholder participation in contract farming: Comparative evidence from five countries', *World Development*, 40(4), pp. 715–30.

Berdegué, J., E. Biénabe, and L. Peppelenbos (2008). 'Innovative practice in connecting small-scale producers with dynamic markets'. Retrieved from: <http://www.regoverningmarkets.org>.

Cai, J., L. Ung, S. Setboonsarng, and P. Leung (2008). 'Rice contract farming in Cambodia: Empowering farmers to move beyond the contract to independence', ADB Institute Discussion Paper No. 109, June.

CIAT (undated). *Agribusiness and Markets—Approaches*, ICRA Learning Resources, CIAT.

Deininger, K. and D. Byerlee (2011). 'The rise of large farms in land abundant countries: Do they have a future?', Discussion Paper, World Bank.

Dixie, G. (2011). 'What does the past teach us about agribusiness investments?' Presentation based on an investigation by Geoff Tyler, ex-head of CDC's Agribusiness Department, World Bank, November.

FAO (2010). 'Cambodia rice value chain study, crop year 2009–2010', prepared for FAO Investment Center by Agricultural Development International, April.

FAO, IFAD, UNCTAD, and World Bank (2010). *Principles for Responsible Agricultural Investment that Respects Rights, Livelihoods and Resources*, Abridged Version, January 25.

Fritschel, H. (2012). 'In Search of a Chain Reaction', *Insights Magazine*, IFPRI, 22 March.

Glover, D. and K. Kusterer (1990). *Small Farmers, Big Business. Contract Farming and Rural Development.* Hampshire and London: MacMillan.

Hazell, P. and M. Evans (2011). 'Environmental, economic and policy aspects of biofuels', in Galarraga, I., M. González-Eguino, and A. Markandya (eds), *Handbook of Sustainable Energy.* Cheltenham, UK and Massachusetts, USA: Edward Elgar.

Hebebrand, C. (2011). *Leveraging Private Sector Investment in Developing Country Agrifood Systems*, The Chicago Council on Global Affairs, May.

Hitchcock, D. (2008). 'Agribusiness and competitive agro-industries in the Asia and Pacific region', LSFM Regional Forum Workshop and 7th Asia DHRRA GA, Session 1a; June, pp. 20–9.

[58] Hebebrand (2011) comes to a similar conclusion when she says that 'the transition from "Corporate Social Responsibility" (CSR) in which efforts [in pursuit of social or wider stakeholder objectives] are pursued separate from profit maximisation to "Created Shared Value" (CSV) where efforts are integral to profit maximisation are not likely to occur automatically' and PPPs will be essential for success.

Leonard, D. K. (1991). *African Successes: Four Public Managers of Kenyan Rural Development.* Berkeley: University of California Press.

McCullough, E., P. Pingali, and K. Stamoulis (2010). 'Small farms and The Transformation of Food Systems: An Overview', in Haas, R., M. Canavari, B. Slee, C. Tong, and B. Anurugsa (eds) (2010), *Looking East, Looking West, Organic and Quality Food market in Asia and Europe*, pp. 47–83. E-book. Retrieved from: <http://www.wageningenacademic.com/eastwest-e?sg=%7BD7EC66C2-3E38-43D5-9196-F1072225D1E9%7D>.

Miller, C. and L. Jones (2010). *Agricultural Value Chain Finance, tools and lessons.* Warwickshire: FAO and Practical Action.

Minten, B., A. Vandeplas, and J. Swinnen (2011). 'The broken broker system? Transacting on agricultural wholesale markets in India (Uttarakhand)', IFPRI, Discussion Paper 01132, December.

Narrod, C., D. Roy, B. Avendano, and J. Okello (2008). 'The impact of food safety standards on smallholders; Evidence from Three Cases', in McCullogh, E., P. Pingali, and K. Stamoulis (eds) (2008), *The Transformation of Agri-Food Systems: Globalization, Supply Chains and Smallholder Farmers*, pp. 355–72. Rome: FAO and London: Earthscan.

Poulton, C. (2009). 'An assessment of alternative mechanisms for leveraging private sector involvement in poorly functioning value chains', FAO AAACP Paper Series No. 8, August 2009.

Poulton, C., G. Tyler, P. Hazell, A. Dorward, J. Kydd, and M. Stockbridge (2008). 'All-Africa review of experiences with commercial agriculture. Lessons for success and failure', background paper for the Competitive Commercial Agriculture in Sub-Saharan Africa (CCA) Study, FAO and World Bank, February.

Reardon T. and C. Barrett (2000). 'Agroindustrialization, globalization, and international development. An overview of issues, patterns and determinants', *Agricultural Economics*, 23, pp. 195–205.

Sartorius, K. and J. Kirsten (2007). 'A framework to facilitate institutional arrangements for smallholder supply in developing countries: an Agribusiness perspective', *Food Policy*, 32 (5–6), pp. 640–55.

Swinnen, J. and M. Maertens (2007). 'Globalization, privatization, and vertical coordination in food value chains in developing and transition countries', *Agricultural Economics*, 37, Issue Supplement s1, December 2007, pp. 89–102.

Tschirley, D., C. Poulton, and P. Labaste (eds) (2009). *Organization and Performance of Cotton Sectors in Africa. Learning from reform experience.* Washington DC: World Bank.

Vorley, B., M. Lundy, and J. MacGregor (2009). 'Business models that are inclusive of small farmers', Chapter 6, in da Silva, C. et al. (eds), *Agro-industries for Development.* FAO UNIDO, IFAD CABI, pp. 186–222.

World Bank (2008). *World Development Report 2008*, Washington DC.

World Bank and FAO (2009). *Awakening Africa's Sleeping Giant: Prospects for Commercial Agriculture in the Guinea savannah Zone and Beyond*, Directions in Development, Agriculture and Rural Development, Washington DC and Rome.

Young, L. and J. Hobbs (2001). 'Vertical linkages in agri-food supply chains: Changing roles for producers, commodity groups and government policy', *Review of Agricultural Economics*, 24 (2), Autumn–Winter 2001, pp. 428–41.

11 A twenty-first-century balancing act: smallholder farm technology and cost-effective research

JOHN K. LYNAM AND STEPHEN TWOMLOW

1. Introduction

'For the first time, we may have the technical capacity to free mankind from the scourge of hunger... within a decade no man, no woman, or child will go to bed hungry,' said Henry Kissinger at the United Nations World Food Conference, 1974. Yet despite more than forty years of research and the Green Revolution in Asia, we are no closer to achieving this goal in the rain-fed, agricultural areas of the tropics, especially in sub-Saharan Africa and the smaller countries of South and Southeast Asia.

Generating sustained and sustainable growth in smallholder productivity under rain-fed conditions has remained one of the major development challenges since the Green Revolution began four decades ago in the 1970s. The Green Revolution in the irrigated areas of Asia vindicated the potential of smallholder technology when adopted at scale, igniting growth multipliers that were the foundations of the current strong overall economic growth rates in Asian economies. Since the 1970s, public-sector agricultural research has been attempting to achieve the same sustained increase in smallholder productivity in rain-fed agriculture, particularly in sub-Saharan Africa. Ex-post impact studies of improved technologies in rain-fed systems continue to demonstrate a positive return on investment in agricultural research, but their adoption is limited spatially and they produce incremental increases in productivity rather than sustained growth, thus reducing the potential of growth multipliers.

Improved technology is now seen as a necessary but not sufficient condition for sustained growth in smallholder productivity. In turn, three central principles now guide the approach to technical change in rain-fed, smallholder agriculture. Firstly the research problem is much more complex, given both the crop diversity that characterises smallholder farming systems and the spatial heterogeneity inherent in rain-fed systems. Secondly, technology adoption and farmer investment strategies are a function of the incentive environment that farmers face, and this in turn is a function of the effectiveness and efficiency of input and output markets, factor markets, and service institutions that support smallholder agriculture. Thirdly, the lack of technology adoption in large parts of rain-fed, tropical agriculture is not well understood and yet is central to the technology design process. Smallholder agriculture in rain-fed areas is thus characterised by multiple biophysical and economic constraints at farming level and by inefficiencies in their institutional environment over which they have little influence. This understanding has in turn resulted in a reinterpretation of the Green Revolution experience, in which the market and institutional context of Asian irrigated agriculture has been seen to be a precondition for the rapid adoption of the high-yielding rice and wheat varieties. The new production possibilities of the improved varieties released what could be considered as the final constraint on the growth potential of the Asian irrigated agriculture, given the investments in market infrastructure, credit, insurance, extension and the range of other services that supported the sector.

Agricultural R&D in developing countries in Asia and sub-Saharan Africa that is focused on rain-fed systems tends to be funded principally by public sources and as a result either implicitly or explicitly tends to focus on smallholder systems. The CGIAR Centres, given that one of their primary objectives is the reduction of rural poverty, also tend to focus on smallholder systems. However, over the past four decades there has not been a consensus on how agricultural R&D can be most effective in ensuring widespread adoption of improved technologies, much less in improving smallholder welfare. This has been reflected in a successive iteration on how best to intersect the supply of new technology with smallholder demand as a means to increase productivity. There has been a consistent critique of traditional National Agricultural Research Institutes (NARI) as 'linear' (research products produced by NARIs but delivered by extension systems) and 'supply driven' (with little correlation to farmers' needs). This, however, underestimates the complexity of the smallholder problem.

The core of the smallholder, agricultural R&D problem is how to balance farm heterogeneity with research cost effectiveness; farm heterogeneity creates the fundamental challenges of meeting farmers' demands while achieving economies of scale in agricultural research. This challenge has been made more complex by a shift in smallholder-productivity research from plot scale to production system and even landscape scale (although this is most clearly articulated in CGIAR research programmes). This shift recognises that income gains from plot-level interventions will not be sufficient to lift smallholders out of poverty and that the sources of productivity gains in these systems can come from multiple sources where the effects on productivity may be synergistic. This shift has complicated the task of making an impact on any reasonable scale, which in turn reduces the return on investment in agricultural research focused on smallholders. Recent approaches such as integrated agricultural research for development (IAR4D) and agricultural innovation systems (AIS) have distilled this challenge to a problem of effective organisation of agricultural R&D.

This complex agricultural R&D challenge has resulted in multiple approaches, but over the last four decades they have adopted a three-stage process: technology design; adaptive research; and dissemination and scaling out. These functions have been separated institutionally, leading to the linear research model. Different institutions tend to be responsible for each of these stages, and shifts in funding priorities across research and extension coupled with ineffective institutional arrangements have resulted in a failure to provide effective articulation between the stages of the R&D process. This has combined with the increasing number of and often conflicting research objectives to reduce the effectiveness of technology supply to smallholders in rain-fed agro-ecologies. This chapter examines technology development for smallholders within an evolving institutional and scientific context, by focusing on these three stages of the R&D process and then assesses the implication for the organisation of agricultural research, often in terms of a hierarchical division of labour, to improve smallholder productivity.[1]

All of this comes at a time of increased accountability for agricultural R&D institutions and significant increases in expenditure on monitoring and evaluation systems to ensure the effectiveness of research investment. It is now not sufficient for the research institutes to judge their performance by such outputs as the number of new varieties or Integrated Pest Management packages. Rather, the focus is on how agricultural research contributes to achieving development outcomes, which in the case of smallholder agriculture includes reduction in rural poverty, increased food security, and improved management of the natural resource base. Given that the agricultural research community alone cannot achieve these outcomes, even more

[1] Innovation systems (World Bank, 2012) are increasingly seen as a conceptual and implementation framework to improve institutional articulation, encourage more demand-driven research, integrate innovation into the dissemination process, and incorporate contextual factors directly into the innovation process. This chapter recognizes the evolving incorporation of innovation systems into the generation of agricultural innovation, but focuses only on the role of agricultural research and technological dissemination. As such the chapter touches on the conceptual thinking that innovation systems can provide in organizing agricultural R&D, but does not use innovation systems as an organizing framework for the chapter. For examples of the latter, see Lynam (2012) and Sumberg (2005).

emphasis is put on organizational linkages that ensure the other preconditions necessary for the adoption/uptake/adaptation of improved technologies. How, then, does agricultural research focused on smallholders ensure technology design and supply are more responsive to farmer demand, and that research itself is more cost effective and able to improve development outcomes? Moreover, does the shift to organizing research around higher systems levels reduce the impact at scale, and does the focus on increased accountability allow for the complexity of the task and the often extensive time lags to produce these outcomes? These are the issues to be explored in this chapter.

2. **Technology design**

The initial evaluation of the Green Revolution generated considerable speculation on whether smallholders would have adopted the high-yielding varieties without the associated improvements in fertilization and water management. Evidence suggests adoption would have still occurred, but at a slower rate; this led to an issue whether the technology design process needed to be tailored specifically to the needs of smallholders? The consensus at that time was that biological technologies were scale-neutral and that agricultural research could focus primarily on improving productivity through the delivery of finished, input-based technologies. However, the challenges of generating large-scale adoption/uptake in rain-fed agricultural systems has over time put more and more emphasis on the importance of technology design in ensuring relevance to the needs of smallholders, which in turn has increased the analysis, understanding, and even participation of smallholder farmers in the technology design process. The design process has in general moved from broad adaption to targeted adaptation, from optimum input application to the most efficient input application, from component research to more integrated systems research, and from assuming markets and policy were independent of technology design to incorporating those requirements as well as agro-ecological constraints. In summary, increasing complexity has been a feature of the technology design process.

This complexity is mirrored in what Byerlee et al. (2010, p. 10) call the adoption puzzle, which they characterize as follows:

> There is, however, an additional problem to be solved in the case of Africa, which is the presumed existence of a backlog of technologies available for adoption that remain unused or inefficiently used, resulting in a large gap between potential and actual yields. This “adoption puzzle” has been the object of much attention in recent years, but it remains largely unresolved.

Is the adoption puzzle due to a lack of appropriate technology or to other constraints on farmers' adoption independent of the technology itself? This question has been exceedingly difficult to answer, and over the last two decades it has moved technology design intended to increase smallholder productivity away from what has been termed a technology supply push approach (Hounkonnou et al., 2012) to a more iterative approach that attempts to understand farmer adoption. However, these feedback loops are rarely established, in large part due to the time lag between technology release and achieving sufficient scale at which to evaluate technology adoption, and also the redesign of technology in an ongoing research process. The result has been a paucity of systematic data on even adoption of inputs such as seed and fertilizer, much less improved crop management techniques; as Byerlee et al. (2010, p.10) note: 'A major effort should thus be made to characterize the state of adoption of major technologies presumed to be lagging such as fertilizers and seeds, and provide simple correlates with conditions of use such as agronomic conditions, relative prices, availability of information, and security of property rights.' The feedback loops between technology design and farmer adoption are rarely made and usually go no further than the adaptive research stage (see the next section).[2]

Crop research involves manipulating biological systems alongside assumptions about farmer management of those systems, whether accessing nutrients, water, and light, avoiding heat or cold stress, and resisting pest and disease attack. Farmer management is in turn influenced by farmer objectives, particularly the balance between cash versus subsistence objectives and risk, and constraints on effective management, such as the inherent quality of the resource base, access to input and output markets and associated relative prices, labour, cash or access to credit, and farm size. Again this reflects the duality between the technology design problem and the adoption problem. The task is then how to incorporate some approximation of farmer management into the technology design process and, given the heterogeneity problem, how to target most effectively different designs based on some understanding of the variation in farmer management constraints. The design problem in increasing smallholder productivity is then usually expressed in

[2] Ex-post adoption studies have tended to focus on farmer characteristics that determine whether a farmer adopted or did not adopt a technology. Such studies provide little feedback directly into the appropriate design of technology. Far fewer studies have focused on understanding the technology adaptation and adoption *process*, which gives insight into the fit of the technology into particular contexts and its plasticity in being adapted under such varying conditions (Douthwaite et al., 2001; 2002). Such studies, however, are rarely done systematically within the heterogeneity of a specified agro-ecology and target range of farming systems. Moreover, when this is done, as for example evaluating alternative striga-control technologies (de Groote et al., 2010), the information may not outweigh the large, prior investments in particular lines of research in favour of a particular approach, especially when different research institutions are involved.

terms of partitioning productivity into G x E x M, where G is the contribution of genetics, E environment, and M management.

Each of these production coefficients has a different spatial scale dimension that structures efficient technology differentiation, testing, and targeting. Spatial targeting at what can be called the meso-scale has tended to include agro-ecology (E), market access (especially the relative price of inputs to farm output prices), and rural population density (as a proxy for farm size and M intensity)—see, for example Omamo et al. (2006). Scale within a heterogeneous target area defines the type of research that is done and the organization of that research. Thus, breeding and varietal testing are organized at the meso-scale. A division of labour between plant breeding, increasingly with molecular screening methods, at an international level and breeding for elite lines at a national level has tended to evolve, although in Africa this is often limited by capacity constraints and an inability to develop the necessary institutional linkages (Lynam, 2011). The more radical strategy, which is encompassed by participatory plant breeding, is to deploy as large amount of genetic diversity as possible and allow farmers to evaluate and select under their conditions and then to reintegrate these selections back into the breeding programme. Organizing breeding programmes that incorporate resistance to critical biotic and abiotic constraints, which often requires sophisticated genetic analysis, and then deploying these resistance traits into diverse agronomic backgrounds, forms the basis of both participatory plant breeding and development of transgenic varieties.

Biotechnology, including its principal elements of marker-assisted selection and transgenics, is often seen as an upstream research methodology in which application is scale-neutral. However, as with plant breeding, the choice of traits and the agronomic background within which those traits are deployed can effectively be used to target smallholder farmers, even more so if those traits can be deployed in locally adapted varietal backgrounds. Because implementing a functional biosafety testing and release capacity has been slow and politically sensitive in most developing countries, apart from large countries like India, China, and Brazil, these ideas have not been widely tested. In East Africa, The International Laboratory for Tropical Agricultural Biotechnology (ILTAB) is testing transgenic cassava lines that incorporate transgenes for resistance to cassava mosaic virus and cassava brown streak virus (actually two species of virus) into local indigenous varieties. The strategy rests on deploying traits that address critical yield constraints in locally adapted varieties that otherwise would meet farmer requirements. However, some crop species are easier and cheaper to transform than others. Also, biosafety frameworks often focus on the transformation event rather than the crop–trait combination and this adds to the costs. Lack of robustness in transformation and the associated costs of production and biosafety approval have skewed the deployment of transgenics to widely adapted, elite varieties with significant market potential, and in general, away from smallholder requirements. Thus, investments in the use of molecular biology to meet smallholder needs have focused on marker-based approaches in which key traits are genetically characterized and very specific markers developed, to be used in breeding and testing programmes, including even participatory breeding. Quantitative traits involving multiple genes, such as for drought tolerance, root characteristics, resistance to bacterial diseases, and insect pests, have added to this complexity and is where biotech research for smallholders is currently concentrated.

Crop management, and natural resource management more generally, is applied at the farm or micro scale, with each farm having its own set of constraints. Yet research cannot be based on conditions in each and every farm. This conundrum has resulted in two contrasting but ultimately complementary approaches. The first, often summarized as the Green Revolution approach, essentially argues that inputs, especially fertilizer and pesticides, provide a first, jump step in productivity that is relatively broadly applicable. However, purchases of inputs by smallholders is linked to well-developed input and output markets and is a function of profitability, which because of underdeveloped transport infrastructure and high transaction costs, especially in sub-Saharan Africa and many lagging areas of Asia, can result in areas where fertilizer use particularly is unprofitable (Guo et al., 2011). With high energy and fertilizer prices, there remain many smallholder areas in the developing world where fertilizer application remains unprofitable. A second approach has focused on more integrated approaches. Management options such as use of organic resources, agroforestry, rotation with legumes, and recently biochar are the basis for more integrated approaches to soil fertility management (Vanlauwe et al., 2010) which can complement timely planting, pest and disease management, intercropping, and residue management. Management can account for up to 70–80 per cent of the yield gap under smallholder conditions (Ken Giller, pc) and have been the focus of research in areas such as soil biology and fertility, agroforestry, manipulation of the agro-ecology for pest management, and conservation agriculture. The research approach centres around understanding processes and principles, developing diagnostic methods, and designing application methods. Dissemination is in the form of decision support systems, models, application principles, and adaptive research methods (see next section).

Technology design in genetic improvement thus strongly contrasts with those technologies where farmer management rather than just farmer choice is critical to improvements in productivity. Farmer knowledge and local adaptation are both critical to soil fertility, integrated pest management, nitrogen fixation, intercropping, and a range of crop-management techniques that support the yield potential of improved varieties while at the same time improving the quality of the natural resource base, including provision of ecosystem services. These differences have in turn led to differences in how

Table 11.1 Characteristics and organizational differences between varietal improvement and crop and resource management research

Varietal Improvement	Crop and Resource Management
Information embodied in seed	Management and information intensive
Potential for wide adaptation	Location and system specificity
Dependent on input markets	Dependent on extension systems
Scale economies in research	Decentralized research systems
Potential for yield take-off	Incremental change
Dependent on market and policy support	Internalized within farming systems

Source: Lynam, 2004.

crop improvement and natural resource management research are organized (see Table 11.1 for a summary), as reflected in the World Bank's meta-evaluation of the CGIAR:

The [CGIAR] System is being pulled in two opposite directions. On the one hand, the CGIAR Centres are not conducting sufficiently coordinated research on the highly decentralized nature of Natural Resources Management (NRM) research, which calls for effective partnerships with National Agricultural research Systems (NARS) to produce regional and national public goods in NRM. On the other hand, the System is not sufficiently centralized to deal with advances in the biological sciences and IPRs, which call for a more unified approach to research strategies and policies.

At the same time, the performance and impact of crop improvement programmes are much more easily evaluated, because of seed systems and the ability to recognize improved varieties in the field. The result has been a perception that investment in crop and resource management is not producing a return on investment, when this research is most needed to close the smallholder yield gap.

The need to forge better integration between plant improvement and crop and resource management has been one factor moving the focus of technology design from component research to farming systems. Other evolving global agendas around climate-smart agriculture, eco-efficiency in water and nutrient use, and payment for ecosystem services all focus on the farming system rather than an individual commodity. Moreover, as Brooks (2010, p. 6) notes in the African case, 'the data on impact suggest that more of the increased income comes from shifts in crop composition than from increased yields'. Evidence from Kenya shows an increase in farm diversification as a response to the market liberalization of the late 1990s. All of this has given impetus to organizing technology design around the production system, as reflected in the three CGIAR research programmes on production systems resulting from the CGIAR reform process.

However, there is little experience of organizing research design where the production system is the unit of analysis. 'System' techniques which integrate a set of principles into a production practice such as conservation agriculture, ICIPE's push-pull technology, and agroforestry systems are in many ways intermediate between the dominance of component research and research on production systems. ILRI's research programme on smallholder dairy systems in the highlands of East Africa comes closest, combining research on animal genetics, forages, animal disease control, livestock management, manure management in relation to staple food productivity and soil quality, and integration into the milk value chain. The smallholder dairy research programme is a harbinger of how production system research programmes might be organized, namely disaggregation into technology components and then integration into more productive farming systems.

Part of the difficulty in organizing research design around production systems essentially derives from the continuing lack of an empirical framework to define representative farming systems, except in a very stylized fashion, for example maize-based farming systems in southern Africa (see Dixon et al., 2001). This example exemplifies attempts to classify rain-fed farming systems by focusing on the principal food staple within a specified agro-ecology. However, the heterogeneity of farming systems still overwhelms such attempts. Thus, maize-based farming systems in highland ecologies of western Kenya have been classified into a typology of five farm types (Tittonell et al., 2010) based on principal livelihood strategies, which are in turn related to land–labour ratios, land quality, and income. Smallholders have multiple objectives which directly affect technology design, namely income optimization, cash flow, nutrition and food security, and risk minimization; these have to be taken into account by researchers along with the increasing number of performance criteria for production systems, such as productivity, quality of the natural resource base, supply of ecosystem services, and system resilience. The researcher's objective of improving the efficiency of system performance in the design process, however, does not translate directly into farmer objectives (Twomlow et al., 2004). Matching farmer objectives with production system performance criteria is a key step in a technology design process organized at the farming-system level, and this is best done at the stage of component integration across a range of farm types.

This increasing complexity has in turn been reflected in changes in the organization of research, involving a number of trade-offs. The movement to systems research has involved organization along agro-ecological or regional criteria at the expense of thematic or disciplinary criteria; it has also required creation of interdisciplinary teams and a focus on economies of scope rather than scale. More local adaptation has required greater decentralization of research, often at the expense of the economies of scale and even scope in larger research stations. Location has become more important in undertaking agricultural research for smallholders; this decentralizing process has been helped by spatial information and modelling technologies, where GIS and modelling allow characterization of the research domain, an understanding of variability in the domain, and improvement in the efficiency of site selection and interpretation of trial results. Multi-site research is critical to effective research design in breeding, crop management, and natural resource management, but this needs to be done as efficiently as possible and the empiricism aggregated through time.

The trend in the CGIAR production system research programmes is to undertake the multi-site G x E x M within large, often multi-country benchmark or sentinel sites and then integrate the components within representative production systems. The core principles of system research, namely a focus on interaction rather than just additive effects, complexity, and system dynamics, particularly in terms of intensification trajectories, are at the heart of this process. In many ways the early attempts in the 1970s and 1980s in organizing technology design around farming systems did not have the necessary base of experience with the component research to make any progress, nor the modelling and spatial tools to characterize and manage heterogeneity in the technology design process.

In summary, the shift of technology design to addressing the constraints of rain-fed farming systems has at the same time supported research methods that focus on the needs of smallholder systems. An understanding of factors causing yield variability, an increased focus on biotic and abiotic stresses, the development of integrated approaches for soil and pest management, and complementing inputs with other management alternatives have all provided a range of options more appropriate for the smallholder farmer working under a wide range of agro-ecological and market conditions. However, we are only starting to understand how crop, livestock, and tree management research are optimally integrated to improve such areas as resource capture and cycling, and thus resource use efficiency, system stability, and resilience. These areas underpin effective adaptation to climate change and provision of ecosystem services, as well as motivating the transition from research on technology components to production systems. However, improved technology design for smallholders has not in itself solved the adoption puzzle.

3. The centrality of adaptive research

The evolution of farming systems research in the 1980s and 1990s towards increased farmer participation and more flexible design of on-farm, experimental trials was central to bridging on-station, applied research with the needs and constraints of smallholder producers. Farming systems research, building on early work at the International Rice Research Institute (IRRI) on cropping systems research, developed the initial methodology for on-farm research, defined the rationale for investment in adaptive research, and has continued to evolve and develop around mutual learning of both scientists and farmers in improving smallholder productivity. One of the initial results that has persisted to today is the very high variability in yield response across farms, even in relatively small geographical areas. This high coefficient of variation was initially seen by agronomists as requiring control of sources of variation in order to understand differences in treatment response. However, it has since evolved into a principal research question driving adaptive research, namely understanding the factors that contribute to this variation in response, which in turn has contributed to methods for farmer selection, trial design and collection of non-treatment data. All of this has emphasized the critical importance of local adaptation in meeting the needs of smallholder farmers under rain-fed conditions, even extending to the point of having to incorporate within farm heterogeneity for such factors as nutrient gradients, soil-borne diseases and pests, and soil type (Tittonell et al., 2005).

Adaptive research in its initial manifestation was seen as a bridge between research and extension, which in turn created some question, if not contest, about whether adaptive research should best be institutionalized in research or extension institutions. Part of this doubt was the purpose of on-farm research itself, and whether the relative focus in trial design and degree of farmer participation should be on scientist learning or on farmer learning. This tension, in fact, resulted in something of an explosion in methods and approaches to adaptive research, which fall into four principal types. The first is farmer-led research, with the local farmer research committees, CIAL (Spanish acronym) developed by the International Centre for Tropical Agriculture (CIAT) being the best example (Ashby et al., 2000). Farmers were given the skills to design their own research, most often within a community or group framework. An external actor provided the skill set for experimental design and interpretation. The approach has not been tracked for any length of time, nor has there been much effort in linking these farmer research groups to broader-scale dissemination pathways. The second approach relies on experimental designs developed by scientists but implemented on farms with variable farmer participation. Mother–baby trials are in many ways a thoughtful approach to combining the complex, multi-factorial trials required by scientists with farmer choice of technology options in simpler, comparative trials involving a significant degree of farmer participation, if relatively limited in terms of the scope of problems being addressed (Snapp et al., 2003).

Research on natural resource management and systems approaches to increasing productivity has spawned a third approach which relies on decision-support systems (DSS) that are employed either by extension or by farmers themselves. DSS go to the heart of conditionality of recommendations in smallholder, rain-fed systems. Such support systems run the gamut from the very simple, such as the Tropical Soil Biology and Fertility Programme's (TSBF) diagnostic criteria on organic matter quality and the associated use of variable quality organic resources under alternative soil fertility conditions, to the more complex. An example of the latter has involved participatory interaction with farmers using crop models such as the Agricultural Production Systems Simulator Model (APSIM), in which alternative scenarios under different rainfall distributions and meteorological forecasts are developed (Dimes et al., 2003). DSS is still relatively knowledge-intensive to apply in terms of developing recommendations, and the farmer then has to decide on changing management practices based purely on simulation results, rather than reliance on empirical evaluation from field trials. This is particularly so for farmers with relatively limited education, a characteristic of most farmers in sub-Saharan Africa and particularly women farmers in South Asia and Africa. These issues have extended some modes of adaptive research into the realm of technology dissemination.

Integrating adaptive research trial results with piloting of dissemination approaches forms the fourth type of adaptive research methodology. Demonstration trials, especially when done at significant spatial scale, such as the N2Africa programme testing grain legume technologies, allow farmers to evaluate the yield advantages of particular technologies, where the trials are often laid out in simple comparative designs, with or without replications (Baijukya, 2011). Appropriately designed across a spectrum of agro-ecological conditions, demonstration trials can provide information of technology response to farmers and at the same time provide data for researchers on technology response under heterogeneous conditions, often involving both genotype and management options, or understanding G x E x M. This allows an evaluation of technology response at different spatial scales from farm to agro-ecology and thus defines how differentiated dissemination strategies need to be, while at the same time providing initial feedback on farmer evaluation of the technology. Alternatively, CIAT has planted a block with a range of different forages from which farmers choose, and the farmers themselves test and adapt the forage and forage combinations to the particular needs of their systems, with researchers focused on understanding why farmers with different resources and market access have adopted particular forages (Roothaert et al., 2003).

Each smallholder, rain-fed farm is unique, yet adaptive research cannot be done on each farm. This has led to methods for classifying types of farming systems and delineating agro-ecological zones, which provide a framework for both efficient selection of adaptive research sites and farm types but also a basis for then designing efficient dissemination strategies. These need to understand how farmers acquire, assess, and act upon new information, whether in the form of physical inputs or improved management practices, and the varying capacity, usually based on education, of farmers to effectively utilize such information. Moreover, in general, there is a negative correlation between the cost per farmer of supplying such information and the intensity and facilitation required for farmers to use that information effectively in changing their farming practices. Thus, farmers proceed through four quite different stages, moving from discovery to actual adoption. These might be classified as sensitization, information/technology sourcing, technology evaluation, and adoption/system integration. Various dissemination methods of varying cost are associated with each stage, e.g. rural radio for sensitization and adaptive trials for technology evaluation, and often these are used together in a phased set of steps. However, the overall objective is to move the farmer to a process of learning by doing, which involves evaluation and decision-making about adoption (Foster and Rosenzweig, 1995).

Optimally, although rarely, adaptive research programmes should be designed to test dissemination strategies simultaneously. Adaptive research programmes focus on the technology evaluation stage of a farmer adoption process, probably the most costly per farmer part of a dissemination programme. Designing communication strategies, developing supply points for accessing the technical inputs, evaluating farmer learning outcomes, and in the end understanding farmer incorporation of the technology into the farming system are critical research areas for adaptive research programmes that feed directly into the design of dissemination programmes at larger scale. Building this bridge between adaptive research and dissemination has often been inhibited by the organizational divide between research and extension. The recent trend towards pluralistic models and away from command and control approaches has helped to close that divide, especially at field level. The effective design of adaptive research programmes while testing options for the design of dissemination strategies requires a broad set of research skills. Moreover, there is little experience with interfacing these results with the needed capacities within a much more differentiated extension capacity, especially in terms of the ability to target adaptive results spatially, to interface them with ICT programmes, to develop delivery systems for the physical inputs, and to collect the relevant data on farmer evaluation and associated co-variants affecting that evaluation.

There have been a number of innovations at the margins in linking adaptive research to scaling-up processes. Much of this builds on the increasing facilitation of farmer groups (social capital), whether through micro-credit programmes, farmer field schools and other extension methods, or market assembly. The social-science literature has analysed the role of social networks, information flows, and social learning, particularly in the adaptive management of natural resource strategies (Leeuwis and Pyburn, 2002; Klerkx et al., 2010). More recently, the economics literature, building on existing practice, has begun exploring the effect of social networks and social learning on the adoption of technology (Foster and Rosenzweig, 2010). The influence of social learning depends on the type of technology, poverty status of the household, the type of network, etc. Moreover, social networks tend to function best if based on kinship or religious group, and most of this work has been done at a community or village level.

Adaptive research tends to build on existing farmer groups and either group or individual experimentation. Rusike et al. (2006) evaluated group methods with differences between whether the researchers or farmers designed and subsequently managed the research trials on their own or together. For a complex technology such as integrated soil fertility management, the findings in Malawi and Zimbabwe suggested that farmers adopted at higher rates when the researchers designed the trials and the farmers managed the trials, at least in Zimbabwe. This approach (researcher-designed, farmer-managed experimental plots) has been used by N2Africa (Baijukya, 2011) in eight countries in reaching over two hundred thousand farmers with improved grain legume technologies (however, the impact on adoption has yet to be evaluated). Nevertheless, a range of adaptive research methods continue to be deployed where there has been little comparative evaluation (e.g. see Johnson et al., 2003) and few have been deployed at any significant scale. It is still assumed that, if there is a certain density of adoption with sufficient connectivity to social networks, the diffusion process will become autonomous, without the requirement of further institutional investment in the testing and dissemination process.

Almost five decades after Rogers' (1995) seminal book and the research on adoption arising from the Green Revolution experience, smallholder adoption of improved technologies under rain-fed conditions still remains poorly understood in practice, although there is a large and growing literature on the subject. Developing efficient adaptive research and dissemination systems will require better understanding of farmer heterogeneity and a greater ability to impart farming learning when using more complex, management-based technologies. Alternative dissemination methods need to be designed to take advantage of social networks and to include poor households and women within them, followed by assessment for their effectiveness and efficiency. As technologies are disseminated, testing needs to move on from pilot projects to a relevant scale.

A fundamental driver of technology adoption is access to efficient markets, whether for inputs, outputs, labour, credit, or insurance. As discussed in the previous section, where formerly technology design incorporated access to markets as a criterion, the prevailing practice over the last decade or so since the period of structural adjustment and market liberalization has been to integrate programmes on technology dissemination with improvements in market efficiency. Interlinked input, output, and credit markets, as best exemplified by warehouse receipt systems, offer most of the preconditions for smallholder technical change (AGRA-ILRI, 2010). A range of projects has attempted to link technology dissemination to improvements or innovations in market efficiency, though these are often linked to individual markets such as fertilizer, credit, or insurance. The Sub-Saharan African Challenge Program adopts an innovative platform approach, at the district level, to integrate market innovations led by private-sector interests with technological dissemination through farmer organizations (Lynam et al., 2010). These approaches are only in their infancy, but offer the potential for interactive improvements in market access and efficiency with technological dissemination at a relevant scale.

Market development, however, favours those smallholders with the land, capital, and education to meet the often strict requirements of urban markets, which can limit access by less endowed smallholders. Neven et al. (2009) in Kenya

analyzed the farm-level impact of supermarket growth on Kenya's horticulture sector, which is dominated by smallholders. The analysis reveals a threshold capital vector for entrance in the supermarket channel, which hinders small, rain-fed farms. Most of the growers participating as direct suppliers to that channel are a new group of medium-sized, fast-growing commercial farms managed by well-educated farmers and focused on the domestic supermarket market. Their heavy reliance on hired workers benefits small farmers via the labour market.

Technical change reinforced by market opportunities can stratify smallholder, rain-fed systems into participants and non-participants in that growth process. Expanded adaptive research programmes that are linked to markets, credit, and organized supply chains could counter the inherent advantages of larger, educated farmers, but can this be done at sufficient scale and appropriate cost?

4. Scale economies in increasing productivity in smallholder, rain-fed agriculture

The challenge remains of how to generate sustained growth in smallholder productivity in rain-fed agro-ecologies that will also generate the growth multipliers and expansion in rural non-farm economic opportunities, creating a buoyant smallholder economy. This is, in part, the framework for Africa's Green Revolution (Rockefeller Foundation, 2006), which focuses on increasing productivity of basic staple food crops, relying primarily on high-yielding varieties and fertilizers delivered by growing agro-dealer networks and input markets. However, even fertilizer turns out to be a relatively complex technology if rain-fed farmers are to use it efficiently and profitably. Given the high cost of fertilizer in sub-Saharan Africa, fertilizer use efficiency is critical to profitable use, and in many areas is not profitable at all (Guo et al., 2011). Nevertheless, scalability of targeted technologies in large areas of sub-Saharan Africa is one the principal criteria used in grant-making by the Bill and Melinda Gates Foundation (Pingali, 2012). And, yet, how to raise smallholder agricultural productivity across a significant spatial scale in rain-fed agriculture continues to be a significant challenge.

For decades the traditional approach to transferring agricultural technology to farmers was through extension systems. In many respects this reached an apogee in the 1980s and 1990s with the funding of Training and Visit (T&V) extension systems by the World Bank. Although research has largely been reorganized into independent parastatals, extension was the preserve of Ministries of Agriculture, which in general had a higher priority for budgetary funding than agricultural research. Part of the objective of T&V was to maximize farmer coverage, and this significantly expanded extension personnel in the ministries just at the inception of structural adjustment and the reigning in of fiscal deficits. In essence T&V systems lost their funding base and retrenched during the 1990s. However, the 1990s were also the period of market liberalization, increasing democratization, and decentralization, leading to an expansion in both civil society organizations and a slow but persistent increase in private-sector engagement in processing, marketing, and logistics. The ineffectiveness of public-sector extension due to budget constraints created a void which was filled by an expanding number of NGOs and by the private sector in animal health and high-value industries such as export horticulture and smallholder dairy. This hybrid, pluralistic system was not effectively coordinated and did not reach anything like universal coverage.

A unique experiment in Uganda with the National Agricultural Advisory Development Service (NAADS) embraced multiple actors and attempted to build an extension system on the basis of farmer fora that identified extension needs and private sector or NGO entities providing advisory services on a contractual basis, initially funded by government but with an increasing share of the costs being borne by farmers themselves. The programme quickly ran into capacity problems in terms of actors that could deliver high-quality extension advice and therefore in the potential returns to farmers in purchasing such advice.

Over the last decade since market liberalization, there has been significant progress in many rain-fed agro-ecologies, particularly in South Asia and East Africa; there have been improvements in smallholder access to input and output markets, in access to rural credit either through micro-credit agencies, rural banks, or village banking schemes, including Savings And Credit

Cooperatives (SACCO), and as crucially in improvements of farmer organizations and farmer associations, thereby giving farmers more voice. These changes provide a more favourable incentive structure for farmer adoption of new technology and reinvestment in farm productivity, such as soil restoration. With expanding capacity in adaptive research on the one hand, and continuing improvements in markets and farmer organization on the other, an extension system acting independently as purely a source of technical information is both ineffective and misses many of the synergies needed to improve smallholder productivity. Thus, instead of evaluating the various actors providing advisory services, more insight is provided by analysing the functionality of different frameworks for scaling up, particularly where there are also objectives of reaching the rural poor and better integrating women into income-generating activities (see Lin, 2012).

Past extension approaches have primarily emphasized demonstration and promotion of improved technologies, without considering much of the context beyond the agro-ecological conditions in which the technology would be adopted. In Asia during the Green Revolution this institutional context was already well in place, including subsidized input supply, bulking and assembly in output markets, rural credit, and extension services. In this regard the production response from the widespread adoption of improved varieties was supported by a range of markets and services that is unavailable in sub-Saharan Africa. During the last decade in the post-market liberalization period, there have been increasing though uneven efforts to develop agrodealer input networks, assembly, and bulking points, especially for perishable produce, price information systems using mobile-phone technology, and rural credit systems, with some exploratory work in crop and livestock insurance. Much of this is being led by increasing private-sector investment and in some cases, as with credit guarantees, by public/private partnerships. However, poor and expensive transport infrastructure, poor contract enforcement with the associated dependence on cash transactions, and high transaction costs through the value chain constrain latent demand for inputs and bias relative, farm-gate prices towards continued planting of staple food crops.

Kydd and Dorward (2004) provide a theoretical framework for what they term the 'coordination' problem of smallholder agriculture in agrarian economies such as those in sub-Saharan Africa. Under such conditions, extension alone has little potential for generating sustained increases in farm productivity.

The potential of such 'integrated' approaches is probably best exemplified by the interlinking of markets and farmer organizations inherent in the cotton industry in West and southern Africa or the smallholder tea industry in Kenya. Both are something of an enclave, export-oriented economy, but it demonstrates the production and productivity potential of interlinked input, output, and credit markets, and the coordination achievable through farmer groups and appropriate governance arrangements in mediating farmer and processor interests at the processing mill. The question is whether and how this coordination can be achieved more broadly across the agricultural sector, especially in contexts where farmers rely on spot markets to sell their produce—the functional delivery of advisory services is a significant contributor to developing improved, interlinked markets. Different approaches for advisory services are briefly evaluated and then reviewed in the context of more integrated frameworks, such as Agricultural Innovation Systems (World Bank, 2012).

Innovation in information and communication technologies (ICT), including rural radio, video, and the virtual saturation of even rural areas with mobile phones, has generated significant work in providing information to farmers, especially smallholders. However, two principal issues limit the use of these technologies to promote adoption, namely the fact that virtually any recommendation in the smallholder, rain-fed system is contextually constrained; it also requires something like a decision support system with associated farm diagnostics looking at pests, diseases, or soil. Mobile phones have been very useful in providing basic factual information such as market prices, names of registered seed producers, meteorological forecasts, and other information useful in farm management. Also, phone cameras with GPS can provide necessary information for disease and pest surveillance, with returned information on control strategies. However, pure information does little more than make farmers aware of problems; farmers still need a framework to access and evaluate technologies, usually through learning by doing.

A far more costly approach compared to the simple provision of information is through building farmer capacity and in many cases associated social capital. This approach goes back to the farmer training units employed in the latter part of the colonial period, but methods have evolved significantly since then. Probably the most recognized approach is that of Farmer Field Schools (FFS), in which a group is given a seasonal curriculum and with a significant focus on farmer learning by doing. The cost of this approach has been criticized in relation to more traditional extension approaches (Feder et al., 2004), but the investment in social capital (compared to merely disseminating extension messages) is not factored into this evaluation, and the expected return over the years in improved farmer social capital is expected to be large. Farmer group formation is also a vehicle for participation of the poor and women, although how this is most effectively done still requires methodological work. Farmer group formation around education and technology testing can also provide the basic building blocks for higher-order farmer associations which organize around marketing—for example, warehouse receipt systems, horticultural bulking, and sorting points—and around access to credit. Farmer associations with the appropriate management and financial skills have the potential to become a coordinating locus for interlinking markets, as discussed earlier, such as raw material supply, bulking and assembly, access to market, credit, and bulk purchasing of inputs. Such farmer collective action is potentially easier in industries like smallholder dairy or smallholder tea, where there is need to organize and coordinate around a rural-based processing capacity, as compared to bulk commodities like food grains. Access to technology and adaptive research thus becomes a complementary service function in improving farm productivity, with the incentives for investment provided by the interlinking markets.

Rather than starting with farmer capacity and organization as an entry point to scaling up, many programmes rely directly on market-based approaches to technology dissemination, even where input markets are relatively thin and underdeveloped. Support strategies in this regard often rely on development of distribution networks, credit guarantees especially for working capital, and agro-dealer training. This approach is based purely on inputs as the key to farm productivity improvement, sometimes with management options provided by agro-dealers. It assumes a significant latent demand for the inputs from farmers and the cash in order to purchase those inputs. For bulky commodities such as fertilizer, distribution costs are high to smallholders, and fertilizer market development initially relies on developing bulk supplies to large farmers, for example tea estates. Smallholder access to inputs through such market-based approaches is not guaranteed, since profitability is not guaranteed. As in much of Asia, there has been a move to institute vouchers or subsidies over the last five years in many parts of sub-Saharan Africa, following the early, quite successful experience in Malawi (Dorward et al., 2008). Scalability of technologies such as legume inoculants, drought-tolerant maize, bio-fortified cereals, or seed coating for striga control are dependent on input market development and private-sector investment, with a tendency for impacts to be biased to more commercial farmers.

Traditional extension and advisory services organized around hierarchical, command, and control models relying on blanket recommendations and repetitive demonstration trials and staffed by diploma-level personnel are now largely an anachronism, incompatible with current needs. How to invest in extension systems remains a subject of debate. There is a move to more pluralistic systems where NGOs and private- and public-sector capacities find some division of labour and attempt to optimally allocate limited resources (Heemskerk and Davis, 2012). However, the 'extension-plus' (Sulaiman, 2012) reinvention of institutional approaches to scaling up will have capacities across ICT, soft facilitation skills in support of farmer group formation and capacity building, decision support systems, innovation brokers, and close links to and understanding of adaptive research within relevant agro-ecologies (Klerkx and Gildemacher, 2012). These functions require a very different type of extension staff with a much wider skill set and greater entrepreneurial skills within a more decentralized organizational structure. Accountability based on farmer evaluation would provide appropriate performance monitoring. The question is whether such functions and skills can be built within existing organizational structures or whether new structures are required, potentially with some functional specialization, such as facilitation skills and farmer organization.

Technology adoption by smallholders and associated dissemination strategies are embedded in a market, organizational, and institutional context that is as heterogeneous as are agro-ecology and farming systems. Initial evaluations of the Green Revolution in Asia focused on the technology and productivity impacts, while the market, organizations, and service delivery were accepted as a given since they were functioning well. This was not the case in sub-Saharan Africa, which has led to the broad debate on how to achieve a Green Revolution on that continent. The idea that agricultural research is embedded in an interacting organizational matrix of markets, farmer organization, agribusiness associations, rural credit and insurance providers, and diverse extension capacities, and that for research to be most successful it should build on those linkages or partnerships, has given rise to the concept of agricultural innovation systems or AIS (World Bank, 2012). In well-developed agricultural market contexts there is usually a self-organizing network of organizational linkages between agro-industrial associations, farmer cooperatives, universities, and government policy and funding agencies.

However, in contexts where agricultural markets do not function efficiently and inhibit private-sector investment, such as in sub-Saharan Africa, research and extension tend to be institutionally isolated, with a lack of funding to support the transaction costs inherent in greater inter-organizational interaction. Moreover, such transaction costs are higher where rain-fed areas are dominated by semi-subsistence smallholders, such as large parts of sub-Saharan Africa and the economically lagging, primarily rain-fed, areas of Asia. In such contexts, where there is a need for an integrated approach to market development, increased productivity and investment in the natural resource base, innovation platforms, as developed by the sub-Saharan Africa Challenge Program, is one experiment in providing effective coordination and organizational linkages. Approaches to achieving scale in raising smallholder productivity in rain-fed agro-ecologies have only increased in complexity, particularly over the last decade.

Four key attributes discussed earlier contribute to this complexity and can be summarized as follows:

1. Given both the heterogeneity and complexity of rain-fed, smallholder systems, place-based adaptation is central to achieving productivity increases that are also resource-efficient and provide the resilience

necessary for adaptation to both climate change and the shocks of intense weather events.
2. A critical threshold of farmer capacity is necessary to attain autonomous diffusion, and this is best achieved through farmer group formation and social capital development.
3. There is an increasing institutional convergence between the organization of agricultural research, expanding on-farm research capacity and the expanded knowledge base and skill set of agricultural extension.
4. Having efficient impact at scale requires locating agricultural research and extension within a larger market, institutional and public/private partnership context, which will develop implementation approaches to agricultural innovation systems.

Taken together, these attributes increase the complexity of agricultural research and argue for significant organizational and institutional innovations in agricultural research systems, especially where the principal clientele is smallholder farmers.

5. Designing R&D institutions to foment growth in smallholder productivity

There are few, if any, agricultural R&D systems that meet the four design criteria described earlier. The factors conditioning the evolving design of agricultural R&D systems in the developing world vary significantly, but probably most in terms of structural change in the overall economy, or what the 2008 WDR termed the three worlds of agriculture (World Bank, 2007). Urbanized economies, such as those in Latin America and increasingly in parts of South-East Asia, tend to have larger farm sizes, a large commercial agro-industrial sector, and market signals biased towards urban consumer food demand and quality characteristics and/or export market development through commercial agro-industry. There is significant private-sector investment in research, and public-sector research tends to focus investments on principal commercial value chains, often within public/private partnerships (see the case of Uruguay in Byerlee, 2012).

Transforming economies, which exist primarily in South-East and South Asia, have industrial sectors that are growing much faster than agriculture. Alongside rural population shifts into urban areas, there is disequilibrium in agricultural growth: agricultural regions with well-developed market and transport infrastructure develop to meet growing urban demand, but large agricultural regions lag economically with much of the rural population still dependent on low-productivity, smallholder farms. This creates a tension in how best to organize agricultural R&D, as the commercial sector and rural poverty in lagging areas require very different approaches and alternative partnerships amid competition for resources and capacities. These tensions can be reduced where R&D is decentralized, usually within a federal structure, by research capacity in universities being improved and better coordinated with public-sector research institutes, and where research policies such as competitive grants programmes support diversified strategies.

The population of agrarian economies located primarily in sub-Saharan Africa is predominantly rural, primarily on smallholder farms, and the agricultural sector is central to overall economic growth, employment, and the balance of trade. Moreover, market liberalization in the 1990s, in the period following almost three decades of state control of agricultural markets and the associated depression of agricultural prices, has taken some time to generate private-sector investment; even then, it is primarily in cash crops and 'commercial' areas, reflecting very high transport costs and significant transaction and coordination costs. Smallholders face high costs of market access, the absence of rural credit and insurance, unintegrated output markets, and lack of enforceable contracts. This has resulted in the need for closer integration between market innovations that improve market efficiency and the design and supply of agricultural technologies within the R&D system. At the same time, investment in agricultural R&D is limited due to other demands on constrained government budgets, often weak capacity, limited operational funds, and the isolation of R&D institutions which tend to operate autonomously.

For the structure and organization of agricultural R&D systems to be effective, they need to undergo a process of transformation and differentiation, as they respond to the rapidly evolving needs of agricultural sectors driven by volatile and changing markets, rapidly advancing science, and associated capacity needs. Moreover, the organization of agricultural R&D is primarily a resolution of the tension between the development or supply of new technology, and defining and responding to the demand for improved technology. In urbanized economies this is primarily resolved by market forces and forging of public/private partnerships between research and an expanding agro-industrial sector. In more agrarian economies, demand arises primarily from a heterogeneous farm population, which creates constraints on realizing economies of scale, scope, and size in agricultural research.

This tension in agrarian economies is in part due to structural rigidity in the supply of agricultural technologies. There is a significant time lag between problem identification and the development of a research programme structure and the longer-term research strategies which guide the development of technology options. This is compounded by differences in the discipline mix depending on the research problem. To shift programme priorities between such options as plant breeding and natural resource management, or biotechnology and integrated crop management, or research on basic staples versus new commercial crops, requires major shifts in staffing, research programme structure, capital and infrastructure, and organizational linkages. This inherent structural rigidity has been responsible for a supply push of new technologies, which in the past was reinforced by large extension systems whose sole responsibility was the dissemination of these technologies. Moreover, accountability was diffused between quite autonomous research and extension organizations. In the literature this organizational model has been characterized and criticized as 'linear', with a lack of effective linkage to farmer demand, insufficient accountability on achieving impact, and little feedback on technology response and adoption (Hounkonnou et al., 2012).

Approaches to reforming this linear model have focused on improving the linkage of supply with demand for new technology. Attempts to link research supply to smallholder demand has primarily focused on introducing the voice of the farmer into either governance structures or funding decisions. For the latter, farmers have been principal decision-makers on competitive grant programmes, such as World Bank-funded programmes in Latin America (World Bank, 2009). However, the projects selected have tended to focus on more adaptive research, using existing technologies that are 'on the shelf', as have the innovation platforms developed under the sub-Saharan Challenge Program (Lynam et al., 2010). These findings reinforce the conclusion that smallholder demand for technology is determined by place and is primarily expressed through adaptive research and integrated dissemination programmes. This argues for a more interactive flow of information and analysis from dissemination and adaptive research back to technology design, an interaction that relies on more effective integration between these three 'stages' and an efficient deployment of adaptive research capacity.

Adaptive research is costly and operational budgets are usually the most limiting constraint on research performance in national systems. Efficient design of adaptive research should both minimize the cost and maximize the value of the information feedback to applied research. Elements in the design of such a system would include an effective characterization of smallholder systems and agro-ecologies within a GIS framework, a rigorous sampling frame for site selection, systematic data on technology response done within efficient multi-locational designs, monitoring of smallholder adoption and productivity change—potentially within farmer panel surveys—and the development of efficient information flows and analysis. Effective and efficient information flows are critical to understanding the interaction between farmer demand for technology as reflected in farmer adoption, technology response under heterogeneous conditions, and next-generation constraints to improving smallholder productivity. What, then, are the implications for the evolving institutional arrangements that will allow such integration and at the same time achieve impact at sufficient scale?

5.1 ORGANIZATIONAL CONSTRAINTS TO LINK TECHNOLOGY DESIGN, ADAPTIVE RESEARCH, AND DISSEMINATION

The development of R&D systems in smaller countries where rain-fed, smallholder agriculture dominates, particularly sub-Saharan Africa, has over the last four decades developed in phases, often involving only one of the three stages discussed in this chapter. The 1970s was a period of development and investment in the National Agricultural Research Institute (NARI). The 1980s focused on the evolution of farming systems research towards its current configuration as adaptive research, while the 1990s saw a primary focus on extension, especially large investments in Training and Visit extension, at the same time as a general decline in investment in agricultural R&D. In many countries, especially in sub-Saharan Africa, extension programmes essentially collapsed under structural adjustment programmes. Over the last decade this vacuum has been filled by improved agricultural extension capacity in domestic and international NGOs and by improvement in farmer organizations. This was supported by shifts in funding of many donors towards civil society and private-sector entities, as well as more downstream projects of CGIAR Centres. Project-based funding of adaptive research across a range of research and development organizations has constrained the development of longer-term capacity as well as systematic information flows.

Rigidity in institutional arrangements, especially between research and extension, and shifting funding priorities have failed to provide effective articulation between the three stages in the R&D process. Over the last decade innovation has come through better integration of adaptive research and extension, but within a decentralized and time-limited profusion of projects. In the process, the links back to applied research have become weaker, reinforcing a supply push of new technology. An obvious but complex missing link is the capacity to design and coordinate information flows. With the increase in project-based monitoring and evaluation (M&E) and the associated investments in baseline surveys, expenditure in data collection has increased, but not in a manner that improves decision-making on technology design or enhances understanding of the adoption puzzle. Harvest Choice is an initiative which collects and systematizes agricultural data for the developing world and has shown the potential of such a capacity. However, it does not coordinate data collection, and the information flows to research are opportunistic rather than systematic. Nevertheless, the increase in studies on global food security and the impacts of climate change are creating more demand for systematic trial data and farm surveys, which may be the driver for forging more effective information flows between adaptive research and research decision-making, whether at national, sub-regional, or regional scales.

5.2 RESEARCH HIERARCHIES AND THE SMALLHOLDER

Agricultural research in support of smallholder development takes place within a nested hierarchy of national, sub-regional, and international research capacities, most significantly the CGIAR research system. The research objectives of CGIAR reflect the priorities of the international aid community, particularly rural poverty alleviation and sustainable development. As such, CGIAR places the smallholder at the core of its research programme. At the same time, donor investments in CGIAR are conditioned on demonstrating impact on smallholder productivity and welfare. Because of capacity constraints (Lynam et al., 2012) in national research and extension systems and a donor tendency to fund more downstream projects, CGIAR Centres find themselves working from technology design through adaptive research to dissemination. Scaling up is the most recent area where CGIAR is developing methodological options (e.g. Lin, 2012). This has resulted in CGIAR working with a wide range of civil society and private-sector partners, but often to the exclusion of national R&D systems.

How to develop a functional R&D hierarchy with a clear division of labour and relevant capacities at all system levels remains an unanswered question, especially for small countries in Asia and sub-Saharan Africa. The benefits to such a division of labour have been expressed by Johnson et al. (2011):

> Whenever research and technologies are shown to be easily transferrable or have large potential spillovers across countries, pooling resources has the potential to reduce R&D costs for individual countries, while helping to improve system wide efficiency by reducing duplication of effort, encouraging greater specialization, and exploiting existing complementarities in research capacities. Additionally, a regional approach can offer greater scope and scale economies than is achievable by individual countries, thereby allowing coverage of a broader range of research topics and generation of the critical mass of human resource capacity needed for success.

This theory has been the basis for the development of sub-regional agricultural research programmes, especially in Africa. However, the theory has worked primarily at a commodity research level, especially organized around plant breeding. The Pan African Bean Research Alliance (PABRA) and WARDA's rice improvement programmes (Lynam, 2011) are the best examples of both a regional division of labour across national breeding programmes, combined with a functional division of labour (population breeding and marker-assisted selection) with CGIAR Centres. There have been attempts to do this with more NRM research programmes, with examples such as Soil Fertility Consortium for Southern Africa (SOFESCA) in southern Africa or TSBF's African Network for Soil Biology and Fertility (*AfNet*) programme. These have not been quite as successful as the breeding networks, partly because of uncertainty over funding. However, at an institutional level, an integrated research hierarchy has yet to be achieved in any of the regions, which in turn complicates the closer integration of technology design, adaptive research, and dissemination.

The recent reform of CGIAR is based on three organizational objectives: the development of CGIAR as an integrated research system rather than an autonomous network of 15 centres; programme development and associated funding organized around 15 global research themes (CGIAR Research Programmes) and implemented by research consortia; and accountability based on results monitoring at the level of strategic outcomes. The institutional arrangements that will link technology design at the CRP level with the dissemination needed to achieve the outcomes are still to be developed. However, the tendency appears to be to move away from the conception of a research hierarchy towards autonomous, place-based project structures with similarities to many of the large Bill and Melinda Gates Foundation (BMGF) projects. Institutional arrangements will depend on which organizations have the capacities to undertake the work, at the expense of building longer-term capacity in dedicated national R&D organizations. With its inherent transaction costs, the research consortia approach will depend on some funding innovations as well. To the extent that funding is also tied directly to results, there will be a tendency to focus on downstream capacity and results, potentially at the expense of technology design work, but also highlighting the point that efficient information flows will be central to programme efficiency and results.

5.3 RESEARCH ORGANIZATION IN RESPONSE TO SYSTEM COMPLEXITY

The organization of agricultural research has primarily been in terms of components of farming systems, whether commodity research programmes, animal production systems, animal health, soils management, agroforestry, or integrated pest management. These partly reflect the disciplinary boundaries which characterize departmental organization within faculties of agriculture as well as much agricultural research organization during the colonial period. The 1970s and the formation of CGIAR were framed within the concept of multi-disciplinary research teams, although still organised around technological components. Multi-disciplinary crop research programmes were the prime example of such approaches. However, over the last five to ten years there has been an increasing trend towards organizing agricultural research—whether formally through programmatic structures or informally through research projects and networks—at what is a higher systems level. Alternative frameworks have evolved around production or farming systems, value or supply chains, and most recently landscapes. This has been a response to the development of new research agendas, on the one hand, and an expanding set of new scientific methodologies on the other. These changes, moreover, are in relatively early phases of exploration and development, and to a significant extent there are inherent trade-offs in the three frameworks—production systems, value chains, and landscapes.

Adaptation to climate change, agro-ecological intensification and eco-efficiency, sustaining ecosystem services, and the continuing challenge of productivity growth in rain-fed, smallholder agriculture are the larger development outcomes that are driving this process of research being organized at higher system levels. All of these are currently being framed within such conceptual frameworks as system resilience, sustainable livelihoods and the associated system vulnerability, and resource efficiency. These frameworks are operational at the level of the farming system or household and position increased productivity within farmer choice of production activities, their interaction, and farmer reinvestment in the system. Moreover, the interaction of farmer management of production activities and the associated impact on a farm's natural resource base have implications for the provision of ecosystem services, but only at scale and mediated by other interacting factors in the landscape. Thus, natural resource management research in areas such as ecosystem services has adopted landscapes as a research framework to understand the interaction between different land use and farming systems, decision-making by a multiplicity of agents, and trade-offs in the objectives of those agents.

On the other hand, organizing technology design around value chains attempts to integrate farm-level research with interventions focused on greater market efficiency and/or development of new markets, such as for biofuel. However, the analysis starts from the evaluation of commodity demand and market structure, and then integrates production focusing on a single commodity and the lower cost production areas. This value-chain framework results in a significantly different programme structure and disciplinary mix than organizing research around production systems, raising the question of which is most appropriate for reaching the majority of smallholder farmers, particularly under heterogeneous production and market conditions. Certainly access to market and improved market efficiency are critical components in smallholder development, and as discussed earlier there are interacting links to the adoption of improved technology. However, there is an inherent bias in the value-chain framework towards more commercial smallholders and areas with better developed market infrastructure.

In many respects the point for linking production system research with marketing options is through the adaptive research process, where organizational innovations in the market sphere can be linked to improving overall productivity of the farming system.

6. **Conclusions**

The challenge of increasing smallholder productivity in the rain-fed, agricultural areas of the tropics, especially in sub-Saharan Africa and the smaller countries of South and South-East Asia, has remained for almost four decades, ever since researchers attempted to move the Green Revolution into rain-fed agriculture. R&D has had to address and adapt to the extraordinary heterogeneity and system complexity which characterizes smallholder agriculture, and more recently it has had to address incomplete, missing, or inefficient markets in both the design and dissemination of new technologies. As a result, smallholder adoption of improved technologies has been primarily limited to single technology components in relatively constrained geographical areas, insufficient to generate the growth multipliers that characterized the Green Revolution. The inability to understand this lack of farmer adoption is in turn mirrored in the uncertainty surrounding the appropriate design of these technologies. One dominant trend in R&D for rain-fed, smallholder agriculture is the increasing centrality in adaptive research to understanding technology response and farmer assessment. That said, there is little sustained institutional capacity to undertake this type of research, nor is it done in a manner that is cost-effective, that efficiently generates information within a well-structured analytical framework, and that provides the critical design features of a dissemination programme at scale. Given the expenditures on adaptive research, the feedback to technology design remains weak or virtually non-existent, and the feed forward to the design of dissemination programmes is literally in its infancy. At the same time, the whole conceptualization of extension is undergoing a major re-evaluation. The challenges facing R&D for smallholder agriculture are more organizational and institutional than either creative or an appropriate application of the science, although those two elements are integrally linked.

REFERENCES

AGRA-ILRI (2010). *Developing Pro-poor Markets for African Smallholder Farmers.* Retrieved from: <http://mahider.ilri.org/bitstream/handle/10568/16491/AGRA-ILRI-Section1.pdf?sequence=5>.

Ashby, J. A., A. Braun, T. Gracia, M. P. Guerrero, C. A. Quiros, and J. I. Roa (2000). *Investing in farmer researchers: Experience with local agricultural research committees in Latin America.* Cali, Colombia: CIAT.

Baijukya, F. (2011). 'N2Africa develops best-fit agronomic practices for grain legumes in Africa', N2Africa Podcaster no. 10. Retrieved from: <http://www.n2africa.org/sites/n2africa.org/files/images/N2Africa%20Podcaster%2010.pdf>.

Brooks, K. (2010). 'African agriculture—What do we not know? What do we need to know? Agriculture for Development—Revisited: Lessons learned from the University of California at Berkeley conference', (mimeo).

Byerlee, D. (2012). 'Producer funding of R&D in Africa: An underutilized opportunity to boost commercial agriculture', Conference Working Paper 4. Agricultural R&D: Investing in Africa's Future. ASTI/IFPRI. Retrieved from: <http://www.asti.cgiar.org/pdf/conference/Theme1/Byerlee.pdf>.

Byerlee, D., A. de Janvry, and E. Sadoulet (2010). 'Agriculture for development—Revisited: Lessons learned from the University of California at Berkeley Conference', (mimeo). Retrieved from: <http://cega.berkeley.edu/assets/cega_events/30/Agriculture_for_Development__Revisited_-_Conference_Lessons.pdf> (22 July 2013).

De Groote, H., B. Vanlauwe, E. Rutto, G. D. Odhiambo, F. Kanampiu, and Z. R. Khan (2010). 'Economic analysis of different options in integrated pest and soil fertility management in maize systems of Western Kenya', *Agricultural Economics*, 41, pp. 471–82.

Dimes, J., S. Twomlow, and P. Carberry (2003). Chapter 6, 'Application Of APSIM in small holder farming systems in the semi-arid tropics', in Bontkes and Wopereis (eds), *Decision support tools for smallholder agriculture in sub-Saharan Africa: A practical guide.* IFDC and CTA, pp. 85–99.

Dixon, J., A. Gulliver, D. Gibbon, and M. Hall (2001). *Farming Systems and Poverty: Improving Farmers' Livelihoods in a Changing World.* Rome: FAO and World Bank.

Dorward A., P. Hazell, and C. Poulton (2008). 'Rethinking agricultural input subsidies in poor rural economies', Future Agricultures Briefing Paper. Retrieved from: <http://www.future-agricultures.org/pdf%20files/Briefing_input_subsidies.pdf> (21 July 2013).

Douthwaite B., J. D. H. Keatinge, and J. R. Park (2001). 'Why promising technologies fail: the neglected role of user innovation during adoption', *Research Policy*, 30, pp. 819–36.

Douthwaite, B., V. M. Manyong, J. D. H. Keatinge, and J. Chianu (2002). 'The adoption of alley farming and Mucuna: lessons for research, development and extension', *Agroforestry Systems*, 56, pp. 193–202.

Feder, G., R. Murgai, and J. B. Quizon (2004). 'Sending farmers back to school: The impact of farmer field schools in Indonesia', *Review of Agricultural Economics*, 26(1), pp. 45–62.

Foster, A. D. and M. R. Rosenzweig (1995). 'Learning by doing and learning from others: Human capital and technical change in agriculture', *Journal of Political Economy*, 103(6), pp. 1176–209.

Foster, A. D. and M. R. Rosenzweig (2010). 'Microeconomics of technology adoption', *Annual Review of Economics*, 2, pp. 395–424.

Guo, Z., J. Koo, and S. Wood (2011). 'Spatial patterns of profitability of fertilizer use in maize production systems in Northern and Central Corridors of East Africa'. Retrieved from: <http://addis2011.ifpri.info/files/2011/10/AgProductivity-Africa_Abstracts1.pdf>.

Heemskerk, W. and K. Davis (2012). 'Pluralistic extension systems', in World Bank, *Agricultural Innovation Systems: An Investment Sourcebook.* Washington, DC: World Bank.

Hounkonnou, D., D. Kossou, T. W. Kuyper, C. Leeuwis, E. S. Nederlof, N. Röling, O. Sakyi-Dawson, M. Traoré, and A. van Huis (2012). 'An innovation systems approach to institutional change: Smallholder development in West Africa', *Agricultural Systems*, 108, pp. 74–83.

Johnson, M., S. Benin, X. Diao, and L. You (2011). 'Prioritizing regional agricultural R&D investments in Africa: Incorporating R&D spillovers and economywide effects', Conference Working Paper 15. *Agricultural R&D: Investing in Africa's Future.* ASTI/IFPRI. Retrieved from: <http://www.asti.cgiar.org/pdf/conference/Theme4/Johnson.pdf>.

Johnson, N., N. Liljab, and J. A. Ashby (2003). 'Measuring the impact of user participation in agricultural and natural resource management research', *Agricultural Systems*, 78, pp. 287–306.

Klerkx, L. and P. Gildemacher (2012). 'The role of innovation brokers in agricultural innovation systems', in *Agricultural Innovation Systems: An Investment Sourcebook.* Washington, DC: World Bank, pp. 211–30.

Klerkx, L., N. Aarts and C. Leeuwis (2010). 'Adaptive management in agricultural innovation systems: The interactions between innovation networks and their environment', *Agricultural Systems*, 103, pp. 390–400.

Kydd, J. and A. Dorward (2004). 'Implications of market and coordination failures for rural development in least developed countries', *Journal of International Development*, 16, pp. 951–70.

Leeuwis, C. and R. Pyburn (eds) (2002). *Wheelbarrows full of frogs: social learning in rural resource management: international research and reflections.* Assen: Koninklijke Van Gorcum.

Lin, J. (ed.) (2012). *Scaling Up In Agriculture, Rural Development, And Nutrition.* Washington, DC: IFPRI.

Lynam, J. (2004). 'Science in improved farming systems: Reflections on the organization of crop research in the CGIAR', Invited paper 4th International Crop Science Congress. Retrieved from: <http://www.cropscience.org.au/icsc2004/symposia/4/2/1333_lynamj.htm>.

Lynam, J. (2011). 'Plant breeding in sub-Saharan Africa in an era of donor dependence', *IDS Bulletin*, 42(4), pp. 36–47.

Lynam, J. (2012). 'Agricultural research within an agricultural innovation system: Overview', in *Agricultural Innovation Systems: An Investment Sourcebook.* Washington, DC: The World Bank.

Lynam, J., N. Beintema, and I. Annor–Frempong (2012). 'Agricultural R&D: Investing in Africa's future: Analyzing trends, challenges, and opportunities: Reflections on the Conference', Washington, DC: ASTI/IFPRI.

Lynam, J., K. Harmsenand, and P. Sachdeva (2010). *Report of the Second External Review of the Sub-Sahara Africa Challenge Programme (SSA-CP).* CGIAR/ISPC Secretariat, Washington, DC Retrieved from: <http://www.sciencecouncil.cgiar.org/publications/reviews/challenge-programs/en/>.

Neven, D., M. M. Odera, T. Reardon, and H. Wang (2009). 'Kenyan supermarkets, emerging middle-class horticultural farmers, and employment impacts on the rural poor', *World Development*, 37(11), pp. 1802–11.

Omamo, S. W., X. Diao, S. Wood, J. Chamberlin, L. You, S. Benin, U. Wood-Sichra, and A. Tatwangire (2006). *Strategic Priorities for Agricultural Development in Eastern and Central Africa.* Washington, DC: International Food Policy Research Institute.

Pingali, P. (2012). 'The Bill & Melinda Gates Foundation: Catalyzing agricultural innovation', in Lin, J. (ed.), *Scaling Up In Agriculture, Rural Development, and Nutrition.* Washington, DC: IFPRI.

Rogers, E. M. (1995) (ed.). *Diffusion of Innovations.* New York: Free Press.

Roothaert, R., P. Horneand, and W. W. Stür (2003). 'Integrating forage technologies on smallholder farms in the upland tropics', *Tropical Grasslands*, 37, pp. 1–9.

Rusike, J., S. Twomlow, H. A. Freeman, and G. M. Heinrich (2006). 'Does farmer participatory research matter for improved soil fertility technology development and dissemination in Southern Africa?', *International Journal of Agricultural Sustainability*, 4(3), pp. 176–92.

Snapp, S. S., M. J. Blackie, and C. Donovan (2003). 'Realigning research and extension to focus on farmers' constraints and opportunities', *Food Policy*, 28, pp. 349–63.

Sulaiman, R. (2012). 'Extension-Plus: New roles for extension and advisory services', in *Agricultural Innovation Systems: An Investment Sourcebook*. Washington, DC: The World Bank.

Sumberg, J. (2005). 'Systems of innovation theory and the changing architecture of agricultural research in Africa', *Food Policy*, 30, pp. 21–41.

The Rockefeller Foundation (2006). *Africa's Turn: A New Green Revolution for the 21st Century*. New York: The Rockefeller Foundation.

Tittonell, P., A. Muriuki, K. D. Shepherd, D. Mugendi, K. C. Kaizzi, J. Okeyoand, and L. Verchot (2010). 'The diversity of rural livelihoods and their influence on soil fertility in agricultural systems of East Africa—a typology of smallholder farms', *Agricultural Systems*, 103(2), pp. 83–97.

Tittonell, P., B. Vanlauwe, P. A. Leffelaar, E. C. Roweand, and K. E. Giller (2005). 'Exploring diversity in soil fertility management of smallholder farms in western Kenya: I. Heterogeneity at region and farm scale', *Agriculture, Ecosystems & Environment*, 110, pp. 149–65.

Twomlow, S. J., J. Rusike, and S. S. Snapp (2004). 'Biophysical or economic performance which reflects farmer choice of legume "Best Bets" in Malawi', in Friesen, D. K. and A. F. E. Palmer (eds), Integrated Approaches to Higher Maize Productivity in the New Millennium: Proceedings of the Seventh Eastern and Southern Africa Regional Maize Conference, 5–11 February, 2002, Nairobi, Kenya: CIMMYT (International Maize and Wheat Improvement Center) and KARI (Kenya Agricultural Research Institute), pp. 480–6.

Vanlauwe, B., A. Bationo, J. Chianu, K. E. Giller, R. Merckx, U. Mokwunye, O. Ohiokpehai, P. Pypers, R. Tabo, K. Shepherd, E. Smaling, P. L. Woomerand, and N. Sanginga (2010). 'Integrated soil fertility management: Operational definition and consequences for implementation and dissemination', *Outlook on Agriculture*, 39, pp. 17–24.

World Bank (2007). *Agriculture For Development: World Development Report 2008*. Washington, DC: The World Bank.

World Bank (2009). *Agricultural Research and Competitive Grant Schemes: An IEG Performance Assessment of Four Projects in Latin America*. Washington, DC: The World Bank.

World Bank (2012). *Agricultural Innovation Systems: An Investment Sourcebook*. Washington, DC: The World Bank.

12 Farmers as entrepreneurs: sources of agricultural innovation in Africa

CALESTOUS JUMA AND DAVID J. SPIELMAN[1]

1. Introduction

Increasingly, images of Africa are characterized not by misery and deprivation, but energized African farmers and other rural entrepreneurs: Rwandan women wash and pack speciality coffee destined for lucrative markets in Europe; Ethiopian farmer cooperatives assemble massive quantities of sesame and coffee to sell on the modern trading floor of the country's commodity exchange; Nigerian women's groups manage sprawling markets to supply the demands of growing urban populations; and Kenyan farmers cultivate a dizzying variety of horticulture crops, while pastoralists in Senegal, Mali, and Ethiopia drive herds of prized cattle, sheep, and goats to meet accelerating demands of sprawling urban centres.

These images are signs of change—signs of entrepreneurship in African agriculture. Throughout the region, innovation is increasingly driven by entrepreneurs (Naudé, 2011). They are men and women who produce, exchange, and use knowledge to create goods and services of social or economic relevance. And throughout the continent, there is proven—yet still untapped—entrepreneurial potential among smallholders and rural communities which can accelerate their role in this transformation process.

Many of these innovation processes are tied to the farm, and there is an ever-increasing stream of innovations designed to improve how farmers till the soil, plant seeds, manage water, apply fertilizer, and tend their livestock. Undoubtedly, these investments in on-farm innovation are critical to increasing agricultural productivity and food security in Africa.

But innovation in African agriculture is also tied to activities that occur off the farm. Rural non-farm employment and entrepreneurial activity are

[1] The authors wish to thank the anonymous referee for helpful comments.

fundamental to agricultural development and economic growth in many countries (Reardon et al., 2000; Pingali and Rosegrant, 1995). The evidence is particularly strong in Asia and Latin America, though no less valid in Africa (Reardon, 1997; Reardon et al., 1994). Yet insufficient attention is given to entrepreneurship in not only African agriculture, but Africa as a whole (Naudé and Havenga, 2005).

Rapid changes in the markets for food, fibre, and fuel are imposing new pressures on African agriculture, with changing consumer preferences at home and abroad driving prices, technology choices, and trade flows. Demographic and environmental pressures are forcing agricultural communities to diversify their livelihoods, migrate to cities, or contend with growing volatility in both climate and markets. With these rapid changes come the intensification of conflict over contested resource claims—food versus fuel, high-yielding intensive production versus low-productivity organic production, short-term gains to satisfy current needs versus long-term strategies to mitigate and adapt to future challenges.

These rapid changes and emerging conflicts strongly suggest that developing countries will need to develop more responsive, dynamic, and competitive agricultural sectors in the short to medium term to benefit from the changing global system. Agricultural innovation will continue to be the order of the day, and developing countries will need innovative policies, programmes, and investments just to keep up.

Thus, the African farmer of the future must be vested with the skills and education needed to experiment, adapt, and compete in a volatile economy where both knowledge and effort are rewarded. In many ways, small-scale farmers actually have a comparative advantage over large commercial farming operations because of their reliance on low-cost household labour, their ability to side-sell in local markets when export markets contract, and their knowledge of local conditions and hazards (Suzuki et al., 2011; Hazell et al., 2007; Poulton et al., 2005).

The future of farming in Africa is synonymous with the future of rural entrepreneurship, a term which can include farmer-entrepreneurs, non-farm business ventures based in rural areas, and small- and medium-scale businesses that provide services or add value to agricultural production.

Unfortunately, policies and investments designed to unlock the entrepreneurial potential of smallholders have received insufficient attention in Africa in past decades. Only in recent years have governments, donors, and investors recognized the need for an enabling environment to unlock this potential. This includes the need for greater human resource capacity, better financial and business development services, and a more effective regulatory environment that encourages growth.

This chapter examines the contribution of rural entrepreneurs to agricultural development and economic growth in Africa, with specific attention given to the policies and investments needed to develop this potential further. Section 1 lays out a conceptual framework that puts the farmer at the centre of the agricultural innovation system. This theory of change is presented in a historical context to illustrate the extent to which the rejection in the early 1950s of the relevance of *The Theory of Economic Development* by Joseph A. Schumpeter shaped development discourse in ways that understated the entrepreneurial function of farmers. The conceptual framework that underpins this chapter emphasizes the role of entrepreneurial capabilities in the context of a wider innovation system. In this respect, Section 2 focuses on infrastructure, technical capacity, and business development as critical requirements for furthering the realization of the entrepreneurial potential of African farmers. The final section outlines the key resources and policy support needed to enable farmers to function as entrepreneurs.

2. Farmers as entrepreneurs: origins of agricultural innovation

Entrepreneurship is emerging as a central feature in economic development. It has been a century since Joseph Schumpeter laid out the theoretical foundations of our current understanding of the role of innovation in economic transformation. In his 1911 classic, *The Theory of Economic Development*, Schumpeter sought to explain the transition from routine production to dynamism as a process of introducing 'new combinations' in the economy (Schumpeter, 1934). He provided a taxonomy of such combinations which included the introduction of 'a new product or a new quality of a new product, a new method of production, a new market, a new source of supply of raw materials or half-manufactured goods, and finally implementing the new organization of any industry' (Hagedoorn, 1996). Schumpeter assigned the act of creating new combinations to entrepreneurs. The entrepreneur is the source of discontinuity in economic systems as well as the agent of change (Schumpeter, 2005).

African farmers do not on the surface fit the largely industrial characterization of entrepreneurship. They are viewed as independent producers who often operate outside the formal markets by producing what they consume. They are perceived to rely largely on social capital embodied in family or local relations (Bauernschuster et al., 2010). The low level of institutional organization and dependence on family labour reinforces the image of farming as activity with low entrepreneurial potential. Farmers are viewed as mere producers whose output becomes part of larger entrepreneurial efforts involving downstream operations. As a result, the bulk of scholarship on entrepreneurship hardly considers farmers as sources of new combinations in the way Schumpeter formulated (Landstrom et al., 2012).

The relevance of Schumpeter's thinking to development was a matter of considerable debate in what later became the foundations of 'development economics'. Wallich (1958) acknowledged that Schumpeter provided 'the most outstanding intellectual performance' because of the internal coherence in his work, but he went on to argue that 'in applying this doctrine to the less developed countries of our day, we find that it does not fit'. The central theme to Wallich's rejection was his view that the role of entrepreneurs as agents of change did apply to less developed countries. Since these economies were largely agricultural, he was in effect stating that agricultural production in these countries did not operate in ways that would lead it to be defined as entrepreneurial in the Schumpeterian sense.

Wallich argued that entrepreneurs hardly existed in low income countries. His position was in contrast to Nurkse (1953), in which he argued that Schumpeter's *The Theory of Economic Development*, 'properly understood, is just what its title says it is: the theory of economic development'. He argued that 'Schumpeter's theory seems to me to provide the mould which we must use, although we may use it with slightly different ingredients'. The debate between the two camps continued and it was generally settled that economic systems that required government could not fit the Schumpeterian mould.

The fact that farmers innovate is not new. In fact, it is in the nature of farming to innovate. What is different from a Schumpeterian perspective is that they create new value through novelty, non-linear adaptations, and constantly manage indeterminate settings (Schumpeter, 2005). The capacity of individuals to innovate is influenced by the individual's interpretation of his or her surrounding environment, the challenges imposed by that environment, and the creativity and commitment that he or she may marshal in response to those challenges (Vandervert, 2003; Shavinina and Seeratan, 2003; Renzulli, 2003). An implication of this is that incentives that encourage innovation and entrepreneurship in Africa must accommodate diverse actors and levels of contribution. There is no standardized approach to unlocking innovative and entrepreneurial capabilities, whether through better education, cheap credit, or new technology.

Another important point is that innovators and entrepreneurs are part of an integral set of interdependent and interactive components within a system that expresses new properties in the form of economic outputs. Often, many diverse actors are involved in the innovation process, influencing and being influenced by interactions with other actors and the formal and informal rules that regulate their practices and behaviours (Lundvall, 1985, 1988; Freeman, 1987, 1988; Nelson, 1988; Dosi et al., 1988; Edquist, 1997). An innovation system embeds the entrepreneur within a larger, more complex system of actions and interactions among diverse actors, social and economic institutions, and organizational cultures and practices. By taking a systems view, farmers are not static components in an economic equilibrium but important sources of novelty.

The performance of an agricultural innovation system hinges on interactions between the research and educational communities and the business and enterprise community, facilitated by a variety of bridging institutions that convey knowledge and information between these communities (Powell and Grodal, 2005; OECD, 1999; Rycroft and Kash, 1999; Nelson and Rosenberg, 1993). Social and economic practices, behaviours, and attitudes, coupled with explicit policies and investments, contribute to economic evolution. The manner in which the various components interact and how they are governed determines the efficiency of the innovation system. But the system itself functions through dynamic and iterative processes of incremental change that engage many different actors, ideas, and actions. It is a twisting, turning, and often convoluted web of processes that somehow drives growth and development.

One implication of this system's perspective is that innovation and entrepreneurship often emerge spontaneously and not necessarily through organized government interventions. This has important implications for planning the growth and development of the agricultural sector, where decades of direct government interventions in all aspects have become the norm in many African countries.

First, the system's perspective forces us to recognize that new inputs or technologies cannot be viewed as innovations themselves, but as embedded information that actors can use in different ways, including those that are used differently from their creator's intentions (Metcalfe, 2000). Second, the approach highlights the need for actors throughout the system to be entrepreneurial in some form and to a significant degree. These actors must be able to learn, or to gather information and use it creatively in response to market opportunities or other social needs (Lundvall, 1999). Learning, in turn, depends on the actor's ability to interact with other system actors in processes of producing or applying knowledge (Fagerberg, 2005; Nelson and Rosenberg, 1993).

Third, the systems approach highlights the fact that innovation is constrained by complexity and the individual and collective capacity of actors to handle complexity. Complexity in process, product, equipment, and instrumentation means that individual actors may not have all the resources they need to innovate fast enough to remain effective or competitive (Powell and Grodal, 2005; Rycroft and Kash, 1999). Hence, the need for collective action in the form of corporations, vertically integrated supply chains, farmers' cooperatives, or small- and medium-sized businesses.

These insights mean that a successful innovation system depends on several key elements: the nature and versatility of the information, the capacity of individuals and organizations to learn and innovate, the nature and character of interactions among innovation agents, and the formal and informal institutions that regulate the agents' interactions. Thus, interventions designed to encourage innovation and entrepreneurship must be dynamic and flexible. A standardized, conventional policy intervention designed to encourage small- and medium-enterprise development will falter if the policy cannot adjust to changes in technology, market structure, business cycles, credit availability, regulations, and a variety of other factors.

3. Support systems and sources of agricultural innovation

3.1 PROVIDING ADEQUATE INFRASTRUCTURE

Infrastructure is vital to the development of a robust innovation system, to agricultural development, and to economic growth. This includes both 'hard' infrastructure—the energy, transportation, and communications networks—as well as 'soft' infrastructure—the rules, norms, and incentives required to ensure their efficient and effective performance in an integrated economy. Increasingly, infrastructure is also associated with the term 'smart', or infrastructure that is engineered simultaneously to address social, economic, and environmental needs. Smart infrastructure includes systems designed with ecological considerations in mind, such as structures and systems that increase carbon sequestration, produce more efficient or off-grid power generation systems that mitigate climate change, or direct rainfall run-off and waste water from roads and buildings into multipurpose uses.

The high social and economic returns on infrastructure development are well documented, particularly in Asia (Fan et al., 2000, 2002); the evidence from Africa, though still emerging, suggests similar returns. Rigorous estimates of the returns on road construction, for example, are documented for Uganda (Fan and Zhang, 2008) and Ethiopia (Mogues et al., 2008), among other countries, with important nuances about road type (primary, secondary, or tertiary) and purpose (primary trade artery, farm-to-market, periphery-to-centre). The evidence suggests a sizeable opportunity for Africa to expand its current road networks—among the lowest in the world in terms of density—to generate strong pro-poor returns. Similar evidence around communications infrastructure has amassed in recent years, with mobile telephony being used in different sections of the agricultural value chain.

And while evidence on the impact of the region's potential in power generation is still unfolding (Gratwich and Eberhard, 2008), most indicators suggest that the opportunities are extensive.

The continental infrastructure challenge is monumental. It is estimated that sub-Saharan Africa needs to invest US$93 billion a year over the next decade to meet its infrastructure needs. Much of this investment should focus on stimulating agriculture. These continental challenges are mirrored at the national level. Take Nigeria as an example. The country's vision to become a middle-income country by 2020 will require sustained investment in infrastructure—nearly US$14.2 billion a year over the next decade, or about 12 per cent of GDP. It currently invests US$5.9 billion per year, 5 per cent of GDP. The country also spends about US$90 billion a year on food imports. Reducing this import bill will require expanding investment in rural infrastructure such as rural energy, transport, telecommunications, and irrigation.

Nigeria's abundant arable land is not being cultivated partly because of poor roads, lack of irrigation, and limited energy supply. Household surveys and spatial data show that between 20 and 47 per cent of rural people live within 2 kilometres of an all-season road. This is well below Nigeria's peer countries. Much of Nigeria's impetus for rural development will come from state-level investments. However, state-run infrastructure suffers most from low maintenance support. Under such conditions, it is not possible for farmers to realize their entrepreneurial potential. They may have individual dynamism, but they can do little with it to create new combinations that add value to the economy.

While national governments, private entrepreneurs, and the community of bilateral and multilateral donors have invested heavily in the development of Africa's infrastructure, it is worth noting that China has entered the picture at a level that may make the difference. China's engagement in Africa's development was set forth in 2000 by the Forum on China–Africa Cooperation (FOCAC), which chalked out an ambitious 'Program for China–Africa Cooperation in Economic and Social Development' (AATF, 2010), and commits China to a range of development assistance ventures including infrastructure and agriculture. These are precisely the types of investments that will play a role in integrating farmers and rural communities into the wider economy, thereby expanding their opportunities for agricultural entrepreneurship.

In short, strategic investments in better road networks, irrigation systems, and power generation have the potential to rapidly integrate farmers and rural entrepreneurs into the economy, providing new and more diversified opportunities beyond the farm. In turn, these opportunities provide farm households and rural communities with new ways in which to manage risk and improve livelihoods. Infrastructure is the backbone—the physical foundation—for entrepreneurship in Africa.

3.2 ENHANCING TECHNICAL COMPETENCE

Upgrading the technical capability of individuals is a key aspect of fostering entrepreneurship (Gries and Naudé, 2011). The path to unlocking the potential of Africa's rural entrepreneurs begins with basic education. Literacy and numeracy skills taught at the primary school level are critical for rural entrepreneurs operating in a modern, commercial economy. Yet primary schooling is a level of education currently unattained by roughly 35 per cent of men and 45 per cent of women in sub-Saharan Africa's low- and middle-income countries (World Bank, 2010). The continued neglect of the region's education sector is a serious stumbling block to fostering greater entrepreneurship in the agricultural sector and, in turn, stimulating broad-based agricultural growth and development.

The potential of Africa's rural entrepreneurs is also dependent on the quality of education. One way of encouraging greater numbers of skilled entrepreneurs to enter the agricultural sector is to teach agriculture as a formal subject in schooling at all levels. From early childhood experiences to university-level education, students can become vibrant entrepreneurs when they are vested with a range of practical skills ranging from technical fields such as agronomy and animal husbandry, to entrepreneurial fields, such as business administration, finance, and economics. These skills are essential to expanding an individual's innovative capabilities.

Many examples illustrate the potential of innovative education designed to encourage entrepreneurship. At the most fundamental level, school gardens and community gardens have been used to motivate students' interest in agriculture and provide them with hands-on exposure to agronomy, food preparation, nutrition, teamwork, and other skills that they will retain throughout their lives (Juma, 2011).

More formal programmes at the higher education level aim to achieve similar goals. For example, the University for Development Studies (UDS) in Ghana's northern region aims to translate higher education and research into knowledge directly relevant to rural communities and their welfare (Juma, 2011). In Mozambique, the Gaza and Manica Polytechnics take a slightly different approach by linking teaching and research to practical and relevant competency development in a practical, hands-on curriculum designed to develop skilled technicians and self-employed entrepreneurs (Davis, 2008). Many other examples exist throughout the continent and beyond, including one that provides a model of particular note—the Escuela de Agricultura de la Región Tropical Húmeda, or EARTH University in Costa Rica. This innovative institution is uniquely designed to provide students with a combination of classroom instruction and learning-by-doing activities conducted in collaboration with local farmers, agroprocessors, and other key players in the agricultural sector (Juma, 2008, 2011; Mitsch et al., 2008; Clark, 2006).

Building technical capacity to support rural entrepreneurship will need to pay particular attention to the role of women. There is emerging evidence that women farmers are starting to define their own technical needs. Some of these efforts are culminating in the creation of new rural universities that focus on the needs of women. An example of such an *in situ* effort is the African Rural University (ARU) for women inaugurated in the Kibaale district of western Uganda in 2011. ARU was incubated by the Uganda Rural Development and Training Program (URDT), a non-governmental organization (NGO) founded in 1987. It is the first African university dedicated to training women. It is also the first African university to be incubated by a rural NGO and shows great promise in the potential for growth among local organizations.

ARU is an innovative model that focuses on building strong female leaders for careers in agriculture and on involving the community in every step of the agricultural value chain. A key feature of the new university is to help young women envision the future they want and design strategies to achieve their goals. Their programming is tailored to meet locally identified needs that value local lifestyles and traditions while allowing the adoption of new technologies and improved production. ARU is building on a long legacy of URDT, which has resulted in better food security, increased educational attainment, raised incomes for families across the district, better nutrition, and strong female leaders who engage in peace-building efforts and community improvements, among others.

A driving factor in this approach is the community–university interaction that focuses on women and agriculture. URDT also has a primary and secondary girls' school that focuses on developing girls' abilities in a variety of areas, including agricultural, business, and leadership skills, and encouraging them to bring their knowledge out to the community.

At URDT Girls' School, students engage in 'Back Home' projects, where they spend some time among their families conducting a project that they have designed from the new skills they learned at school. Such projects include creating a community garden, building drying racks to preserve food in the dry season, or conducting hygiene education. Parents also come to the school periodically to engage in education and to help the girls design the Back Home projects. School becomes both a learning experience and a productive endeavour; families are therefore more willing to send children, including girls, to school because they see it as relevant to improving their lives.

URDT focuses on agriculture and on having a curriculum that is relevant for the communities' needs. It has an experimental farm where people can learn and help develop new agricultural techniques, as well as a Vocational Skills Institute to work with local artisans, farmers, and businessmen who have not had access to traditional schooling. A local radio programme is designed to share information with the broader community. URDT also runs an Appropriate and Applied Technology programme that allows people from the community to interact with international experts and scientists to develop new methods and tools to improve their lives and agricultural productivity.

There are also examples of innovative approaches to extension and advisory services in Africa designed to encourage rural entrepreneurship. Farmer field schools and other innovative extension approaches aim to provide farmers with technical skills through experiential learning modalities that are appropriate to adult learners (Davis et al., 2011; Davis, 2006, 2008). Rural radio and community radio programmes are helping to provide farmers with greater information on both technical and business topics, allowing them to engage more effectively in commercial markets for their produce (Pinkerton and Dodds, 2009).

Unfortunately, the challenges to strengthening agricultural education and extension in Africa are immense due in part to the historic legacies that must be overcome (World Bank, 2007). Many educational institutions and extension services are still steeped in formal colonial systems designed to create administrators whose main task is to administer rural populations. In some countries, efforts designed to develop more responsive, demand-driven agricultural education and extension systems have been constrained by inadequate resources and poorly incentivized staff. As a result, few countries have succeeded in creating cadres of motivated scientists, extension agents, and entrepreneurs who, in turn, drive the demand for motivated entrepreneurs who supply value-added agricultural goods and services.

The missed opportunity is particularly acute among women engaged in agriculture in Africa. Women provide 70 to 80 per cent of the labour for food cultivation in Africa and make up 48 per cent of the African labour force. Yet the proportion of women in higher education, business, and science is extremely low. By focusing on building the capacity of women in all of these fields, Africa will include an important set of actors who similarly drive agricultural entrepreneurship.

There are no easy solutions to the structural and cultural constraints facing agricultural extension and education in Africa, including the exclusion of significant numbers of potential women entrepreneurs at all levels. Neither grand 'centres of excellence' nor ready-made approaches will provide solutions. Educational approaches and learning philosophies need to accommodate different types of individuals, and alternative menus of learning opportunities must be expanded to accommodate diverse innovative capabilities. In effect, more practical approaches are needed, with emphasis placed on mobilizing greater political support for continuous public investment in education and extension, designing incentives that attract motivated professionals, strengthening both formal and non-formal education opportunities, and encouraging strong linkages between education, extension, and research, on the one hand, and business, on the other. Creative and context-specific approaches to achieving these goals are vital to unlocking the innovative capabilities of rural entrepreneurs.

3.3 FOSTERING BUSINESS DEVELOPMENT

Though armed with innovative capabilities, entrepreneurs are nothing without capital and expertise. Entrepreneurs are often the individuals willing to take risks and launch new commercial ventures, but are rarely the actual financiers of these ventures. For the entrepreneur's financing needs, individual investors, banks, venture capital firms, philanthropies, and governments have a major role to play. For the entrepreneur's business development needs, other types of services have a similar role to play. Lifting the fundamental economic constraint facing rural entrepreneurs—the scarcity of resources with which to innovate—is vital to encouraging greater innovation in African agriculture.

Unfortunately, rural entrepreneurs face an uphill battle given the risky nature of agriculture and its dependence on the vagaries of both markets (that is, price risk) and weather (production risk). Banks and other financial institutions have played a negligible role in promoting entrepreneurship among small- and medium-sized businesses (SMEs) in African agriculture. Venture capitalists—including private equity funds, investment groups, non-governmental organizations, and commercial funds—are rarely found outside the modern edge of South Africa's economy. Government programmes to finance rural entrepreneurs have often been constrained by weak mechanisms to insure risk, weak administrative systems, and a tendency towards rent-seeking behaviour, while donor programmes often demonstrate short-term success at the expense of sustained impact.

There are scattered pools of capital and investors willing to stake their claim on Africa's agricultural sector, to transfer risk, and encourage entrepreneurship. For example, South Africa, Nigeria, and Kenya are emerging as financial hubs that are providing new agricultural investments. Seed companies, food and feed processors, breweries, and exporters of high-value horticulture and floriculture are common destinations for investors looking for new profit opportunities.

Investments made by such players as the Acumen Fund (Acumen Fund, 2012) and the Alliance for a Green Revolution in Africa (AGRA) in local companies are testaments to the ability of venture capital to support entrepreneurship. Since 2001, Acumen Fund has invested more than US$65 million in enterprises that provide access to water, health, alternative energy, housing, and agricultural services for low-income customers in South Asia and Africa. AGRA, established in 2006, supports an integrated portfolio of investments in Africa's agricultural sector: its African Seed Investment Fund (ASIF) alone has a total capitalization of US$12 million with the aim of financing 30 seed enterprises over 8 years, while Kilimo Biashara, a partnership with Equity Bank in Kenya, marshals US$50 million in low-interest loans for smallholder farmers and small- and medium-sized agricultural businesses (AGRA, 2011).

In the immediate future, the widening and deepening of venture capital in Africa's agricultural sector will depend on our understanding of the geographical, technological, and institutional dimensions of the business. Geography plays a key role in constraining the ability of finance to move into areas of profitability. The solutions—massive investment in transportation, energy, and communications networks—are well documented in terms of their social and economic returns.

But overcoming geography is only half the battle: viable technology and supportive institutions are also necessary. The difficulty of developing and delivering technologies that fit the needs of smallholders is enormous, and influenced by multiple factors affecting adoption, including farm-level heterogeneity (farm size, soil quality, slope, irrigation, and rainfall), heterogeneity among farmers (gender, age, education, risk preferences), and social dynamics such as access to extension, learning from others, and related network learning effects. These difficulties are further complicated by the operational complexities of markets, especially for smallholders who must make decisions about what to cultivate, consume, or sell given their expectations of prices, information asymmetries, and market distortions.

There are many viable and innovative solutions in the making in Africa. The Ethiopian Commodity Exchange (ECX), for example, provides buyers and sellers of selected agricultural commodities with an organized marketplace where they can trade in a manner that provides assurances of quality, quantity, payment, and delivery, thereby reducing information asymmetries that otherwise constrain efficient market exchanges. ECX includes not only a modern trading floor, but also a warehouse receipts system, grading laboratories, rural price tickers, training services for traders and members, and an advanced communications infrastructure. In 2010, the ECX traded 502,000 metric tonnes of coffee, as well as significant volumes of sesame, maize, and white pea bean, all of which are vital sources of income for many smallholder households (ECX, 2011).

Other solutions are based on a social enterprise model and serve as bridging institutions between entrepreneurs, market opportunities, and the agricultural sector. For example, since 1991 over 133,000 families in Africa have created microenterprises using basic technology and know-how provided by Kickstart, a non-profit organization based in Kenya. Kickstart supported these families by identifying business opportunities that serve the poor, designing products such as hand and treadle pumps, building supply chains, and developing markets (Kickstart, 2012).

Similarly, CNFA, a non-profit venture working in 16 African countries, approaches the challenge by working to strengthen agricultural input supply chains for smallholders. CNFA builds networks of village-level agro-dealerships or one-stop-shops that serve as a key link in the value chain by offering smallholder farmers access to agricultural inputs, services, financing, and output marketing. Their work has reached out to some 280,000 smallholders with market creation activities and helped stimulate US$150 million in sales of improved inputs to smallholders in Africa (CNFA, 2008).

The One Acre Fund takes on a similar challenge in Kenya and Rwanda by providing farmers with the tools they need to improve their harvests and livelihoods. Their emphasis is on distributing appropriate technologies in a manner that supports sustained adoption. A combination of in-kind loans of seed and fertilizer, peer group membership, in-field training, market linkages, and a flexible service model allows the One Acre Fund to create not only farmer-entrepreneurs, but also field officers who themselves are rural innovators. As of 2012, the One Acre Fund was serving over 130,000 farm families and 520,000 children in three countries (One Acre Fund, 2012).

4. **Implications for public policy and institutions**

The entrepreneur also requires an enabling policy environment within which to operate. Regulatory and legal aspects of entrepreneurship are the frame conditions that foster or impede innovation. They cover public policies on science, industry, business, labour, and agriculture, as well as the informal institutions that govern the behaviours, practices, and attitudes that condition the ways in which individuals and organizations act and interact.

Often, too little attention is paid to the role of 'innovation policy' because rarely is there a single governmental body—a ministry, agency, or directorate—mandated to address policy issues relating to the enabling environment for innovation. Thus, the policy levers of change are highly dispersed and conflicting, making it difficult to improve the prospects for entrepreneurial growth through well-designed interventions. In an effort to focus our thinking on innovation and entrepreneurship, several ongoing attempts have been made to measure discrete components of an innovation system in an effort better to identify key areas of policy leverage over innovation. They include the World Bank Ease of Doing Business Index (DBI) and the World Economic Forum's Global Competitiveness Index (GCI).

The 2011 DBI ranks countries based on the ease of doing business, with rankings ranging from 1 (the easiest country in which to do business, which is Singapore) to 183 (the most difficult country, which is Chad).[2] The 2011 DBI finds that a large number of governments in Africa have implemented

[2] More specifically, a high ranking on the DBI means the country's regulatory environment is more conducive to the starting and operation of a local firm. The DBI averages the country's percentile rankings on nine separate topics made up of a variety of indicators, and gives equal weight to each topic.

regulatory changes designed to improve the ability of firms to start up and operate. This represents a significant improvement in the enabling environments that existed just eight years ago (World Bank, 2011).

Similarly, the 2011 GCI provides a scoring of countries based on assessments of the institutions, policies, and factors that determine the level of productivity of a country. Scores range from 1 to 7, with the lowest scores allocated to worst-performing countries (Chad, with the lowest score of 2.87) and higher scores allocated to best-performing countries (Switzerland, with the highest score of 5.74).[3] The 2011–12 GCI finds similar evidence of progress in creating a supportive enabling environment in Africa, but cautions that the region still lags behind most other developing and industrialized countries (WEF, 2011).

These measurement efforts are often useful in informing policymakers about the need to reform policies in support of enterprise development (much like sports league tables encourage low-performing teams to change their game plan). And the bottom line emerging from these efforts for many is that policy reform efforts in much of Africa still move too slowly to stimulate rapid growth in entrepreneurship.

Part of the problem lies in the absence of a coherent understanding of how policy processes occur in individual countries. Indeed, there is a growing consensus that our theories of policy change are falling short, and that alternative approaches are needed to strengthen the role of evidence-based policymaking in Africa and other developing countries (Young, 2005; Court and Maxwell, 2005; Court and Young, 2004; Pannell, 2004). Three important lessons emerge from the recent literature and experience on policy processes in Africa.

First is the importance of context. Efforts to create conducive policy environments for entrepreneurial growth must be more cognisant of local, country processes of institutional innovation, and their historical contexts, and must rely less on formula-based prescriptions for agriculture in Africa (Omamo, 2003). Here, efforts to encourage innovation and entrepreneurship demand an understanding of institutional innovation, institutional context, and historic path-dependency, but also demand asking the tough questions. What are the alternative policy options available, and how can they be designed and implemented to foster greater entrepreneurship?

Second is the importance of analysis and evidence. Good policymaking requires good information and analysis. This rather simple idea underlies the

[3] More specifically, the GCI scores combine a range of indicators to measure a country's potential for productivity growth and, ultimately, international competitiveness. The GCI measures and indexes 89 indicators, covering institutions, infrastructure, the macroeconomy, health and primary education, higher education and training, market efficiency, technological readiness, business sophistication, and innovation.

design of knowledge support systems for managing and streamlining information and analysis on the agricultural and rural economies of developing countries. For example, initiatives such as the Strategic Analysis and Knowledge Support Systems (SAKSS) seek to address these information gaps by providing location-specific analysis of rural development and food security strategies. In turn, these analyses are used to inform policymakers on optimal solutions that are sufficiently grounded in evidence—in hard facts, figures, and cross-country analysis. A key challenge among SAKSS and other such programmes is to link up successfully with key policymaking processes to ensure a more systematic demand for and supply of knowledge support, and to provide knowledge products that are appropriate for this demand.

Third is the power of cross-country cooperation. Africa is chalking up some important successes in bringing neighbouring countries together to address common, regional problems relating to science, trade, governance, and commerce. A watershed event was the formulation of the Comprehensive Africa Agriculture Development Programme (CAADP) in 2003, an undertaking of the New Partnership for Africa's Development (NEPAD), itself established in 2001 to assert African ownership over the region's growth and development agenda with leadership from the Africa Union. CAADP's aim is to improve food security, enhance nutrition, and increase rural incomes in Africa by increasing public investment in agriculture to 10 per cent of national budgets per year and raising agricultural productivity by at least 6 per cent per year, a substantial increase on past levels.

There is still much to be done to achieve these goals. Between 2000 and 2008, Africa's real GDP grew by about 4.9 per cent, but agriculture represented around 12 per cent of this growth. The commitments and targets set forth under CAAPD may help accelerate this growth and, when combined with the explicit African ownership and leadership of NEPAD and CAADP, will play a key role in putting agriculture squarely on the regional and national development agendas.

Cross-country cooperation has also supported several unprecedented policy changes that may usher in a new generation of regional entrepreneurial opportunity. The harmonization of seed trade regulations across the Common Market for Eastern and Southern Africa (COMESA) region, for example, opens the door for new private investments in cultivar improvement by crop science companies ranging from small local seed cooperatives to large multinational companies. COMESA represents 19 states in the region, with a total population of 420 million and a total GDP of more than US$450 billion. Intra-COMESA trade was estimated at US$14.3 billion in 2008, and as trade barriers come down, the COMESA region stands to become an even larger and more lucrative market in the near future.

In addition to promoting local sources of economic growth, Africa is moving rapidly to foster regional integration aimed at creating larger continental markets. The most advanced of such efforts is the June 2011 launch of negotiations for a Grand Free Trade Area (GFTA) stretching from Libya and Egypt to South Africa. The proposed GFTA would merge three existing blocs, including the Southern African Development Community, the East African Community (EAC), and the Common Market for Eastern and Southern Africa.

The GFTA will include 26 countries with a combined GDP of over US$1 trillion and an estimated consumer base of 700 million people. This significant market will appeal to foreign as well as domestic investors. Local industrial and agricultural development will take centre stage, but many inputs will come from abroad, and talks on developing this tripartite free trade area are already under way.

Larger trading blocs facilitate the economic growth that in turn enhances the expansion of the middle class. It is estimated that the free trade area initiatives of the three existing regional blocs in Africa led exports among the 26 member states to increase from US$7 billion in 2000 to over US$32 billion in 2011. These efforts build on ongoing integration efforts in the EAC, including a customs union, common market, common currency, and political federation. The five member countries (Burundi, Kenya, Rwanda, Tanzania, Uganda) count 135 million people with a total GDP (at current market prices) of about US$80 billion, representing a powerful consumer base. These efforts will have far-reaching implications for farmers whose potential for regional trade will expand.

For many seed companies, the market opportunities are backed by organizations that will also contribute to leveraging global and regional advances in both science and science policy. Science networks, programmes, and organizations—the Biosciences eastern and central Africa (BecA) Hub, the Forum for Agricultural Research in Africa (FARA), the Association for Strengthening Agricultural Research in Eastern and Central Africa (ASARECA), and the Consultative Group on International Agricultural Research (CGIAR)—will all play a role in enabling private entrepreneurs to develop better products that cater to the demands of farmers throughout the COMESA region. Companies such as Western Seed Company in Kenya and Victoria Seeds in Uganda are already benefiting from many of these organizations and opportunities.

5. Conclusions

Entrepreneurs are the drivers of innovation. They play a vital role in organizing the transmission of adaptation information, products, and processes.

Yet in Africa's agricultural innovation system, farmers as entrepreneurs continue to operate without adequate support to enable them to realise their full potential. This will not change unless public policies and organizational cultures change to give farmers the same level recognition and support that is accorded to industrialists. First, 'smart' improvements in Africa's hard and soft infrastructure must be the target of forward-thinking public policy and investment. Second, improvements in technical education—at all levels and in all forms—are needed to unlock individuals' innovative capabilities. Third, resourcing strategies are needed to provide individuals with the financing and expertise needed to launch and sustain a successful enterprise. Finally, public policies, regulations, laws, and norms are needed to create a more enabling environment for innovation and entrepreneurship.

These conclusions should not be interpreted to suggest that African entrepreneurs—including its farmers—are a population in stasis, relying on outdated methods of production or simple barter systems of exchange. Too often, this is the dismal picture painted of the region and its people. In fact, many farmer-entrepreneurs are innovators who skilfully adapt their practices to changing market signals, rainfall patterns, price fluctuations, unforeseen shocks, and commercial opportunities. Nonetheless, improvements in infrastructure, education, resourcing, and the wider enabling environment can expand the choices available to farmer-entrepreneurs, thereby offering greater opportunity to manage risk, increase their incomes, and improve their livelihoods.

Finally, we conclude with a call for greater investment in the evaluation of experiences in fostering rural entrepreneurs in African agriculture. Many of the examples and illustrations provided are well known because they have captured our imagination, generated high profiles, or been 'talked up' within the development community. In fact, few are the subject of systematic documentation and rigorous evaluation to demonstrate that they have generated a positive influence on entrepreneurial capabilities or innovation systems. Regardless of the type and level of evidence, the key point is that successes in agricultural development—and failures, too—need to be systematically documented, examined, and shared so that others can learn lessons, adapt them to different circumstances and contexts, and avoid similar pitfalls.

REFERENCES

AATF (African Agricultural Technology Foundation) (2010). *A Study on the Relevance of Chinese Agricultural Technologies to Smallholder Farmers in Africa.* Nairobi, Kenya: African Agricultural Technology Foundation.

Acumen Fund (2012). Retrieved from: <http://www.acumenfund.org> (27 September 2012).

AGRA (Alliance for a Green Revolution in Africa) (2011). *Reports and Publications.* Retrieved from: <http://www.agra.org/what-we-do/> (27 September 2011).

Bauernschuster, S., O. Falck, and S. Heblich (2010). 'Social capital access and entrepreneurship', *Journal of Economic Behavior & Organization*, 76(3), pp. 821–33.

Clark, N. (2006). 'Application of the innovation systems perspective in the African higher education sector: Experiences and challenges', paper presented at the Innovation Africa Symposium, Kampala, Uganda, 21–23 November.

CNFA (2008). Agrodealer Model. Retrieved from: <http://www.cnfa.org/wp-content/uploads/2012/05/Core-Capability-Input-supply-and-farm-services_nov.16.pdf> (9 September 2012).

Court, J. and S. Maxwell (2005). 'Policy entrepreneurship for poverty reduction: bridging research and policy in international development', *Journal of International Development*, 17, pp. 713–25.

Court, J. and J. Young (2004). *Bridging Research and Policy in International Development: An Analytical and Practical Framework.* Research and Policy in Development Programme Briefing Paper No. 1, Overseas Development Institute, United Kingdom.

Davis, K. (2006). 'Farmer field schools: A boon or bust for extension in Africa?', *Journal of International Agricultural and Extension Education*, 13(1), pp. 91–7.

Davis, K. (2008). 'Extension in Sub-Saharan Africa: Overview and assessment of past and current models, and future prospects 2008', *Journal of International Agricultural and Extension Education*, 15(3), pp. 15–28.

Davis, K., E. Nkonya, E. Kato, D. A. Mekonnen, M. Odendo, R. Miiro, and J. Nkuba (2011). 'Impact of farmer field schools on agricultural productivity and poverty in East Africa', *World Development* (in press).

Dosi, G., C. Freeman, R. Nelson, G. Silverberg, and L. Soete (eds) (1988). *Technical Change and Economic Theory.* London: Pinter.

Edquist, C. (ed.) (1997). *Systems of Innovation Approaches: Technologies, Institutions and Organizations.* London: Pinter.

ECX (Ethiopian Commodity Exchange) (2011). *The Ethiopian Commodity Exchange.* Retrieved from: <http://www.ecx.com.et/> (9 September 2012).

Fagerberg, J. (2005). 'Innovation: A guide to the literature', in Fagerberg, J., D. C. Mowery, and R. R. Nelson (eds), *The Oxford Handbook of Innovation.* Oxford and New York: Oxford University Press.

Fan, S. and X. Zhang (2008). 'Public expenditure, growth and poverty reduction in rural Uganda', *Africa Development Review*, 20(3), pp. 466–96.

Fan, S., P. Hazell, and S. Thorat (2000). 'Government spending, agricultural growth, and poverty in rural India', *American Journal of Agricultural Economics*, 82(4), pp. 1038–51.

Fan, S., L. Zhang, and X. Zhang (2002). *Growth, Inequality, and Poverty in Rural China: The Role of Public Investment*, IFPRI Research Report no. 125, Washington, DC: IFPRI.

Freeman, C. (1987). *Technology Policy and Economic Performance: Lessons from Japan.* London: Pinter.

Freeman, C. (1988). 'Japan: A new national system of innovation', in Dosi, G., C. Freeman, R. Nelson, G. Silverberg, and L. Soete (eds), *Technical Change and Economic Theory.* London: Pinter.

Gratwich, K. N. and A. Eberhard (2008). 'An analysis of independent power projects in Africa: Understanding development and investment outcomes', *Development Policy Review*, 26(3), pp. 309–38.

Gries, T. and W. Naudé (2011). 'Entrepreneurship and human development: A capability approach', *Journal of Public Economics*, 95(3–4), pp. 216–24.

Hagedoorn, J. (1996). 'Innovation and entrepreneurship: Schumpeter revisited', *Industrial and Corporate Change*, 5(3), pp. 883–96.

Hazell, P., C. Poulton, S. Wiggins, and A. Dorward (2007). 'The future of small farms for poverty reduction and growth', 2020 Discussion Paper No. 42. International Food Policy Research Institute (IFPRI), Washington, DC.

Juma, C. (2008). 'Agricultural innovation and economic growth in Africa: Renewing international cooperation', *International Journal of Technology and Globalisation*, 4(3), pp. 256–75.

Juma, C. (2011). *The New Harvest: Agricultural Innovation in Africa.* New York: Oxford University Press.

Kickstart (2012). About Kickstart. Retrieved from: <http://www.kickstart.org/about-us/> (9 September 2012).

Landstrom, H., G. Harirchi, and F. Astrom (2012). 'Entrepreneurship: Exploring the knowledge base', *Research Policy* (in press).

Lundvall, B. A. (1985). *Product Innovation and User-Producer Interaction.* Äalborg, Denmark: Äalborg University Press.

Lundvall, B. A. (1988). 'Japan: A new national system of innovation', in Dosi, G., C. Freeman, R. Nelson, G. Silverberg, and L. Soete (eds), *Technical Change and Economic Theory.* London: Pinter.

Lundvall, B. A. (1999). 'Technology policy in the learning economy', in Archibugi, D., J. Howells, and J. Michie (eds), *Innovation Policy in a Global Economy.* Cambridge, UK: Cambridge University Press.

Metcalfe, S. J. (2000). 'Science, technology and innovation policy in developing economies', Paper prepared for the Enterprise Competitiveness and Public Policies workshop, Barbados, 22–25 November 1999, and revised following that presentation.

Mitsch, W. J. et al. (2008). 'Tropic wetlands for climate change research, water quality management and conservation education on a university campus in Costa Rica', *Ecological Engineering*, 34(4), pp. 276–88.

Mogues, T., G. Ayele, and Z. Paulos (2008). 'The Bang for the Birr: Public Expenditures and Rural Welfare in Ethiopia', IFPRI Research Report no. 160. Washington, DC: IFPRI.

Naudé, W. A. (2011). 'Entrepreneurship in not a binding constraint on growth and development in the poorest countries', *World Development*, 39(1), pp. 33–44.

Naudé, W. A. and J. J. D. Havenga (2005). 'An Overview of African Entrepreneurship and Small Business Research', *Journal of Small Business Entrepreneurship*, 18(1), pp. 101–20.

Nelson, R. R. (1988). 'National systems of innovation: Institutions supporting technical change in the United States', in Dosi, G., C. Freeman, R. Nelson, G. Silverberg, and L. Soete (eds), *Technical Change and Economic Theory*, pp. 309–29. London: Pinter.

Nelson, R. and N. Rosenberg (1993). 'Technical innovation and national systems', in Nelson, R. (ed.), *National Innovation Systems: A Comparative Analysis.* New York: Oxford University Press.

Nurkse, R. (1953). *Problems of Capital Formation in Underdeveloped Countries.* New York: Oxford University Press.

OECD (Organisation for Economic Co-operation and Development) (1999). *Managing National Innovation Systems.* Paris: OECD.

Omamo, S. W. (2003). *Policy research on African agriculture: trends, gaps, challenges.* ISNAR Research Report No. 21, International Service for National Agricultural Research, The Hague.

One Acre Fund (2012). One Acre Fund: Farmers First. Retrieved from: <http://www.oneacrefund.org/> (9 September 2012).

Pannell, D. J. (2004). 'Effectively communicating economics to policy makers', *Australian Journal of Agricultural and Resource Economics*, 48(3), pp. 535–55.

Pingali, P. and M. Rosegrant (1995). 'Agricultural commercialization and diversification: Processes and policies', *Food Policy*, 20(3), pp. 171–85.

Pinkerton, A. and K. Dodds (2009). 'Radio geopolitics: Broadcasting, listening and the struggle for acoustic space', *Progress in Human Geography*, 33(1), pp. 10–27.

Poulton, C., A. Dorward, and J. Kydd (2005). 'The future of small farms: New directions for services, institution, and intermediation', in *The future of small farms: Proceedings of a research Workshop*, Wye, UK, 26–29 June 2005, International Food Policy Institute, Washington, DC. Retrieved from: <http://www.ifpri.org/events/seminars/2005/smallfarms/sfproc.asp> (9 September 2012).

Powell, W. W. and S. Grodal (2005). 'Networks of innovators', in Fagerberg, J., D. C. Mowery, and R. R. Nelson (eds), *The Oxford Handbook of Innovation*. Oxford and New York: Oxford University Press.

Reardon, T. (1997). 'Using evidence of household income diversification to inform study of the rural nonfarm labor market in Africa', *World Development*, 25(5), pp. 735–48.

Reardon, T., E. Crawford, and V. Kelly (1994). 'Links between nonfarm income and farm investments in Africa households: adding the capital market perspective', *American Journal of Agricultural Economics*, 76(5), pp. 1172–6.

Reardon, T., J. E. Taylor, K. Stamoulis, P. Lanjouw, and A. Balisacan (2000). 'Effects of nonfarm employment on rural income inequality in developing countries: an investment perspective', *Journal of Agricultural Economics*, 51(2), pp. 266–88.

Renzulli, J. S. (2003). 'The three-ring conception of giftedness: Its implications for understanding the nature of innovation', in Shavinina, L. V. (ed.), *The International Handbook on Innovation*. Oxford: Elsevier Science.

Rycroft, R. W. and D. E. Kash (1999). *The Complexity Challenge: Technological Innovation for the 21st Century*. New York: Cassell.

Schumpeter, J. A. (1934, 1980). *The Theory of Economic Development*. London: Oxford University Press.

Schumpeter, J. A. (2005). 'Development', *Journal of Economic Literature*, XLII (March), pp. 108–20.

Shavinina, L. V. and K. L. Seeratan (2003). 'On the nature of individual innovation', in Shavinina, L. V. (ed.), *The International Handbook on Innovation*. Oxford: Elsevier Science.

Suzuki, A., L. S. Jarvis, and R. J. Sexton (2011). 'Partial vertical integration, risk shifting, and product rejection in the high-value export supply chain: The Ghana pineapple sector', *World Development*, 39(9), pp. 1611–23.

Vandervert, L. R. (2003). 'The neurophysiological basis of innovation', in Shavinina, L. V. (ed.), *The International Handbook on Innovation*. Oxford: Elsevier Science.

Wallich, H. (1958). 'Some Notes towards a theory of derived development', in Agarwala, A. N. and S. P. Singh (eds), *The Economics of Underdevelopment*. London: Oxford University Press.

WEF (World Economic Forum) (2011). *The Global Competitiveness Report 2011–2012*. Geneva: WEF.

World Bank (2007). *Cultivating Knowledge and Skills to Grow African Agriculture: A Synthesis of an Institutional, Regional, and International Review*. Washington, DC: World Bank.

World Bank (2010). *World Development Indicators*. Washington, DC: World Bank.

World Bank (2011). *Doing Business 2012: Doing Business in a More Transparent World*. Washington, DC: World Bank.

Young, J. (2005). 'Research, policy and practice: why developing countries are different', *Journal of International Development*, 17, pp. 727–34.

Part III

Enhanced Livelihood Opportunities for Smallholders

13 The changing rural world and livelihood options for resource-poor rural people

EDWARD HEINEMANN[1]

1. Rural areas are changing

Intrinsic to the process of economic development is structural transformation—the evolution of an economy characterized by four interrelated processes: a declining share of agriculture in GDP and employment; rural-to-urban migration that stimulates the process of urbanization; the rise of a modern industrial and service economy; and a demographic transition from high rates of births and deaths to low rates of births and deaths (Timmer, 2009). This structural transformation involves the evolution of an economy from a traditional one centred almost exclusively on agriculture to a modern one in which the agricultural sector plays a modest or minor role within total economic activity. It is the route described by the World Bank (2007), from being an agriculture-based country to become a transforming and then urbanized and eventually high-income country. Yet agriculture is critical to the success of that evolution, particularly in its early stages: as agricultural productivity rises, it supplies—and reduces the prices of—food to the urban economy; it releases labour for employment in the industrial sector; it provides raw materials for agro-industry; and it offers a market for agricultural inputs and consumer products from the urban sector. This broad path of structural transformation is a 'stylized fact of history' (Losch et al., 2011) that has been observed across all regions, albeit with significant variations in different contexts.

Today, that broad path has specific dimensions that are different from the one traced 50–100 years ago: this is partly because the environment and range of policy possibilities that national governments face is different; partly

[1] The author is grateful to his colleagues Carlos Sere, Bettina Prato, and Gary Howe for their guidance and comments provided on draft versions of the chapter.

because development today, and in particular the economic opportunities that are available to agricultural-based countries, are conditioned by already developed regions; and partly because of changed conditions in agricultural systems themselves. In the next section we describe seven factors that are shaping the rural sector in different degrees the rural sector, and the prospects for poor rural people to pull themselves out of poverty in the developing world today.

1.1 ACROSS MUCH OF THE DEVELOPING WORLD, AGRICULTURAL LAND ISSUES ARE DRIVING CHANGE

Two distinct factors are at play, with important implications for poor rural people. The first is the fragmentation of land, the process by which, through inheritance and the sharing of land between heirs, family holding sizes become smaller and smaller with each generation. Fragmentation has resulted in a rapid decline in average farm sizes over the past 50 years, particularly in agriculture-based and transforming countries. In parts of eastern and southern Africa, cultivated land per capita has halved over the last generation and, in a number of countries, the average area today amounts to less than 0.3 hectares per capita (Jayne et al., 2011). In India, the average farm size fell from 2.6 hectares in 1960 to 1.4 hectares in 2000 and it is still declining (IFAD, 2008). Such trends drive rural people either to farm part-time, producing only for their own food needs, and look for income-generating opportunities elsewhere, or, where market opportunities permit, to adopt more intensive, commercialized production systems.

Second, and in response to the surge in global food and fuel prices that kicked in from 2007, there has been an intensification of commercial pressures on agricultural land for both food and biofuel production. There are two dimensions to this. The first, and most discussed, are the large-scale investments. Although estimates have been made of how much land has been acquired by large-scale investors (see for example Cotula et al., 2009; Anseeuw et al., 2012), it is still very unclear how much land has actually been developed or used. The jury is also still out on the impacts of large-scale investment on rural communities. Deininger and Byerlee (2011) argue that it can potentially bring a variety of benefits, while Anseeuw et al. (2012) assert that in practice, rural people typically lose out through loss of access to natural resources, and that the jobs created through the investments are limited in number, low-paid, and insecure.

The other, and almost certainly more pervasive, trend is the changing patterns of land ownership and use resulting from *local* land transactions—both sales and different types of rental contracts. It is unclear just how widespread such transactions are, though even 20 years ago—prior to the new interest in land, a study of 6 countries in East and West Africa found that 16 per cent of smallholder plots had been acquired through the market (Migot-Adholla et al., 1991). Today, there is growing competition for land in many countries in sub-Saharan Africa—particularly high-value land—usually steered by more powerful actors, and this is leading to a gradual individualization of rights and accentuating wealth differentiation (Cotula, 2007; Sjaastad, 2003). Similar forces are in play in some countries in Asia, where land markets are contributing on one hand to a concentration of land ownership and on the other to an increase in rural landlessness. In Cambodia, for example, the rate of landlessness increased from 13 per cent in 1997 to 20 per cent in 2004 (World Bank, 2004), with female-headed households twice as likely to have sold their land as male-headed households, and significantly more likely to be landless (STAR Kampuchea Organization, 2007).

1.2 ENVIRONMENTAL DEGRADATION, CLIMATE CHANGE, AND THEIR IMPACTS

Across large parts of the developing world, the natural resources from which poor rural populations derive their livelihoods are being degraded or depleted through overuse, poor land management, or soil nutrient mining, and climate change is multiplying the impact of this. Population growth combined with extreme poverty pushes people into more marginal areas, and compels them to overuse the fragile resource base; the results include deforestation, soil erosion, desertification, and reduced recharge of aquifers. As much as 5 to 10 million hectares of agricultural land are lost each year to severe degradation, and in large areas of China, South Asia and the Middle East, and North Africa, water use is unsustainable at current rates of extraction (World Bank, 2007). This not only has a direct negative impact on agricultural productivity, it also leaves the land more vulnerable to extreme weather patterns.

Many rural people already have to cope with increased climate variability, which exacerbates existing processes of degradation, leads to increasingly unreliable growing seasons and ever-more frequent extreme events, and creates ever-greater uncertainty and risk for small farmers—all factors undermining agricultural productivity. Agriculture is highly sensitive to climate change, and in the longer term, increasing temperatures will further increase uncertainty and exacerbate weather-related disasters, and lead to biodiversity loss, drought, the emergence of new pest and disease patterns, and scarcities of water and arable land. Growing seasons are likely to be shorter and more erratic, and in some cases entire farming systems will have to change. In some locations existing staple crops will have to be replaced by others that are more drought resistant, in others arable systems will have to be replaced by livestock production (IFAD, 2012). Poor rural people everywhere will be the least equipped to adapt and the most affected because of their reliance on predictable weather patterns, and climate change may have a substantial and negative impact on levels of hunger, particularly in Africa (Parry et al., 2009). In some areas, this will result in the displacement of sizeable populations, and many poor rural people, individually or in groups, will migrate in search of resources or opportunities; and this may in turn fuel conflict between different groups with competing claims to the same resource—be it within or between rural communities and even among nations.

1.3 AGRICULTURAL MARKETS HAVE CHANGED PROFOUNDLY IN THE LAST 30 YEARS, AND THERE IS AN EXPANDING AND INCREASINGLY SEGMENTED DOMESTIC DEMAND IN DEVELOPING COUNTRIES FOR A GROWING RANGE OF FOOD PRODUCTS

This expansion is a result of, and associated with, rapid growth in urban populations, higher incomes, and the emergence of a sizeable middle class with new lifestyles and consumption patterns; and it is true both in large cities and in smaller urban centres. Overall, there has been increasing demand for food, and in most countries the fastest-growing demand has been for higher-value produce such as vegetables, fruit, meat, and dairy, and for foods of higher and assured quality. In response, made possible by widespread market liberalization in the 1980s, there has been rapid restructuring of national food markets, substantial new private investments have occurred in processing and retailing, and new market arrangements and standards have emerged to varying degrees across countries. Supermarkets have grown rapidly across the developing world, starting in Latin America, where supermarkets now typically account for 60 per cent or more of retail food sales, and more recently spreading to Asia and then Africa (Onumah et al., 2007).

Despite this trend, in all regions traditional markets continue to play an important role for consumers and for producers, particularly poorer ones; indeed, in most cases, traditional marketing arrangements are substantially more important than modern markets (Losch et al., 2011). Traditional markets typically don't offer high prices to producers—the result of inefficiencies and high levels of risk and transaction costs along the supply chain—though they are more easily accessed. Having said that, in many countries it remains only a minority of farmers who actually sell food and agricultural products (Jayne et al., 2011)—typically those close, or with easy access, to urban markets. As a result, the proportion of households participating in agricultural markets can vary considerably in different parts of the same country (Losch et al., 2011), with weaker market integration in more remote areas.

For many smallholder farmers—though certainly not the poorest, who enter the market principally as food buyers—these domestic markets, modern and integrated, traditional and atomized, offer a new range of economic opportunities. They also offer new threats and opportunities for employment. On one hand, the upgrading and integration of agricultural value chains and the associated concentration of processors, wholesalers, and retailers have displaced many small rural businesses—from brokers to retailers, particularly in Latin America. On the other hand, where value is added, they can create a wide range of opportunities, not only in production, but also up- and downstream in the value chain, from input supplies and distribution, to on-farm labour, to transportation and processing of agricultural products.

1.4 MOBILE PHONES HAVE IMPROVED LIVES AND STIMULATED A WHOLE RANGE OF ECONOMIC ACTIVITIES IN RURAL AREAS, AND OTHER EMERGING TECHNOLOGIES WILL ALMOST CERTAINLY FOLLOW

The use of mobile phones is expanding exponentially, and many poor rural people now either own handsets or are able to access them. While rural women are less likely to own mobile phones than men, studies suggest that phone ownership can offer them important gains in terms of economic empowerment (IFAD, 2010). For smallholder farmers, mobile phones have greatly reduced market transaction costs and risks, and there is a whole range of SMS-based services now on offer, covering all stages of the agricultural production cycle. They allow farmers to find out seed prices, get weather forecasts and news about disease outbreaks, receive technical advice, find out market prices for their produce, contact buyers, transfer money, and arrange loans. While concerns remain about the scope for the poorest and marginalized to benefit from such services, a series of studies suggests that—in different contexts—mobile phone coverage can induce the market participation of farmers who are located in remote areas and produce perishable crops; enable farmers to obtain higher prices; reduce grain price dispersion across markets and reduce intra-annual price variation; and allow farmers to sell their produce at more distant wholesale markets, rather than at the farm gate (see for example Fafchamps and Minten, 2012; Baumüller, 2012).

Banking services too are now supplied through mobile phones. According to Cobert et al. (2012), there are now more than one hundred telecom companies and banks that have launched mobile-money services in developing countries, with examples including M-PESA in Kenya, MTN Uganda, Vodacom Tanzania, FNB in South Africa, and GCASH and Smart Money in the Philippines. M-PESA, which offers savings, domestic money transfers, and other services through local agents on commission, is one of the best-established schemes. With more than 14 million customers, it now provides mobile-banking facilities to more than 70 per cent of the country's adult population (IMF, 2011). In addition, however, a whole range of businesses is now using M-PESA functions for a variety of financial transactions in sectors and businesses of relevance to rural Kenyans: in agriculture, health delivery, water supplies, microfinance, and social protection.[2]

Just as the number of mobile phones now far exceeds the number of fixed lines in many countries, so the future in rural areas may be one in which electrical supplies no longer need to be connected to a national grid and fuel supplies no longer need to be transported. Wind, solar and hydro power, methane, and biodiesel all offer substantial promise for off-grid energy generation. In particular, the price of solar panels is falling rapidly, as is that of light-emitting diode (LED) lighting, and the technology is close to being within the reach of a large number of rural people. Such a future could play a critical role in providing new rural employment opportunities in the industries growing up around power generation, in generating power on which rural industries can be based, and in improving the living conditions and prestige of rural areas.

1.5 THE RAPID INCREASE IN MIGRATION FROM RURAL AREAS, AND THE REMITTANCES THAT THE MIGRANTS SEND HOME, ARE CHANGING RURAL HOUSEHOLDS AND THEIR LIVELIHOOD STRATEGIES

Migration has become a widespread and important livelihood strategy in most regions. Worldwide, there are estimated to be over 200 million international migrants—countries as varied as Mali, Mexico, and Morocco all have around 10 per cent of their population living abroad (Losch et al., 2011). Recorded remittance flows to developing countries from international migration have increased rapidly from around US$100 billion in 2002 to US$350 billion in 2011, and they are expected to continue growing. Yet international migration represents only a tiny fraction of total migration in each country—perhaps as little as 1 per cent (Anríquez and Stloukal, 2008). Most migration is domestic, and most of it is rural-to-urban migration. So while the level of remittances sent home by individual domestic migrants is undoubtedly many times lower than that sent home by international migrants, the sheer numbers of domestic migrants, plus the fact that all their remittances go to the rural areas (as against only 30 to 40 per cent of remittances from international migrants), mean that domestic rural-to-urban migration is almost certainly a more significant contributor to the rural economies of most developing countries.

[2] <http://www.thinkm-pesa.com/2012/03/how-m-pesa-is-transforming-kenyas.html>.

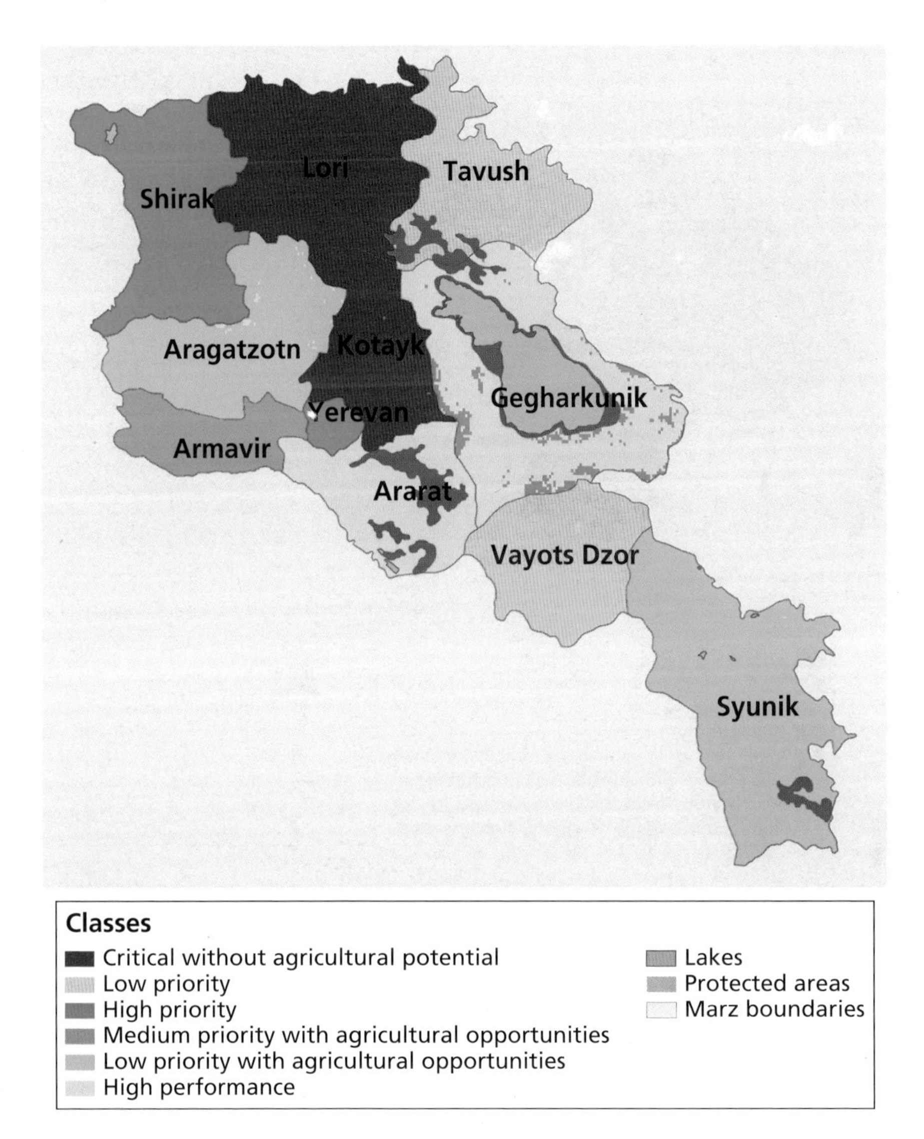

Figure 6.14 Armenia—typology combining efficiency, potential, and poverty in seven categories

Note: For Armenia, we have only six classes. We do not have medium priority without agricultural opportunities.

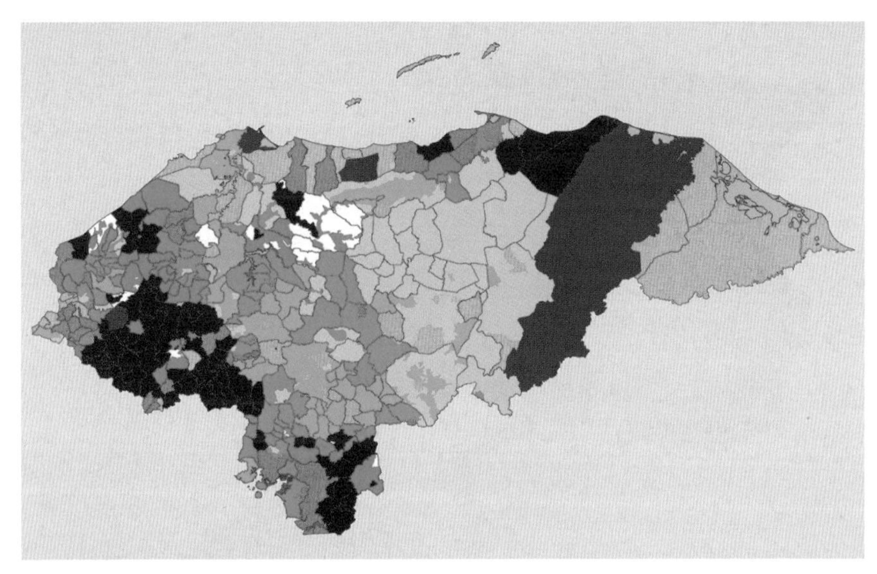

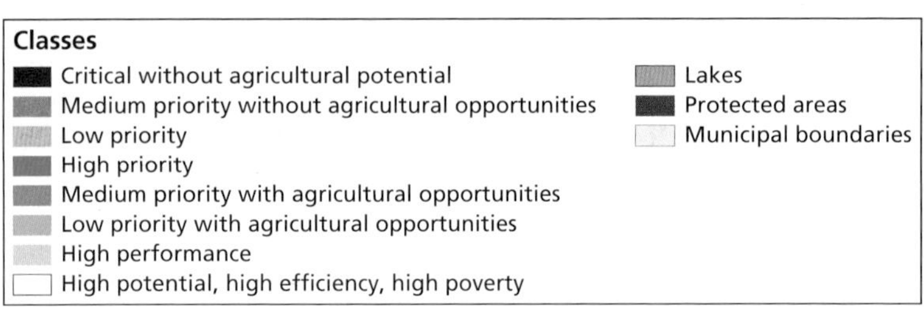

Figure 6.15 Honduras—typology combining efficiency, potential, and poverty in seven categories

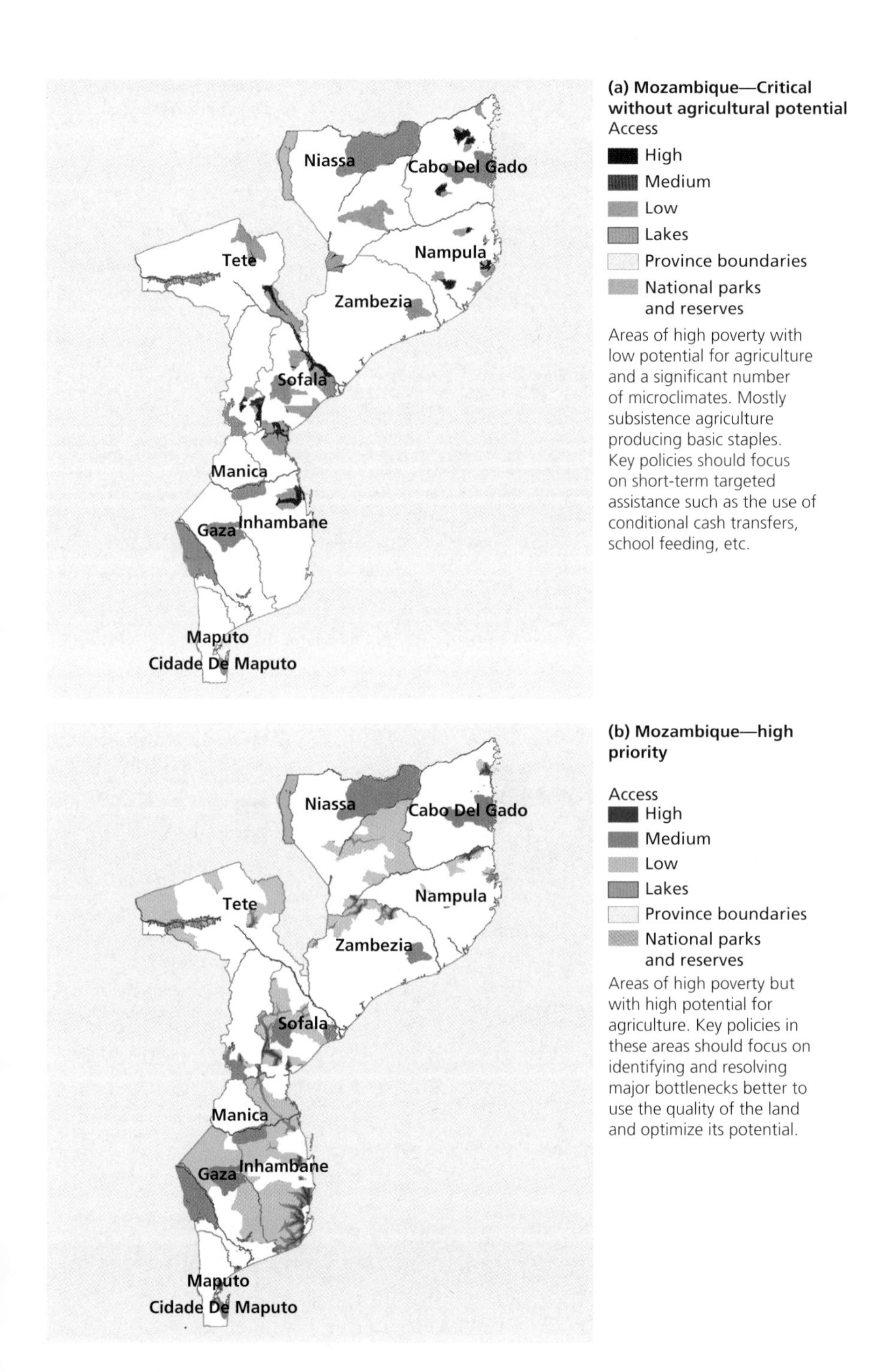

Figure 6.16 Mozambique—four key areas and types of public investment needed

(c) Mozambique—
High performance

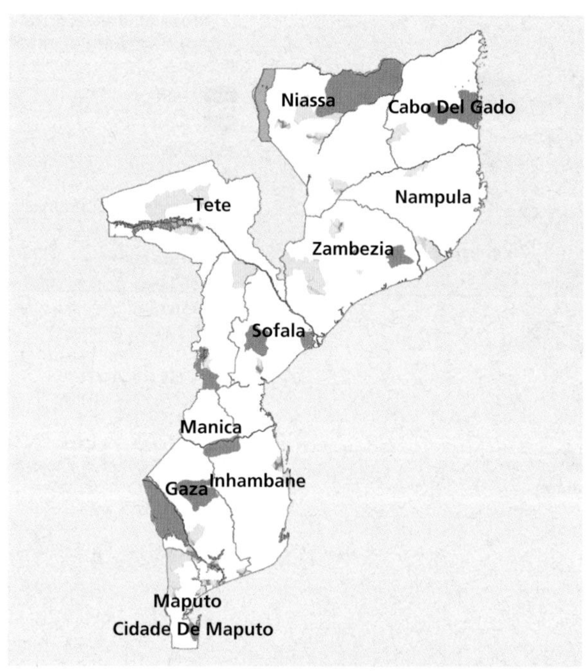

Access

- High
- Medium
- Low
- Lakes
- Province boundaries
- National parks and reserves

Areas of low poverty rates with significant pontential and high efficiency. Given their high performance, these regions could be used to learn from and replicate best practices.

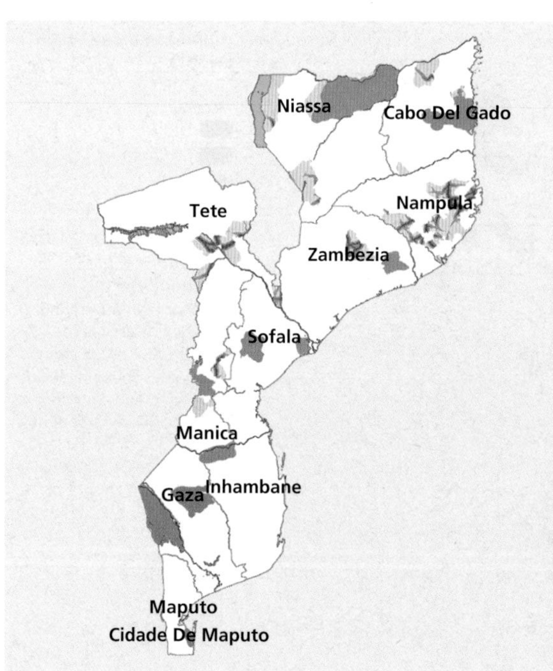

(d) Mozambique—
Medium priority without agricultural opportunities

Access

- High
- Medium
- Low
- Lakes
- Province boundaries
- National parks and reserves

Areas with moderate poverty rates, low potential, and medium and high efficiency. Despite the high efficiency, agricultutal potential is low and as a result there are signigicant levels of poverty. Policies in these areas should focus on non-farm activities or activities that could give the capabilities for them to migrate as a way to increase households' income.

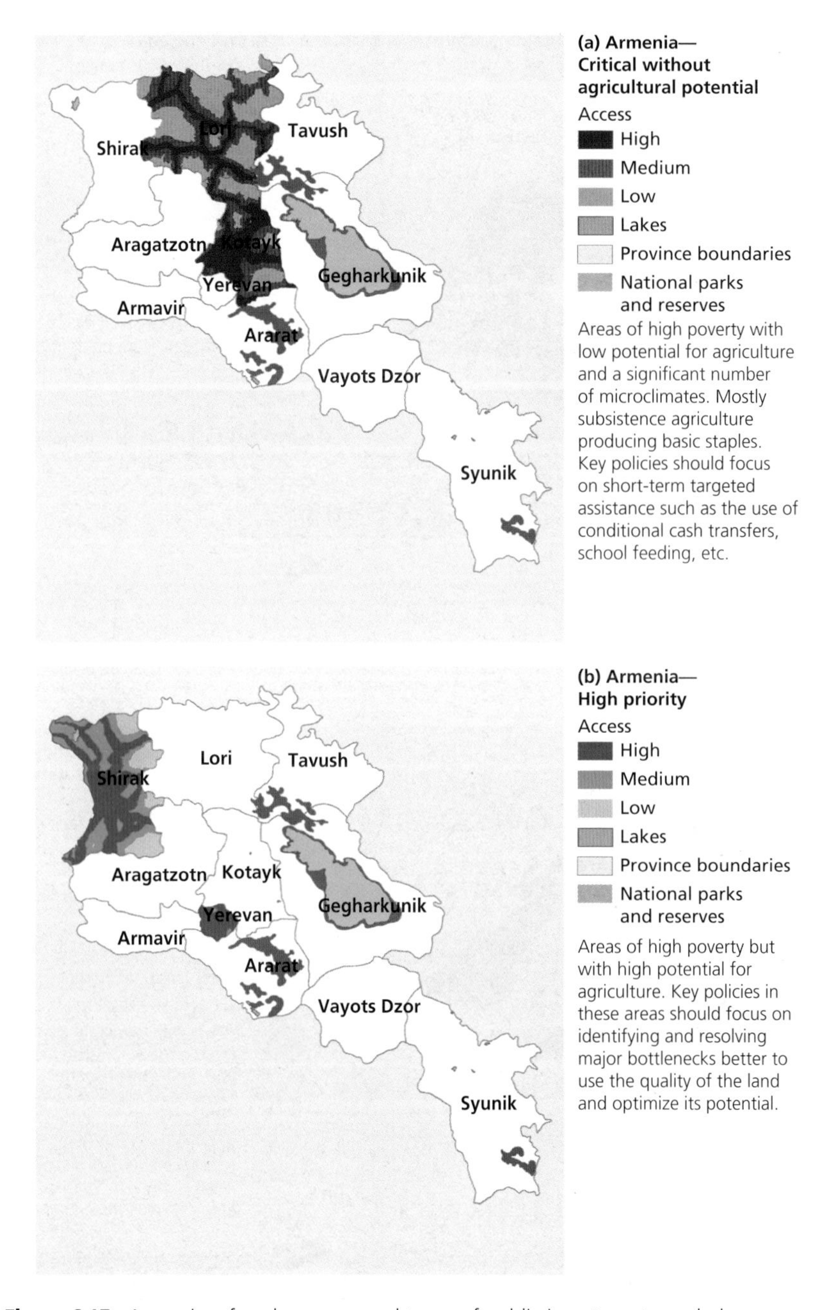

Figure 6.17 Armenia—four key areas and types of public investment needed

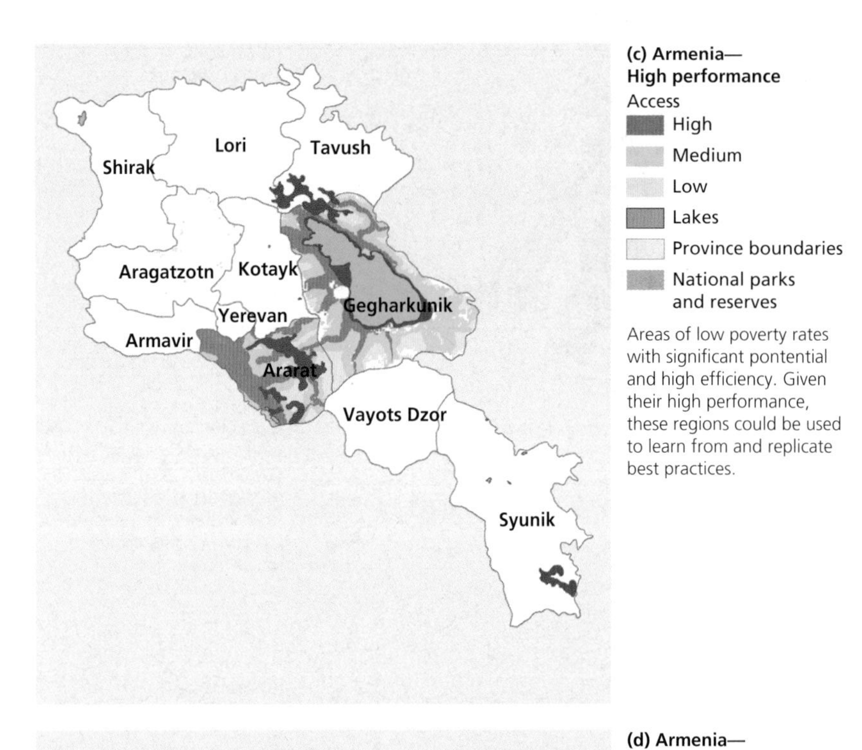

(c) Armenia—High performance

Areas of low poverty rates with significant pontential and high efficiency. Given their high performance, these regions could be used to learn from and replicate best practices.

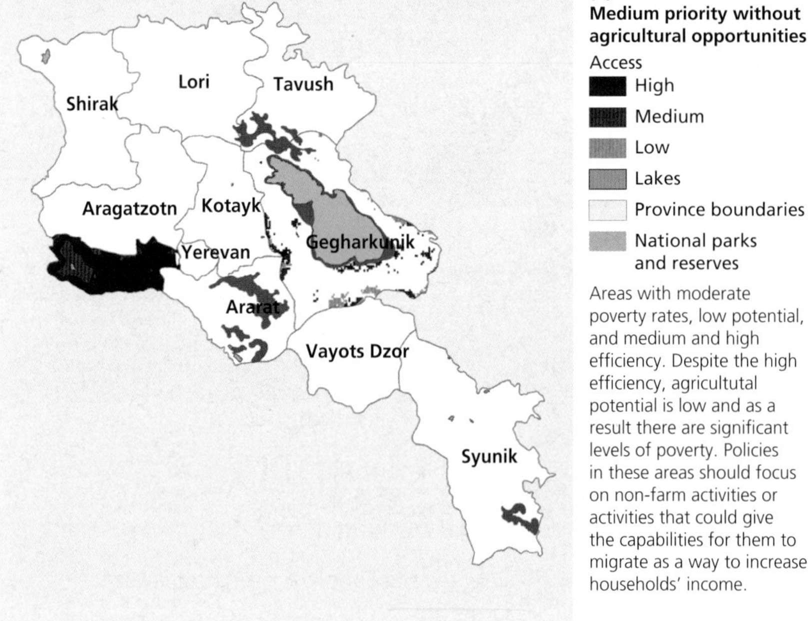

(d) Armenia—Medium priority without agricultural opportunities

Areas with moderate poverty rates, low potential, and medium and high efficiency. Despite the high efficiency, agricultutal potential is low and as a result there are significant levels of poverty. Policies in these areas should focus on non-farm activities or activities that could give the capabilities for them to migrate as a way to increase households' income.

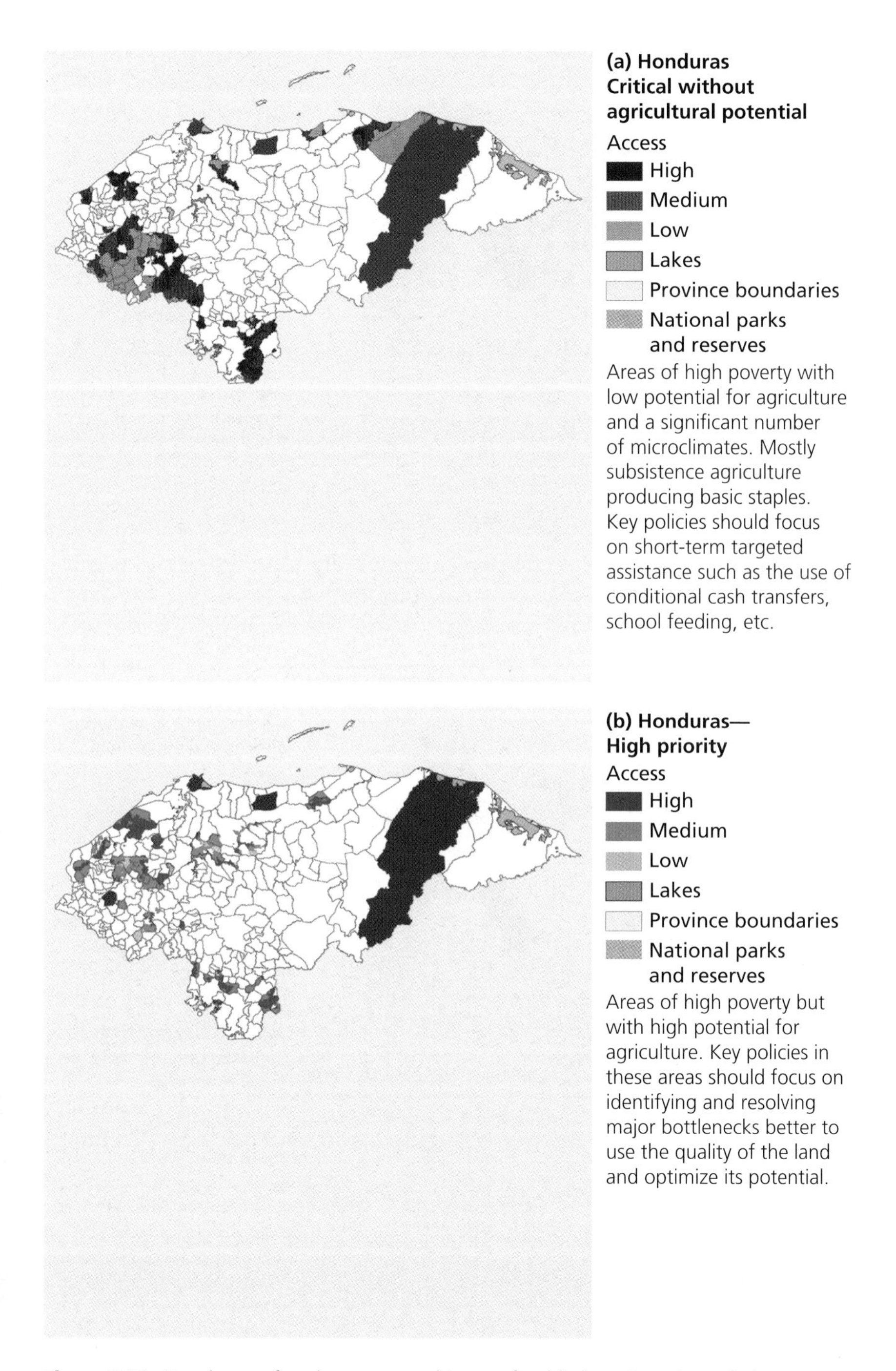

Figure 6.18 Honduras—four key areas and types of public investment needed

(c) Honduras—High performance

Areas of low poverty rates with significant pontential and high efficiency. Given their high performance, these regions could be used to learn from and replicate best practices.

(d) Honduras—Medium priority without agricultural opportunities

Areas with moderate poverty rates, low potential, and medium and high efficiency. Despite the high efficiency, agricultutal potential is low and as a result there are signigicant levels of poverty. Policies in these areas should focus on non-farm activities or activities that could give the capabilities for them to migrate as a way to increase households' income.

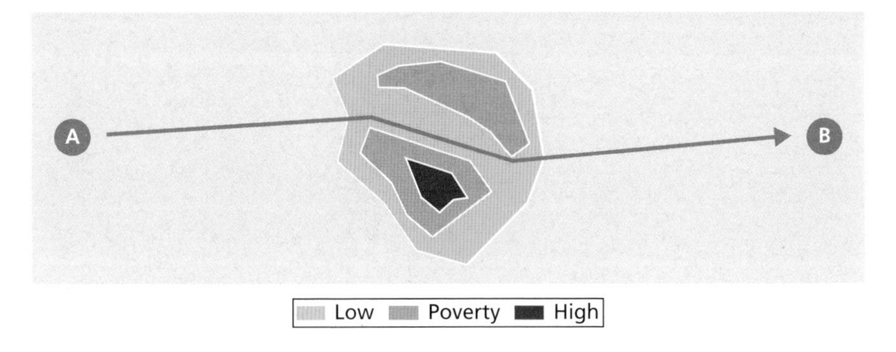

Figure B6.1 Friction surface between points A and B

•	15	15	15	15	15	15
1	15	15	50	50	15	15
1	15	15	50	50	15	•
1	15	15	50	50	50	15
1	1	15	15	50	50	15
1	1	15	15	15	15	15
90	1	1	1	1	1	1
1	90	1	1	1	1	1
1	1	90	90	90	90	90

Figure B6.2 Values indicating the difficulty of crossing a ‘cell’

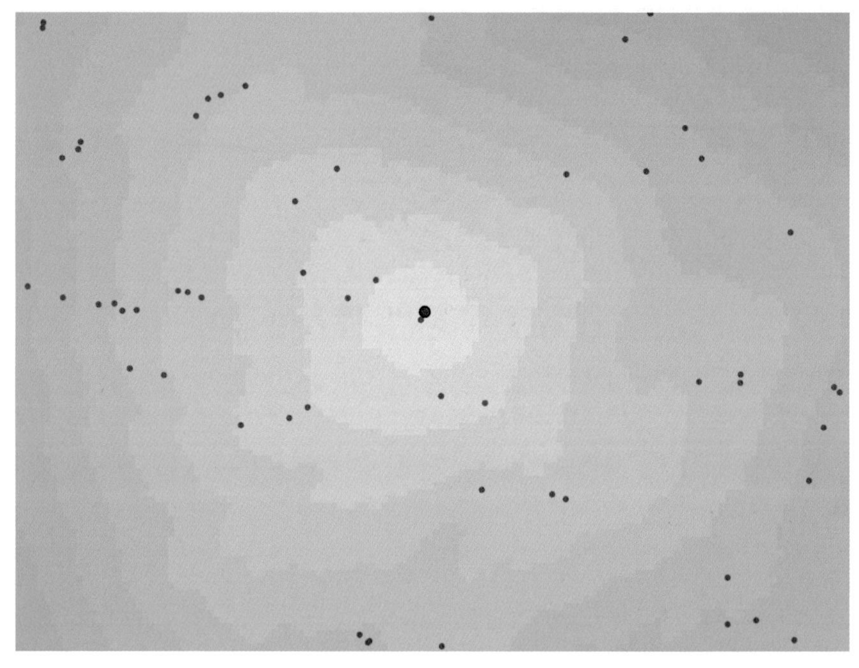

Figure B6.4 Times calculated only with the off-path walking velocity

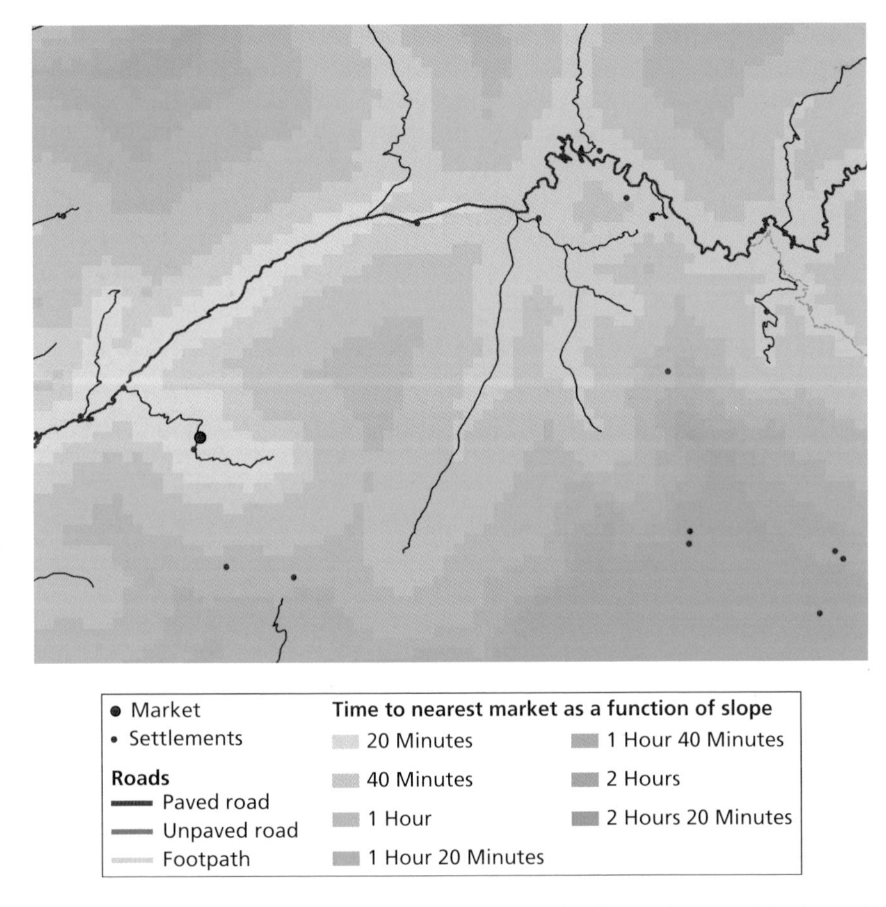

Figure B6.5 Times calculated with the three variations of walking velocity and fixed speed road classification

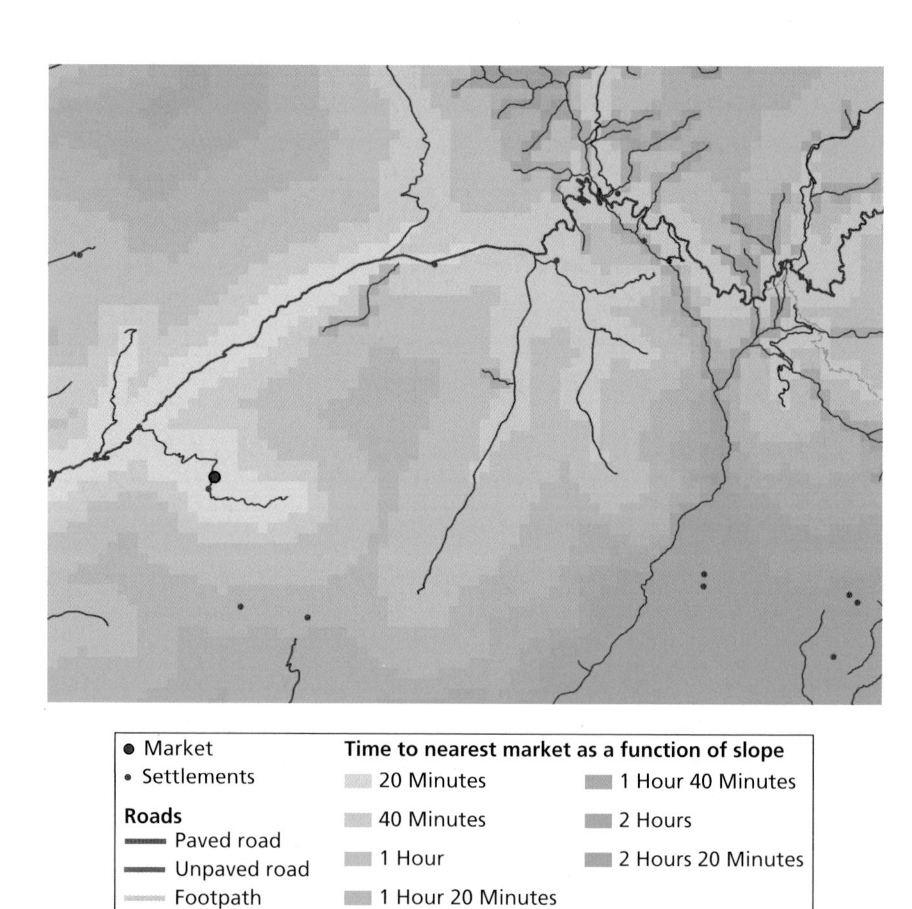

Figure B6.6 Friction surface map

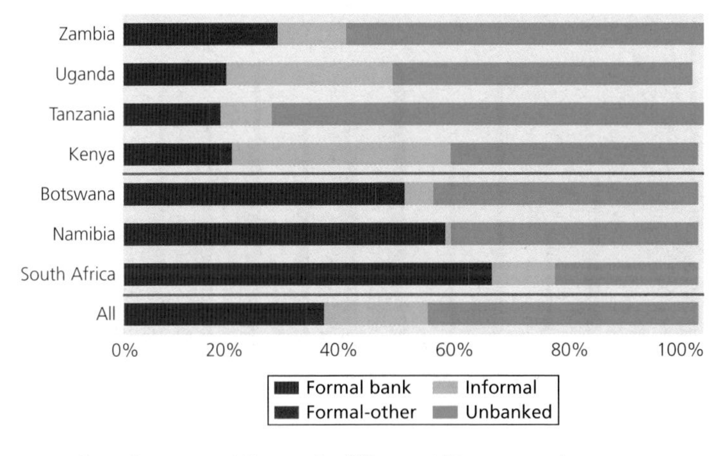

Figure 8.1 Informal sources of finance in different African countries

Source: FinScope (2007).

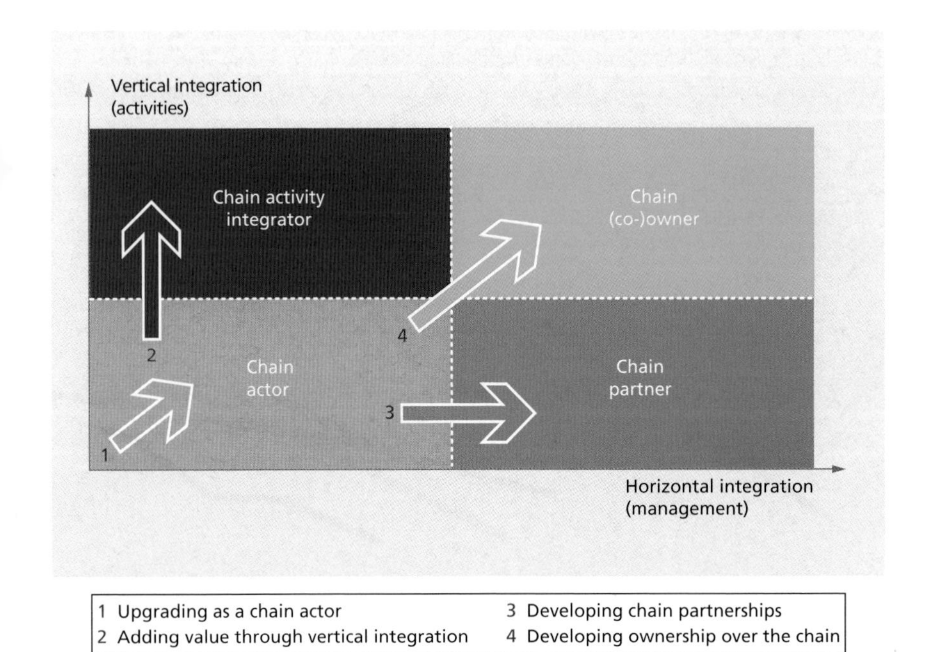

Figure 10.2 Strategies for empowering farmers

Note: Adapted from CIAT (undated).

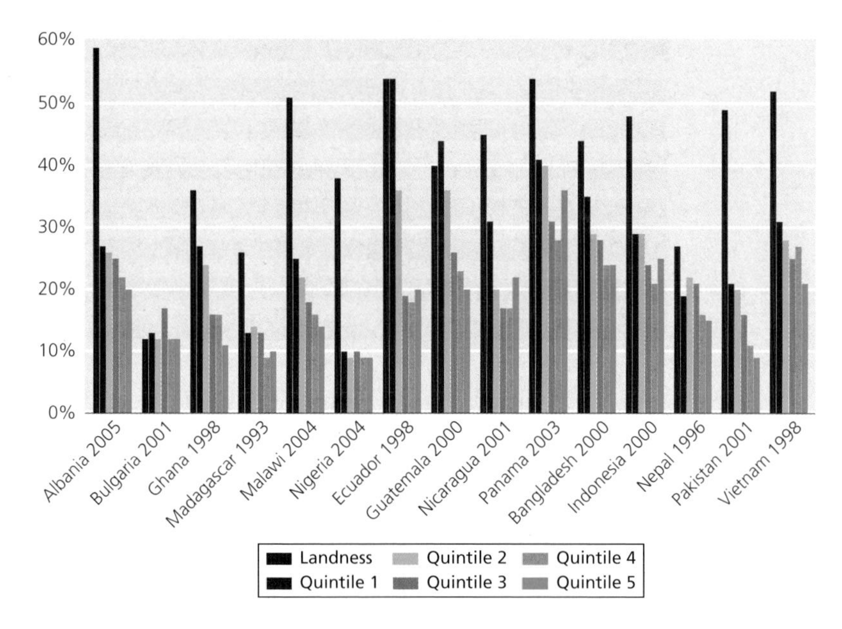

Figure 17.3 Share of household incomes from non-farm sources, by access to land

Source: Carletto et al. (2007). Constructed from data Table A8 reporting statistics for 15 countries carried out under national living standards measurement surveys.

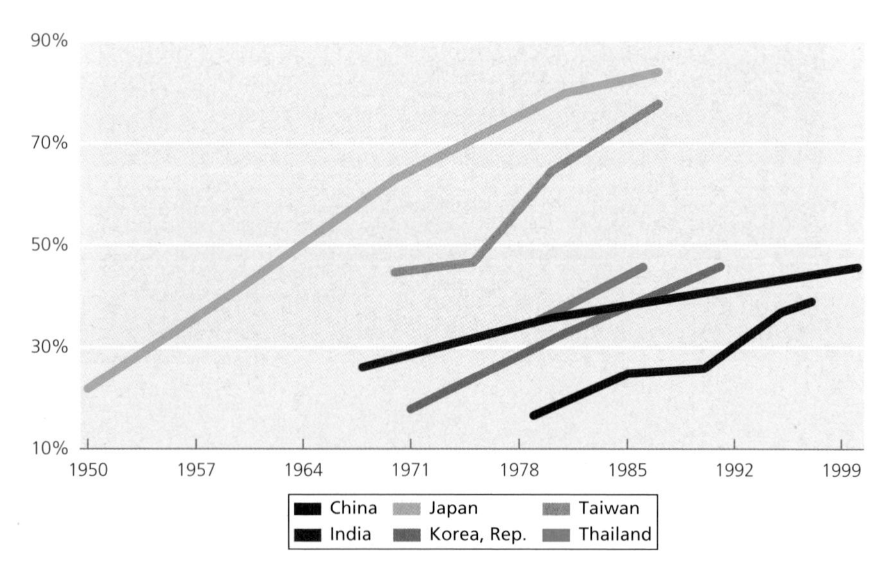

Figure 17.4 Share of farm household incomes from non-farm activities, Asia

Source: Haggblade et al. (2007). Table 4.1. India statistics are for share of rural incomes, rather than farm households alone.

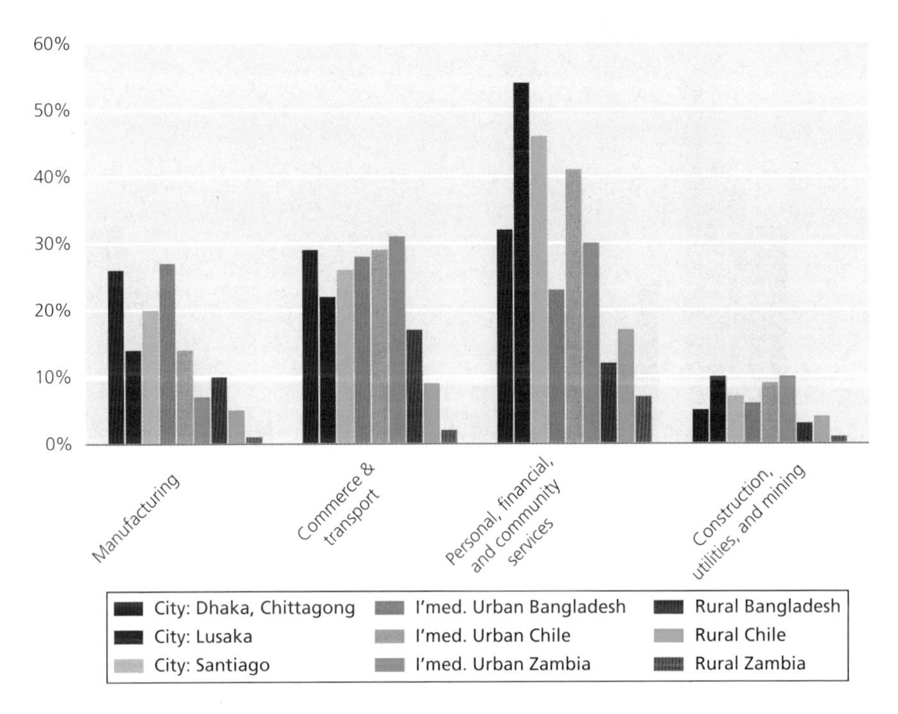

Figure 17.7 Non-farm activity by share of employment for city, intermediate urban and rural areas in Bangladesh, Chile, and Zambia (share of all jobs, %)

Source: Haggblade et al. (2007). Constructed from data in Table 1.6. Original sources: Bangladesh (2003), Banco Central de Chile (1986, 2002), and Zambia (2003).

Notes: Data for Bangladesh and Zambia are 2000, Chile from 1984. The total of jobs for any given location do not sum to 100, since agricultural employment has been omitted from the chart.

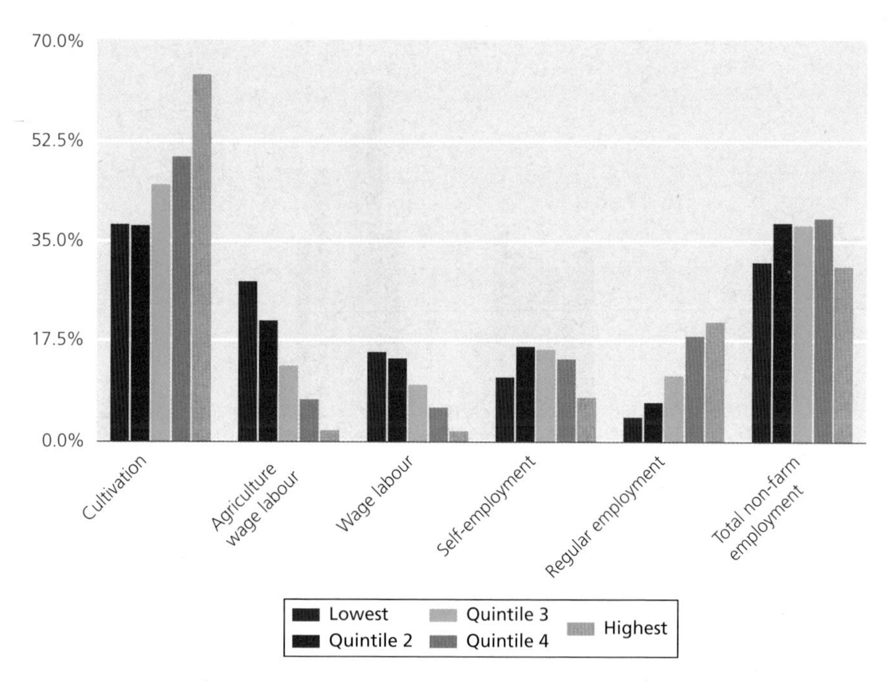

Figure 17.8 Sources of income in rural India by per capita income quintile, 1993–4 (%)

Source: Lanjouw (2007), Table 3.2a. Original source: Lanjouw and Shariff (2004).

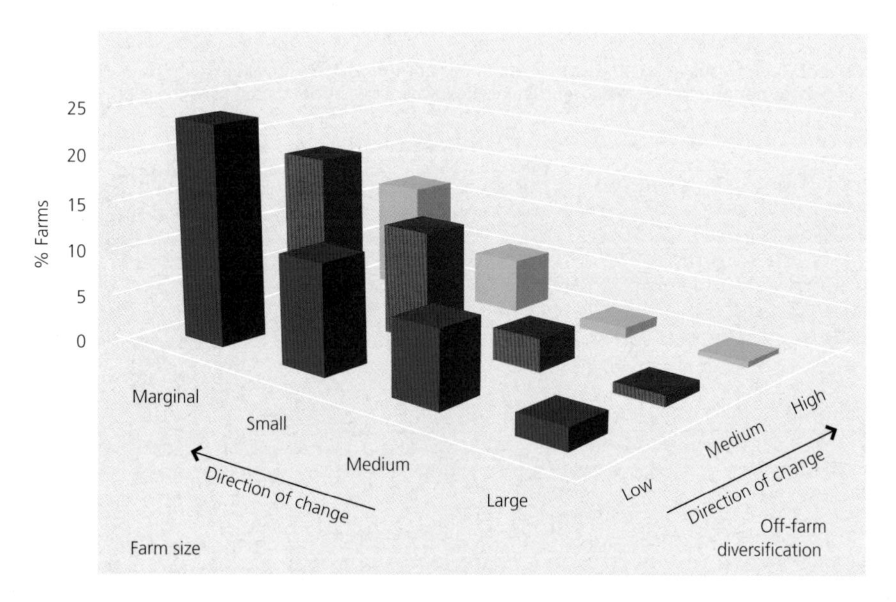

Figure 18.1 Emerging distribution of farm households by farm size group and degree of off-farm income diversification

These substantial levels of remittances going to the rural areas have multiple and diverse impacts. First, remittances sent home by migrants are extraordinarily important for entire rural societies and economies: almost 60 per cent of rural households in Panama, almost 80 per cent in Malawi, and over 80 per cent in Indonesia receive remittances (IFAD, 2010), and for those households remittances may play an essential role in providing a supplementary—or for some, the only—source of income. Second is the impact of migration on agriculture, which is mixed and highly contextual. In some cases, remittances foster on-farm investment and increased agricultural production. However, perhaps because migration rates are usually highest from the poor regions of countries where opportunities are most limited, it seems that more often they are invested in non-agricultural income-generating activities, and that they actually accelerate a move out of agriculture by the household (Vargas-Lundius et al., 2008). Third, remittances are associated with increased household investments in children's education and in health care (Kugler, 2005). In essence, remittances can help to build the human capital that is a precondition for rural people to access economic opportunities. And fourth, beyond the household, remittances can also create job opportunities and stimulate economic growth. In South and South-East Asia, each migrant was found to create an average of three jobs, on- and off-farm, through remittances (IFAD, 2010). In Mexico, remittances have been found to create second-round income effects that favour poor people in and beyond the rural economy, and several studies show that remittances provide capital to small, credit-constrained entrepreneurs, which may be assumed to create jobs (Vargas-Lundius et al., 2008).

1.6 IMPROVED INFRASTRUCTURE, COMBINED WITH GLOBALIZATION, IS STARTING TO CHANGE THE MAKE-UP AND STRUCTURE OF RURAL ECONOMIES

In a globalized world, characterized by the free movement of capital and goods, there has been a weakening both of those linkages between rural and urban areas that, historically, contributed to the process of structural transformation (UNRISD, 2010), and of the correlation between agricultural growth and the growth of non-farm income and employment (Wiggins and Hazell, 2008). Yet improved infrastructure has made possible the emergence of new sorts of economic linkages between the urban—and even global—economy and rural areas. Particularly where patterns of urbanization are dispersed, characterized by dynamic regional towns and small cities rather than a more limited number of large cities, a range of market opportunities may emerge—many of them agriculture-related—for products, services, and labour, which can be accessed by rural households living within their orbit.

Globalization has meant that in many poor countries, the products of traditional or artisanal manufacturers—of household goods, clothes, etc.—are unable to compete with mass-produced, low-cost imports. At the same time, though, improved transport and communication linkages between rural and urban areas, often combined with new markets created as a result of globalization, offer new opportunities for rural households, particularly in transforming and urbanizing economies. In South-East Asia and China, high population densities and low transport costs have led to labour-intensive manufacturing for export markets being subcontracted to rural industries (Wiggins and Hazell, 2008). In Mexico too, urban centres create manufacturing and service employment opportunities up to 150 kilometres around them (World Bank, 2008). In India, rapid rural non-farm growth is occurring along transport corridors linked to major urban centres, largely independent of their agricultural base, and Indian villages close to towns and cities have a better record of reducing poverty than others (Bhide and Mehta, 2006) and this is common in other countries too. Improved transportation means that rural-to-urban commuting has become a reality: in some densely settled Asian and Latin American countries it has become a significant phenomenon (Haggblade et al., 2010).

To date, globalization and urban-led transformation are proving most powerful in driving rural growth in densely populated, transforming economies such as China and India, where the rural non-farm economy is well developed and may make up more than half of total rural incomes. They are less effective as catalysts of change in the poorest, agriculture-based countries, where the rural non-farm economy represents a smaller part of the rural economy as a whole, and for the poorest people, since they typically lack the skills to take advantage of these opportunities and are likely to live far from urban centres.

1.7 AS THE WORLD AROUND THEM IS CHANGING, SO ARE RURAL POPULATIONS THEMSELVES

First, in all developing regions, the rural population is falling as a proportion of the total population. In 2010, around 55 per cent of the total population of the developing world was defined as rural, but by 2020–25 this figure will fall to less than 50 per cent. The relative decline of the rural population results from the continuous migration of large numbers of people from the rural to the urban areas; yet natural population growth means that the rural population of the developing world as a whole—currently 3.1 billion—is still increasing, albeit at a declining rate, and shortly after 2020 it will peak at around 3.2 billion and then start to fall. There is of course enormous variation across regions. At one extreme, in Latin America and the Caribbean and in East and South-East Asia, rural populations are already in decline, and in

Latin America and the Caribbean, only 20 per cent of the population is still rural. At the other, in South-Central Asia and sub-Saharan Africa the populations are still over 60 per cent rural, and rural populations are still growing (United Nations, 2011).

Second, rural areas are becoming less poor. Certainly, poverty rates are considerably higher in rural than urban areas: of those living on less than US$1.25/day, an estimated 1 billion—around 70 per cent—live in rural areas (IFAD, 2010), and a majority of the world's poor will live in rural areas for many decades to come.[3] However, those 1 billion people represent just under 35 per cent of the total rural population of developing countries, down from around 54 per cent in 1988 (IFAD, 2010). This decline in rural poverty is mainly due to massive reductions in China above all, and in South-East Asia. By contrast, rural poverty rates are considerably higher in South Asia (45 per cent) and sub-Saharan Africa (over 60 per cent); and while the rate is declining in both, in South Asia over 500 million rural people still live on less than US $1.25 a day, while in sub-Saharan Africa the figure currently stands at around 300 million people, and it is still growing—albeit slowly (IFAD, 2010).

Third, the demographic profile of rural populations is changing in response to changes in natural population growth rates and migration. However, the key point is that, in all developing regions, dependency ratios—the shares of the total population under 15 and over 60 years old—are higher in rural than in urban areas (Anríquez and Stloukal, 2008), which is the result of higher fertility rates in rural areas as well as rural-to-urban migration among the 15–60 years age group. The implication is that rural populations everywhere are intrinsically, and by definition, less productive than urban populations, a greater proportion of which is made up of those in a more productive age group. Dependency ratios are highest—though falling—in the least developed regions, and sub-Saharan Africa has the highest rural dependency rates. The fact that 44 per cent of the population there is under the age of 15 means that there will be a rapidly growing labour force for the foreseeable future, which will almost certainly not be absorbed in the urban/industrial sector (Valdés et al., 2008). Finding expanded employment opportunities for this growing rural labour force will be a priority for all governments.[4]

Fourth, the countries in which poor rural people live are becoming less poor. Sumner (2011) shows that, as many developing countries have progressed from low-income country (LIC) status to become middle-income countries (MICs) (essentially going from being agriculture-based to

[3] In early 2012 the World Bank (Chen and Ravallion, 2012) estimated that by 2008 the number of people in the world living on less than US$1.25 a day had declined to 1.29 billion, suggesting that the number of rural people living on less than US$1.25 a day is likely to have declined to close to 900 million.

[4] Outside agriculture-based countries, by contrast, rural populations are ageing: a phenomenon that has important implications for agricultural productivity and innovation and for future farm sizes.

transforming countries), so the proportion of the world's poor (those on less than US$1.25 a day) living in LICs has declined from 94 per cent in 1990 to only 24 per cent in 2007–8.[5] Fully three-quarters live in countries now classified as MICs—and 60 per cent of them in just five of them: China, India, Indonesia, Nigeria, and Pakistan. He also finds that less than a quarter of the world's poor live now in fragile and conflict-affected countries, split evenly between LICs and MICs. So most of the world's rural poor live today in stable, middle-income countries, characterized by growing levels of inequality. On one hand, they are increasingly those unable to take advantage of, and excluded from, existing patterns of growth; yet on the other, they live in countries where their governments have the resources to invest in programmes to promote broad-based economic growth and can take advantage of a growing array of models for social protection policies to help rural people move out of poverty.

These seven factors combine to create the drivers for what has been described as a 'new rurality', characterized by changing relations between the rural and urban spaces, and by growing interconnections—of goods, services, and people—between the two. Yet these factors are not present to the same degree everywhere: rural areas are changing at different paces and in different ways, not only in agriculture-based, transforming, and urbanized countries, but also *within* individual countries; and while some rural areas are becoming ever-more closely linked to the urban areas, others are left behind, and indeed may even become increasingly marginalized from processes of development. Effectively, there is a shifting rural geography, with the destinies of different territories shaped by these interlinked factors. To a greater or lesser extent, for the rural populations within them, new opportunities, new risks, and new processes of integration into, or marginalization from, the modern economy are being created. As we will see, the ways in which those populations, and in particular the households within them, respond to these forces vary considerably, according to the assets the households have at their disposal, and their assessment of the relative opportunities and risks associated with different alternatives in terms of livelihood strategies.

2. **What do these changes mean for the poor rural people?**

For large numbers of poor rural people in the developing world, the playing out of some or all of the seven factors of change referred to—and of course

[5] While Sumner is not looking at *rural* poor per se, there is no reason to assume that his conclusions do not broadly hold true for this group.

others—has created an environment for them that is profoundly different from that faced by their parents. New or growing areas of risk include the ever-more limited availability of land, the growing value of (and interest in) that land, and the growing insecurity with which they access or own it, including the threat of dispossession. They include the deterioration of the natural resource base—declining soil fertility, erosion or salination of soils, declining water availability, etc.—as well as climate change, bringing new uncertainties and multiplying existing risks. By contrast, growing food and agricultural markets, improved communications and new technologies, and expanding employment in the growing rural non-farm economy all provide important growth opportunities; however, without enabling policies and appropriate support, many of these are likely to be beyond the reach of many poor rural people. Not only do they face possible exclusion from such opportunities, they may even end up worse off as others expand their asset base to respond to the opportunities.

Rural people's decisions about how to allocate and use the limited cash, land, and labour at their disposal are a function both of the opportunities available to them, and of the need to minimize the possibility of shocks throwing them into poverty, preventing them from moving out of it, or reducing their ability to spend on their primary needs. In many cases, however, the need to minimize these possibilities undermines people's ability to seize opportunities, which generally come with a measure of risk. Poor households have fewer buffers to fall back on than less poor households, and so it is critical for them to adopt strategies that reduce their risks to the greatest extent possible. Diversification is a key strategy for reducing risk: highly diversified cropping or mixed farming systems reduce the risk of crop failure; non-farm activities complement and reduce the risks attached to farming—and vice versa. Commercial production under contract is an important element of a risk-reduction strategy for those farmers who are able to access such opportunities. Asset accumulation—including money, land, livestock, and other assets—is also critical to build a buffer against shocks, and a crucial component of household risk-management strategies. Yet in minimizing risk, households often have to pass up opportunities that could help them increase their incomes. Some studies estimate that average farm incomes could be 10–20 per cent higher were it not for the need to avoid risks (IFAD and WFP, 2010).

Exposure to major illnesses, market volatility, failed harvests, natural disasters, or conflict are the sorts of shocks that are a major factor pushing people rapidly into poverty or keeping them there, and those most likely to fall are the least resilient—typically large households with high dependency ratios and without sufficient assets to cushion the fall. When shocks do occur, it is this group that is the most likely to have to resort to coping strategies that involve incurring debt, selling assets, or foregoing on education opportunities for children and youth—all of which leave them deeper in poverty and that much more vulnerable to future shocks. Yet people do not resign themselves to poverty: they are repeatedly taking initiatives to exploit the opportunities open to them and improve their lot (Narayan et al., 2009), though moving out of poverty is a slow process, based on a successful enterprise (agricultural or non-agricultural) or employment. Assets such as land and livestock are important factors associated with moving out of poverty, as are education, participation in non-agricultural wage labour, and the share of income generated from non-agricultural self-employment. Effective risk management, and avoidance of the shocks associated with the risks (e.g. continued good health), is a precondition for moving out of poverty.

The consequence of this interplay between human enterprise and debilitating shocks is that there are many rural households that fall into poverty at some stage, while there are many others that are able to move out of poverty, and there are some that may repeatedly move out of and fall back into poverty. Indeed, it is very typical for 10 to 20 per cent of the population to fall into, or move out of, poverty within a period of five to ten years (IFAD, 2010). Dercon and Shapiro (2007) find that in countries as varied as Argentina, Bangladesh, Chile, China, Egypt, Ethiopia, Indonesia, Iran, and Uganda there are more people who are sometimes poor than always poor. The degree of movement in and out of poverty, and the speed with which people's conditions change, are remarkable; and it reminds us that 'the rural poor' is not a group separate and distinct from others in society.

Of course, rural poverty is not just a consequence of shocks; it is above all the asset base of rural households, and their economic, social, and political relations with the world around them, that determine whether or not they live in poverty. Valdés et al. (2008) found that in all regions poor households have on average significantly larger households, and in most, a higher share of dependents; they have substantially smaller plots of land than rich households—or are landless—and in most contexts they also own less livestock; they have significantly fewer years of education; and are less likely to have access to running water and electricity. Rural poverty also has a strong geographical dimension, with the highest rates of rural poverty often found in remote, marginal, or weakly integrated areas, characterized by a combination of an unfavourable natural resource base, poor infrastructure, weak state and market institutions, and political isolation (Chronic Poverty Research Centre, 2004). Above all, rural poverty is rooted in historical factors and in economic, social, and political relations within societies. It may be reflected in a variety of forms of exclusion, discrimination and disempowerment, and unequal access to and control over assets, all of which limit the opportunities of particular groups of people (in different rural societies women, youths, and indigenous peoples all face this marginalization) to improve their livelihoods and undermine their efforts to do so, and increase the risks they face. These issues of power and relations all contribute to the multi-dimensionality of rural poverty; indeed, in Latin America and some parts of Asia it can be defined primarily in terms of non-income deprivations.

3. Livelihood strategies of the rural poor

In response to the opportunities they see open to them, their perceptions of the risks they face, as well as their own asset profiles and characteristics, rural households make choices about how they can derive their incomes. There is a range of possible sources, including: their own on-farm production (crops and livestock); common property resources to which they have access (forests, fisheries); employment (agricultural and non-agricultural, private sector and government); self-employment; and transfers, including remittances from migrant relatives and social transfers. Diversified income sources are common among rural households—if less so among individual household members (World Bank, 2007), reflecting their strategies to reduce and manage risks of failure in any single income source. In most of the fifteen countries studied by Valdés et al (2008), between 30 and 60 per cent of rural households depended on at least two sources of income to generate 75 per cent of their total income. However, there are variations across regions and countries. On-farm production remains a particularly important income source in agricultural-based countries—particularly in sub-Saharan Africa—and less so in the transforming and urbanized countries of Asia and Latin America.

The 2008 World Development Report (World Bank, 2007) distinguishes five distinct livelihood strategies practised by rural households, according to where those households put most emphasis. Some farm households derive most of their incomes from agricultural production explicitly oriented towards the market (*market-oriented smallholders*). Others depend primarily on farming for their livelihoods, but use the majority of their produce for home consumption (*subsistence-oriented farmers*). Others still derive the larger part of their incomes from wage work in agriculture (on-farm, marketing, processing, and retail) or the rural non-farm economy, or from non-agricultural self-employment (*labour-oriented households*). Some households may choose to leave the rural sector entirely, or depend on household members who have migrated (*migration-oriented households*). Finally, in recognition of the fact that these are not mutually exclusive options, *diversified households* combine income from farming, off-farm labour/self-employment, and migration.

It is important to recognize that none of these livelihood strategies is set in stone: they reflect in part the existing opportunities open and available to those households; and in part their capacity to take advantage of the opportunities and their assessment of the risk associated with each. And as these change, so the strategies may change. Thus, for example, a new, remunerative, and low-risk market for agricultural produce may encourage some rural households to allocate more resources to market-oriented agriculture; increasingly irregular rainfall may lead some households to invest less into their on-farm production and put more emphasis on non-farm sources of income; expropriation of the household's land on behalf of a large-scale investor may push the household, or members of it, to migrate to the urban areas; while a rural employment programme that guarantees a minimum income can provide a springboard that enables a household to focus on market-oriented production, or on a non-agricultural micro-enterprise, or even to send one household member to the city. As a result, rural households can—and many do—use agriculture, the rural non-farm economy, and migration as alternative or complementary routes out of poverty, sometimes simultaneously, sometimes at different stages in their life cycles (IFAD, 2010). So there is no single or predetermined set of livelihood activities associated with moving out of poverty.

Implicit here is the recognition that the role of agriculture as a vehicle for poverty reduction is not a constant one, applicable in all circumstances. For many rural households, commercialized agriculture, linked into modern value chains, may be their route out of poverty, and available household resources—financial, managerial, and labour—are allocated to that activity accordingly. But for many others, including some better-off rural households, agriculture may be considered rather as a way of assuring the household's food consumption requirements: a safety net that allows for diversification of activities and incomes and the allocation of resources to these. Among poorer rural households, farming is by necessity usually subsistence-oriented; yet there are also situations where poor households are able to focus on market-oriented production: landless women dairy farmers in southern India provide an example (Rao et al., 2002). So the assumption that poorer rural households are subsistence-oriented while better-off rural households are market-oriented is not one that universally holds; indeed it may be an inaccurate and misleading generalization.

Having said that, typically agriculture continues to play a key role in the economic portfolios of rural households: Valdés et al. (2008) found that about 80 per cent of rural households in 11 of the 15 countries studied engage in farm activities of some sort. Virtually everywhere, it was the poorer rural households that derived the highest proportion of their incomes from farming and agricultural labour, while the better-off households derived the most from non-farm activities. Income gains at the household level were found to be associated with a shift out of agriculture towards more non-agricultural wage and self-employment income. But this is not a universal truth: a range of other studies throw up diverse and context-specific results: in some cases the poor rely more on non-farm activities than do the better off, while in other cases the opposite is true.[6] This should not be surprising: to repeat, much depends on the range of opportunities—agricultural and non-agricultural—available in the local economy, as well as their relative profitability and how risky and accessible they are to different groups.

At the same time, rural non-farm income sources can be important for all types of rural household. They can be a critical part of the livelihood portfolio of wealthier households, and they can play key roles in the risk-mitigation and risk-management strategies of poorer households. These sources include both non-farm wage employment (agricultural and non-agricultural, private and government) and non-farm self-employment, and they cover a highly diverse collection of activities, including trading, commercial and service activities, agro-processing, construction, and manufacturing. Often highly seasonal, many non-farm businesses operate according to the rhythms set by the agricultural season.

Wage employment is typically more important than self-employment as a source of household income in transforming and urbanizing countries, with the service sector and government services providing significant employment opportunities in some countries (IFAD, 2010). In sub-Saharan Africa self-employment is more important: Losch et al. (2011) find that there are areas in Kenya and Senegal where it has become the backbone of rural livelihoods. Yet within countries, there may be differences in the non-farm economy, according to differing natural resource endowments, population density, labour supply, location, infrastructure, and culture. Non-farm enterprises perform better in densely populated areas, where demand is higher (World Bank, 2007); their composition is often a function of this. Deep in the rural areas, the non-farm economy may be limited to small retailers, farm equipment repair services, and input supply firms, while in small towns, there may be demand for a broader range of services, both public and private.

The opportunities open to different groups are not the same, of course. Typically, education is key to accessing good employment opportunities in the non-farm economy. Poor people dominate unskilled wage labour, particularly in agriculture, as well as many of the low-return activities such as cottage industries and small-scale trading. The poor are more likely to be in casual rather than regular wage labour, while their businesses are likely to be labour-intensive and small-scale. For rural women, the rural non-farm economy is generally more important as a source of employment than agricultural

[6] Jayne et al. (2010) find that in Kenya, Ethiopia, Rwanda, Mozambique, and Zambia, it is the poorest quintile of households that derive the highest proportion of their incomes from off-farm sources, while the richest quintile derive the lowest. Of six studies covering countries in Asia, Africa, and Latin America, cited by Valdes et al. (2008), some found that poor rural households derived a higher share of their total income from non-farm sources, others that it was richer households that derived a higher proportion of their total income from non-farm sources.

labour markets, except in South Asia (World Bank, 2008). Women make up between 10 and 40 per cent of those employed in the rural non-farm economy, with the highest shares in sub-Saharan Africa and Latin America (Wiggins and Hazell, 2008), though women frequently face poorer conditions of employment and lower wages for the same work compared to men.

Of particular significance is the fact that in *all* developing regions of the world, the rural non-farm economy has become more important. In periods of ten years or less between the late 1980s and early 2000s, the share of non-farm income in total rural household income increased by between 10 and 20 per cent in 7 out of 13 countries in Africa, Asia, and Latin America (Valdés et al., 2008). Only in one country did the share decline, and this was by less than 5 per cent. In India too (not one of the 13), the rural non-farm sector has grown steadily during the past 25 years, with some acceleration between the mid-1990s and mid-2000s (Himanshu et al., 2011). Such an evolution follows the West European experience where, by the early 2000s, more than 50 per cent of professional farms in the six countries studied were actively involved in activities that go beyond conventional agricultural production,[7] and in Germany more than 50 per cent of total farm incomes was found to come from these activities (van der Ploeg and Roep, 2003).[8]

Rural-to-urban and international migration is an important element of the overall livelihood strategy for many rural households, and one that is rapidly growing in importance. In general, migration is driven by the need to manage risk through diversification of income sources at the household level, especially when there are limited diversification opportunities nearby. Not surprisingly, migration rates are typically highest in the poorest regions of countries, where opportunities are most limited. Those who migrate are typically not the poorest (Beegle et al., 2008), and being able to migrate may be conditional on education, having an existing network of migrants with whom to relate, being able to meet up-front costs, and not being constrained by family duties at home—duties that typically bear more heavily on women. For rural households, the migration of a family member can provide opportunities for more diverse and secure incomes, and although wealthier households generally gain more in absolute terms, poor households count

[7] These activities are not in their entirety non-agricultural, though they go beyond what is termed 'conventional agriculture'. They include activities categorized as 'broadening' (agri-tourism, new farm activities, diversification, and nature and landscape management); 'regrounding' (new forms of cost reduction, and off-farm income); and 'deepening' (organic farming, high-quality production and regional products, and short supply chains).

[8] It is important to note that this important shift took place during a sustained period of low food prices and limited market opportunities for smallholder farmers. Higher food prices over the past five years suggest this trend in rural incomes may have slowed, or in specific contexts even started to reverse, though it is an area in which further research is needed.

remittances as a vital component of their income and a key element of their risk-mitigation and coping strategies.

While it is true that the livelihood strategies described earlier are not set in stone, it is also the case that for the poorest rural households the options open to them are extremely limited, by their lack of resources—particularly land and capital—by their lack of education and skills, and by the premium that they have to place on avoiding risk. Such households are those most likely to be stuck in, and unable to escape from, livelihoods based around subsistence farming and/or casual agricultural labour, with the need to earn money through the latter often undermining their efforts to increase their own agricultural production. Particularly in agriculture-based countries, these households may make up a significant proportion of the rural population. Identifying appropriate support mechanisms to enable these households to expand their livelihood options, increase their incomes, and contribute to rural economic growth is a major challenge for governments and their development partners alike.

4. **Supporting the livelihood options of resource-poor rural people**

As we have seen, in a rapidly changing and increasingly complex world, the economic realities confronting rural people are shaped by a range of factors which, while typically having global origins, play out in very different ways in different local contexts. As a consequence, in their efforts to move out of poverty rural households respond in different ways, according on one hand to their assessment of the economic opportunities that are open to them, the potential returns that these may offer and the risks they present, and on the other their capacity to take advantage of the opportunities and manage the risks, according to their own asset base and characteristics.

There are many routes out of poverty, and most of them depend on a diversified set of household activities. The precise nature and blend of activities is constantly shifting, according to changing opportunities and risks and to changes in the household's profile. A livelihood strategy that works for a household today may no longer be appropriate for it ten years on. We have also seen that agriculture is often an important part of rural livelihood strategies, and that while many of the poorest households are by necessity 'subsistence-oriented' farmers, there are also many less poor households that *choose* to remain so—producing enough for their food needs, but preferring to invest their surplus capital and labour in non-agricultural activities. Commercialized agriculture linked to markets represents a specific choice about how best to allocate household resources that can work for some, including some of the very poorest, but may not be appropriate for all; it is likely to depend on the adequate availability of household land, labour, and capital, *as well as* markets to produce for, *as well as* the absence of alternative non-farm activities that are more remunerative and less risky.

What does all this mean for national governments and an international development community concerned with assisting rural people to overcome poverty? Above all, it is evident that there are no 'silver bullets'—no single technology, no single set of policy prescriptions, no single sector investment programme—that can offer a comprehensive solution. Instead, we argue, a development agenda for pro-poor rural growth needs to do four things. First, it needs to reflect and respond to the full range of livelihood strategies—agricultural and non-agricultural—that rural households adopt in their efforts to move out of poverty; and it must support their efforts in each of these areas—commercial and subsistence agriculture, labour and self-employment in the rural non-farm economy, and migration—and seek to create economic opportunities that they can access. Second, it needs to assist poor rural people to develop the skills, the knowledge, and the organization that they need to take advantage of such opportunities. Third, it needs to reduce the risk that poor rural people face in allocating resources to any single activity rather than a diversified range of activities, and assist them to manage the diverse risks they face more effectively, so as to avoid falling deeper into poverty. And fourth, while development policies and investments must be determined in large part by an assessment of the available drivers of pro-poor growth, the fact is that opportunities—and the barriers to achieving them—are shaped by global, national, and local factors, all of which are evolving and can vary enormously within and across the same country. This means that policies and investments also need to recognize and reflect the specificity of context, and focus particularly on expanding opportunities and responding to constraints that are locally defined and current.

Whatever the livelihood strategy of the rural household, a focus on agriculture is critical. The role of agriculture as a driver of growth, and in the larger process of structural transformation, is perhaps not as clear-cut as in the past. Yet against a backdrop of higher food prices, as well as a weakened natural resource base and environmental degradation, energy scarcities, and climate change and its impacts, there is growing recognition that support for agriculture needs to be substantially scaled up. Moreover, there is also an emerging consensus as to the need for a particular approach to agriculture: sustainable intensification—an approach that, while taking advantage of modern technologies for enhanced productivity, gives emphasis to better preserving or restoring the natural resource base, better exploiting agro-ecological processes and maximizing synergies within the farm cycle, and increasing the resilience of farming systems to climatic variation and change.

Since the full benefits of many sustainable practices accrue over several years rather than immediately, secure tenure that provides the incentive for farmers to invest their labour and capital is vital for their success: a land-tenure regime that offers such security, in practice as well as in policy, is a prerequisite. So too is increased investment in research, with a particular focus on developing varieties that are resilient to increasing temperatures and extreme weather events. Market-smart subsidies, including payments for the environmental services that sustainable intensification can provide, can also contribute to an incentive framework that encourages farmers to make the change to more sustainable practices (FAO, 2011). At the same time, to move towards these sorts of production systems, smallholder farmers must develop the skills to combine their experience and knowledge with modern science-based approaches, and develop effective solutions to their problems. This requires strengthening agricultural education in schools, reforming advisory services, and fostering greater collaboration, innovation and problem-solving among smallholders, researchers, and service providers. It also requires building of coalitions, sharing responsibilities, and creating synergies among governments, civil society, the private sector—and above all—farmers and their organizations.

For some—particularly those who are subsistence-oriented or oriented principally towards non-farm activities—that support may be aimed primarily at enabling them to produce enough for their own household consumption needs and reduce the losses they suffer post-harvest. For the poorest households, productive safety nets may be a necessary complement to investments in agricultural service delivery, as they can offer those households the safety net they need to invest their scarce time or resources in a new enterprise, technology, or practice.

For others—those who are already market-oriented or those who have the potential and interest to become so—the focus on production needs to be supported by efforts to enable households to become better linked up to markets in more efficient and equitable value chains. Strengthening their capacity to organize is a key requirement to participating in markets more efficiently and to reducing transaction costs for them and for those with whom they do business. Infrastructure is also important—particularly transportation, and information and communication technology—for reducing costs and uncertainty, and improving market information flows. Contracts can help, as they often build trust between smallholders and agribusiness, as well as facilitating farmers' access to input credit and other financial services. And although smallholder farmers engage at only one point in value chains, there is a need to intervene along them to support the development and expansion of efficient value chains in which increased numbers of smallholder farmers can engage on more equitable terms.

Yet while a focus on agriculture—and on enhancing its productivity, sustainability, and resilience—is critical, this alone is clearly insufficient. Particularly in sub-Saharan Africa, where the proportion of the rural population under 15 years old is highest and where, for many, extremely small land sizes make it difficult for smallholder agriculture to provide a way out of poverty, there is at the same time a pressing need to find decent, non-agricultural employment opportunities for rural youth. But in all regions, attention needs to be given to the growing rural non-farm economy—both wage employment and self-employment—and to creating opportunities that poor rural people can access. This requires investment in rural infrastructure and services such as energy, communications and transportation, and better governance.

Prerequisites for encouraging private investments include improving the business climate by reducing red tape and corruption, and providing business services suited to the needs of both men and women small entrepreneurs. For firms, the possibility of acquiring a labour force with appropriate skills is crucial. For rural workers, an improved environment is one in which they find decent employment opportunities and their rights and ability to organize are recognized, and in which efforts are made to address the prevalence of poorly paid, insecure, and unregulated jobs—taken up predominantly by women—in the informal sector. Governments have a key role to play in strengthening the capabilities of rural people to take advantage of opportunities for decent employment or entrepreneurial activity in the rural non-farm economy, as well as in creating an improved environment for the rural non-farm economy, including facilitating and catalysing initiatives taken by others such as firms or rural workers' organizations.

In some rural areas, where local economic opportunities are limited for individuals or households, migration may be the only route out of poverty. Yet many governments have policies aimed at lowering migration to urban areas, even if they are often ineffective and merely distract from the need to plan for the inevitable future urban expansion (UNFPA, 2011). Instead, governments can recognize the reality of rural–urban migration and support migrants: they can ensure that migrants gain access to services, including information on their rights and on available labour opportunities, and that they are able to claim the same rights and entitlements as non-migrants. They can facilitate, or at least not hinder, organization of migrant workers, and they can monitor and punish labour trafficking. Given the value of remittances to rural people, this is an important area for greater efforts by governments, in partnership with other actors—MFIs, other financial institutions, and providers of banking and communication technology. Further efforts are needed to reduce the costs and risks of transferring remittances to poor rural areas and to harness the benefits of remittances through improved financial services, including savings and insurance. Although some innovations have emerged in recent years, there is still a great need to invest in more effective and efficient technology solutions to reduce transfer costs, and to link remittances to effective financial services and profitable investment opportunities.

Livelihood options, agriculture, the non-farm rural economy, and job-seeking beyond the rural areas are all increasingly knowledge-intensive, and in all of these areas having relevant and up-to-date knowledge and a range of skills is usually a prerequisite both for identifying and taking advantage of economic opportunities and for individual success. In the rural economy as a whole, productivity, dynamism, and innovation all depend on there being a skilled, educated population. Advancing individual capabilities is critical, and it needs far more attention in the rural development agenda. Investment is particularly needed in post-primary education, in technical and vocational skills development, and in re-oriented higher-education institutes for agriculture; and attention is needed at all levels to ensure that syllabuses reflect and respond to the current needs of the agricultural and rural economy, and help people to develop relevant and up-to-date skills.

Beyond the individual level, strengthening the collective capabilities of rural people can give them the confidence, security, and power needed to overcome poverty. Membership-based organizations have a key role to play in helping rural people reduce risk, learn new techniques and skills, manage individual and collective assets, and market their produce. They also negotiate the interests of people in their interactions with the private sector or government, and can help to hold them accountable. While many organizations have problems of governance, management, or representation, they usually represent the interests of poor rural people better than any outside party can. They need strengthening to become more effective, and more space needs to be made for them to influence policy.

If managing risk is central to the livelihood strategies of poor rural people, and the grasping of new economic opportunities requires that they take on new and additional levels of risk, then reducing the risks that poor rural people face and helping them to improve their risk-management capacity needs to become a central, cross-cutting element within a pro-poor rural development agenda. It needs to drive support both to agriculture—and sustainable intensification reflects this concern—and to the rural non-farm economy. On the one hand, it involves strengthening the capacity of rural women, men, and youths to manage risk by supporting and scaling up the strategies and tools they use for risk management and for coping, and helping them to gain the skills, knowledge, and assets to develop new strategies. On the other hand, it requires that the conditions they face be made less risky, be it in terms of markets, health care and other essential services, natural environment, or security from conflict, and it involves developing or stimulating the market to provide new technologies and services for smallholders and poor rural people that reduce, transfer, or help them manage the risks they face.

There is growing recognition of the importance of risk management for public policy, and there are more and more experiences with policy instruments to draw upon. These include: social protection measures such as conditional cash transfers and employment guarantee/public works schemes that can help poor households to build their assets, reduce risks, and more easily invest in profitable income-generating activities; promoting the expansion and deepening of a range of financial services to poor rural people, including (but not limited to) insurance products such as weather index insurance that can help farmers cope with catastrophic events and crop failure; and strengthening community-level organizations and assisting them to identify new mechanisms of social solidarity.

Finally, we started this chapter by highlighting a number of areas in which the rural areas are changing, and how these changes impact on poor rural people. Yet their impact is different in different places, and the realities that poor rural people actually face are shaped by national and local factors as well as global, and all of these can vary enormously, even within the same country. Such heterogeneity points to the need for policies that recognize the importance, and accommodate the specificity, of context, and permit an implementation approach that is varied and prioritizes the expansion of local opportunities and responses to local constraints and risks.

REFERENCES

Anríquez, G. and L. Stloukal (2008). 'Rural population change in developing countries: Lessons for policymaking', ESA Working Paper No. 08–09. Rome: FAO.

Anseeuw, W., L. A. Wily, L. Cotula, and M. Taylor (2012). *Land Rights and the Rush for Land.* Rome: International Land Coalition.

Baumüller, H. (2012). 'Facilitating agricultural technology adoption among the poor: The role of service delivery through mobile phones', ZEF Working Paper Series 93, University of Bonn.

Beegle, K., J. De Weerdt, and S. Dercon (2008). 'Migration and economic mobility in Tanzania: Evidence from a tracking survey', Policy Research Working Paper, WPS 4798, World Bank, Washington, DC.

Bhide, S. and A. K. Mehta (2006). 'Correlates of incidence and exit from chronic poverty in rural India: Evidence from panel data', in Mehta, A. K. and A. Shepherd (eds), *Chronic Poverty and Development Policy in India.* New Delhi: Sage Publications.

Chen, S. and M. Ravallion (2012). 'An update to the World Bank's estimates of consumption poverty in the developing world', Development Research Group, Briefing Note, World Bank, 3 January 2012.

Chronic Poverty Research Centre (CPRC) (2004). *Chronic Poverty Report 2004–05.* Manchester, UK: CPRC.

Cobert B., B. Helms, and D. Parker (2012). *Mobile Money: Getting to Scale in Emerging Markets.* McKinsey & Company. Retrieved from: <http://mckinseyonsociety.com/downloads/reports/Economic-Development/Mobile-money-Getting%20to-scale-in-emerging-markets.pdf>.

Cotula, L. (ed.) (2007). *Changes in 'Customary' Land Tenure Systems in Africa.* London: International Institute for Environment and Development/Rome: Food and Agriculture Organization of the United Nations.

Cotula, L., S. Vermeulen, R. Leonard, and J. Keeley (2009). *Land Grab or Development Opportunity? Agricultural Investment and International Land Deals in Africa.* London: International Institute for Environment and Development/Rome: Food and Agriculture Organization of the United Nations and International Fund for Agricultural Development.

Deininger, K. and D. Byerlee, with Lindsay, J., A. Norton, H. Selod, and M. Stickler (2011). *Rising Global Interest in Farmland: Can It Yield Sustainable and Equitable Benefits?* Washington, DC: World Bank.

Dercon, S. and J. S. Shapiro (2007). 'Moving on, staying behind, getting lost: Lessons on poverty mobility from longitudinal data', in Narayan, D. and P. Petesch (eds), *Moving Out of Poverty: Cross-disciplinary Perspectives on Mobility.* Washington, DC: World Bank and New York/ Basingstoke, UK: Palgrave Macmillan.

Fafchamps, M. and B. Minten (2012). 'Impact of SMS-Based Agricultural Information on Indian Farmers', World Bank Economic Review, Article in press available online, 27 February 2012.

FAO (2011). *Save and Grow: the Policymaker's Guide to the Sustainable Intensification of Smallholder Crop Production.* Rome: FAO.

Haggblade S., P. Hazell, and T. Reardon (2010). 'The rural non-farm economy: Prospects for growth and poverty reduction', *World Development,* 38(10), October 2010, pp. 1429–41.

Himanshu, P. Lanjouw, A. Mukhopadhyay, and R. Murgai (2011). 'Non-farm diversification and rural poverty decline: a perspective from Indian sample survey and village study data', Asia Research Centre Working Paper 44. London: London School of Economics & Political Science.

IFAD (2008). *IFAD Policy on Improving Access to Land and Tenure Security.* Rome: IFAD.

IFAD (2010). Rural Poverty Report: *New Realities, New Challenges: New Opportunities for Tomorrow's Generation.* Rome: IFAD.

IFAD (2012). *Climate-Smart Smallholder Agriculture: What's Different?* Rome: IFAD.

IFAD and World Food Programme (WFP) (2010). *The Potential for Scale and Sustainability in Weather Index Insurance for Agriculture and Rural Livelihoods.* Rome: IFAD.

IMF (2011). *Regional Economic Outlook: Sub-Saharan Africa: Sustaining the Expansion.* October 2011. Washington DC: International Monetary Fund.

Jayne, T. S., D. Mather, and E. Mghenyi (2010). 'Principal challenges facing smallholder agriculture in sub-Saharan Africa', *World Development,* 38(10), October, pp. 1384–98.

Jayne T. S., M. Muyanga, and J. Chamberlin (2011). 'Land constraints in smallholder agriculture: Towards the identification of appropriate agricultural commercialization strategies for densely populated rural areas', Presentation at seminar organised by the International Food Policy Research Institute at the ILRI Campus, Addis Ababa, Ethiopia, 4 November 2011.

Kugler, M. (2005). *Migrant Remittances, Human Capital Formation and Job Creation Externalities in Colombia.* September. Borradores de Economia 370, Banco de la Republica de Colombia. Retrieved from: <http://ideas.repec.org/p/bdr/borrec/370.html#biblio>.

Losch, B., S. Fréguin-Gresh, and E. White (2011). *Rural Transformation and Late Developing Countries in a Globalized World: A comparative analysis of rural change.* Washington, DC: World Bank.

Migot-Adholla, S., P. Hazell, B. Blarel, and F. Place (1991). 'Indigenous land rights systems in sub-Saharan Africa: a constraint on productivity?', *World Bank Economic Review,* 5(1).

Narayan, D., L. Pritchett, and S. Kapoor (2009). *Moving Out of Poverty: Vol. 2, Success from the Bottom up.* Washington, DC: World Bank/Basingstoke, UK: Palgrave Macmillan.

Onumah, G., J. R. Davis, U. Kleih, and F. J. Proctor (2007). 'Empowering smallholder farmers in markets: Changing agricultural marketing systems and innovative responses by producer organizations', Working Paper 2, Empowering Smallholder Farmers in Markets (ESFIM), Wageningen University and Research Centre, The Netherlands.

Parry, M., A. Evans, M. W. Rosegrant, and T. Wheeler (2009). *Climate Change and Hunger: Responding to the Challenge.* Rome: World Food Programme.

Rao, S. V. N., S. Ramkumar, and K. Waldie (2002). 'Dairy farming by landless women in southern states of India', in Morrenhog, J., V. Ahuja, and A. Tripathy (eds), *Livestock services and the poor: papers, proceedings and presentation of the international workshop*, organized in Orissa in collaboration with the Department of Fisheries and Animal Resources Development of the Government of Orissa as a joint venture of the Indo Swiss NRM Programme, Orissa, Global Initiative of Livestock Services, and the Poor and the Pro-poor Livestock Initiative.

Sjaastad, E. (2003). 'Trends in the emergence of agricultural land markets in sub-Saharan Africa', *NUPI Forum for Development Studies, No.1 2003*, Norwegian Institute of International Affairs, Oslo, Norway.

STAR Kampuchea Organization (2007). *Landlessness and land conflicts in Cambodia.* Retrieved from: <http://www.landcoalition.org/sites/default/files/legacy/legacypdf/07_r%5bt_land_cambodia.pdf?q=pdf/07_r[t_land_cambodia.pdf>, 5 March 2013.

Sumner, A. (2011). 'Poverty in Middle-Income Countries', The Bellagio Initiative, Briefing Summary, November, Institute of Development Studies, London.

Timmer, P. (2009), 'A world without agriculture? The historical paradox on agricultural development', in American Enterprise Institute for Public Policy Research, No. 1, May.

UNFPA (2011). *Population Dynamics in the Least Developed Countries: Challenges and Opportunities for Development and Poverty Reduction.* New York: United Nations Population Fund.

United Nations (2011). *World urbanization prospects: the 2011 revision.* New York: United Nations Department of Economic and Social Affairs, Population Division.

UNRISD (2010). *Combating poverty and inequality.* Geneva: United Nations Research Institute for Social Development.

Valdés, A., W. Foster, G. Anríquez, C. Azzarri, K. Covarrubias, B. Davis, S. DiGiuseppe, T. Essam, T. Hertz, A. P. de la O, E. Quiñones, K. Stamoulis, P. Winters, and A. Zezza (2008). 'A profile of the rural poor', Background paper for the IFAD Rural Poverty Report 2011.

van der Ploeg, J. D. and D. Roep (2003). 'Multifunctionality and rural development: the actual situation in Europe', in van Huylenbroeck, G. and G. Durand, *Multifunctional Agriculture: A New Paradigm for European Agriculture and Rural Development.* Hampshire, England: Ashgate Publishing Limited, pp. 37–53.

Vargas-Lundius R., G. Lanly, M. Villarreal, and M. Osorio (2008). *International Migration, Remittances and Rural Development.* Rome: IFAD.

Wiggins, S. and P. Hazell (2008). 'Access to rural non-farm employment and enterprise development', Background paper for the IFAD Rural Poverty Report 2011.

World Bank (2004). 'Regional Study on Land Administration, Land Markets, and Collateralized Lending East Asia and Pacific Region', 15 June, Washington, DC.

World Bank (2007). *World Development Report 2008: Agriculture for Development.* Washington, DC: World Bank.

World Bank (2008). *World Development Report 2009: Reshaping Economic Geography.* Washington, DC: World Bank.

14 Securing land rights for smallholder farmers

KLAUS DEININGER[1]

1. Background and motivation

While efforts to secure the land and property rights of smallholder farmers have long been recognized as essential for economic development, land values have been significantly increased (Anseeuw et al., 2012) by a combination of population growth, soil degradation, and urban expansion, recently accentuated by land appreciation in response to rising global demand for agricultural commodities and environmental services (Deininger et al., 2011b). Demand is disproportionately high in Africa (Arezki et al., 2011), where productivity on currently cultivated land is well below the land's potential, with potential food security implications.[2] The fact that most of these lands are used by locals without formal documentation and many are legally state land, creates a risk that non-transparent transfers of these lands may foster speculation and result in a 'resource curse' and loss of livelihoods rather than benefit from technology and access to credit and infrastructure. Having rights recognized, boundaries demarcated, and, if land is held by groups, internal decision-making clarified will strengthen local users' ability to negotiate with outsiders and allow their integration into value chains by a variety of mechanisms.

Economic development and structural change entail moving labour from agricultural to non-agricultural sectors. This is less likely to happen if those who join the non-farm economy fear losing their land or if land remains idle because the cost of registering transfers is too high. Impediments to land market functioning will also undermine the ability to use land as collateral in financial markets and make it more difficult for entrepreneurs, small or large,

[1] This chapter benefited from discussions with and comments by H. Binswanger, G. Feder, P. Hazell, and S. Holden as well as an anonymous referee. The views expressed in this chapter are those of the author and do not necessarily represent those of the World Bank, its Board of Executive Directors, or the countries they represent.

[2] With the exception of South Africa, none of the African countries where interest by agricultural investors has recently materialized achieves 25 per cent of its potential productivity on currently cultivated land (Deininger et al., 2011b).

to access land for developing entrepreneurial activities.[3] Resolving these issues in a context of rapid urbanization and increasing land values requires institutions that can cope with new demands so as to avoid corruption and to facilitate planning and the availability of serviced land. Incentives for production of environmental amenities also require land rights to be identified.

This multi-faceted relevance of land to development is recognized by policymakers at global (Food and Agricultural Organization of the UN, 2012) and regional level (African Union, 2009). As land issues are technically complex and often politically sensitive, with responsibility distributed across a range of ministries (from agriculture and environment to urban and local governments), the challenge has been to translate abstract high-level commitments into action on the ground (Place, 2009). This chapter aims to cover not only the 'why' but also the 'how' of securing land rights and setting in motion a process of policy reform in the land sector. It argues that moving beyond individual examples requires casting land issues in the broader framework of good land governance and establishing a constituency for outcome-based reform at the national level.

We first recap why securing land rights is important, in particular because (i) secure land rights increase incentives for land-related investment and stewardship; (ii) they allow low-cost operation of land markets and the associated development of financial markets based on the ability to use land as collateral; and (iii) having secure rights empowers asset-holders in the long term. We also highlight key reasons why, in practice, many interventions failed to realize these benefits. These include: (i) a neglect of distributional issues that resulted in interventions having unanticipated negative effects or being unviable politically; (ii) a failure to appreciate and build on the benefits of existing arrangements (including the fact that rights are normally linked to obligations) that led to efforts at wholesale 'replacement' of rights rather than their gradual evolution along a continuum of rights; and (iii) a focus on unaffordable solutions, the high cost of which could not be sustained, causing reversion to informality.

Addressing these requires putting land into a framework of good governance in at least three areas. First, existing rights need to be recognized and effectively enforced. Second, there is a need for provision of comprehensive and current information on rights to interested parties at low cost. Finally, the ways of exercising public interest in land, including land-use planning, the acquisition of land for public purpose, and the divestiture of state land to private interest are clearly defined and transparently implemented. We highlight examples of effective interventions (and how their effect has been demonstrated) in the area of communal rights mapping and recognition,

[3] The difficulty of accessing land for enterprise development has emerged as one of the main complaints by private-sector operators in a large number of enterprise surveys in African countries.

registration of individual rights, improving the efficiency of land administration institutions, and improving gender equity.

We conclude by arguing for a three-step approach to sustainably improving land governance at country level, building on the land governance assessment framework. The first step is a technical assessment of land governance compared to international best practice and prioritization of policy actions supported by key stakeholders. Second, as in many contexts, adjustments in regulations, institutional structures, and processes will be needed to move towards better land governance in practice, and continued feedback from pilot schemes and rigorous evaluation of new activities will be critical. This will provide the basis for assessing progress with land governance on a larger scale in a way that is comparable globally.

2. Conceptual issues

A large literature documents how crucial secure land rights are to motivate productivity-enhancing investments, provide a basis for the functioning of land and financial markets, and assign political and decision-making power within the household. However, past interventions often failed to achieve their full potential due to (i) neglect of distributional impacts and the political dimension of land rights; (ii) failure to appreciate and acknowledge the plurality of rights and responsibilities in existence; and (iii) use of unsustainable high-cost technology which undermined broad access and affordability.

2.1 WHY SECURING LAND RIGHTS IS IMPORTANT

Development economists have long highlighted the central role of institutions (i.e. socially imposed constraints on human interaction that structure incentives in any exchange) in shaping growth and the distribution of its gains among the population (Greif, 1993; North, 1971). Property rights are social conventions, backed by the enforcement power of the state (at various levels) or the community, allowing individuals or groups to lay 'a claim to a benefit or income stream that the state will agree to protect through the assignment of duty to others who may covet, or somehow interfere with, the benefit stream' (Sjaastad and Bromley, 2000).

Since in most contexts, land and associated real estate are one of households' most important assets, societies have, from the earliest days of recorded history, developed customs and laws on how to define land rights, and many set up registries to make public the assignment of rights and their transfer among private parties (Powelson, 1988). The creation and maintenance of such property rights systems are an important public good that reduces the need for landholders to expend resources (e.g. in hiring private armies) to protect their rights.

Investment incentives

Secure property rights affect economic outcomes most immediately by reducing the risk of land loss, increasing investment incentives and reducing the need for individuals to spend resources on protecting their rights (Besley and Ghatak, 2010). Historically, land rights emerged at the transition from the hunter-gatherer stage when investment in land becomes important (Binswanger et al., 1995). The prospect of being able to enjoy the fruits of their labour encourages owners to make long-term land-related investments and manage land sustainably (Besley, 1995). Positive impacts of land-tenure security on investment in rural areas have been documented in China (Jacoby et al., 2002), Thailand (Feder et al., 1988), Latin America (Bandiera, 2007), eastern Europe (Rozelle and Swinnen, 2004), and Africa, where weak ownership rights, often for disadvantaged groups or outsiders, lead to significant reductions in fallowing that then reduced yields (Deininger and Jin, 2006; Fenske, 2010, 2011; Goldstein and Udry, 2008). In urban areas, efforts to enhance tenure security increased levels of self-assessed land values (Lanjouw and Levy, 2002), investment in housing (Field, 2005), female labour market participation, and reduced child labour due to a reduced need to stay at home and guard land assets (Field, 2007).

If there is widespread insecurity of property rights, clarification of such rights through systematic adjudication and registration of land rights can be a cost-effective way to increase tenure security. The magnitude and distribution of the associated benefits will depend on the reduction in enforcement effort afforded by formal recognition, the increment in security afforded by the intervention (which will depend on the legitimacy and legality of existing arrangements and the level of disputes) and the availability of investment opportunities. Benefits will be higher if the increment in tenure security is large, e.g. if land tenure had been insecure or conflict-ridden before, if new arrangements enjoy wide legitimacy and if returns from land-related investment are high.

Land transfers and financial markets

Economic development normally involves specialization and a move of part of the labour force out of the agricultural sector. Such movement creates heterogeneity in the population and increases the scope for efficiency-enhancing land transfers. Institutions allowing such transactions at low cost and removing the fear of those transferring use rights that they might lose their land, can increase the productivity of land. Land rental allows labour to move out of agriculture without foregoing the benefits (e.g. the social safety net function associated with land ownership), encouraging transfers through rental rather than sale. Initially they are likely to involve community members. High transaction cost, caused by unclear rights or institutional inefficiencies, can reduce the number of such transactions or drive them into informality, with potentially negative impacts on long-term economic development (Libecap and Lueck, 2011).

Asymmetric information and risk have long been shown to lead to credit rationing and the use of collateral as one way of reducing such credit rationing (Stiglitz and Weiss, 1981). Land's immobility and relative indestructibility make it ideal collateral. However, banks will use it for this purpose on a large scale only if they have access to low-cost means to make reliable inferences on ownership and the absence of other encumbrances. Such information is normally provided by land registries. If it is reliable and comprehensive, it can eliminate the need for physical inspection of the land in question or inquiry with neighbours, thus reducing the transaction cost of exchanging land in impersonal markets and creating the preconditions for using it as collateral to secure loans. While this provides the conceptual foundation for credit impacts from land titling or registration, such effects would be expected only if there is already latent and unsatisfied demand for credit (i.e. a portfolio of viable projects), registry information is comprehensive and remains up to date over time, third parties, such as mortgage lenders, can access reliable registry information at low cost on a routine basis, and the prospect of foreclosing on collateral in case of default is credible.

Compared to the overwhelming empirical support for investment impacts, evidence of credit impacts from land titling, although not entirely missing (Feder et al., 1988), is surprisingly limited. They may accrue only to wealthy producers (Carter and Olinto, 2003), and in a number of cases expectations for property rights reform to improve credit access (de Soto, 2000) failed to materialize (Field and Torero, 2006). One reason is that better access to information on land ownership will affect credit supply only if other impediments are absent, i.e. if agents have been credit-constrained before and are endowed with sufficient levels of illiquid wealth that can be foreclosed upon at reasonable cost (Besley and Ghatak, 2010). Lack of investment opportunities, risk aversion, and political, social, or economic restrictions on land market liquidity that make foreclosure difficult are identified in the literature as key reasons for the limited attractiveness of rain-fed agricultural land to lenders.

Power relations

The limited overall availability of land implies that, especially in settings where land is the main asset, how access to and use of land are organized becomes highly political (Boone, 2007). While this has long been documented qualitatively (Binswanger et al., 1995), a growing number of studies now provide quantitative evidence of long-term impacts of land institutions on outcomes, such as public good provision and educational attainment in India (Banerjee and Iyer, 2005; Iyer, 2010), human capital formation and democratic development in Central America (Nugent and Robinson, 2010), transparency and governance in Brazil (Naritomi et al., 2009), and financial-sector development across US counties (Rajan and Ramcharan, 2011). In all of these places, land institutions and the way they have changed had sustained impacts on economic outcomes.

In addition to affecting the distribution of political power, land access is also a key determinant of bargaining power within households. The far-reaching 'downstream' impact of female property rights is shown in studies documenting that women's access to assets can affect girls' survival rates (Qian, 2008), their anthropometric condition (Duflo, 2003), and, for some groups, investment in (girls') schooling (Luke and Munshi, 2011). Yet, in many contexts, women can access land through male relatives only and their ability to inherit or hold on to it in case of widowhood or divorce is limited. Such constraints affect not only intra-household bargaining power (Behrman, 1990; Strauss et al., 2000) but also efficiency of land use (Udry, 1996), women's participation in non-farm opportunities (Quisumbing and Maluccio, 2003), and their being affected by land conflict (Deininger and Castagnini, 2006; Joireman, 2008). Greater attention to gender in design and evaluation of land-related interventions will thus be critical.

2.2 CHALLENGES FOR EFFORTS TO STRENGTHENING LAND RIGHTS

While the literature highlights the scope for beneficial effects from efforts to strengthen land rights, it also documents instances where such efforts either failed to achieve their stated objective or may even have had the opposite effect. Key reasons include insufficient appreciation of (i) the distributional aspects of land ownership and access; (ii) the complexity of land rights that co-exist in practice and the advantages of traditional systems in managing them; and (iii) the resource requirements (in terms of money and human capital) of conducting and maintaining land administration systems that require high-precision ground surveys and the fact that in many instances the corresponding gains are rather modest.

Distributional aspects

Although interventions to recognize land rights can, by encouraging investment and improving transferability, make everybody better off, the fixed stock of land implies that these can easily degenerate into a zero-sum game. It is not unusual for such efforts to set off speculative land acquisition by powerful and well-informed individuals (Benjaminsen and Sjaastad, 2002; Peters, 2004). In cases where there is no intensive dissemination effort to bring everybody to the same level of knowledge, the informational advantages of the rich and well-connected (Peters, 2004) may formalize land access by local elites and bureaucrats while disempowering the poor (Jansen and Roquas, 1998).

The potential for adverse distributional consequences is particularly pronounced if the rights of certain groups do not enjoy legal recognition. Rights by 'migrants' or non-nationals, which may have been held for generations, often enjoy limited legal recognition (Colin and Ayouz, 2006; Fenske, 2010). Failure either to ensure that such rights enjoy proper legal protection before trying to register rights or to adopt adjudication process that will ensure such rights are upheld on the ground often led to negative social consequences and even conflict, as in Côte d'Ivoire due to inappropriate laws. If overlapping rights exist, clear and widely accepted rules of priority will need to be used in adjudicating such rights to prevent such programmes being viewed as thinly veiled attempts to formalize illegitimate land acquisitions.

While distributional issues relating to pre-existing private rights are one area of concern, land registration programmes can also legitimize or even encourage encroachment on communal or public land, including areas such as forests which are highly valuable from an environmental perspective. This includes land sales by chiefs who perceive themselves as landlords rather than custodians of a vital community asset (Berry, 2009) and pocket receipts (Lavigne-Delville, 2000).

Land transfers pose specific issues in two sets of circumstances. On the one hand, economy-wide distortions may mean that land markets will not bring about outcomes in line with efficiency and poverty reduction, as is well documented in the case of Brazil (Rezende, 2006).[4] On the other hand, at low levels of development, land is not only a productive asset but also performs important functions as a social safety net and old-age insurance. With credit-market imperfections, introducing transferability may give rise to distress sales or myopic transactions that may have a negative social impact (Andolfatto, 2002). This has often prompted communities to adopt rules that limit the scope for individual land alienation to outsiders. Such rules are unlikely to be harmful provided they are the product of a conscientious and participatory choice, arrived at by weighing associated cost and benefits, and ways to transact with outsiders as a group are available. If traditional social ties loosen or the efficiency gains from allowing sales increase, groups can move towards gradual individualization and sales to outsiders at their own pace.

[4] While technical innovations may weaken the efficiency-advantage of owner-operated small farms, most large farms emerged due to their ability to deal with market imperfections (Deininger and Byerlee, 2012).

Benefits from existing arrangements

Failure to appreciate and build on the benefits of existing systems has often resulted in elements of traditional systems that were functioning rather well being replaced with land registration of greater complexity, cost, and risks. The rights and responsibilities associated with land use or ownership are context- and culture-specific and evolve in response to needs. Recognizing existing rights and using them as a starting point is likely to be more effective and sustainable than trying to import concepts for which there is no demand or that may not be understood. Plural arrangements adapted to different conditions and a continuum of rights are preferable to a mistaken quest for uniformity and wholesale 'replacement' of traditional with seemingly modern institutions (Meinzen-Dick and Mwangi, 2008).

Efforts to 'formalize' tenure have often eliminated or weakened secondary or communal rights. Women's rights, especially those in informal or polygamous unions, are often neglected. Finally, secondary and often temporary rights for use of arable land for grazing by pastoralists after the harvest or for fuel wood collection by the poor may simply be eliminated. In each case, failure to recognize such rights can lead to negative social impact and conflict. Where they exist, legal recognition of such rights is a pre-condition for any attempt at recording rights. Even with legal recognition, ways to protect and document such rights during implementation need to be found. In the case of secondary rights, elaboration of land-use plans may be more cost-effective than efforts to record rights individually.

It is now also recognized that formalizing through individual title is not always the most appropriate solution; in fact, rather complex arrangements to define rights and obligations by the individual relative to groups, e.g. through condominium associations, are frequent in modern economies. In situations of high risk, well-governed communal tenure arrangements can provide more flexibility and an important safety net (Baland and Francois, 2005). Efforts to individualize such arrangements may reduce the flexibility of risk-management options by the poor and possibly leave them worse off.[5] Attempts to replace local institutions that functioned reasonably well with 'better' ones that fail to materialize can actually increase conflict (Deininger and Castagnini, 2006).

[5] In Mexico, less than 15 per cent of *ejidos*—mostly those in peri-urban areas where land use had already lost communal links—made use of the opportunity fully to individualize their land, suggesting that, even at relatively high levels of per capita income, the spatial reach of insurance mechanisms to replace the safety net function of (communal) land ownership remains more limited than is often thought (Zepeda, 2000).

Economically unaffordable solutions

Land registration and registry operations incur cost. They will be desirable from an economic perspective only if the total current cost of establishing and maintaining them, in the light of expected workloads, are commensurate with the benefits in terms of increased investment incentives and the scope for efficiency-increases through land transfers. Once overall costs and benefits are assessed, it will be possible to discuss how these should be distributed between the public sector and private users. It is often argued that since it establishes a critical piece of infrastructure, first registration is something that should be supported by public funds.

Two critical components of the cost of recording land rights are those related to boundary demarcation and the establishment of the institutional structure. Regarding the former, many observers confuse high-precision boundary surveys with greater security of rights, with major cost implications.[6] This, together with the fact that lower-cost options became widely available only recently, implied that boundary surveys with a cost of US $20–60 per land parcel were a key cost component in efforts systematically to register individual land rights (Burns, 2007). Image-based solutions that allow quick coverage of large areas and simple updating can be cheaper and will be desirable in most situations where achieving quick coverage of large areas is critical. A second cost element relates to the institutions to maintain land rights at local level. In areas with limited transaction volumes, it will make sense to have existing institutions perform some land-related functions at the local level rather than trying to establish entirely new ones.[7]

The impact of having cost-effective institutions can be immense; for example, in Madagascar the cost of traditional surveying approaches was well beyond what would have been affordable or justifiable in light of the expected benefits and thus it would never have achieved broad coverage. However, benefits from secure land rights were estimated to exceed the cost of image-based approaches (Jacoby and Minten, 2007), making broad coverage viable. At the same time, many studies suggest that a high cost of registering subsequent transactions can quickly result in obsolescence of systems that have been established at high cost (Atwood, 1990; Pinckney and Kimuyu, 1994). Even in a case where land titling had generated significant investment benefits, as in the case of Argentina (Galiani and Schargrodsky, 2010), the high cost of formalizing transactions prompted a process of 'deregularization'

[6] As a rule of thumb, the cost of survey increases exponentially with precision (Dale and Mclaughlin, 2000).

[7] Unaffordable institutional designs are widespread in practice and undermine implementation, as in Uganda (Hunt, 2004), where more than a decade after the passage of the 1998 Land Act not a single certificate of customary ownership had been issued.

where subsequent transfers were not registered, thereby eroding the sustainability of these benefits (Galiani, 2011). Historically, lack of attention to costs made broad coverage with a sustainable system of land registration impossible in the Philippines (Maurer and Iyer, 2008) and resulted in the obsolescence of such systems (Barnes and Griffith-Charles, 2007).

3. **Embedding smallholders' rights in a system of good land governance**

To improve land governance in a consistent way that allows scaling up within a reasonable timeframe, attention to three areas is critical (Deininger et al., 2011c). First, there is need for a legal and institutional framework that recognizes existing rights, enforces them at low cost, and allows users to exercise them in line with their aspirations and in a way that is transparent, equitable, and benefits society as a whole. Second, reliable and complete information on land and property rights needs to be broadly accessible, comprehensive, reliable, current, and cost-effective in the long run. Finally, transparent management of the public-good aspect of land requires both regulations to avoid negative externalities from uncoordinated action by private parties and clear limits to, and rules for, the way the state exercises its right of pre-emption and manages or disposes of public land.

3.1 SECURING, ENFORCING, AND ALLOWING TRANSFERABILITY OF RIGHTS

Because failure to recognize existing rights will impede tenure security, curb investment in land, increase the potential for conflict, and divert resources that can be more productively deployed elsewhere in the defence of property claims, the legal recognition of existing land rights is a key element of good land governance. Failure to identify clearly or define different types of land rights—including individual, secondary, usufruct, customary, or other types of group rights—will make land transactions more costly, thereby blocking the movement of land to more efficient uses, which is very important in times of far-reaching social and economic change (Baland et al., 2007), and possibly its use as collateral.

Within any given country or jurisdiction, different rights can coexist (legal pluralism) and evolve over time. This implies a need for flexibility in the types of rights that can be recognized and in the ways rights can be upgraded. In traditional systems, rights held by women, children, and vulnerable groups such as migrants or herders are often insufficiently protected, come under threat as land values increase, or are in danger of being appropriated by the better-off or the well informed. Special safeguards are thus needed to ensure that such rights are protected. It is important that registries of rights be administered transparently and cost-effectively and in ways that have local legitimacy. For example, communal rights are not inferior to individual ones providing the choice in favour of collective or communal ownership arrangements is made by users on the basis of careful and informed consideration of advantages and disadvantages of different arrangements, rules are clear, and the scope for elite capture is minimized. On the contrary, if internal structures of accountability exist and if arrangements can be revisited as circumstances change, the identification and registration of community boundaries can be a very cost-effective way to cover large areas within limited time and resource availability.

To allow for effective protection against competing claims, legal recognition needs to be backed by right holders' ability to identify land boundaries unambiguously and call on the powers of the state to defend their rights if challenged. If, with land becoming more valuable, increased frequency of transfers makes it more likely for competing claims to arise, registration to put rights and transfers on public record is often worth the effort, especially if low-cost locally accepted mechanisms are used. Although all rights along the continuum can be registered in principle, the fact that adjudication and registration of rights are costly will imply that in practice, only some types of rights will need to be formalized. Participatory land-use planning can help record secondary rights (such as those by herders) that extend across large areas in a way that is more cost-effective than individual registration. If rights are assigned to groups, then regulations will need to address how such groups set rules for themselves, interact with outsiders, and call on external agencies to adjudicate disputes.

Although putting rights to land on record will not miraculously transform an economy through a sudden emergence of credit markets, mechanisms to formalize rights in a way that respects existing arrangements (and the continuum of rights they imply) are often justified if land values and the frequency of transactions increase. However, if existing land rights are unclear or weak, using a sporadic or an on-demand approach for the first-time registration of rights will often carry a significant risk of land being concentrated in the hands of well-connected and powerful elites. Systematic registration can reduce these dangers by including ways to (i) inform all potential claimants about the processes and criteria used to decide between competing rights, (ii) require claimants to come forward, and (iii) adjudicate rights at one point in time. It can thus achieve a much lower unit cost than sporadic registration.

Although user groups or society at large can impose limits on the types of rights or ways in which rights can be exercised by individuals, such limits should be based on a careful assessment of the cost and benefit of different options, should aim to achieve effects (environmental, health, security, or other) at low cost, and should not disproportionately disadvantage certain groups of right holders. Restrictions on land use such as minimum plot sizes or building standards that are not affordable for large portions of right holders usually result in increased informality and reduced respect for the law. Despite their negative impact (e.g. through evasion and discretionary enforcement), removal of such laws may be opposed by vested interests.

3.2 PROVIDING EFFECTIVE AND LOW-COST ENFORCEMENT (INCLUDING DISPUTE RESOLUTION)

Public-sector functions related to land are normally performed by different institutions. Where local capacity is available, routine tasks should be decentralized. Unclear or overlapping mandates and functions increase transaction costs and can create opportunities for discretion that undermine good governance and can push users into informality. They can also create confusion or parallel structures that threaten the integrity and reliability of the documents and information provided by land-sector institutions, rendering policy implementation difficult. The legitimacy of land-sector institutions and the actions they perform depend on the extent to which the policy framework guiding institutional activities is backed by social consensus rather than by the perception of it being captured by special interest groups. Land policy is thus most appropriately developed in a participatory and transparent process that clearly articulates policy goals, identifies different institutional responsibilities, and includes an assessment of the resources needed for quick and effective implementation. It is important to define ways to measure progress towards achieving land policy goals and to assign clearly the responsibility for monitoring and publicizing progress towards meeting those goals—in ways that can be understood by those affected and that can feed into the policy dialogue. Responsibilities should be assigned on the basis of administrative information by relevant ministries or by an independent institution.

Inconsistencies between legal provisions and their enforcement or overlapping institutional mandates are a key source of weak land governance. The desire to facilitate cost-effective provision of land information through registries has traditionally been at the heart of titling and registration programmes. Although such programmes have often contributed to positive outcomes, they have also often failed to meet their objectives in terms of outreach, equity, and sustainability. Those instances point to a need to ensure that registries provide broad access to comprehensive, reliable, up-to-date information on land ownership and relevant encumbrances. A regular assessment of the extent to which programmes provide that information can be an important element in improving land governance.

Because potential investors cannot be sure whether any gaps in the data available from the land registry could be relevant to their interests, they will derive few benefits from registries that do not provide complete, geographically exhaustive, and reliable information. If unsure, they will need to check land ownership information on a case-by-case basis. The most extreme form of ensuring reliability of land registries is for the state to indemnify anybody for losses suffered from deficient information in the registry, an institutional characteristic that is often (but not always) associated with systems of title registration. However, what matters is not the legal guarantee but the quality of the underlying information and, provided good-quality information is available, the ability of the state to honour any promise of indemnification. The fact that some deeds systems include a promise of indemnification for errors suggests that the type of system is not the only variable to be considered.[8]

Even if documenting land rights and boundaries has clear benefits, the sustainability of land registries[9] and their ability to reach out to those with limited resources will depend on this being done in a low-cost manner. Registries that were too costly to operate or burdened with high stamp duties or informal fees often had difficulty in achieving full coverage or remaining current and instead reverted back to informality if users were unable or unwilling to pay. Ensuring efficient operations, fee schedules in line with perceived benefits, and transparent collection are preconditions for registries serving all sectors of society. Rather than trying to squeeze operating costs to unrealistically low levels—which could create governance challenges of its own, low-cost technology, especially for ground surveys, seems more appropriate.

While avoiding land conflicts may be impossible, it will be important they are handled consistently rather than on an ad hoc basis and that the institutions to resolve disputes and manage conflict are legitimate, locally accessible, effective, and empowered to resolve issues conclusively, subject to basic considerations of equity and avenues for appeal. Gaps in any of these areas can lead to 'forum shopping', whereby those with better knowledge or connections choose channels more likely to yield outcomes favourable to them, or simultaneously pursue conflicts in multiple fora. Festering disputes can

[8] With modern technology, deeds systems can also provide reliable information that allows checking for possible pre-existing claims and includes protection against malfeasance.

[9] Though systematic studies of the extent to which systematic titling campaigns are followed by informal transactions that would undermine what has been achieved, available studies suggest that the magnitudes involved can be large (Barnes and Griffith-Charles, 2007). More rigorous quantitative assessment of this issue would be highly desirable.

impose huge costs, not only on individuals, but also on society as a whole, e.g. by precluding much-needed investment from happening on land whose status is disputed. While customary institutions can effectively deal with internal conflicts in ethnically relatively homogeneous communities though often in ways that are biased against women), their ability to fairly resolve problems between communities or with 'migrants' is less clear.

3.3 IMPARTIALITY MANAGING PUBLIC LAND

Identification and recording of land rights are essential to provide management incentives. At the same time, rights come with responsibilities and obligations, and there is a clear social interest in having land used in a way that cost-effectively provides public goods and avoids negative externalities. Moreover, especially at low levels of development when other revenue sources are limited, land taxation can help support decentralization and encourage effective land use.

Public land ownership is justified if public goods (such as infrastructure or parks) are provided or if land is used by public bodies (such as schools, hospitals, defence, or state enterprises). How state land is managed, acquired, and divested often poses serious governance challenges. To minimize such risks, it is important that (i) state land with economic value be clearly identified on the ground; (ii) acquisition of state land through expropriation is implemented promptly and transparently with effective appeals mechanisms, and used only as a last resort to provide public goods if direct negotiation is not feasible; and (iii) divestiture of state land is done transparently, does not negatively affect existing local rights, and helps to generate (local) public revenue. Effective management of public land is impossible if there is no inventory of such land or if boundaries are unclear. Lack of a public land inventory provides opportunities for well-connected individuals to capture state assets through squatting, often with negative environmental impacts. Information about revenues from public lands and the costs incurred to manage public lands should be open for public scrutiny.

Although expropriation can be justified to prevent attempts by private owners trying to extract resources by unreasonably withholding their consent, the ability to use it beyond a very narrow set of circumstances raises the risk of public officials using their powers to promote private rather than public interests, which can encourage rent seeking and political meddling. Expropriation procedures should therefore be clear and transparent, with fair compensation in kind or cash at market values made available expeditiously. Rather than using expropriation, land transfers to private parties should be based on users' voluntary and informed agreement which normally can be gained if they are provided with a fair share of the proceeds. If expropriation is needed, independent valuation of land assets should thus be the norm, where possible, when market systems for valuation have been established. Those whose land rights are affected will need access to mechanisms for appeal that can provide authoritative rulings quickly and in an independent and objective way. Land rights under consideration should extend beyond landowners to tenants and dependents (that is, spouses and children) and should also include those affected by re-zoning and land-use changes (that is, farmers affected by expansion of urban boundaries). Maintaining a given standard of living for those affected should be a key compensation objective.

In many countries most of the land not formally registered by individuals is considered to be property of the state. With increased demand from outside investors, the way in which such land is handed out has become one of the most egregious forms of bad governance, outright corruption (for example, bribery of government officials to obtain public land at a fraction of market value), and squandering of public wealth, often resulting in local resentment and long-term conflict. Avoiding such outcomes will require clear, transparent, and competitive processes that are based on recognition of existing occupancy rights and negotiation with current land users, publicizing contract terms (including payments at different levels) and payment collection, and conducting regular and independent audits.

While land-use planning is justified to allow effective provision of public goods consistent with available resources, land-use plans and regulations should be designed to utilize unused potential from land. It should be done in a participatory fashion, (i) to cope with future land demands, (ii) avoid unrealistic standards that would force large parts of the population into non-compliance and thus leave the formal system, and (iii) consider the affordability, the resources needed for enforcement, and the availability of mechanisms to merge new with existing regulations. Changes in zoning or infrastructure development should provide benefits to society at large, rather than to specific groups, they should be transparent, and be implemented in a way that holds decision-makers accountable. The way in which land for urban growth is acquired often deprives rural dwellers of their rights and may provide insiders with large rents. It should therefore be transparent, with broad participation, build on market forces, and be combined with measures (such as capital gains taxes or betterment levies) that would allow the public to capture a significant part of the generated surplus, and compensation paid if there are land-use restrictions. Frequently the same government institution or individuals impose land-use plans, hear appeals, and even act as ultimate landowners, creating conflicts of interest that should be avoided. As slow and opaque processes can hinder investment and economic development by imposing uncertainty and costs, routine requests for building and development permits should be handled promptly and predictably.

The ability to raise revenue and decide on desired levels of service provision at the local level are key features of effective decentralization, and taxes on land or property are among the best sources of self-sustaining local revenues if (i) local governments are allowed to retain a large part of the property tax revenue they collect, (ii) they are provided with the technical means (for example, cadastres) to do so, and (iii) clear principles are established for evaluating and updating valuation rolls to avoid arbitrariness. In addition, land taxes are very effective in curbing speculative acquisition of land and rent seeking often associated with such transfers.

4. **Promising approaches to enhancing smallholders' rights**

To illustrate the broader context for land governance, as well as the potential for different entry points, modalities, and the impact of improving land governance at country level, we draw on examples for: (i) registration of communal as well as individual rights; (ii) improving the functioning and effectiveness of land registries; and (iii) empowering women through legal change and specific interventions. Each of these examples also highlights complementarities between different types of interventions and the longer-term focus of land reforms.

4.1 REGISTRATION AND MAPPING OF COMMUNAL LAND

In land-abundant settings (e.g. Mozambique, Tanzania, Zambia), delimiting community boundaries but leaving internal management to communities can be more relevant and cost-effective than individual titling. It allows large areas to be covered quickly and a focus on institutional aspects, e.g. formulation of bylaws to govern ways in which communities interact with outsiders or how conflicts are resolved (Knight et al., 2012); these have been found to be more critical determinants of perceived tenure security and the ability to effect efficiency-enhancing land transfers. If authoritative information on community rights and representatives is available in a public registry, the transaction costs for outside investors in search of land can be reduced and the transparency of transactions greatly enhanced. If combined with legal measures, this can help reduce the neglect of existing rights that is often noted to have accompanied recent processes of land-related direct investment (Alden-Wily, 2010; Anseeuw et al., 2012). Recognition of community rights and demarcation of relevant boundaries will be a precondition for establishment of mechanisms to compensate land users for positive externalities and global public goods that emerge from their land-use decisions (e.g. leaving land forested).

The fact that many African countries made provisions to recognize customary tenure and communal land creates the legal basis for delimiting community boundaries while leaving registration and management of individual plots to community institutions, with the possibility of making the transition to more formal or individualized systems as the need arises.[10] Mexico's *ejido* reform, implemented in the late 1990s, illustrates the potential of such intervention, but also the need for institutional support in three areas. First, dissemination and assistance helped achieve close to full participation in a voluntary process, setting the stage for cost-effective implementation and achieving the network effects it entails (e.g. availability of registry information). Second, a critical precondition for mapping to be possible and meaningful was setting up internal community institutions to manage resources transparently and in a way that flexibly responds to local needs and can resolve conflicts. Third, establishment of a registry to provide public access to updated spatial and textual information was critical.

Ejidos are communal settlements of land reform beneficiaries who had received non-transferable use rights in a process that started in 1917 and reached its peak in the 1930s. With *ejidos* covering about half of Mexico's agricultural land, 70 per cent of its forests, and two-thirds of the land needed for urban expansion, there were three main motivations for reform. First, without land markets (and inheritance to only one heir), many old *ejidatarios* were unable to use their agricultural land effectively while the young could not access land. Second, powerful members were able to appropriate large common property resources and forests for use as a source of patronage, hurting the poor and indigenous people. Finally, urban expansion was based on informality, land invasions, and corruption rather than proper planning, leading to conflict and a high cost of service provision.

Reforms comprised legal changes, institutional reform, and a systematic programme of land regularization (*PROCEDE*).[11] They aimed to empower communities to choose the property rights regime most suitable to their needs (i.e. either communal, individual, or mixed), increase transparency, tenure security and investment through issuance of land ownership certificates, facilitate land transfers, reduce the cost of verifying land rights, and deal with a backlog of long-standing conflicts. To strengthen self-governance, the law recognized *ejidos*' legal personality and mandated creation of three bodies (the assembly, a vigilance committee and a secretariat) to ensure separation of

[10] For a discussion of institutional options see Fitzpatrick, 2005, and for a detailed example of legislative arrangements to put this into practice see Government of Mexico, 2000.

[11] PROCEDE is the acronym for ***Pro****grama Nacional de* ***Ce****rtificacion de* ***De****rechos* ***E****jidales y Solares Urbanos.*

powers and internal checks and balances. The assembly audits other *ejido* organs and can, with a majority of at least 75 per cent of all members, approve contracts with outsiders or decide to individualize land holdings.

Three public institutions were created to support the process. First, 42 tribunals plus an appeals tribunal were tasked to resolve the immense backlog of land conflicts in ways that encouraged settling out of court and access by the poor. During 1992–9, they dispensed with about 350,000 conflicts. Second, a registry (*Registro Agrario Nacional*; RAN) with delegations in each state was created to document rights to individual and collective land and identify individuals empowered to make decisions on behalf of the *ejido*. Finally, the office of the ombudsman (*Procuraduria Agraria*; *PA*), with local presence in each state, was created to provide para-legal assistance and oversight of the regularization process to prevent elite capture and help small farmers assert their rights.

These elements fed into a programme of voluntary land regularization which, in a 12–18-month process, clarified property rights for individual *ejidos* in a number of steps. A first step of boundary assessment, demarcation, and conflict resolution entailed an independent review of legal documents (i.e. the documents establishing the *ejido*, any modifications in area or membership) to identify ambiguities or conflicts that required prior resolution. Once external boundaries were determined, the PA launched a dissemination campaign to explain the programme, culminating in a formal decision on programme participation by the assembly (with a 50 per cent quorum). A positive decision led to formation of a committee responsible for the next steps: identification of boundaries of different types of land (urban plots, parcelled and common lands); preparation of sketch maps; and establishment of a complete inventory of rights once pending conflicts had been resolved or registered separately. These materials were presented to a second assembly for formal approval to clear the way for formal demarcation of lands and public display of the results for at least two weeks before a third assembly; with a quorum of at least 75 per cent of all community members, this approved the outcome in the presence of a public notary and a representative of the PA who prepared minutes and certified that proper procedures were adhered to. This allowed results to be published in the registry and certificates to be issued.

The quantitative accomplishments achieved by *PROCEDE* have been impressive. In less than a decade, close to 100 million hectares (larger than Spain and France together) were measured and mapped, and 2.9 million households received certificates to individual, common, and housing land. The programme increased land access for about 1 million households which previously had no rights. It also improved governance; over eighteen thousand *ejidos* formalized internal by-laws through assembly representatives, and the well-regarded process helped improve household welfare by increasing participation in the non-agricultural economy. Contrary to initial fears, it did not cause a wave of land sales, but instead provided a basis for numerous contracts and joint ventures. It prompted an increase in migration, which increased productivity by allowing more able farmers in high-potential areas to expand cultivated area (de Janvry et al., 2012) and welfare through remittances from international migration (Valsecchi, 2010).[12] Although technology constraints at the time the programme was implemented implied that costs were high by current standards, net benefits were positive (World Bank, 2002). With current technology, implementation cost could be significantly reduced.

4.2 REGISTRATION OF INDIVIDUAL RIGHTS

As the most densely populated country in Africa, Rwanda in the late 1990s recognized the importance of clarifying land issues to overcome a history of land-related conflict and tribal division, end gender discrimination in access to assets, and provide a framework to encourage optimum use of available land resources. While customary land-tenure systems had traditionally provided high levels of tenure security, outdated processes and misuse, e.g. by officials invoking the state's power to expropriate land (eminent domain) for private benefits, had weakened the system's ability to cope with far-reaching social, economic, and political changes. Competition for land in a setting where non-agricultural income opportunities had expanded only slowly led to illegal land sales and 'land grabbing', which exacerbated inequality, landlessness, dispute, and social tensions. Land-related conflicts contributed to the 1994 genocide (Andre and Platteau, 1998). To rise to the challenge, the country embarked on a strategy of legal and institutional reform and eventually nation-wide implementation of a land-tenure regularization programme.

First, a number of far-reaching legal and institutional changes were embarked upon. As is common in customary systems, women in Rwanda had land-use rights only through their husbands whose lineage controlled the land, implying that their right to own or inherit land was severely compromised. In fact, widows were unable to inherit their deceased husband's property and at most were allowed to use it until male children grew up. Those without children lost even use rights to family land unless they maintained family ties by marrying one of their husband's brothers (Republic of Rwanda, 2004). The 1999 inheritance law changed this by advancing in three areas, namely: (i) granting daughters and sons equal rights to inherit their parents' property; (ii) protecting women's property rights under legally registered marriages

[12] The political impact from such a reform is more complex (de Janvry et al., 2011).

subject to the provisions of the family law; and (iii) requiring spousal consent for transactions (e.g. sale, mortgage, or exchange) of matrimonial property by any of the partners.

This was followed by the 2004 Land Policy that put forward principles for efficient and sustainable use of land, to be detailed in the 2005 Organic Land Law (OLL). The OLL vests land ownership with the state, but provides landholders with long-term usufruct rights (up to 99 years, depending on land use) that can be sold, passed on to heirs, mortgaged, leased, or otherwise transferred. It also defines procedures for expropriation (with compensation) in the public interest, and makes first-time registration and recording of any follow-up land transfers compulsory.

To allow implementation, the institutional infrastructure is critical. Land commissions were formed at national and district levels to oversee OLL implementation. The National Land Center (NLC), which includes the registrar's office, was established as a technical agency in charge of activities related to land administration, land-use planning, management, and OLL implementation. At district, municipal, and town level, District Land Bureaux (DLBs) were established, complemented by land committees at sector and cell levels to serve as focal points for land registration and land-use planning, and thus facilitate a decentralized and participatory implementation process.

To clarify existing land rights on all of the country's estimated 10–11 million land parcels, the NLC developed and is now implementing a land tenure regularization programme (Sagashya and English, 2010). Given the lack of successful models that could be drawn upon and the heterogeneity of situations across different parts of the country, it was decided to develop and fine-tune the methodology through a pilot,[13] implemented in 2007/8 in four cells chosen to reflect typical situations in the country, before embarking on a national roll-out. Operationally the pilot declares an area subject to adjudication and conducts stakeholder sensitization programmes, clearing the way for locally trained para-surveyors to conduct a field-based adjudication process. This consists of identifying parcel boundaries in the presence of landowners and adjoining neighbours, marking them on an aerial photo to create a graphical record, and, for undisputed parcels,[14] issuing a claim receipt

[13] This pilot resulted in demarcation and adjudication of 14,908 parcels with a total area of 3,448 hectares, owned by 3,513 households.

[14] Parcels where ownership claims are disputed are recorded in a separate dispute register. Disputed claims can then be pursued separately through either administrative or judicial channels with the possibility of civil society playing an important role in moving the process along. While it is hoped that adjudication and registration of land rights will help to increase activity in land markets and consolidate plots, no efforts to this end are included in the adjudication process. Similarly, as the goal was to establish a model for a nationwide roll-out in campaign style, areas covered by the pilot land adjudication did not receive preferential access to other services such as extension (Ali et al., 2011b).

signed by neighbours. Information from this receipt, in particular the names of all persons, including women and minors, with claim to the property, is transferred to a registry book, digitized, and displayed publicly. If no objections are raised within a two-week public display period, the information is formally registered, creating the precondition for award of a formal certificate upon payment of a nominal fee.[15]

An evaluation of the pilots using a survey conducted in 2009, i.e. not too long after pilot implementation, points towards significant benefits from land-tenure regularization (Ali et al., 2011b). First, those whose parcels were registered, in particular female-headed households, were more likely to invest in soil conservation (bunds, terraces, check dams) on their land. For affected households, the likelihood of such investment is more than double the propensity to invest in the control group (10 per cent) and the estimated increase in the propensity to invest is 19 percentage points for female-headed households, supporting the notion that low levels of female tenure security before the programme had precluded investment by this group. Clarification and documentation of rights reduced uncertainty over who would inherit land, with substantial benefits for female children who might otherwise have been discriminated against. Finally, legally married women were significantly more likely to have their informal ownership rights documented and secured after registration. But women who were not legally married saw diminished property rights, in accordance with the law, suggesting that rights by women who are not legally married may need special protection, an issue that has in the meantime been taken up by the Government. In addition, it suggests that areas where policy is unclear, ambiguous, or at variance with practice on the ground, will need to be addressed; these include high fees for registration of subsequent transactions, unclear provisions regarding subdivision of land, and institutional arrangements that can guarantee updating and sustainability at low cost.

The case of Rwanda illustrates a number of lessons. First, a clear policy framework and extensive piloting were essential to create the preconditions for a massive roll-out. As documented, some five years were spent developing the policy framework and two years refining the process in small pilots (15,000 plots in total). Once processes were worked out, nearly 10 million parcels were demarcated in about two years. Second, a side effect of doing a rigorous impact evaluation is that, if properly designed and integrated into the programme, it can provide information that will be critical to adjustments of specific elements during implementation (e.g. to strengthen women's

[15] The fee of RWF 1000 (US$1.84 at the 2008 rate) per parcel or about RWF 4000 (US$7.36) per household taking the national average of four parcels per household. This compares to a cost of approximately US$9–11 in the pilot—since then reduced to about US$5 per parcel.

rights). Finally, while the programme achieved impressive results in terms of demarcation, operation of registries at local level (in a context of decentralized governance) remains a challenge that needs to be resolved quickly to ensure sustainability.

In Ethiopia, an earlier effort to record rights shows that: (i) if accompanied by dissemination and outreach, low-cost mechanisms based on elected local institutions offer considerable potential to reduce conflict, increase investment, and empower women without suffering from elite capture; (ii) even with a low-tech approach and if market transactions are limited, comprehensive records as well as ways of updating of records are critical; and (iii) policy issues such as limits on market transactions and unclear scope for expropriation compensation may limit further benefits.

Based on results from issuing land-use certificates to about 630,000 households in Tigray in 1998–9, other Ethiopian regions embarked on a large-scale certification effort, issuing land-use certificates to more than 6 million households (18 million plots) in 2003–5. The process starts with local awareness creation, sometimes with the distribution of written material, followed by elections of land-use committees in each village. After receiving some training, these committees, if needed with elders' assistance, systematically resolve existing conflicts, referring cases that cannot be settled amicably to the courts. This is followed by demarcation and surveys of undisputed plots in the presence of neighbours with subsequent issuing of land-use certificates which, for married couples, include names and pictures of both spouses but no sketch map or corner co-ordinates.

Assessment based on a nationwide survey suggests that decentralized implementation over a time period long enough to sort out conflicts allowed the programme to adapt to local conditions while still making rapid progress overall. Implementation was not biased against the poor; procedures were adhered to and recipients appreciate the certificates received (Deininger et al., 2008). While there is scope for improving the process, including coverage of common property resources (CPRs), clarity of policy guidelines, providing access to legal and policy information in written form, and ensuring better participation by women and, in some regions, their inclusion on certificates, the overall record of first-time registration is impressive. This is particularly notable given that the programme has been implemented at a cost well below what is reported anywhere else in the literature. While this allowed for a very low-cost process with an estimated cost of about US$1 per plot, addition of a map to be marked on an aerial photo would be possible with a modest increase of costs, and pilots to test this are currently underway (Holden et al., 2009).

A more detailed impact assessment in Ethiopia's Amhara region finds that the programme significantly increased tenure security, investment, and renting out by landlords with a rough estimate of programme-induced benefits (based on the net increase in productivity due to investment)[16] pointing towards a favourable cost-benefit ratio (Deininger et al., 2011a). At the same time, although certification has a positive effect, some policies, in particular the threat of uncompensated expropriation for urban expansion, leave considerable tenure insecurity. To realise the full potential of first-time registration, action will be needed to (i) put in place arrangements for low-cost updating and publicity of registry information; (ii) consistent inclusion of common property resources (CPRs), possibly in combination with land-use planning and assignment of group rights, to arrest threats of encroachment and resource degradation; (iii) exploring options for systematic addition of a graphical record; and (iv) policy measures to compensate in case of expropriation, protect contracts and provide security against arbitrary redistribution, and allow land-use rights to be transferred for longer time periods (World Bank, 2011).

4.3 MAKING LAND INFORMATION MORE WIDELY AVAILABLE AND IMPROVING INSTITUTIONAL EFFICIENCY

While securing rights can significantly increase investment incentives, the cost of registering property or accessing reliable information will affect the extent of market transfers and the scope of using land as collateral for loans. High costs of registering land transactions either drive them into informality or leads to land lying idle, as in the case of Albania (Deininger et al., 2012a). If accessing information on land ownership is too costly (or registry information is judged incomplete or unreliable), banks may be forced to use more costly ways of securing loans, with a possible decline in credit supply.

There are three main elements of such transaction costs, namely (i) the operational cost of the registry, possibly including a need to involve lawyers in transactions, (ii) the cost of surveys, determined by the degree of precision required, and (iii) stamp duties levied on land transfers. Taken together, these costs can be sizeable. In the 183 countries included in the World Bank's 2012 'Doing Business' study, the mean official fees to transfer a property is 5.9 per cent of the property value, with the process taking an average of 55 days. The fees vary widely between countries: in 40 of them it is 2 per cent or less of property value, in 94 it amounts to 5 per cent, but in 39 it is 10 per cent or more. Informal fees and the need to use brokers can increase these costs, affecting not only users' ability to access information but also confidence in

[16] Given the policy-induced limits to the functioning of land, labour, and credit markets in Ethiopia, further productivity-impacts of land certification through these channels will be small at best. To obtain a conservative estimate, we focus on investment-induced productivity effects only.

the land registry.[17] The most viable short-term options to reduce costs are improving operational efficiency of registries or reducing stamp duties.

An example for the former is Georgia, a country that in the early 2000s established a single national land administration agency, made all information publicly available on the Internet, established free legal consultation service, put licensed private surveyors (rather than bureaucrats) in charge of conducting surveys, drastically cut staff from 2,100 to 600, and increased salaries eightfold. To keep the registry financially independent, the registry law was revised and the fee structure adjusted. This provided the basis for a decrease in the cost of property registration from 2.4 per cent to 0.6 per cent of property values, and a reduction in the time required from 39 to 9 days. Land market and mortgage activity increased significantly (Dabrundashvili, 2006).

The Indian state of Karnataka illustrates the potential for using IT to reduce informal and eventually also formal fees. Over less than two years, it computerized land registration under a public–private partnership, a step that is estimated to have saved users $16 million in bribes (Lobo and Balakrishnan, 2002), largely by eliminating discretion in land valuation. The automated and predictable process greatly improved demand for formal registration, allowing cuts to stamp duty from 14 to 8 per cent (which is still very high), while quadrupling tax revenue from $120 to $480 million (World Bank, 2007).

In most contexts, changes in record management are adopted jointly with improvements in the quality of the underlying information, making it difficult to identify the impacts of each. Computerization of records in India is an exception; it does not alter the nature or quality of underlying information, but reduces the cost of keeping registries up to date and, especially if it includes online access, makes third-party use of the data easier, allowing banks to ascertain ownership status and existence of pre-existing liens on property that might be offered as collateral reduced banks' cost of extending credit, and allowing credit supply and the volume of registered mortgages to expand. In the state of Andhra Pradesh, variation in the date of shifting to fully computerized operations across the state's 387 sub registry offices (SROs) between 1999 and 2005 provides a test of this hypothesis (Deininger and Goyal, 2012). Credit access is measured by the total volume of credit disbursed by all scheduled commercial banks, based on banks' mandatory reports to the Reserve Bank of India on a quarterly basis for the 1995 to 2007 period.[18] Data on the volume of registered land transactions by SRO allow us to test whether, as credit effects would lead one to expect, computerization affected numbers of mortgages but not other types of land transactions.

[17] In India, a recent study estimated that bribes paid annually on land administration amount to $700 million (Transparency International India, 2005).

[18] To allay concerns that pre-existing differences in the growth of credit could drive results, we test whether the timing of computerization is related to this variable (Deininger and Goyal, 2012).

Results point towards significant credit impacts in urban but not in rural areas, consistent with underlying institutional parameters,[19] with a 10.5 per cent increase in credit volume on average. The effect increased over time, from some 3–5 per cent in the two years after computerization to 15 per cent in the long run. This is in line with data on the volume of registered land transactions, implying that in urban SROs the number of registered mortgages increased by 18 per cent in the short term and by 32 per cent in the long term, suggesting that the intervention helped expand coverage. Computerization had no effect on the volume of registered land sales or that of non-monetary land transfers such as gifts, inheritances, and leases, supporting the notion of reduced transaction cost for institutional credit providers in urban areas being a main benefit from computerization. Although institutional constraints limited the size of effects in this case, if record coverage is good, innovative ways of facilitating access to them offer scope to reduce transaction costs and, over time, also improve the quality of information.

4.4 GENDER EQUITY

Given the importance of land assets and the far-reaching impacts of asset access for intra-household decision-making, there were concerns that land regularization efforts might weaken women's rights (Deere and Doss, 2006) and seriously reduce the benefits from such programmes. Indeed, the share of females whose land rights are formally recognized remains low. For example, in Tanzania, only 5 per cent of residential licences (RLs) or short-term intermediate (and non-transferable) documents certifying occupancy to urban properties carry a woman's name either as a sole or a co-owner.

To explore demand (and willingness to pay) for formal titles and reasons for such gender imbalances, a random experiment with two interventions was conducted in informal settlements in Dar es Salaam (Ali et al., 2011a). A cadastral survey for 1,000 randomly selected plots in two neighbourhoods (one with and one without infrastructure upgrading) was conducted to be able to issue a Certificate of Right of Occupancy (CRO), the most secure form of document available.[20] Residents were offered the opportunity to acquire

[19] Two factors explain this result. First, in urban areas, land registries are the only available form of documentary evidence on land ownership. By contrast, land records maintained at village level are more accessible than land registries. The fact that these records were not computerized would limit potential impacts of registry computerization even under ideal circumstances. Second, the limited liquidity of rural land markets due to restrictions on the transferability of land, as well as mandated levels of lending to the rural sector and repeated episodes of credit forgiveness, would limit the scope for credit effects in rural areas.

[20] Although formally the CRO is a lease for 33 to 99 years, it is fully transferable and routinely accepted as collateral by banks, thus being equivalent to a title in other settings. We therefore use CRO and title interchangeably.

the CRO at a subsidized price that was randomly varied between TSh 20,000 and TSh 100,000 to trace out the demand curve. Moreover, while all participants were exposed to training on the importance of having females as co-owners of land, a randomly selected group was offered an additional discount if they were willing to list a female as a (co)-owner.

Results are interesting in two respects. First, with a median willingness to pay about TSh 50,000 (US$36), demand for CROs among poor informal households is surprisingly large, both in light of their overall income and compared to the cost of alternatives.[21] Second, both awareness-raising and a relatively minor discount were effective in fostering female co-ownership: in the baseline survey, only 13 per cent considered women to be default landowners and 25 per cent expressed an intention to include a woman on a formal title. After awareness-raising, over 60 per cent indicated their intention to include a woman on the title and over 90 per cent of those who received the modest conditional subsidy chose to include a woman as a co-owner.[22] While the impact in terms of female empowerment remains to be explored, this suggests that programme design can have significant impact on gender outcomes.

The extent to which any female empowerment effects from land registration can be sustained will depend on how inheritance is distributed between sons and daughters, a variable affected by legal provisions as well as social norms.[23] While interactions among these remain poorly understood, impacts of legal change on gendered access to assets and productivity provide an opportunity to understand better these issues, as the case of India illustrates. Although India's constitution requires equality before the law, inheritance was biased against females as practices followed the 1956 Hindu Succession Act (HSA), which limited inheritance of joint property, in particular land, to males. In the late 1980s and early 1990s, a number of states amended this Act (Andhra Pradesh in 1989, Tamil Nadu in 1989, and Maharashtra and Karnataka in 1994) to make women's status equal to that of men. Information on changes in actual inheritance for three generations (parents, siblings, and children of the household head) in a nationally representative survey of more than eight thousand rural households allows us to assess the impact of this

[21] In the study area, mean per capita monthly income is TSh 80,000 (US$67), but sample households are poor. RLs could be acquired for between TSh 4000 and 6000. Obtaining a CRO presently costs about US$400–500; systematic issuance could reduce costs to below US$100.

[22] Households offered a discount on a title conditional on including a woman were no less likely to purchase a land title than those offered unconditional discounts, allaying fears of excluding households where women may benefit the most from titling.

[23] Recall that in the case of Rwanda discussed earlier, female-headed households were significantly more likely to transfer their land to sons rather than daughters, despite legal provisions to the contrary (Ali et al., 2011b).

legal change on women's asset ownership and socio-economic outcomes (Deininger et al., 2012b).[24]

Three results are of interest. First, the reform led to an increase in the likelihood of females inheriting land at the time of the reform and a continued growth in this variable thereafter. It also had a positive impact on the total value of assets women received (i.e. there was no full substitution), the share of household land they received (i.e. they did not just receive token amounts), and their level of land ownership at the time of the survey (i.e. effects persisted). Second, beyond effects on assets, they note an increase in the age of marriage among other variables, suggesting that eligibility for asset transfers affected women's status more broadly. Finally, girls but not boys whose education decisions were made after the legal change had significantly higher levels of primary education (by 0.37 years) than those born earlier. Still, the fact that, almost two decades after its passage, the Act had not fully equalized inheritance between males and females suggests that ways to increase awareness of rights under the new inheritance regime could be an effective way of empowering women, especially in view of the Amendment's nationwide expansion in 2005.

5. Conclusions and next steps

This points towards better appreciation of the scope for securing smallholders' land rights, and the far-reaching and multi-faceted impacts that can arise as a result of doing so. There is also a growing body of good practice that can be drawn upon to inform such policies. To ensure broad and sustained impact, general principles need to be translated into country-level reality, put in the context of a longer-term vision, and linked to ways to track progress towards longer-term goals. We conclude the chapter with thoughts on how recent agreements on the importance of land governance could be used to: (i) assess the status of land governance across countries in a comparable manner and build a constituency for reform; (ii) drawing on global good practice, identify the most immediate steps to improve governance in the land sector in ways that can benefit the poor and evaluate the impact of doing so; and (iii) track progress in key outcome variables over time in a way that allows input by all stakeholders.

[24] We assess impacts in two ways, namely (i) within-household differences between males and females for households in reform states under the original compared to the new legal regime, and (ii) outcomes for females only between reform and non-reform states. Gender and age as well as household fixed effects, together with other controls and tests for non-Hindus whose inheritance was not affected by the reforms, are used to check the robustness of results.

In situations where land issues are dealt with in a fragmented fashion or if there is little consensus on priority actions, a broader assessment of a country's position in terms of good land governance can create support for follow-up action. To help with this, the land governance assessment framework (Deininger et al., 2011c) draws on good practice for 21 indicators in 5 key areas (legal and institutional framework for recognition of rights; land-use planning, management, and taxation; management of public land; access to land information; and dispute resolution and conflict management).

Implementation is overseen by a country coordinator with extensive knowledge of the sector who is well respected and impartial. Following a preparation phase,[25] a tenure typology and institutional map as well as background reports and panel briefs are prepared. About five to eight panels consisting of three to six subject-matter experts and users are then constituted. Their aim is to arrive at a consensus ranking of the country's status with respect to each of the dimensions based on statements derived from global practice with implementation. Areas where the country is lagging are then translated into agreed policy priorities. Discussions are recorded and material from all panels is synthesized to form a country report that can be shared, reviewed by outside experts, and validated in a national workshop and dialogue with policymakers to define priority policy actions, timeframes, and monitoring indicators. Use of a consistent framework across countries removes arbitrariness and provides a basis for learning and transfer of good practice across countries.

Applications to date identified issues in three areas. First, legal and institutional frameworks may have gaps, lack clarity, or be duplicative. Issues relating to institutional capacity and efficiency of operation (e.g. fee structures, staffing, operational procedures) are easier to address in the short to medium term, but may require more specific policy action. Finally, in many of the settings, cost-effective processes to adjudicate land rights were deemed necessary and a need for pilot schemes was identified. Carefully assessing demand for this by different segments in the population (including women), selecting pilot sites to reflect situations likely to be encountered in practice, elaboration and documentation of specific steps, and rigorous evaluation of longer-term impacts can provide the basis for subsequent scaling-up. Some of the impact evaluations discussed highlight the scope for exploring design options, examining how demand responds to exogenous variations, and deriving parameters that sustainable programmes need to satisfy within

[25] The specific activities to be performed in this phase include: (i) review of the LGAF implementation manual to identify any areas where customization to country conditions may be needed, as well as potential data availability; (ii) identification of the team, in particular expert investigators and panel members from a wide range of sectors, and formulation of a time plan; and (iii) government buy-in as evidenced by the appointment of a person to liaise with relevant ministries and departments, and agreement to make specific data available to the team.

a given institutional context, thus providing a feedback loop to link different strands.

In addition to providing actionable follow-up at country level, linking to data that should be generated by the system on a routine basis[26] can facilitate tracking of progress at higher levels (regional or global). If used to implement global initiatives, such as the Voluntary Guidelines on Tenure of Land and Natural Resources or the African Land Policy Initiative, this could allow progress in securing smallholders' land rights to be tracked. It will also monitor the effectiveness with which these rights are exercised to set in motion a process that can help improve land governance in response to changing requirements and to the benefit of the broader rural economy.

REFERENCES

African Union (2009). *Land policy in Africa: A Framework to Strengthen Land Rights, Enhance Productivity and Secure Livelihoods.* Addis Ababa: African Union and Economic Commission for Africa.

Alden-Wily, L. (2010). 'Whose land are you giving away, Mr. President?', paper presented at the Annual Bank Conference on Land Policy and Administration, Washington DC, 26 and 27 April 2010.

Ali, D. A., K. Deininger, D. Stefan, J. Sandfur, and A. Zeitlin (2011a). 'Are poor slum-dwellers willing to pay for formal land title? Evidence from Dar es Salaam', World Bank Policy Research Paper. Washington, DC, World Bank.

Ali, D. A., K. Deininger, and M. Goldstein (2011b). 'Environmental and gender impacts of land tenure regularization in Africa: Pilot evidence from Rwanda', World Bank Policy Research Paper 5765. Washington, DC, World Bank.

Andolfatto, D. (2002). 'A theory of inalienable property rights', *Journal of Political Economy*, 110 (2), pp. 382–93.

Andre, C. and J. P. Platteau (1998). 'Land relations under unbearable stress: Rwanda Caught in the Malthusian Trap', *Journal of Economic Behavior and Organization*, 34(1), pp. 1–47.

Anseeuw, W., L. Alden Wily, L. Cotula, and M. Taylor (2012). *Land Rights and the Rush for Land. Findings of the Global Commercial Pressures on Land Research Project.* Rome: International Land Coalition.

Arezki, R., K. Deininger, and H. Selod (2011). 'What drives the global land rush?' IMF Working Paper WP/11/251. Washington DC: International Monetary Fund.

Atwood, D. A. (1990). 'Land Registration in Africa: The impact on agricultural production', *World Development*, 18(5), pp. 659–71.

[26] Good candidates are (i) the number of registered land transfers of different types (sales, mortgages, gifts, inheritances, etc.); (ii) the share of land transactions or parcels (by value or size) registered in women's names; (iii) total receipts of land tax revenue; (iv) the breakdown of privately/community owned or state land identifiable on maps; (v) numbers and value of expropriation cases; (vi) numbers of land-related conflicts in the courts. With modifications of existing systems, these could be generated routinely at national and regional levels in many countries and compiled by a multilateral institution such as FAO or IFAD.

Baland, J. M. and P. Francois (2005). 'Commons as insurance and the welfare impact of privatization', *Journal of Public Economics*, 89(2–3), pp. 211–31.

Baland, J.-M., F. Gaspart, J. P. Platteau, and F. Place (2007). 'The distributive impact of land markets in Uganda', *Economic Development and Cultural Change*, 55(2), pp. 283–311.

Bandiera, O. (2007). 'Land tenure, investment incentives, and the choice of techniques: Evidence from Nicaragua', *World Bank Economic Review*, 21(3), pp. 487–508.

Banerjee, A. and L. Iyer (2005). 'History, institutions, and economic performance: The legacy of colonial land tenure systems in India', *American Economic Review*, 95(4), pp. 1190–213.

Barnes, G. and C. Griffith-Charles (2007). 'Assessing the formal land market and deformalization of property in St. Lucia', *Land Use Policy*, 24(2), pp. 494–501.

Behrman, J. R. (1990). 'Intrahousehold allocation of nutrients and gender effects: A survey of structural and reduced form estimates', in Osmani, S. R. et al. (eds), *Nutrition and poverty*. Oxford: Oxford University Press.

Benjaminsen, T. A. and E. Sjaastad (2002). 'Race for the prize: land transactions and rent appropriation in the Malian cotton zone', *European Journal of Development Research*, 14(2), pp. 129–52.

Berry, S. (2009). 'Property, authority and citizenship: Land claims, politics and the dynamics of social division in West Africa', *Development and Change*, 40(1), pp. 23–45.

Besley, T. (1995). 'Property rights and investment incentives: Theory and evidence from Ghana', *Journal of Political Economy*, 103(5), pp. 903–37.

Besley, T. and M. Ghatak (2010). 'Property rights and economic development', in Rosenzweig, M. R. and D. Rodrik (eds), *Handbook of Economic Development Vol 5*. Oxford and Amsterdam: Elsevier.

Binswanger, H. P., K. Deininger, and G. Feder (1995). 'Power, distortions, revolt and reform in agricultural land relations', *Handbook of development economics*, 3B, pp. 2659–772.

Boone, C. (2007). 'Property and constitutional order: Land tenure reform and the future of the African state', *African Affairs*, 106(425), pp. 557–86.

Burns, T. A. (2007). 'Land Administration: Indicators of success and future challenges', Agriculture & Rural Development Department, Washington, DC: World Bank.

Carter, M. R. and P. Olinto (2003). 'Getting institutions "Right" for Whom? Credit constraints and the impact of property rights on the quantity and composition of investment', *American Journal of Agricultural Economics*, 85(1), pp. 173–86.

Colin, J. P. and M. Ayouz (2006). 'The development of a land market? Insights from Côte d'Ivoire', *Land Economics*, 82(3), pp. 404–23.

Dabrundashvili, T. (2006). 'Rights registration system reform in Georgia', Paper presented at the Expert Meeting on Good Governance in Land Tenure and Administration. Rome: FAO.

Dale, P. F. and J. Mclaughlin (2000). *Land Administration (Spatial Information Systems)*. Oxford and New York: Oxford University Press.

de Janvry, A., K. Emerick, M. Gonzalez-Navarro, and E. Sadoulet (2012). 'Certified to migrate: Property rights and migration in rural Mexico', Working Paper. Berkeley, CA: University of California.

de Janvry, A., M. Gonzalez-Navarro, and E. Sadoulet (2011). 'Why are land reforms granting complete property rights politically risky?' Working Paper. Berkeley, CA: University of California.

de Soto, H. (2000). *The Mystery of Capital: Why Capitalism Triumphs in the West and Fails Everywhere else*. New York: Basic Books.

Deere, C. D. and C. R. Doss (2006). 'The gender asset gap: What do we know and why does it matter?' *Feminist Economics*, 12(1–2), pp. 1–50.

Deininger, K., D. A. Ali, S. Holden, and J. Zevenbergen (2008). 'Rural land certification in Ethiopia: Process, initial impact, and implications for other African countries', *World Development*, 36(10), pp. 1786–812.

Deininger, K., D. A. Ali, and T. Alemu (2011a). 'Impacts of land certification on tenure security, investment, and land market participation: Evidence from Ethiopia', *Land Economics*, 87(2), pp. 312–34.

Deininger, K. and D. Byerlee (2012). 'The rise of large farms in land abundant countries: Do they have a future?' *World Development*, 40(4), pp. 701–14.

Deininger, K., D. Byerlee, J. Lindsay, A. Norton, H. Selod, and M. Stickler (2011b). *Rising global interest in farmland: Can it yield sustainable and equitable benefits?* Washington, DC: World Bank.

Deininger, K., G. Carletto, and S. Savastano (2012a). 'Land fragmentation, corpland abandonment, and land market operation in Albania', *World Development*, forthcoming.

Deininger, K. and R. Castagnini (2006). 'Incidence and impact of land conflict in Uganda', *Journal of Economic Behavior & Organization*, 60(3), pp. 321–45.

Deininger, K. and A. Goyal (2012). 'Going digital: Credit effects of land registry computerization in India', *Journal of Development Economics*, forthcoming.

Deininger, K., A. Goyal, and H. K. Nagarajan (2012b). 'Women's inheritance rights and intergenerational transmission of resources: Evidence from India', *Journal of Human Resources*, forthcoming.

Deininger, K. and S. Jin (2006). 'Tenure security and land-related investment: Evidence from Ethiopia', *European Economic Review*, 50(5), pp. 1245–77.

Deininger, K., H. Selod, and A. Burns (2011c). *Improving Governance of Land and Associated Natural Resources: The Land Governance Assessment Framework*. Washington, DC: World Bank.

Duflo, E. (2003). 'Grandmothers and granddaughters: Old-age pensions and intrahousehold allocation in South Africa', *World Bank Economic Review*, 17(1), pp. 1–25.

Feder, G., Y. Chalamwong, T. Onchan, and C. Hongladarom (1988). *Land Policies and Farm Productivity in Thailand*. Baltimore and London: Johns Hopkins University Press.

Fenske, J. (2010). 'L'Etranger: Status, property rights, and investment incentives in Côte d'Ivoire', *Land Economics*, 86(4), pp. 621–44.

Fenske, J. (2011). 'Land tenure and investment incentives: Evidence from West Africa', *Journal of Development Economics*, 95(1), pp. 137–56.

Field, E. (2005). 'Property rights and investment in urban slums', *Journal of the European Economic Association*, 3(2–3), pp. 279–90.

Field, E. (2007). 'Entitled to work: urban property rights and labor supply in Peru', *Quarterly Journal of Economics*, 122(4), pp. 1561–602.

Field, E. and M. Torero (2006), 'Do property titles increase credit access among the urban poor? Evidence from a nationwide titling program', Group for Development Analysis (GRADE) and International Food Policy Research Institute (IFPRI) working paper. Retrieved from: <https://econ.duke.edu/uploads/media_items/fieldtorerocs.original.pdf>.

Fitzpatrick, D. (2005). '"Best practice" options for the recognition of customary tenure', *Development and Change*, 36(3), pp. 449–75.

Food and Agricultural Organization of the UN (2012). *Voluntary Guidelines on the Responsible Governance of Tenure of Land, Fisheries and Forests in the Context of National Food Security*. Rome: FAO.

Galiani, S. (2011). 'The dynamics of land titling regularization and market development', Working Paper. Helsinki: United Nations University.

Galiani, S. and E. Schargrodsky (2010). 'Property rights for the poor: Effects of land titling', *Journal of Public Economics*, 94(9–10), pp. 700–29.

Goldstein, M. and C. Udry (2008). 'The profits of power: Land rights and agricultural investment in Ghana', *Journal of Political Economy*, 116(6), pp. 980–1022.

Government of Mexico (2000). 'Marco legal agrario', *Edicion commemorativa—reforma agraria 1915–2000*. Mexico, DF: Procuraduria Agraria.

Greif, A. (1993). 'Contract enforceability and economic institutions in early trade: the Maghribi traders' coalition', *American Economic Review*, 83(3), pp. 525–48.

Holden, S. T., K. Deininger, and H. Ghebru (2009). 'Impacts of low-cost land certification on investment and productivity', *American Journal of Agricultural Economics*, 91(2), pp. 359–73.

Hunt, D. (2004). 'Unintended consequences of land rights reform: The case of the 1998 Uganda Land Act', *Development Policy Review*, 22(2), pp. 173–91.

Iyer, L. (2010). 'Direct versus indirect colonial rule in India: Long-term consequences', *Review of Economics and Statistics*, 92(4), pp. 693–713.

Jacoby, H. and B. Minten (2007). 'Is land titling in Sub-Saharan Africa cost effective? Evidence from Madagascar', *World Bank Economic Review*, 21(3), pp. 461–85.

Jacoby, H. G., G. Li, and S. Rozelle (2002). 'Hazards of expropriation: Tenure insecurity and investment in Rural China', *American Economic Review*, 92(5), pp. 1420–47.

Jansen, K. and E. Roquas (1998). 'Modernizing insecurity: The land titling project in Honduras', *Development and Change*, 29(1), pp. 81–106.

Joireman, S. F. (2008). 'The mystery of capital formation in sub-Saharan Africa: women, property rights and customary law', *World Development*, forthcoming.

Knight, R., J. Adoko, S. Siakor, A. Salomao, T. Eilu, A. Kaba, and I. Tankar (2012). *The Community Land Titling Initiative: International Report*. Rome: International Development Law Organization.

Lanjouw, J. O. and P. I. Levy (2002). 'Untitled: A study of formal and informal property rights in urban Ecuador', *Economic Journal*, 112(482), pp. 986–1019.

Lavigne-Delville, P. (2000). 'Harmonising formal law and customary land rights in French-speaking West Africa', in Toulmin, C. and J. Quan (eds), *Evolving Land Rights, Policy and Tenure in Africa*. London: DFID/IIED/NRI.

Libecap, G. D. and D. Lueck (2011), 'The demarcation of land and the role of coordinating property institutions', *Journal of Political Economy*, 119(3), pp. 426–67.

Lobo, A. and S. Balakrishnan (2002). 'Report card on service of bhoomi kiosks: An assessment of benefits by users of the computerized land records system in Karnataka', Working Paper, Bangalore: Public Affairs Centre.

Luke, N. and K. Munshi (2011). 'Women as agents of change: Female income and mobility in India', *Journal of Development Economics*, 94(1), pp. 1–17.

Maurer, N. and L. Iyer (2008). 'The cost of property rights: Establishing institutions on the Philippine frontier under American rule, 1898–1918', Working Paper 14298, Cambridge, MA: National Bureau of Economic Research.

Meinzen-Dick, R. and E. Mwangi (2008). 'Cutting the web of interests: Pitfalls of formalizing property rights', *Land Use Policy*, forthcoming.

Naritomi, J., R. R. Soarez, and J. J. Assuncao (2009). 'Institutional development and colonial heritage within Brazil', Working Paper, Rio de Janeiro: Department of Economics—PUC.

North, D. C. (1971). *Structure and Change in Economic History*. New York: W. W. Norton.

Nugent, J. B. and J. A. Robinson (2010). 'Are factor endowments fate?' *Revista de Historia Economica*, 28(1), pp. 45–82.

Peters, P. (2004). 'Inequality and social conflict over land in Africa', *Journal of Agrarian Change*, 4(3), pp. 269–314.
Pinckney, T. C. and P. K. Kimuyu (1994). 'Land tenure reform in East Africa: Good, bad or unimportant?', *Journal of African Economies*, 3(1), pp. 1–28.
Place, F. (2009). 'Land tenure and agricultural productivity in Africa: A comparative analysis of the economics literature and recent policy strategies and reforms', *World Development*, 37(8), pp. 1326–36.
Powelson, J.P. (1988). 'The Story of Land: a World History of Land Tenure and Agrarian Reform', Cambridge, MA: Lincoln Institute of Land.
Qian, N. (2008). 'Missing women and the price of tea in China: The effect of sex-specific earnings on sex imbalance', *Quarterly Journal of Economics*, 123(3), pp. 1251–85.
Quisumbing, A. R. and J. A. Maluccio (2003). 'Resources at marriage and intrahousehold allocation: Evidence from Bangladesh, Ethiopia, Indonesia, and South Africa', *Oxford Bulletin of Economics and Statistics*, 65(3), pp. 283–327.
Rajan, R. and R. Ramcharan (2011). 'Land and credit: A study of the political economy of banking in the United States in the early 20th century', *The Journal of Finance*, 66(6), pp. 1895–931.
Republic of Rwanda (2004). 'National Land Policy', Kigali: Ministry of Lands, Environment, Forests, Water, and Mines.
Rezende, G. C. D. (2006). 'Labor, land and agricultural credit policies and their adverse impacts on poverty in Brazil', Paper for Discussion No. 1180, Rio de Janeiro.
Rozelle, S. and J. F. M. Swinnen (2004). 'Success and failure of reform: Insights from the transition of Agriculture, *Journal of Economic Literature*, 42(2), pp. 404–56.
Sagashya, D. and C. English (2010). 'Designing and establishing a land administration system for Rwanda: Technical and economic analysis', in Deininger, K., C. Augustinus, P. Munro-Faure, and S. Enemark (eds), *Innovations in Land Rights Recognition, Administration, and Governance*. Washington, DC: World Bank.
Sjaastad, E. and D. Bromley (2000). 'The prejudices of property rights: on individualism, specificity and security in property regimes', *Development Policy Review*, 18(4), pp. 365–89.
Stiglitz, J. E. and A. Weiss (1981). 'Credit rationing in markets with imperfect information', *American Economic Review*, 71(3), pp. 393–410.
Strauss, J., G. Mwabu, and K. Beegle (2000). 'Intrahousehold allocations: a review of theories and empirical evidence,' *Journal of African Economies*, 9, pp. 83–143.
Transparency International India (2005). *India corruption study 2005*. New Delhi: Transparency International.
Udry, C. (1996). 'Gender, agricultural production, and the theory of the household', *Journal of Political Economy*, 104(5), pp. 1010–46.
Valsecchi, M. (2010). 'Land certification and international migration: Evidence from Mexico', Working Papers in Economics No 440. Gothenburg: University of Gothenburg.
World Bank (2002). *Mexico—Land Policy: A Decade after the Ejido Reforms*. Washington, D.C.: The World Bank, Rural Development and Natural Resources Sector Unit.
World Bank (2007). *Doing business 2008*. Washington, DC: World Bank, International Finance Corporation and Oxford University Press.
World Bank (2011). *Ethiopia: Options for strengthening land administration*. Addis Ababa: World Bank, Africa Region Sustainable Development Department.
Zepeda, G. (2000). *Transformación Agraria. Los Derechos de Propriedad en el Campo Mexicano bajo el Nuevo Marco Institucional*. CIOAC, Mexico.

15 Empowering women to become farmer entrepreneurs: case study of an NGO-supported programme in Bangladesh

MAHABUB HOSSAIN and W. M. H. JAIM[1]

1. Introduction

The importance of gender as an issue in developing countries was re-emphasized at the World Conference on Women in Beijing in 1995 (United Nations, 1996). The Conference recognized women's work in the productive and social sectors as a key aspect of development.

Women's empowerment is defined as 'the capacity of women to be economically self-reliant with control over decisions affecting their life options and freedom from violence' (Rao and Kelleher, 1995, p. 70). Women suffer from different types of powerlessness in social and economic spheres of life. This lack of empowerment is reflected in lower educational levels, lower incomes, and less control over their own incomes. It is also evidenced by less bargaining power in selling their labour and produce, less access to production inputs and resources, and inequality in opportunities for employment in the labour market compared with men, as well as decision-making on their own reproductive health. Development practitioners are concerned to raise women's empowerment levels, which will make women capable of challenging their dependency or oppressive situations in the family and society (Basu and Basu, 2001).

Gender specialists suggest that economic development programmes could pave the way to improving the economic status of women and thereby their overall status in the family and the community (Kandiyoti, 1988; Buvinic, 1987). They argue that participation in economic development programmes

[1] The authors thank the anonymous referees for their helpful comments.

improves skills and knowledge, as well as enhancing exposure to those outside the family sphere, and interactions with peers through participation in meetings of village organizations promoted by NGOs. These changes may weaken traditional values and norms, and could lead to the path of greater empowerment. Studies in South Asia find that economic empowerment has been the entry point for overall empowerment of women if they are organized under a common platform (Carr et al., 1996). There are contrary views, however. According to Mayoux (1998), women's empowerment is more than simply marginal increases in incomes; it requires a transformation of power relations within households, markets, communities, and national and international economies. Against this backdrop, it is important to examine whether women's involvement in agriculture not only helps to increase household incomes, but also contributes to their economic and social empowerment.

This chapter provides an overview of women's engagement in agricultural activities based on previous studies and assesses recent trends in women's engagement in productive work. It looks at the drivers of women's participation in agricultural activities based on analysis of unpublished data from a longitudinal national-level sample survey conducted by one of the authors, and evaluates the impact of efforts to promote women's engagement in agriculture by providing access to credit and training. The first section, entitled 'Women's Participation in Agriculture: Prevailing Knowledge', provides a review of existing knowledge of women's engagement in economic and agricultural activities based on previous studies. The second section, entitled 'Findings of Longitudinal Surveys', presents findings from a longitudinal sample survey on trends in women's engagement in agricultural activities and drivers of engagement. The impact of a special project to promote women's entrepreneurship in crop farming through provision of credit and extension is assessed in the third section entitled 'Case Study of a Targeted Project'. The final section, entitled 'Concluding Remarks', concludes the chapter with a recapitulation of key findings and implications for policies.

2. Women's participation in agriculture: prevailing knowledge

Until recently, women in Bangladesh hardly participated in agricultural activities outside the home (Abdullah and Zeidenstein, 1982; Westergaard, 1993; Bose et al., 2009; Hossain and Bayes, 2009). A number of studies conducted in the 1980s and 1990s found that women's contributions to socio-economic development were not visible due to a set of social norms that enabled men to dominate women (Abdullah and Zeidenstein, 1982; Ahsan et al., 1986; Begum, 1983; Chowdhury, 1986; Farouk, 1979, 1983; Halim and McCarrthy, 1985; Westergaard, 1983, Jaim and Rahman, 1988).The studies found that women work longer hours than men, particularly in low-income households, more in agricultural than in non-agricultural economic activities, and more as unpaid family labourers than as farm managers. Even if women do most of the work, men mostly control the decision-making and the income generated from such work. Women's economic activities were confined to homestead production and post-harvest operations. White (1992), however, found some evidence of women's marginal engagement in the marketing of agricultural produce within the village. Women rarely went to the marketplace, which is the domain of their male relatives, but women did conduct minor exchanges within their villages with poultry, eggs, and goats, and provided small loans to other women from their savings.

A rigorous analysis of primary data from the 1983–4 Agricultural Census conducted by Safilios-Rothschild and Mahmud (1989) as part of the UNDP-coordinated *Agriculture Sector Study* found marginal involvement of women in agricultural activities outside the homestead. The study noted that 43 per cent of women had agriculture as a primary occupation outside their domestic work, and another 12 per cent had agriculture as a secondary occupation. The rate of engagement in agriculture was found to be higher among the landless and the marginal landowning households than among those with medium and large farms. The analysis also revealed a high participation rate of women as waged agricultural labourers among landless households. In female-headed landless households, two-thirds of women worked as agricultural labourers.

In her study *New Technology in Bangladesh Agriculture: Adoption and Its Impact on Rural Labour Market*, Rushidan Islam Rahman found that adoption of improved agricultural practices and intensive cultivation had little impact on women's participation in the labour market but it increased their workloads due to additional harvests (quoted in Westergaard, 1993). For the majority of women from small households, the technology adoption increased their workload in post-harvest processing; for instance drying and seed selection activities are often done exclusively by women within the homestead. Women from large farm households, however, could avoid the additional load by employing hired female labour from the landless and marginal landowning households.

Westergaard (1983) looked into women's control over resources and participation in decision-making. Although women are entitled to inherit 33 per cent of parental property after the father dies, the majority of women did not claim their share. They perceived that 'it is no good to claim the share from the brothers because if they do, they cannot claim any support from their brothers in case of widowhood or divorce'. The study reported that women often had ownership rights to livestock and poultry, and they could keep the sales proceeds of such assets. This observation is confirmed in the study on women's role in agriculture conducted by Safilios-Rothschild and Mahmud, 1989.

Several studies focused on women's empowerment in terms of degrees of participation in decision-making, which could improve women's status in the family and thereby reduce gender inequality (Westergaard, 1983; Hannan, 1986). The studies used indicators such as decision-making by husband and wife (or jointly) on such things as choice of crops and varieties, use of inputs, marketing of products, sale of assets, education and health care of their children, and their subsequent marriages. Westergaard's study (1983) revealed that women participate very little in decisions regarding how much rice to sell, whether or not to send the rice for husking in the mill, or whether or not to hire female labourers. These decisions were mostly taken by the husband alone, in consultation with in-laws. In rare cases the decision was taken jointly by husband and wife. However, decisions with regards to treatment of poultry and livestock and planting of trees and vegetables in the home garden were taken either by the wife alone or by the wife and the husband together (Hannan, 1986). Only in a few cases did women take decisions alone regarding taking loans or maintaining household finances. The studies also looked into whether women could decide on how to spend the income they earned through employment. It was noted that employment for wages did not lead to female autonomy or empowerment. This finding is not surprising because most women who find waged employment outside their own house (*bari*) come from extremely poor economic backgrounds and are forced to take up such employment for survival. Women in such jobs were paid in kind with rice or other food items that they shared with the family (Westergaard, 1983).

Gender-related indicators across countries also show that Bangladeshi women are less empowered despite the efforts of large NGOs channelling credit through women in order to generate employment opportunities, provide training to augment skills, increase literacy, and raise awareness about their rights. With regards to the Gender-Related Development Index (GDI), Bangladesh ranks 120 out of 156 countries. According to the Gender Empowerment Measure (GEM), which takes into account gender inequality in economic and political spheres, Bangladesh ranked 81 out of 93 countries (UNDP, 2007). Thus attempts to transform the traditional roles of women in society have had limited success due to long entrenched cultural traditions (ADB, 2001; Hossain, 1998).

Micro finance is generally thought to be an effective means of improving women's status and overall household welfare that will also contribute to empowering women economically and socially. Since micro-lending programmes target women and the poorest section of the population, expansion of micro-credit should increase women's participation in agricultural activities. But because of limited ownership of land, such participation may be confined to home-based activities like homestead gardening, poultry and livestock rearing, etc. Studies documenting the impact of micro-credit on women's involvement in agriculture are rare. Hashemi et al. (1996) show that participation in credit programmes is positively associated with a woman's level of empowerment, which is a function of her relative physical mobility, economic security, ability to make various purchases on her own, freedom from domination and violence within the family, political and legal awareness, and participation in public protests and political campaigning (ADB, 1997, p. 15). This implies that women's participation in economic activities can automatically increase the overall status of women and empower them. But there are contrary views. Some observers (Rahman, 1998), however, note that women are usually used as a font for supply of micro-credit to the household, as women share the money borrowed with their male relatives. Only in a small fraction of cases do they themselves engage in economic activities. If the women refuse to share the money with the male relatives it may lead to more domestic violence and deepen the misery of these women. Micro-credit, however, has led to an increase in women's participation in poultry, livestock, and home gardening.

3. Findings of longitudinal surveys

National-level information on labour force participation is available from the reports of the Labour Force Survey conducted occasionally by the Bangladesh Bureau of Statistics. The latest survey taken in 2006 reports a civilian labour force of 49.5 million, 25 per cent of whom are women. Female participation in the labour force has, however, been increasing rapidly from a low base. Between 1999 and 2006 female workers increased by 42 per cent compared with a 16 per cent increase in the number of male workers. The data on gender disaggregation of agricultural workers, as obtained from recent Labour Force Surveys, are reported in Figure 15.1. The trend towards the 'feminization' of agriculture is clearly evident from the data. In the period from 1999 to 2000, 19 per cent of agricultural workers were women. By 2005–6 the women's share of the agricultural labour force had increased to 34 per cent.

However, in-depth information about the nature and extent of women's participation in agriculture is not available from the Labour Force Survey. There have also been changes in the definition of women workers over time. For deeper analysis in this section, we use unpublished data from a longitudinal rural household survey for 1988–2008 conducted by the first author (see Hossain and Bayes, 2009 for details of the survey). The benchmark survey

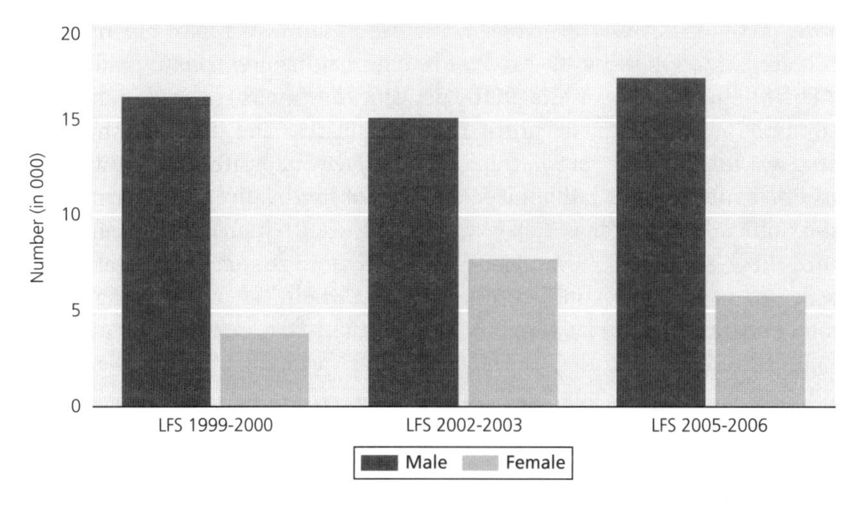

Figure 15.1 Trend of female participation in agriculture compared with male: Bangladesh
Source: Bangladesh Bureau of Statistics (2008).

was conducted in 1987–8 to assess the impact of the 'Green Revolution' on income distribution and poverty (Hossain et al., 1994). A multi-stage random sampling was used to draw a national-level representative survey. A sample of 1,240 households was drawn from 62 randomly selected villages covering 57 districts (out of 64 districts) in the country. A new sample of 1,860 households was drawn in 1999–2000 from the same villages, covering the old households and their offshoots for a study on the determinants of change in rural livelihoods conducted by the International Rice Research Institute. The same households were visited again in 2008 to assess the impact of increases in food prices on income distribution and poverty. A module of time budget was used in all three surveys. The respondents were asked to report time allocation to different activities for 12 hours from 6 a.m. to 6 p.m. for workers above 14 years of age, for 4 days preceding the date of the survey. The following information is based on the analysis of the data.

We define economic activities as those that generate income for the household or save household expenditures. These include employment in agricultural and non-agricultural labour markets and also unpaid work for the household in crop production activities, homestead gardening, livestock and poultry rearing, fishing, cottage industries (handicraft production), house repairing and construction work, transport operation, business, trade and shop keeping, and personal services. There are many other activities done by women that are quasi-economic in nature, including food processing and preparation of meals for family members, child care, care of old and sick family members, and tutoring of children. If the household had hired workers to do these jobs, it would involve some expenditure. We have not included such activities, following the practice in national income accounting.

The findings of the surveys on the per cent of workers engaged in different economic activities can be noted from Table 15.1. The time allocation on these activities is reported in Table 15.2. It may be noted that women are heavily involved in raising poultry, animal husbandry, home gardening, and crop cultivation such as post-harvest operations, drying, and seed selection. Since these are mostly homestead-based activities, it is convenient to carry them out in between conducting domestic duties. The activities in which women are involved relatively full time are non-farm services. In contrast, major economic activities for men are crop farming, non-farm services, business and shop keeping, and transport operations.

The importance of cultivation in generating employment is on a downward trend because of the continuous reduction in farm size under population pressure. The opportunities for job creation in the rural non-farm sector is on the rise due to expansion of rural roads and the generation of a marketable surplus from agriculture (transport and trade services), expansion of supply of micro-credit (generating self-employment in trade and business), and expansion of education and health services in rural areas. Labour has also moved out of cottage industries with the expansion of rural roads and electrification. In recent years farming has attracted more labour due to increased farm profitability as a result of an upward trend in agricultural prices.

Table 15.1 Employment of adult men and women in agriculture and non-agriculture over time

	Percentage of adult workers employed in the activity					
	1988		2000		2008	
Activity	Men	Women	Men	Women	Men	Women
Agriculture	**83.8**	**59.2**	**59.5**	**57.6**	**65.3**	**66.4**
Crop cultivation	67.7	15.7	47.9	6.4	52.6	3.9
Livestock and poultry	28.2	29.5	23.9	50.8	34.5	68.9
Homestead gardening	1.5	9.9	2. 6	9. 1	2.4	18.0
Fisheries	5.1	1.0	5.6	0.4	3. 7	0.5
Non-agriculture	**36.6**	**14.3**	**45.1**	**8. 1**	**43.7**	**8.4**
Industry/processing	2.9	8.1	3.8	1. 4	3.2	1.1
Transport operations	3.0	0.0	5.3	0.0	5.9	0.0
Construction work	10.0	3.8	3.7	1.1	4.2	1.6
Business/trade	12.6	0.8	16.6	0.4	14.8	0.4
Services	11.5	3.2	17. 4	5.5	16.9	5.7
Total employed	**96.8**	**66.0**	**91.9**	**64.3**	**94.5**	**71.4**
Total (multiple responses)	**120.0**	**73.5**	**104.9**	**65.7**	**109. 0**	**74. 8**

Table 15.2 Trend in time allocation of adult men and women in agricultural and non-agricultural activities

	Duration (Hours per day per worker) of work for those employed					
	Men			Women		
	1987	2000	2008	1987	2000	2008
Activity	(N = 1756)	(N = 2464)	(N = 2472)	(N = 1563)	(N = 2419)	(N = 2614)
Agriculture	**5.02**	**2.94**	**3.65**	**1.33**	**1.11**	**1.28**
Crop cultivation	4.30	2.27	2.92	0.57	0.11	0.16
Livestock and poultry	0.56	0.48	0.55	0.64	0.84	0.91
Homestead gardening	0.02	0.04	0.03	0.11	0.14	0.19
Fisheries	0.14	0.15	0.15	0.01	0.02	0.02
Non-agriculture	**1.97**	**3.06**	**2.47**	**0.42**	**0.36**	**0.22**
Industry/processing	0.15	0.26	0.21	0.21	0.07	0.04
Transport operations	0.20	0.36	0.44	0.00	0.00	0.00
Construction work	0.42	0.19	0.21	0.08	0.03	0.02
Business/trade	0.74	1.10	1.05	0.02	0.02	0.03
Services	0.47	1.15	1.13	0.10	0.24	0.30
Economic activities	**6.99**	**6.00**	**6.69**	**1.75**	**1.47**	**1.67**
Domestic activities	**1.27**	**1.25**	**1.14**	**7.16**	**5.84**	**5.87**
Total	**8.26**	**7.24**	**7.83**	**8.91**	**7.30**	**7.54**

Over the last two decades women's involvement in crop cultivation has sharply declined from about 23 per cent in 1988 to about 3 per cent in 2000, and 4 per cent in 2008. This is mainly because of the fact that the work normally done by women in post-harvest operations, particularly rice processing (winnowing, drying, parboiling, husking/milling, etc.) has been mechanized. At present, women are involved mostly in livestock and poultry production activities rather than crop production activities. Participation of adult women in livestock and poultry production activities increased from 43 per cent in 1988 to 51 per cent in 2000 and further to 69 per cent in 2008. Involvement of women in homestead gardening has also increased in recent years. Credit support from NGOs has largely facilitated involvement of women in livestock and poultry rearing as well as in homestead gardening in rural Bangladesh. Participation of women in fisheries activities was found to be negligible.

Allocation of time for men working in agriculture was mainly for crop production activities, while for women it was for livestock and poultry rearing activities as revealed from all the surveys in 1987, 2000, and 2008. However, allocation of time for crop production activities for men has reduced from 4.30 hours per day in 1987 to 2.27 hours per day in 2000, and 2.92 hours per day in 2008. On the other hand, allocation of time per day for women in livestock and poultry-rearing activities has increased from 0.64 hours in 1987

Table 15.3 Participation of wage labourer by gender in agriculture and wage rate over time

	1988		2000		2008	
Gender	Percentage of labour	Wage rate per day (USD)	Percentage of labour	Wage rate per day (USD)	Percentage of labour	Wage rate per day (USD)
Female	2.5	0.54	1.0	0.59	1.1	1.07
Male	24.6	0.73	22.3	1.02	23.1	1.76

to 0.84 and 0.91 hours respectively in 2000 and 2008. Allocation of time for women has also slightly increased in recent years for homestead gardening.

Very few women participate in the agricultural labour market (Table 15.3). In 1988, only 2.5 per cent of women workers sold their labour in the agricultural labour market compared with 25 per cent among male workers. Women's participation in the agricultural labour market further declined to 1.1 per cent in 2008. Participation of women in the agricultural labour market is poverty induced, and since poverty has improved substantially over the last two decades, women's participation in the agricultural labour market has also declined. The other notable finding from these surveys is the continuing gender disparity in the wage rate. The wage rate in the agricultural labour market has increased substantially, particularly during the period between 2000 and 2008 (Table 15.3). The wage rate per day has increased by 73 per cent during this period for men and by 81 per cent for women. In 1988, women's wages were 27 per cent lower than those of men. In 2008, the wage disparity further widened to 40 per cent.

In order to understand the drivers of women's participation in agricultural activities, we ran a logit regression model (qualitative dependent variable). The dependent variable was measured by a dummy variable with the value of 1 for households where women allocated more than one hour a day to agricultural activities and a value of 0 otherwise. The explanatory variables included: the wage rate for agricultural labour and for non-agricultural labour at the village level; land owned by the household (hectares); cultivated land with access to irrigation (hectares); the number of years of schooling of the woman worker; whether the worker is a member of a microfinance organization; the age of the worker and the age square (to capture the non-linear effect of age); the distance of the village from the nearest bus stop; and whether the village has access to electricity.

The results, reported in Table 15.4, indicate that agricultural wage rates have a positive effect on participation in agriculture, while the non-agricultural wage rates have a negative effect. Women's participation might increase if tightening of the labour market increased the agricultural wage rate. Developed infrastructure facilities (electricity and proximity to transport facilities) will

Table 15.4 Determinants of women's participation in agriculture—estimates of logit function: 2008

Dependent variable (1 for worker allocating at least one hour per day to agricultural work, 0 otherwise)	Odds ratio	Z value	Probability of Z value
(X_1) Age of female workers (years)	1.275	13.3	0.000
(X_2) Square of age	0.997	−12.8	0.000
(X_3) Household-owned land (ha)	0.968	−0.54	0.590
(X_4) Cultivated land, irrigated (ha)	1.318	2.37	0.018
(X_5) Years of schooling of worker (years)	0.944	−3.75	0.000
(X_6) Access of the village to electricity (with electricity = 1, no electricity = 0)	4.399	3.92	0.000
(X_7) NGO membership (membership = 1, no membership = 0)	1.441	3.08	0.002
(X_8) Distance of bus stop from the village (km)	0.356	−11.8	0.000
(X_9) Square of distance of the village from bus stop	1.062	16.45	0.000
(X_{10}) Agricultural wage rate in village (Taka/day)	1.030	11.64	0.000
(X_{11}) Non-agriculture wage rate in village (Taka/day)	0.903	−87.9	0.000
Degrees of freedom: 2,614 Chi square = 668, Pseudo R Square = 0.20			

induce women to participate more in agriculture. Higher education has a negative effect on participation. Similarly, access to credit through NGOs contributes to greater participation in agricultural activities. The size of land ownership has a neutral effect, but irrigated land increases participation through an increase in agricultural production that requires women to engage in post-harvest operations. Older women participate more in agricultural activities than younger women, but the positive effect slows down as women become older. A major policy implication of the findings is that supply of credit will encourage more women to engage in agriculture. Also investment in infrastructure development may promote participation in agriculture by facilitating linkages with markets.

4. Case study of a targeted project

4.1 PROJECTS FOR WOMEN'S ENGAGEMENT IN AGRICULTURE

In Bangladesh, a number of micro-credit projects have targeted women to encourage greater involvement in crop production activities. Prominent among these are: (a) the Small Farmers Development Project funded by IFAD channelled through the Palli Karma Sahayak Foundation (PKSF) and implemented by the Department of Agricultural Extension and several small NGOs; (b) the Food Security at Household level (FoSHol) project implemented by the International Rice Research Institute in collaboration with a number of NGOs; and (c) the Northwest Crop Diversification Project

(NCDP) funded by the Asian Development Bank and implemented by the Department of Agricultural Extension in Partnership with BRAC, RDRS, and other NGOs.

The Northwest Crop Diversification Project (NCDP) funded by FAO and IFAD and implemented through NGOs (RDRS, BRAC, etc.) and PKSF are amongst the few development projects which targeted small farmers (with special focus on female farmers) in order to provide credit and training to extend knowledge of improved technologies. NCDP is provided to farmers through NGOs at a subsidized interest rate (13 per cent) in order to achieve the following objectives:

- Diversification of cropping patterns away from rice;
- Engage small farmers, especially women, to grow high-value crops (HVCs) including some vegetables and spices;
- Create women entrepreneurs;
- Link small farmers to markets.

This section assesses the effectiveness of NCDP attempts to feminize agriculture by turning women from unpaid family help to farm managers, women's performance compared to men in conducting farm enterprises, and the effect of the project on economic and social empowerment of women.

4.2 DATA AND METHODOLOGY

For this study, two NGOs (namely BRAC and RDRS) were deliberately selected, as their coverage in implementing the project is more comprehensive than other NGOs. Sherpur Upazila of Bogra district for BRAC–NCDP and Thakurgaon Upazila of Thakurgaon district for RDRS–NCDP were selected randomly. From each of these two locations, five female and five male small farmers' groups were selected randomly. A list of all the female and male members in these groups under NCDP (each group consists of 15–20 members) were collected from the concerned officer of BRAC and NCDP. Thus, a list of roughly 75–80 female and male farmers from each of the locations was collected from BRAC and RDRS regional offices. Taking this list as the sample frame, 15 female and 15 male farmers from each location (Sherpur and Thakurgaon) were selected using a systematic random sampling method. Thus the total number of samples from the two locations was thirty both for female and male farmers. However, two male farmers were dropped at the time of data analysis, as they were found to belong to the large farm category.

Before collecting field-level data using a structured questionnaire, focus group discussions with the female and male farmers' groups were held by the authors to assess the survey design and selection of samples. Primary data was collected by trained investigators from the selected farmers during October 2010.

To assess the nature of participation in economic activities (EAs), all activities related to agriculture and non-agriculture were classified into nine categories. Farmers were asked to report the degree of engagement on a 3-point rating scale: 0 for never participating, 1 for occasionally participating, and 2 for frequently participating. Thus a respondent's score might range from 0 to 18 for nine EAs. Frequency counts of responses for each of the EAs were used to measure the Participation Index (PI) for each of the respondents.

For estimating determinants of participation, the Probit qualitative dependent variable model was used. The dependent variable was measured using the Participation Index (PI). If the PI was less than or equal to 4, the respondent was marked as not actively participating in the economic activities (assigned value = 0), and those with PI values of 5 and above were considered as participants (assigned value = 1). The dependent variables used are farm size, total family members in the active age group, area under high-value crops, education of the head of the household (years of schooling), and education of the female farmers.

A Women's Empowerment Index (WEI) was constructed following the methodology employed by Bose et al. (2009) using the data on women's participation in decision-making in agricultural and non-agricultural activities. We assigned the value 1 when the decision is taken by male members alone (least empowered). When the decision is taken jointly by husband/male and female we assigned value 2 (equally empowered). The highest value 3 is assigned when the decision is taken by a female alone (most empowered). We considered twelve inter-household decision-making indicators: seven indicators for agriculture-related decisions and five indicators for non-agriculture-related ones. The seven agriculture-related indicators are: (1) selection of crops and variety, (2) management of production activities, (3) purchase of agricultural inputs, (4) rearing of cattle and poultry, (5) marketing of crops and other agricultural produce, (6) homestead gardening, and (7) post-harvest operations. The five non-agriculture-related indicators are: (1) management of family budget, (2) children's education, (3) purchase and sale of land and other fixed assets, (4) travel and recreation, and (5) voting in local and national elections.

WEI was calculated separately for both agricultural and non-agricultural activities as well as for all economic activities, by dividing the total value of the scores with the number of indicators. As the score ranged from 1 to 3, the average value was 2.0 for each of the decision-making indicators. Therefore, to assess the individual empowerment status and position, women with average scores of 2.0 or more were labelled 'empowered'.

4.3 RESULTS AND DISCUSSIONS

Socio-economic profile of NCDP farmers

According to the selection criteria of NCDP, only farmers holding 0.2 to 1.2 hectares should be included in the project. Our survey results show mistargeting by 27 per cent for female farmers and 4 per cent for male farmers. Many of the farmers covered in the project had farms below the minimum ceiling of 0.2 hectares, as a considerable proportion of women from vulnerable groups had been included in the sample. Only 7 per cent of female farmers and 21 per cent of male farmers had land above the maximum ceiling of 1.2 hectares. The field staff of the NGOs explained that while forming farmers' organization for a village, they did not find enough interested female farmers in the target group required to provide the service in a cost-effective way, so they had to go beyond the target group at the lower end of the farm size scale.

Average farm size was 0.51 hectare for female farmers and 0.85 hectare for male farmers (Table 15.5). The average farm size in Bangladesh is 0.65 hectare. It may also be noted that due to religious and social systems in Bangladesh, daughters inherit land from their parents only half of that inherited by their brothers. In many cases, the sisters do not take their share in order to keep good relations with their brothers for social protection during difficult times. So most of the land in the family is owned by the male head of the household.

The female tenant farmers recorded the rented land (locally called land taken under 'agreement') in their own name in order to qualify for NCDP credit. This study found that 70 per cent of the female farmers were either pure tenants (27 per cent) or part-tenants (43 per cent) while the majority of male farmers were owner farmers (53 per cent).

Table 15.5 Socio-economic background of the respondent farmers

Socio-economic characteristics	Female farmers	Male farmers
Average age (in years)	38	40
Education status:		
Had formal schooling (%)	60	89
Average years of schooling	3	6
Family size (no. per household)	4.9	4.4
Full-time engagement in agriculture	71	44
Average farm size (hectare)	0.51	0.85
Farm size distribution (% of total)		
Below 0.20 hectare	27	4
0.21 to 1.20 hectare	66	75
1.21 to 2.00 hectare	7	21
Tenancy status (% of farmers)		
Owner farmer	30	53
Part tenant	43	36
Pure tenant	27	11

The average family size of female farmers' households was found to be 4.9 while it was 4.4 for the male farmers' households. The participation of female farmers in agriculture was found to be significantly different. Full-time engagement in agriculture for female farmers was found to be 71 per cent while the corresponding percentage for male farmers was only 44 per cent, as male farmers also engaged in many rural non-farm activities. A large proportion of the female farmers (40 per cent) were illiterate compared with male farmers (11 per cent). In Bangladesh nearly 40 per cent of the adult population is illiterate. It implies that the participants in the project are drawn from relatively educated farmers. The average number of years of schooling was three for females and six for male farmers.

Project intervention

(a) *Provision of credit*: About three-quarters (73 per cent) of the female farmers joined NCDP under BRAC in 2002, while in 2009, two-thirds (67 per cent) of the female farmers joined NCDP under RDRS. The male farmers joined NCDP under both BRAC and RDRS in 2005. The amount of loan received increased substantially with repeat loans. The average size of loan for first-time female borrowers was Tk 5700 (range Tk 2000 to Tk 10,000), which increased to Tk 17,200 (range 5000 to 50,000) in 2010. This indicates that there is large unmet demand for credit from women to engage in agriculture. For male farmers the average size of loan increased from Tk 5000 to Tk 19,000 between 2005 and 2010.

Besides NCDP credit, 23 per cent of the female farmers received credit from other NGOs. For male farmers, the only source of credit was NCDP, as most NGOs provide micro-credit only to women. The male farmers did not have loans from commercial banks and specialized agricultural credit agencies either. Only 7 per cent of the male farmers received credit from the banks. The female farmers who did not take credit from other sources besides NCDP were asked to specify reasons for it. Half of them reported that there was no need for additional loans, while 27 per cent reported that the interest rate was high, and another 18 per cent reported problems with attending weekly meetings and paying weekly instalments (for NCDP the instalment is collected on a monthly basis). With the exception of one female farmer and two male farmers, no participant in the project reported any problems related to NCDP credit. For RDRS, which collects instalments on a weekly basis, farmers suggested making provision for paying monthly instalments or payment after the harvesting of the crop.

(b) *Provision of training*: NCDP also provided training on crop management for project clients. The training is organized by the concerned NGOs, but the trainers are from the Department of Agricultural Extension under the

Ministry of Agriculture. The survey showed that 100 per cent of both male and female farmers under BRAC received such training, while in the case of RDRS, 53 per cent of female and 64 per cent of male farmers received such training. The duration of training was for one day only and it mostly pertains to the production of maize followed by potato and some vegetables. More than 80 per cent of the farmers commented that one-day training is inadequate since very little can be learnt from this short course. They also wanted occasional refresher training, as they face many problems when they practise the knowledge received from the first training.

Farmers were also asked to specify the areas of training they want under NCDP. The majority of female farmers gave the highest priority to the production of maize (37 per cent), followed by vegetables (26 per cent), potato (11 per cent), and rice cultivation (11 per cent). For male farmers the stated priority is maize (31 per cent) and potato (19 per cent), followed by rice (15 per cent) and vegetables (12 per cent). Some (12 per cent) are also interested in getting training in livestock raising (Table 15.6).

Impact of the project

(a) *Crop Diversification*: In order to diversify crops, the NCDP encouraged farmers to grow high-value crops (HVCs) such as vegetables, potato, spices, and maize, through provision of credit and training facilities supplied by selected NGOs. Distribution of cropped area showed that non-rice crops occupied 31 per cent of the cropped area for female farmers and 35 per cent for male farmers (Table 15.3). In Bangladesh only 26 per cent of the cropped land is allocated to non-rice crops (Government of Bangladesh, 2011). Thus the project had some impact on crop diversification, but not to a large extent. The NCDP also targeted the expansion of areas under spice cultivation (i.e. onion, garlic, ginger, etc.). But the allocation of land under spice production

Table 15.6 Fields of training demanded by farmers

	Female farmers		Male farmers	
Area/subject where more training is demanded	No. of farmers	Percentage of total	No. of farmers	Percentage of total
Maize cultivation	10	37.0	8	30.8
Potato cultivation	3	11.1	5	19.2
Rice cultivation	3	11.1	4	15.4
Vegetable cultivation	7	25.9	3	11.5
Wheat cultivation	1	3.7	1	3.8
Poultry rearing	1	3.7	1	3.8
Livestock rearing	1	3.7	3	11.5
Cultivation of spices	1	3.7	1	3.8
All	27	100.0	26	100.0

was found to be insignificant (only 1 per cent) both for male and female farmers. The reason for this insignificant production was due to the low profitability of spices compared with other crops, in part because spice crops did not benefit from the technological progress experienced by rice, wheat, and maize. The allocation of land to pulses and oil seeds was also found to be negligible (Table 15.7). The findings indicate that the provision of incentives in the form of subsidized credit and extension services is not a sufficient condition to promote the crops unless they are more profitable compared with other crops for which technological progress is necessary.

Cropping intensity of the female farmers was 192 per cent while it was 204 per cent for the male farmers. Cropping intensity for both female and male farmers' households under NCDP was found to be higher than the national average of 191 per cent (Government of Bangladesh, 2011). It shows that credit, training and extension services, and intensive supervision can help increase cropping intensity.

(b) *Linking to markets*: In the case of rice, female farmers marketed (sold) 44 per cent of *Aman* paddy and 51 per cent of *Boro* paddy. The remaining portion was kept for home consumption. The corresponding numbers for male farmers were 56 per cent and 63 per cent, respectively. For maize, 100 per cent of the produce was marketed by both female and male farmers. For potato, 78 per cent of the product was marketed by female farmers and 97 per cent by male farmers. In the case of vegetables, more than 90 per cent of the products were marketed by both female and male farmers. The findings imply that although the small farmers are mostly subsistence in nature, they also supply substantial proportions of produce to the market. It shows that higher agricultural prices benefit both small and large farmers.

Table 15.7 Distribution of land for different crops amongst NCDP farmers

	Percentage distribution of cropped area	
Crops	Female farmers	Male farmers
Rice: Aman	39.26	39.35
Boro	29.58	23.40
Aus	0	2.76
Wheat	7.07	8.59
Maize	9.90	9.23
Potato	3.33	6.60
Pulses	.45	0.10
Mustard	1.46	0.81
Jute	3.18	5.18
Vegetables	4.93	2.79
Spices	0.84	1.19
Total	100.00	100.00

The majority of female farmers (77 per cent) and all the male farmers reported that they did not face any problems marketing their crops. Those who reported problems mentioned low prices offered to them by marketing agents. They mentioned that for maize the wholesalers directly procure the crop from the field after harvest, while for other crops the produce could be sold for a reasonable price at the local market. In areas with good road communication, farmers with large marketable surpluses can sell the product directly at the secondary market. The price disadvantage for selling in the local market was small. Only 27 per cent of female members mentioned receiving lower prices in the local market compared with those in the secondary markets when the transport costs must be taken into account.

(c) *Gender difference in participation*: The study found substantial gender differences in the participation in economic activities (Table 15.8). Women were more heavily involved in poultry and livestock rearing activities and in the production of vegetables and spices compared with their male counterparts. Their participation in the production of crops grown in the field was marginal, which was the domain of male farmers. Women's participation in pond aquaculture and fishing in the flood plains was rare. Similar findings were reported in many recent studies (Rahman and Naoroze, 2007; Hoque and Itohara, 2008; Sultana and Thompson, 2008). NCDP did not have any effect in inducing female farmers to participate in marketing of produce, which was done by male relatives. Only 2 out of 30 women participated in non-farm activities, compared to 17 out of 28 male farmers. A few female farmers participated in the agricultural labour market as waged labour, but these women mostly belong to poverty-stricken households. However, the wage rate of female farmers was considerably lower (ranging from Tk 100 to Tk 135) compared with male farmers (ranging from Tk 160 to Tk 250). Bose et al. (2009) reported similar findings of discrimination of women in the

Table 15.8 Time allocation of female and male farmers in different economic activities

	Female farmers		Male farmers	
Economic activities	No. of farmers	Hours per day	No. of farmers	Hours per day
Production of field crops	25	2.56	28	3.75
Production of spices and vegetables	17	2.82	17	3.59
Poultry rearing	24	1.25	0	0
Goat rearing	19	1.58	10	1.0
Cattle rearing	23	2.30	24	2.0
Aquaculture	0	0	5	2.2
Waged labour	3	6.67	4	8.0
Marketing products	0	0	13	1.0
Non-agricultural activities	2	2.00	17	6.5

Note: The average amounts of time spent are for those participating in the activities.

labour market. The findings imply that women's work remained focused on more home-based activities while men's work concentrated more on market-oriented non-farm activities.

The time allocation of male and female farmers in different economic activities is reported in Table 15.8. The numbers show that women are engaged in economic activities on a part-time basis (except in waged labour), but the intensity of engagement is higher than that of men in poultry and livestock raising. Men are engaged relatively full-time in agricultural waged labour and non-farm activities, but their engagement in agriculture is also relatively part-time.

(d) *Impact on Social Empowerment*: Mobility outside the boundary of the home and interaction with people outside the family circle improves self-confidence among women. The mobility of female farmers in Bangladesh was facilitated by the NGOs in Bangladesh by the very nature of the delivery of credit. Women go to the NGO offices to deposit savings and attend meetings of village organizations to repay loan instalments. The numbers in Table 15.9 show that the most frequent movements of female farmers outside the home were due to depositing loan instalments with the concerned NGOs. About 87 per cent of female farmers go outside the home frequently for this purpose, while the percentage of those who occasionally move outside home for this purpose was reported at 13 per cent.

As the NCDP credit delivered to female farmers was for crop production activities, this required purchases such as seed, fertilizer, insecticides, etc. from local or secondary markets. The majority of female farmers (about 67 per cent) reported that they never go to markets to purchase inputs such as seed, etc., and male members of the family do this work. Almost all female farmers (97 per cent) reported that they never went to market to sell products, while a negligible percentage (3 per cent) goes outside the home for this purpose. Many women, however, go outside their villages to visit relatives. Mobility in order to participate in NGO activities helped women to build

Table 15.9 Mobility of NCDP female farmers outside the home

	Degree of mobility (percentage of female farmers)		
Reasons for mobility	Never	Occasionally	Frequently
Purchasing inputs from markets	66.7	30.0	3.3
Selling products to markets	96.7	3.3	nil
Attending meetings/workshop of NGOs	13.3	70.0	16.7
Depositing loan instalments at village organization meetings	nil	13.3	86.7
Visiting relatives, travelling to towns, shopping for goods for personal care, etc.	3.3	86.7	10.0

social capital through interactions with other women in their group. This has made them somewhat more socially empowered than before.

(e) *Participation in decision making*: Women's participation in intra-family decision-making processes indicates their level of social empowerment. The economic independence of women and their share of contribution to the family income are considered major factors which influence the participation of women in intra-family decision-making processes. However, there are other social and cultural factors that influence the participation of women in household decision-making.

The degree of participation has been assessed by asking female farmers whether the decisions are taken by males alone or males and females jointly or females alone with respect to each of the activities presented in Table 15.8. The Women's Empowerment Index (WEI) was constructed using the method explained in the methodology section.

The findings reported in Table 15.10 showed that in the majority of cases, decisions were taken jointly by husband and wife. This is in contrast to findings of earlier studies. Presumably, women participating in NCDP received some orientation from the concerned NGOs (BRAC and RDRS) regarding the need for women's empowerment. Providing access to credit and extension may also have a confidence-building effect on women and strengthened their bargaining power in decision-making in the household. Indeed, in

Table 15.10 Participation of women in household decision-making processes

	Decision is taken by			
Decision-making parameters	Husband alone	Jointly by husband and wife	Wife alone	Average Women's Empowerment Index (WEI)
Agriculture-related indicators	**1.98**			
Selection of crops variety	3	27	0	1.90
Crop management	3	27	0	1.90
Marketing of inputs	2	28	0	1.93
Cattle and poultry rearing	1	27	2	2.03
Sale of crop products	0	30	0	2.00
Homestead gardening	0	27	3	2.10
Post-harvest operations	1	29	0	1.97
Non-agriculture-related indicators	**1.91**			
Cash management	2	28	0	1.93
Education of children	1	27	2	2.03
Land purchase and sale	8	22	0	1.73
Travelling and recreation	2	28	0	1.93
Casting votes in elections	2	28	0	1.93
All indicators	**1.95**			

Number of female farmers' household = 30: Empowerment status: Less empowered, if $WEI_i \leq 2.0$ and highly empowered if $WEI_i > 2.0$

a few cases such as cattle and poultry raising and homestead gardening, the decision was taken by the wife alone. At the other end of the spectrum, the decision was taken by the husband alone in a few cases for crop production activities, cash management, travelling, and casting votes in elections.

The Women's Empowerment Index presented in Table 15.10 clearly indicates that women are empowered at the above-average level in overall decision-making activities in both agricultural and non-agricultural arenas, as well as in the cases of agricultural and non-agricultural activities separately. This was due to awareness-raising activities of the NGOs through which the project was implemented and the effect of channelling credit through women for entrepreneurship development in agriculture. In the absence of such mobilization, the value of the WEI could have remained very low, as found in other studies (Bose et al., 2009).

5. **Concluding remarks**

Earlier studies on gender roles show that in Bangladesh, women rarely engage in economic activities outside the homestead. In agriculture, women are engaged in activities within the homestead such as livestock and poultry rearing, homestead gardening, and post-harvest operations such as threshing, drying, seed selection, and food preparation as unpaid family labour. Engagement in such activities is part-time and is done in between performing domestic duties. Women working outside the homestead for wages are rare, and those who do, belong to poverty-stricken households. Recent data show that feminization of agriculture has been growing, albeit slowly, and is driven by the migration of male members to rural non-farm activities and to urban areas in search of better economic opportunities.

Recently several projects have targeted women to facilitate the process of feminization of agriculture with provisions for credit and training, and linking them to markets. The Northwest Crop Diversification (NCDP) project funded by the Asian Development Bank and IFAD is one such project. Unlike micro-credit provided by different NGOs, credit from the NCDP is an exception, as it focuses on engaging women in agricultural activities, particularly in crop cultivation in which women are rarely engaged. This study aimed to assess the effectiveness of NCDP in reaching women with credit and training, and its effect on crop diversification and the empowerment of women.

The demand for credit from both male and female farmers has increased over time since joining NCDP. Compared with the first year of joining the project, the amount of credit received per female farmer has increased three times, while the amount has increased four times for male farmers. Despite credit given for crop production activities, home-based economic activities for women got priority compared with field-level activities, which appears to be a reflection of the entrenched social system and cultural values that confine women within homesteads even when they conduct economic activities.

Gender differentials in participation in economic activities continue despite the intervention. Although female farmers participated in crop production activities, their involvement was less than male farmers. Female farmers were more involved in livestock and poultry-rearing activities compared with male farmers. Participation in non-farm activities was also found to be dominated by males. Only a few female farmers worked as wage labour, and the wage they received was considerably lower than that for their male counterparts.

The study found that the mobility of women outside the home has increased because of the requirement to attend village organization meetings and the repayment of credit instalments at such meetings. Such mobility, along with handling money and participation in decision-making in agricultural and non-agricultural arenas, has contributed to confidence-building amongst participating women. The farmers participating under NCDP are found to be more empowered than women in general, as revealed by recent studies on women's empowerment.

REFERENCES

Abdullah, T. A. and S. Zeidenstein (1982). 'Village women in Bangladesh: prospects for change', A study prepared for the International Labour Office within the framework of the World Employment Programme. Oxford: Pergamon Press.

Ahsan, R. M., S. R. Hussain, and B. J. Wallace (1986). *Role of Women in Agriculture.* University of Dhaka, Dhaka: The Centre for Urban Studies.

ADB (1997). *Microenterprise Development: Not by Credit Alone.* Asian Development Bank.

ADB (2001). 'Women in Bangladesh: Bangladesh Country Briefing Paper'. Retrieved from: <http://www.adb.org/documents/women-bangladesh-country-briefing-paper> (24 July 2013).

Bangladesh Bureau of Statistics (2008). *Report of the National Labour Force Survey 2005–06*, Planning Division, Ministry of Planning, Government of the People's Republic of Bangladesh.

Basu, S. and P. Basu (2001). 'Income generation program and empowerment of women—A case study in India', Charles Sturt University, Bathurst, NSW 2795, Australia. Retrieved from: <http://crawford.anu.edu.au/acde/asarc/pdf/papers/conference/CONF2001_03.pdf>.

Begum, S. (1983). 'Women and rural development in Bangladesh', Master's thesis (mimeographed), Cornell University.

Bose, M. L., A. Ahmad, and M. Hossain (2009). 'The role of gender in economic activities with special reference to women's participation and empowerment in rural Bangladesh', *Gender, Technology, and Development*, Volume 13, No. 1, January–April 2009.

Buvinic, M. (1987). 'The psychology of donor support', *World Development*, Vol. 17, No. 7–12, pp. 1045–57.

Carr, M., M. Chen, and R. Jabvala (1996). *Speaking Out, Women's Economic Empowerment in South Asia.* London: IT Publications.

Chowdhury, N. (1986). 'Revaluation of women's work in Bangladesh', *The Bangladesh Journal of Agricultural Economics*, 9(1), pp. 8–15.

Farouk, A. (1979). *Time use of rural women.* Bureau of Economic Research, Dhaka: University of Dhaka.

Farouk, A. (1983). *The Hardworking Poor.* Bureau of Economic Research, Dhaka: University of Dhaka.

Government of Bangladesh (2011). *Yearbook of Agricultural Statistics of Bangladesh, 2011*, Bangladesh Bureau of Statistics, Ministry of Planning, Dhaka.

Halim, A. and F. E. McCarthy (1985). 'Women labourers in rice producing villages of Bangladesh', *Women in Rice Farming*, proceedings of a conference on Women in Rice Farming Systems, The International Rice Research Institute, Manila, Gower Publishing Company Ltd, Aldershot and Gower Publishing Company, Brookfield, pp. 242–54.

Hannan, F. H. (1986). *Resources Untapped, an Exploration into Women's Role in Homestead Agricultural Production System.* Women's Desk, BARD, Comilla.

Hashemi, S. M., S. R. Schuler, and A. P. Riley (1996). 'Rural credit programmes and women's empowerment in Bangladesh', *World Development*, Vol. 24, No. 4.

Hoque, M. and Y. Itohara (2008). 'Participation and decision-making role of rural women in economic activities: A comparative study for members and non-members of the micro-credit organizations in Bangladesh', *Journal of Social Sciences*, 4(3), pp. 229–36.

Hossain, M. (1998). 'Credit for Alleviation of Rural Poverty: The Grameen Bank of Bangladesh', *IFPRI Research Report 65*, International Food Policy Research Institute, Washington DC. Retrieved from: <http://www.ifpri.org/pubs/abstract/65/rr65.pdf> (14 January 2008).

Hossain, M. and A. Bayes (2009). *Rural Economy and Livelihoods: Insights from Bangladesh.* Dhaka, Bangladesh: A. H. Developing Publishing House.

Hossain, M., M. A. Quasem, M. A. Jabbar, and M. M. Akash (1994). 'Production environments, modern variety adoption and income distribution in Bangladesh', in David, C. C. and K. Otsuka (eds), *Modern Rice Technology and Income Distribution in Asia.* Boulder and London: Lynne Reiners Publishers.

Jaim, W. M. H. and M. L. Rahman (1988). 'Participation of Women and Children in Agricultural Activities—A Micro Level Study in an Area of Bangladesh', *The Bangladesh Journal of Agricultural Economics*, Bureau of Socio-economic Research and Training, Bangladesh Agricultural University, Mymensingh, Vol. XI, No. 2, pp. 30–5.

Kandiyoti, D. (1988). *Women and Rural development Policies: the Changing Agenda.* Brighton, UK: Institute of Development Studies.

Mayoux, L. (1998. 'Women's Empowerment and Micro-finance Programmes: Approaches, Evidence and Ways Forward', The Open University Working Paper No. 41.

Rahman, M. H. and K. Naoroze (2007). 'Women empowerment through participation in aquaculture: Experience of a large-scale technology demonstration project in Bangladesh', *Journal of Social Science*, 3(4), pp. 164–71. Retrieved from: <http://www.biomedsearch.com/article/Women-empowerment-through-participation-in/182288520.html> (24 July 2013).

Rahman, R. I. (1998). *New Technology in Bangladesh Agriculture: Adoption and Its Impact on Rural Labour Market.* Bangkok: ILO/ARTEP.

Rao, A. and D. Kelleher (1995). 'Engendering Organizational Change: The BRAC Case', *World Development*, Vol. 24, No. 1.

Safilios-Rothschild, C. and S. Mahmud (1989). 'Women's Roles in Agriculture: Present Trends and Potential for Growth', *Bangladesh Agriculture Sector Review*, UNDP and UNIFEM, Dhaka.

Sultana, P. and P. M. Thompson (2008). 'Gender and Local Floodplain Management Institutions—A Case Study from Bangladesh', *Journal of International Development*, 20, pp. 53–68. New Jersey: John Wiley & Sons, Ltd.

UNDP (2007). *Human Development Report, 2006*. Washington, DC: United Nations Development Program.

United Nations (1996). *Report of the Fourth World Conference on Women, 4–15 September, Beijing, China*. New York. Retrieved from: <http://www.un.org/womenwatch/daw/beijing/pdf/Beijing%20full%20report%20E.pdf> (24 July 2013).

Westergaard, K. (1983). *Pauperization and Rural Women in Bangladesh: A Case Study*. Comilla, Bangladesh: Bangladesh Academy for Rural Development (BARD).

Westergaard, K. (1993). 'Review on Women and Gender Issues', in Asaduzzaman, M. and K. Westergaard (eds), *Growth and Development in Rural Bangladesh: A Critical Review*. Dhaka, Bangladesh: University Press Limited.

White, S. (1992). *Arguing With the Crocodile: Gender and Class in Bangladesh*. London: Zed Books.

16 Securing a future for smallholder farmers in an era of climate change

CAMILLA TOULMIN[1]

1. Introduction

'Smallholder agriculture' covers an enormous range of producers in terms of size, crop and land use mix, income diversification, and market engagement. There has been a long-running debate in policy circles and academia about the merits of small and large farms but, increasingly, the case for support to smallholders has been holding ground. Such a case relies on the relative neglect to date of smallholder agriculture and the consequent unmet potential from increased investment in their livelihood systems. Smallholders usually exhibit more intensive use of labour and higher yields per area than extensive agricultural holdings. Equally, smallholder agriculture tends to make use of a more diverse range of crops and activities, thereby maintaining wider biodiversity. There are also powerful arguments, in favour of smallholders, related to social justice and equity. Evidence shows that growth originating in agriculture is much more effective at benefiting the lowest income groups than growth generated from industrial or service sectors, so investing in smallholders should increase output and reduce rural poverty.

Adapting to climate change has until recently been seen as requiring large-scale concrete investments, such as in flood protection measures or irrigation systems. And, while this hardware has a role to play, it needs to be complemented by investment in the software of development, in the form of institutions, governance, and accountability systems. Institutional structures are invisible to the naked eye, but no less vital to building more resilient social and economic systems.

[1] The author acknowledges, with thanks, the contributions of IIED Colleagues Simon Anderson, Muyeye Chambwera, and Charlotte Forfieh. She also thanks the anonymous referee and Atiqur Rahman for their helpful comments and suggestions.

Despite their huge number, smallholders throughout the world face a fundamental hurdle thrown up by their lack of political and economic weight. This means that they face an asymmetry of power vis-à-vis government and corporations in relation to their access to land, water, and natural resources. Equally, the diverse nature of smallholder production means that engagement with high-value markets can be very risky as well as costly (Vorley et al., 2012). The global economy is undergoing major shifts, with a rapid increase in demand for key commodities and the resources needed to grow them. Global markets are also changing in the face of consumer and other pressures. This presents smallholders with an increasingly complex world, both of opportunities but also rising uncertainties. The impacts of climate change and greater variability will exacerbate such difficulties (Godfray et al., 2010).

2. **Main climate threats by region and type of farm**

2.1 DIRECT IMPACTS

Agricultural development faces an array of challenges in low-income nations, due to pressure on resources, falling farm size, and changes in market structure. The impacts of climate change, more variable rainfall, and greater volatility in prices and markets will exacerbate such difficulties. Recent extreme events in Africa and Asia illustrate the devastating impacts of too much or too little rainfall on agriculture and rural systems. While not definitively linked to climate change, the floods in Pakistan in August 2010 brought widespread loss of human life and devastation to the irrigated rice fields of the Indus valley, roads, canals, towns, and villages. It is reckoned that the Indian monsoon has already weakened somewhat, due to the atmospheric brown cloud that sits in a haze over South Asia and Indian Ocean (Lenton, 2011). The floods in 2007 in sub-Saharan Africa also led to agricultural losses, infrastructural damage, and homelessness, while harsh droughts in 2010–11 in the Horn of Africa led to millions displaced by hunger and famine. Global climate change is expected to be associated with unexpectedly heavy rainfall and more extreme storm events, since global temperature rise will generate warmer, more heavily moisture-laden air. Rising temperatures will speed up evaporation from plants and soils, and could lead to a fall in yields from rain-fed agriculture in North Africa of up to 50 per cent by 2020, due to a reduction in the growing season and increased heat stress on plants (Agoumi, 2003).

Looking forward, the impacts of climate change on different agricultural systems depend on the assumptions made as regards the rise in levels of greenhouse gases (GHG) in the atmosphere, and how this affects global, regional, and more local weather patterns. Projections from different climate models used by the Intergovernmental Panel on Climate Change (IPCC) present a range of pathways to 2100, depending on how sharply emissions growth is controlled over the next few decades (IPCC, 2007). Most observers agree that there is now little that can be done to avoid a 2°C rise in global average temperatures by 2050, and further increases in temperature depend on the speed at which we can start bringing down global emissions today. The window for keeping global warming to only 2°C seems to be closing fast, and analysis of commitments made at Copenhagen and Cancun shows that a 4° world is far more probable (Anderson and Bowes, 2011; World Bank, 2012). It is also clear that average warming temperatures mask a wide range of likely outcomes, such that a global average of 4°C warming is projected to bring terrestrial temperatures of more than 4° in most places, especially at the Poles. Predicted temperatures for northerly parts of Russia, Europe, and the Americas of 7–10°C, a likely consequence of a 4° world, would bring catastrophic levels of sea-level rise from melting of polar ice sheets, plus de-frosting of methane stocks in Siberia (Lenton, 2011). Thus, unless we bring down GHG emissions in the very near future, climate change will get worse in terms of rising temperatures and increasingly volatile rainfall patterns.

The higher the level of greenhouse gases in the atmosphere and associated shifts in temperature and rainfall, the greater the impacts. In the event of exceeding 2°C significantly, there will be ever-more marked consequences for survival in different regions, especially those heavily reliant on rainfall and natural resources. A massive increase in environmental migration has been flagged as a likely outcome in a 4°+ world, as in situ adaptation to such great changes will have its limits (Thornton et al., 2011).

A large number of assessments exist for how climate change will impact on agricultural systems around the world. Smallholder farmers are particularly at risk because of their high dependence on agriculture, limited reserves, poor access to credit, insurance, and other financial services, and the fact that they operate in areas with already high levels of risk and uncertainty, and in countries with weak governance. The last of these points means that smallholders face insecure rights of access to and ownership over the land and natural resources on which their livelihoods depend. Customary rights to farmland and collective resources—grazing, woodlands, water bodies—are often not recognized by statutory law. Crops are often growing at the margins of their ecological range and, hence, have little room for further adaptation. In global terms, an International Food Policy Research Institute (IFPRI) study shows the greatest predicted impacts on crop yield are in irrigated wheat and rice systems, where yields could fall by 30 and 15 per cent, respectively, by 2050, with farmers in South Asia faring worst (IFPRI, 2009). However, the smallholders of sub-Saharan Africa are seen as being most vulnerable, due to their more limited resources and lack of access to government systems of support. Particular difficulties are faced in areas where sea-level rise threatens high potential farming zones, such as the Nile delta south of Alexandria and the urban belt along the coast of West Africa.

Climate models demonstrate that farming systems in southern Africa and the coastal fringe of north Africa are particularly vulnerable to a fall in the length of the growing season, with the former region likely to see a shortening of 20 per cent. Most models show that the East African region will receive more rain, but it is likely to come in more intense bursts. There is much less certainty about rainfall trends in West Africa, where the models show a wide range of possible outcomes, from 20 per cent drier to 20 per cent wetter (Conway, 2009). A recent study for Ethiopia has assessed the costs of climate change for a range of scenarios, both wet and dry (World Bank, 2010a). In both cases, there is an adverse impact on levels of GDP, ranging from 2 to 12 per cent, in the first case because of flood damage to infrastructure, and in the second from failed harvests and grazing resources due to drought.

Modelling of rainfall and temperature in India shows high levels of vulnerability to floods and drought, running in a broad swathe across the centre of the country from Gujarat and Rajasthan in the west to the Bay of Bengal in the east. Pakistan and Nepal face a combination of heightened risks of flooding from rapid snow melt and increased aridity in dry areas. Bangladesh is particularly vulnerable to sea-level rise and storm surges, alongside increased flooding from the Brahmaputra river system. Many dry areas throughout the region have already become drier, and major rivers and associated irrigation systems are vulnerable to excessive flooding, as was seen in August 2010. These areas are most vulnerable because of high reliance on the monsoon, and on irrigation from snowfall in the Himalayas. Water stress is likely to grow significantly, with less available from groundwater sources, given continued depletion of supplies. This will have a major impact on irrigated rice and wheat, and highlights the enormous importance of water-storage structures to capture and store rainfall, for use later in the season. However, as was seen with the Pakistan floods in August 2010, a very large volume of rain falling in a very short period may be impossible to accommodate, whatever the storage available.

Given the uncertainties and complexities around modelling and the results, several authors have identified 'hot spots' of vulnerability as a means to focus attention on adaptation challenges, and implications for action in particular areas. For example, Ericksen et al. (2011) bring together different aspects of food security with expected climate changes of relevance to agriculture, such as length of growing season, reliable crop growing days, and variability in rainfall. Each region exhibits a combination of impacts; within sub-Saharan Africa, specific regions of concern include south-west Niger, central Chad, Rwanda, Burundi, and the Ethiopian Highlands, while a global view flags up the extreme vulnerability of Bangladesh, Nepal, Vietnam, and the south-east coast of China.

2.2 INDIRECT CLIMATE-CHANGE IMPACTS

Climate change brings not only direct impacts on farm and livestock production from changes to temperatures and the length of the growing season, and the amount and distribution of rainfall; it will also affect people's welfare and livelihoods through a number of other processes, such as rising sea level, shifts in pests and diseases (IAMP, 2010), and changes to resource availability. For example, as water becomes scarcer, this will place heavier burdens on women and girls, who traditionally have the task of fetching domestic water supplies, as well as their taking on more responsibility for farming.

There are also a number of indirect policy-related impacts which, combined with other powerful global forces, are already unleashing a big rise in demand for land and natural resources. Well-known examples include the rapid increase in demand for biofuels, following the agreement of legal directives to increase their use in the EU, US, Brazil, and China, and the search for extra-territorial land to purchase or lease for growing food crops by countries with a high dependence on food imports (Cotula et al., 2009). In most cases, these new forces have generated adverse impacts for smallholders, since they have put further pressure on land availability in contexts where the rights of local land users are poorly respected by government. To date, this rush to acquire land has been very damaging for smallholder farmers, due to their limited legal protection and weak negotiating position (ILC, 2011).

2.3 OVERALL COSTS OF CLIMATE-CHANGE IMPACTS

Discussion of adaptation of agriculture to climate change is often couched in the language of 'win-win-win', with investment in 'sustainable intensification' bringing improvements in productivity and livelihoods, reduced greenhouse gas emissions, and greater resilience. There may indeed be examples where there are clear positive gains to be made of this nature, and it is important to search for them. However, such optimism needs to be tempered by the real, tangible costs identified from studies of current and future impacts, and the unequal distribution of impacts across the planet. It also needs to be tempered by remembering the formal responsibility held by the major emitters of GHG under the UNFCCC, of compensating those countries and communities damaged by global warming. At a global level, the UNFCCC and World Bank have produced estimates of the costs of adaptation which range from US$41 billion to 171 billion/year by 2030. Stern states that 'adaptation is so broad and cross-cutting—affecting economic, social and environmental conditions and vice versa—that it is difficult to attribute costs clearly' (Stern, 2009). Parry et al. suggest that these figures may underestimate the costs to a significant extent, and that actual costs are likely to be bigger by a factor of 2 to 3 (Parry et al., 2009). The higher the rise in average global temperatures, the greater will be the cost and risk of a breakdown in life support systems.

3. Adapting to climate change

3.1 BUILDING RESILIENCE TO DEAL WITH RISK AND UNCERTAINTY

People face a range of different risks relating to production, prices, and personal circumstances. Dealing with risk involves assessing a set of outcomes, with known probabilities attached, based on the past being a good guide to the future. This is the basis on which cover is provided by insurance companies. Uncertainty is harder, as it involves an unknown distribution of different outcomes, preventing the past being an accurate guide to the future, and insurance is not so readily available. One current example is the rapid change in pension provisions for people in Europe, given the recognition that people are living longer than anticipated and consequently the sums that have been set aside are no longer enough to provide cover at the anticipated level. Climate change is another example which generates uncertain outcomes, given that the past no longer provides a good guide to the weather patterns of today and tomorrow. Small island states have had difficulty in getting insurance for infrastructure, and hotels for more than a decade because of fears of sea-level rise and more intense storms. A further area of risk and uncertainty relates to state fragility and poor governance. IFAD notes that 'unaccountable public authorities and institutions introduce an element of unpredictability into public life that can significantly increase transaction costs associated with market investments and contracts, access to services and utilities, and practices of citizenship. These affect particularly poor rural people due to their lack of power' (IFAD, 2010).

There are various responses to uncertainty, which include diversification into less weather-prone activities, seeking out information about likely risks and returns of different activities, and building a more resilient system. Resilience refers to the ability of a region, country, city, village, or household to protect itself from adverse impacts and recover from damage. Debate surrounds the best ways of ensuring resilience, with some arguing for more intense market links, while others see market engagement adding further risks. Thus, for example, increased income and economic development are often presented as the best means to adapt to climate change, since having a higher income provides a wider range of opportunities. Others argue that agriculture and rural development should seek greater self-reliance, reduced use of external inputs, re-use and recycling of resources, and stronger local organizations (Jones et al., 2010). As will be seen, some combination of these strategies seems the best answer, depending on context.

Diversification: Strategies to address risk and uncertainty surrounding farm production include diversification of assets and activities, mixing crop and livestock production, pursuit of off-farm incomes, and so on (IFAD, 2010). Understanding household survival in risk-prone areas has often taken an approach based on portfolio analysis, farm households protecting themselves by investing in a range of different assets (Toulmin, 1992; Scoones, 1998). Such protection works best when each asset or activity is subject to risks which are not closely correlated. For example, growing crops of different cycle length provides greater assurance of a harvest when distribution of rainfall within the growing season is highly variable. Planting a combination of different crops, and raising a mix of cattle, sheep, and goats achieve the same purpose. Combining crops and livestock not only generates co-benefits from provision of traction, dung, and fodder, but different risks are associated with each. Livelihood diversification has long been a strategy pursued by many smallholders in all parts of the world. In Mali, farming households combine cultivation of two millet varieties with groundnuts and sesame, cattle, sheep and goats, a shop-keeping business, craftwork, migration of young men and women to town to earn cash, and individual ways of earning cash (hunting, fortune-telling, petty trade, cloth dyeing). In other marginal farming areas, such as the north-west highlands of Scotland, crofting smallholders mix off-farm work and running a Bed & Breakfast with fishing, a few head of sheep on the hill, and reliance on a plot of vegetables. Rural dwellers also try to improve their options in the longer term by, for example, getting some of their children into school, marrying their children into better-off families, and establishing part of the household in town.

Farm and crop contracts: People also face other risks, such as those linked to market price volatility, and may seek to reduce their exposure to this risk by contract farming, establishing reserves and storage for later sale, or use of forward contracts. A range of farm contracts exists, such as sharecropping, which enables landowners and land users to share the risks associated with growing a particular crop and associated investments. Well-documented examples are those of *abuna* and *abusa* sharecropping contracts amongst the Akan of Ghana and Côte d'Ivoire, which have been at the heart of the expansion in cocoa plantations through the twentieth century (Lavigne et al., 2002). The terms of these contracts have changed over time to accommodate shifts in the relative scarcity of labour and land. A range of other contracts has been used for cash crops such as tea, cotton, and coffee, through which the buyer provides a series of inputs on credit against delivery of the crop at harvest. Each contractual form offers a different balance of risks, rewards, and voice to the parties concerned (Vermeulen and Cotula, 2010). In the last few years, rising uncertainty and scarcity in basic commodities have led many large agri-food companies, wanting to guarantee themselves assured supplies, to agree long-term contracts with farmers. Equally, where land ownership by foreign investors is not an option, alternative contractual forms which do not involve land purchase must be sought, such as joint ventures, with shared equity.

Insurance: In developed countries, farmers can access a wide range of insurance packages to cover against loss or damage from weather, pests, diseases, and price changes. A Google search for the UK found 95 different companies offering a range of products. In the event of a claim being made, this often requires costly time-consuming loss assessments through farm visits to see the damage being claimed, a cost which is bearable if the farm is large. However, many insurance schemes have some level of government subsidy; for example, in the US, it is estimated that the Federal Government covers 59 per cent of the cost of the main crop insurance programmes.

There are many examples of agricultural insurance in Asia and Latin America, but sub-Saharan Africa has been very poorly served to date, with less than 0.5 per cent of the total market (Mahul and Stutley, 2010). Lower-cost models are needed for insurance to be of value to poor farmers. New weather-related insurance considers rainfall data, rather than actual yield, which is much less costly to monitor. It also eliminates the need for farm visits. As a reliable and independently verified index, it can be reinsured into the international market, but does require a dense, high-quality network of weather stations. However, there are concerns that the historic rainfall record may provide insufficient guidance to future conditions and, hence, the premium to be paid (Hazell et al., 2010). Such insurance policies are often bundled into input packages and farm credit deals. One of the drawbacks is that policies do not cover all of the risks farmers face regarding production, such as price fluctuations and impacts of pests. In many cases, the relationship between farmer and institution offering the insurance is vital, for farmers to feel confidence that their losses will be covered fairly in the event of rainfall failure (United Nations, 2007). Such trust takes time and experience to build up.

Over the last decade, a wider range of crop insurance products has been developed (see Rahman and Smolak, Chapter 8 of this book). For example, in India the Hyderabad-based micro-finance company ICICI launched the first index-based weather insurance for two hundred groundnut and castor farmers, which has now spread to some eleven thousand farmers. Equally, an Indian seed company has attached free weather insurance to its cotton-seed packages in Maharashtra. Syngenta Foundation has experimented with insurance in Kenya, aimed at covering maize crops from adverse weather in western Kenya. Growing from 200 farmers in 2009 to 11,500 in 2010, the aim is rapidly to scale up further. Payments are made based on weather data recorded at a particular set of weather stations, and there are no farm visits, cutting very substantially the costs of administration. Payment is made through M-Pesa mobile telephone banking. Farmers can buy the insurance at a 5 per cent premium when purchasing farm inputs (Ferroni, 2010). In India, Pepsico provides weather insurance for farmers contracted to supply potatoes for its chip business, and Nyala Insurance in Ethiopia has extended its support to include farmer education and awareness-building (see Rahman and Smolak, Chapter 8).

Collective insurance mechanisms: People have various collective means by which to protect themselves against risk, of which the family is often the foundation. Individuals and households invest in relationships which assure some level of mutual support. In 'normal' times, these may be pretty effective, as when a single family or village is badly hit by crop losses. In such cases, people will know they can go for help to others in the same or neighbouring communities. They may be much less able to cope where losses are extensive and heavy, such as when drought or conflict affects a whole region. Individuals also face risks linked to health and welfare; strengthening ties with family, friends, social networks, and relationships is one means by which to ensure support in case of need.

There is much that can be learnt for adaptation strategies from the last 30 years' analysis of drought and famine, in terms of household resilience, why some are more vulnerable than others, and the form taken by successful interventions. These lessons demonstrate the importance of protecting the assets of poorer households in times of drought and famine to avoid loss of essential productive capital, such as cattle, land, ploughs, and work-oxen. Evidence from disaster-preparedness shows it is far cheaper to invest in effective preparation and protection of assets, than to restore damage ex-post (IFAD, 2010).

3.2 ADAPTING RURAL SYSTEMS TO CLIMATE-CHANGE RISKS

Climate-change adaptation encompasses, according to the IPCC, 'not just adjustments in ecological, social or economic systems in response to actual or expected climate stimuli and their effects', but also adjustments to reduce harm from, or to benefit from, current climate variability as well as anticipated climate change. It can be a specific action, a systemic change in livelihoods, or an institutional reform which provides better management and protection of natural resources. Adaptive capacity varies between individuals and communities according to parameters such as wealth, status, gender, and access to livelihood alternatives.

Adaptation is likely to involve a range of strategies, pursued by actors at different scales, some market-based and others relying more on social and institutional networks. At an individual level, farmers and livestock keepers often adapt their strategies to address shifts in weather or market conditions. Farmers can adjust to changes in rainfall by changing the planting date, seeking different seeds with greater resilience to drought or flood, and managing their water and soils in different ways. Where rainfall is declining, farmers can invest in irrigation and more careful management of water, such as by constructing planting pits to capture and hold moisture more effectively (Reij et al., 1996). Herders can adjust the date they move their herds to different pastures, and shift the composition of animals held, towards more drought-hardy species.

Governments have an important role to facilitate adaptation at community level. Work by local governments, rural councils, and municipal authorities is of great importance in building an effective response that can make the most of local knowledge and priorities. The International Institute for Environment and Development's (IIED) engagement with national and local governments in Nepal and Kenya is aimed at exploring the best way for government to support the diverse means by which different households and communities are adapting their livelihoods. First findings include the need for local adaptation plans tailored to the demands of different agro-ecosystems and livelihoods; establishing a means by which local communities can exert demands on government and research bodies to call for the kind of help they need; taking account of the fact that people differ substantially in their assets and ability to cope; and firm support by the state to decentralized governance structures and decision-making.

The debate over what makes for effective adaptation to climate change has evolved considerably over the last five years, from an initial focus on large-scale infrastructure, such as building flood defences, to recognition that adaptation needs to promote bottom-up processes of change based on support for community-based action. This demands that the 'software' of development and change is addressed, especially issues around power, accountability, and governance. Investment in physical infrastructure may be essential, but choices made here need to consider how this infrastructure will protect and benefit different groups. For example, there is currently renewed enthusiasm for investment in major dam projects in many parts of Africa, to fulfil multiple purposes, such as hydro-power, irrigated agriculture, and domestic water supply. Yet those local people most affected by construction and changes to river flows are usually the last to be consulted, and rarely compensated for their loss of land and assured water supply (Skinner et al., 2009). Smaller scale systems of water capture and use may generate far broader benefits to local people than building a few large dams, yet the latter option is usually preferred by governments and donors as it is seen as 'modern', as well as generating contracts and revenue for those in government and the private sector.

IIED has championed the concept of Community Based Adaptation (CBA) as the best means to enhance capacity amongst NGOs and other civil society groups working with the most vulnerable groups. It starts from the basis of recognizing the knowledge and skills of local people, their different skills and capacities, and facilitates their own analysis and identification of forms of adaptation that make sense in their own context. Over the last six years, an annual international workshop has brought together practitioners from across the world to learn and share from each other on how best to support CBA. There is a growing body of evidence from the field of many diverse adaptation activities in practice (Reid and Huq, 2010). Work has also been carried out to support local innovation systems, through Participatory Innovation Development, Participatory Learning and Action, Climate Field Schools, and other approaches, drawing on the legacy of earlier approaches to support sustainable rural development (PROLINNOVA, 2010).

Box 16.1 outlines some of the practices used by pastoralists in Ethiopia and Niger to address the impacts of climate change. This recognition of the value of indigenous knowledge and innovation capacity is vital in strengthening local responses. But it needs also to admit there are limits to adaptation, and value to be gained by complementary investments in infrastructure, knowledge, markets, and policy which can further strengthen people's ability to cope with change. A combination of measures, operating at different levels, is the best response, with local and national governments seeking to support community-based processes. While this is the ideal scenario, there are no guarantees of this approach being followed. Instead, governments faced by what they perceive as a climate change 'crisis', may adopt a much more top-down approach. Equally, elite groups may seize their chance to exploit new opportunities, such as by grabbing land or tapping into funds designed for Reduced Emissions from Deforestation and Forest Degradation (REDD).

BOX 16.1 PASTORALIST RESPONSES TO CLIMATE CHANGE IN ETHIOPIA AND NIGER

Developing cut and carry fodder systems, to conserve fodder.
Investment in water points to ensure secure access for people and herds.
Closer relationships with people in town, often former pastoralists.
Shift in herd composition to more hardy species, especially donkeys for transport.
Diversification of livelihood sources.
Strengthening traditional institutions to re-establish control over key resources, and enable negotiations with neighbouring groups.
Reliance on mobile phones for access to information and social contact.
Many of these are collective responses, given strong traditional structures in pastoral groups.

Source: GebreMichael et al. (2010).

Much of the debate around adaptation to climate change presents an optimistic picture in which there are significant co-benefits from combining mitigation and adaptation actions. However, it must be recognized that adaptation has its limits. It is very hard to adapt to some shifts in rainfall and temperature. Equally, the higher level of volatility in weather patterns brings its own challenges. Such difficulties will increase, the higher the level of warming. There are likely to be thresholds and disruptions to production systems, particularly in a world of 4°C and beyond. We do not know where these thresholds might be, so it would be prudent to stay below such planetary boundaries (Rockstrom et al., 2009).

4. New opportunities from climate-change policy and markets

Three main market opportunities have been generated by climate-change policy: a global cap on emissions, production of biofuels, and funding for reducing deforestation and forest degradation (REDD). The agreement of a global cap on carbon emissions has created a market for permits to emit carbon, and a set of new activities and potential market opportunities. The European Emission Trading System and Kyoto Protocol Clean Development Mechanism are examples of this, although current carbon prices have been weakened by low levels of commitment to mitigation. Given that the agricultural and land-use sectors are reckoned to contribute up to 30 per cent of global emissions, agriculture is expected over the next few years to be brought into arrangements to curb greenhouse gas emissions by setting targets for their reduction. The principal sources of GHG from agriculture are carbon dioxide associated with ploughing and land clearance, methane from livestock emissions and irrigated rice growing, and nitrous oxide from use of chemical fertilizer.

While agriculture and land use are a major source of GHG, land is also a very large sink for absorption of greenhouse gases, through sequestration of carbon in soils and vegetation. Estimates vary for the incremental carbon that could be sunk into agricultural landscapes through changes in land management, but these may be as high as 5Gt of carbon per year, equivalent to one-sixth of global emissions (Steer, 2010). In theory, if land users could gain revenue from sequestering carbon, and reducing emissions from methane and nitrous oxide, this could provide a significant source of additional income. This could be achieved through changes to land-use practices, such as 'no-till' conservation agriculture, making biochar, managing manure, grazing management, re-greening through natural regeneration of vegetation, and tree-planting.

Table 16.1 Potential for carbon sequestration in different soil management and cultivation systems

Management Practice	Mean change in tCO_2e/ha/year
Vegetation cultivation	9.39tCO_2e/ha
Avoided land cover/land use charge	0.40tCO_2e/ha
Grazing management	2.16tCO_2e/ha
Fertilization	1.79tCO_2e/ha
Fire control	2.68tCO_2e/ha

Source: Tennigkeit and Wilkes (2008).

Table 16.1 shows the potential for carbon sequestration in different soil management and cultivation systems. A recent review of 22 carbon sequestration projects in agriculture around the world shows huge diversity in scale and returns, with levels of sequestration spanning from 1.37 to 140tC/ha per year and prices of C varying from US$10–180/tC (FAO, 2010). The average price paid was US$20/t of C, with almost all projects financed through the voluntary carbon market.

Thus, in theory, there is a range of carbon market options for smallholders to pursue. In practice, however, experience to date has been with niche voluntary markets. Access to the formal carbon market may be harder to establish for smallholders, and is currently offering very low returns. For markets to work well, a clear specification of the product being transacted is required. In the case of sequestered carbon in farming landscapes, there are several hurdles to jump, not dissimilar to those encountered for a range of other goods and services, such as organic or fair trade produce and for REDD. The current science of understanding, monitoring, and measuring changes in the status of soil carbon is limited, especially in tropical soils. As with REDD, there are questions around persistence of soil carbon pools and their vulnerability to shocks (such as heavy rain bringing soil erosion), leakage due to intensified pressure in neighbouring areas, and issues around who can claim rights over the resources able to provide this service, such as soils or forests (Robbins, 2011). There are also significant costs for producers of these services in engaging in markets, because of the transaction costs associated with certification and quality assurance. Hence, if carbon markets are to bring significant benefits for smallholders in practice, they need to design ways to minimize their transaction costs. This would require use of a simple proxy for C sequestration, and bundling of C services to reduce costs of contracts and market engagement.

A second area where smallholders could potentially gain from climate change-related market development concerns production of feedstocks for biofuels. The market in biofuels is driven very largely by government policy measures around the world. The EU, for example, has proposed a mandatory target by 2020 for 10 per cent of all member states' transport fuels to come from biofuels, and similar targets have been set by the US, Brazil, India, and China. As a result, global demand has increased by around 20 per cent per year up to 2011, reaching more than 90 million metric tonnes, and will expand further. The expected mismatch between global demand and supply means that feedstock production and international trade in biofuels are expected to grow very rapidly in the coming years. African countries have a very limited market share in biofuel production and trade at present, but there has been great interest from governments and investors seeking to establish an industry in a number of countries. One reason is the prospect of attracting investment, for example through accessing some funds from carbon trading systems (e.g. the Clean Development Mechanism). Another is that many developing countries located in tropical and subtropical areas have, or may develop, a comparative advantage for feedstock production, such as sugar cane and oil palm (OXFAM, 2007; Dufey et al., 2007). This creates an opportunity for African nations to develop a new export market for their agricultural produce and to increase export revenues. Biofuel production has been presented as potentially able to improve agricultural employment, incomes, and livelihoods. Such potential may be attained by large-scale plantations, but also when cultivation involves small-scale farmers, particularly for meeting their own energy needs. In the case of large-scale production, there are several very serious concerns about which land these crops will be grown on, the risks of eviction for thousands of smallholders, the number of jobs created and urgent need for consultation, arbitration, and compensation in the event of land deals being struck which displace smallholders.

A third field for potential income for smallholders involves access to REDD funding, in exchange for reducing emissions from deforestation and forest degradation. There is significant potential for benefit, but concerns have been raised for how it might work out in practice. The focus to date has been on tropical moist forests, in countries such as Brazil, Indonesia, Guyana, and the Democratic Republic of the Congo (DRC). The basic assumption driving a REDD scheme is that forests need to be more valuable standing than felled. But it needs to be asked—'more valuable to whom?' Standing forest may well be more valuable to forest dwellers than seeing their land cleared, but they are rarely in a position to decide. Conversely, standing forest may be worth much less to a forestry official than when it is felled, since he can gain money from the issue of timber-felling permits. This difference in interests and incentives, combined with ineffective governance and oversight of forest management, are serious weaknesses that limit the effectiveness of forest carbon markets.

REDD could also offer revenue for savanna and dryland regions. Recent work in the West African Sahel shows that these drylands can act as significant carbon sinks, because of the very extensive areas involved. A conservative estimate for carbon sequestration in the mixed Acacia woodlands of the Sahel gives a figure of 20 tonnes of carbon per hectare, based on the tree mass. Including the below-ground carbon and grassland elements would increase the volume associated with the restoration of woodlands in the Sahel. If carbon is priced at US$10/ton of carbon, this would generate the equivalent of US$200/hectare. Reij reckons an area of 5 million ha in Niger has already undergone this improvement, equal to US$1 billion worth of carbon sequestered (Reij, 2008). None of this activity has received any financial compensation for the carbon to date. It is interesting to ask how much more might be achieved were such a financial mechanism in place, and the heightened risk of land grabs associated with making land more valuable in this way.

There are many challenges to building a strong and equitable market-based REDD policy (Saunders and Nussbaum, 2007), given the weak institutions responsible for managing forests and woodlands, and associated systems of governance. Serious concerns have been raised about the distribution of funding from such schemes, and the share to be gained by national and local government coffers, as opposed to local people. Equally, forest land will inevitably become more valuable as a result of REDD payments, and more powerful groups will likely displace forest-dwellers in order to reap the REDD rewards. In a global context where commodity prices are highly volatile, it is unclear at what level the REDD payment should be pitched. If it is too low, relative to the gains to be made from soya bean and oil palm, then it will not act as an incentive to stop further forest clearance. If it is too high, it will prompt further land seizures by elites.

While many governments are keen to acquire REDD money, they are much less keen to share this with local people whose actions ensure the maintenance of forest cover. There is a mismatch between REDD as a mechanism for reducing emissions and the best ways of ensuring sustainable management of forests for local people's benefit. REDD strategies need to base themselves on building rights, capacity, and incentives for good forest management. Reliance on a market mechanism alone delivers neither the carbon nor local livelihoods (Mayers, 2010; IIED 2012).

5. **What mix of measures will help smallholders adapt and prosper?**

An array of measures could support adaptation to climate change amongst smallholder farmers. These include technology, securing property rights, investment in infrastructure, and support for local organizations and knowledge systems. Each of these elements has something to offer, but no single one can provide all that is needed. Rather, a combination of measures is required, the particular mix depending on context and opportunity. A focus on market opportunities should not diminish Annex 1 nations' commitment to support adaptation as a legally binding component from the UNFCCC treaties.

Research and technology: Much agricultural research to date has focused on breeding new varieties for drought or flood tolerance as the best means to address climate change. There are considerable opportunities here from screening of crops for drought tolerance and selective breeding of livestock and crop varieties in relation to changes in the amount and distribution of rainfall, length of growing season, incidence of pests and diseases, and frost conditions. However, it is important also to think about improving the broader cultivation system, such as ways of managing soil and water in the wider landscape, and building more effective nutrient recycling systems. In the case of sub-Saharan Africa, a key constraint to sustainable farming is the low availability and use of nutrients across the continent. The use of chemical fertilizer averages 6–8 kg/ha in comparison with 100kg/ha globally and more than 200kg/ha in China. While global usage of chemical fertilizers clearly needs to be cut, by improving nutrient use efficiency, farmers in Africa will need to increase their inputs of key soil nutrients if they are to increase their crop yields (Pretty et al., 2011).

One of the drawbacks of the international agricultural research system has been the privatization of most research capacity in many developed countries, which had formerly been a major source of innovation for adaptation and take-up in other parts of the world. Since the 1980s, state support for this research has fallen, so that research by large agri-food corporations now dominates the landscape. While this corporate concentration provides strong incentives for development of products which can be commercialized, there is no incentive for research into the generation of public goods, nor changes in agronomic practice which might lead to farmers using fewer, rather than more, inputs. In many developing countries, especially in Africa, spending on agricultural research and extension was cut back in the structural adjustment programmes of the 1980s and '90s. Some more positive changes are now underway, with substantial public investment in agricultural research in major middle-income countries, such as Brazil, China, India, and Turkey. There has also been renewed commitment by donors to the global network of the Consultative Group on International Agricultural Research (CGIAR) research centres, and some pilots are underway to test out private–public partnerships that could generate crop varieties of value to poor countries and communities. Building resilience to climate change for the smallholder sector needs to be the focus for a renewed programme of research supported by public funds (CCAFS, 2010).

Information and communication technologies: The revolution in information and communication technologies (ICT) is opening up access to external knowledge among even the poorest. The growth in the number and spread of mobile phones has been particularly striking in the last decade, despite little government support. Investment in transmission masts, the use of 'pay as you go' contracts, and sharing of handset costs have brought mobile phones within reach of many relatively remote communities in low-income countries. This has facilitated greater connectivity, access to extension services, and reduced market transaction costs. There is evidence, for example, that the introduction of mobile phones in Niger reduced the volatility of prices in local markets and the differences in prices between markets (Aker, 2008). Mobile phone technology may also facilitate greater transparency and better governance, through local people being better informed about their rights and expectations of government. In sub-Saharan Africa, subscriptions have risen from 16 million in 2000 to over 500 million in 2009, an astonishing 30-fold increase, which demonstrates the receptiveness of many poor people to new technology when it fits their priorities and the price is affordable. Because of the enormous growth in the market for phones at village level, a number of financial service providers are now designing their services for smallholders, such as M-Pesa banking in Kenya.

Recognizing local resource rights: Justice demands that local people have their rights recognized and respected by government. In addition, investment in land, soils, water, and trees requires confidence that the benefits reaped tomorrow and next year will repay effort laid out today. Effective adaptation to climate change will be much more effective where local people have their rights recognized, since this will encourage them to improve the resilience of their systems, and gain some benefit from sale of land assets should they decide to leave farming. In the case of land and other natural resources, such as water and forests, there are often poorly defined rights at both individual and collective levels, with overlapping and inconsistent claims to these resources. Many countries exhibit pluralistic legal systems, where customary rights continue alongside more formal statute law. Uncertainty surrounding land rights may discourage agricultural investment, through fear of contest and eviction. Land rights are particularly insecure for groups with little political weight at local or national levels, such as women, pastoral herders, and migrants. Equally, tenancy and sharecropping arrangements can lead to sub-optimal levels of investment, if the land-user feels a risk of land being taken away. Attributing firmer rights to land-users and clarity regarding the terms of tenancy can bring substantial yield gains through increasing the incentive to invest in longer-term improvements. This is of growing importance given rising pressures on land in many developing countries (Anseeuw et al., 2011; World Bank, 2010b; Cotula et al., 2009; Lavigne et al., 2002). As IFAD notes, 'land dispossession has been a continuous process over centuries. Yet the new attractiveness of agriculture resulting from higher commodity prices and subsidies from biofuel production is leading to increases in domestic and transnational demand for agricultural land, bringing new risks for poor rural people' (IFAD, 2010). These risks are only likely to grow further in the decades to come. The government of Brazil has imposed a moratorium on large land purchases by foreigners, but many other governments are keen to attract incoming agricultural investment by making land available.

Strengthening of local rights to land and other natural resources needs to combine legal measures, administrative provisions, and practical tools that enable people to get formal recognition of often unwritten rights. Formal land titling programmes have often missed their goal because they have proved slow, inaccessible to rural dwellers, and costly. Equally, women tend to miss out when land rights are formalized, since under customary tenure, they gain access through a male relative. In a few cases, women are now getting their rights recognized either as independent land rights holders, or as joint holders with their husbands. Although many African countries now give attention to gender equality in constitutional provisions, in practice in rural areas, women's position remains largely unaffected by such legislation (Joireman, 2007).

Weak property rights also generate difficulties in the management of collective resources, such as common grazing land, forests, and fisheries. Without well-defined and recognized institutions for management, the pursuit of short-term interests by individuals often results in a race for harvesting what can be taken, with adverse consequences for the long-term viability of the resource in question. This problem also applies to ecosystem services, such as those associated with air, soil, and water quality, and the lack of private incentives to invest in maintaining and enhancing their value. Worldwide, two-thirds of forest land is claimed by government, allowing state agencies to grant logging concessions without consulting local people. The assertion by many African governments of rights over all trees and forests in rural areas, including those on farmland, and an inability to exercise management responsibility in practice, due to inadequate staff and other resources, open up a vacuum in which local people feel no ownership of these resources and the state can exert no effective authority. Building more resilience in smallholder systems demands that local rights to manage and control access to land and key natural resources be recognized. Such systems of registration need to be simple and accessible, so that they offer effective security, encourage smallholder investment, and protect less powerful groups (women, migrants) from eviction by others.

Bridging local knowledge and modern science: The understanding and application of modern science offer large benefits for food production, when linked to local knowledge of soils and ecosystems. Blending insights from different knowledge systems can help improve productivity and efficiency as well as produce more resilient food production systems by engaging local farmers in defining research priorities, and assessing new methods and practices (Dryzek, 2000). It is particularly vital to integrate a better understanding of gender and inequality into research and extension by recognizing women's expertise and the diverse roles they play in the farm and food economy. Women usually have greater responsibility for family food production and processing, whereas men have greater engagement in market-oriented production and off-farm income. Women also have responsibility for preparing food for their families, which means they may have marked preferences for certain crop varieties, in terms of ease of cooking, or pounding into flour. Women are often more involved in vegetable and fruit production around the homestead, which rarely receives as much attention as the main grain or cash crops. Research shows that a better understanding of the different roles played by women and men in the agricultural system and related businesses leads to better interventions for boosting productivity and building adaptive capacity. It is reckoned that reducing inequalities in access to inputs between men and women farmers could bring about an increase in agricultural productivity of 10–20 per cent, as well as addressing a fundamental inequity (Meizen-Dick et al., 2010). Building climate resilience means bridging local and outsider knowledge in ways which enable an effective dialogue, and which recognize the differing needs and resources available to rich and poor (Scoones and Thompson, 2009).

Investment in social infrastructure and social learning lies at the heart of building more resilient social systems, by strengthening the collective capabilities of people to work together (IFAD, 2010). Traditional extension has been centred on the linear transfer of knowledge from an expert to the farmer (Foresight, 2011). Methods which turn the extension system upside-down, such as social learning groups, collective action, and general empowerment, can be more effective in spreading knowledge and uptake of new practices. Examples of more 'bottom-up' methods of facilitating knowledge transfer in the absence of major public or private financing are provided by 'farmer field schools' and related initiatives (Pretty et al., 2011). Originating as a movement in Indonesia to reduce reliance on agrochemicals, farmer field schools have become an international movement, incorporating many of the ideas of group membership, collective action and local research, aided by trained facilitators who have now become part of modern extension practice.

Strengthening the voice and power of farmer associations is a vital means of addressing the range of challenges faced by small-scale farmers, whether for issues of the environment, market access, or innovation. In Uganda, women have organized into groups to process and sell cassava. In Nigeria, aquaculture entrepreneurs are raising and selling fish, while others concentrate on producing and selling feed. In Kenya, the extension system encourages farmers to form common interest groups for generating business activities (Pretty et al., 2010).

Market engagement: There has been considerable effort to think through how markets might 'work for the poor', through fair-trade and other mechanisms. Some major companies have sought to show they are drawing some part of their produce from smallholders (such as fair-trade roses from Kenya). Yet evidence also shows that the bottom third of the farm population will always find it much harder to access new market opportunities and require specific attention if they are to reap the benefit. IIED's analysis of markets for ecosystem services shows that the rules for participation in many of these markets tend to favour the better-off. A review of eleven years of Costa Rica's payments for ecosystem services, a role model for initiatives worldwide, illustrates the difficulties in making such payment systems accessible to the poorest farmers. A very clear filtering process is needed if more payments are to reach the hands of smallholders (Porras et al., 2008). Reasons why poorer groups find it difficult to access markets include information, transaction costs, and the burden of demonstrating compliance, and the overall return from engagement in formal markets is often very low (Shames et al., 2012).

As markets become ever more connected and complex, those producing food for sale must learn how to link to markets, identify market niches and consumer requirements, and navigate the complexities of national and international regulations involving food safety, food quality, and environmental sustainability. Individuals need the basic training to cope with business challenges and the skills to make use of information sources—both economic (market prices for example) and agronomic (such as up-to-date weather forecasts). The notion of 'farming as a business' has become a theme in the national agriculture strategies of several African countries, including Uganda and Ethiopia, and has been shown to increase rural incomes when communities have sufficient training to be able to capitalize on commercial opportunities (State of Science Review, 2010). In India and Vietnam, evidence shows the important role to be played by market intermediaries, in helping farmers gain information and guidance in new market conditions (Vorley et al., 2007). While market opportunities are hugely important in enabling farmers to improve their incomes and savings, attention is also needed to the institutional and social infrastructure through which such market opportunities can best be channelled.

A broader question concerns whether market engagement is a good means for smallholders to adapt and protect themselves from risk, or if it generates greater vulnerability. IIED's analysis shows that it rather depends on the 'markets' in which people engage, and how they are structured and governed. Smallholders face dualistic markets; on the one hand, high-value export crops require a high level of traceability, quality control, and certification. Transaction costs for market engagement are high and there are significant risks associated with the high returns, as was seen, for example, from the impact of the volcanic ash cloud on air freight in April 2010. There are also consumer concerns around buying air-freighted produce because of its higher carbon footprint, which may make demand more volatile. On the other hand, many local and regional markets have few if any barriers to entry, but returns are much lower, especially in markets for coarse grains. In general, it is likely that smallholders find it easier to access local and domestic markets than global supply chains. Over time, however, this dualism is blurring, with the transformation of food markets in many middle- and low-income nations, and the arrival of large-scale wholesale and retail buyers. This will put a tighter squeeze on local farmers, unless they can demonstrate that they can meet the rising standards required for sale into more formal markets (Murphy, 2010).

The 2007–8 food and commodity crisis also showed the limits to reliance on global markets to assure food supplies at times of shortage. While there were various causes for the escalation in food and commodity prices, including drought, cereal export bans, and the rapid rise in oil prices, the globalization of commodity and financial markets now means hot money can flow into a set of assets which include foodstuffs and land. Even those food importing countries with plenty of money found, in practice, that it was much harder to secure stocks than they had assumed, generating a strong interest in securing farmland elsewhere so they could control their own food production.

Investment in physical infrastructure: In many poorer countries, agricultural production and market access may be severely limited by lack of transport infrastructure, energy, or irrigation facilities. In sub-Saharan Africa, public investment in infrastructure fell from 6.3 per cent of government budgets in 1980 to 3.7 per cent in 2005 (Fan et al., 2009), bringing in its wake low levels of irrigated agriculture, poor road density and inadequate maintenance, and minimal electrification in rural areas. Levels of irrigation development are particularly low in sub-Saharan Africa, at only 4 per cent of cultivable land, in contrast to the world average of 18 per cent and 34 per cent in Asia. Yet yields from irrigated crops tend to be up to three times higher than those on rain-fed lands (Foresight, 2011). Cheaper, quicker transport between farm and market is also likely to increase levels of investment and production, since the gap between farm gate and market price is reduced, and perishable crops can reach the buyer in a better condition. Amongst road infrastructure investments, it is estimated that rural and feeder roads have a larger impact than higher-quality roads (Foresight, 2011). Investment levels in sub-Saharan Africa are also very low in water storage and irrigation. There are clear choices to be made in terms of the kind of infrastructure to build. For example, if you choose to build one large dam rather than twenty small dams, there will be very different implications for the wider society and landscape. If building resilience of smallholder farmers is the chief priority, then the latter option may be better, while if electricity generation for the capital city is the prevailing concern, then the former may count as a better investment. The African Union's Programme for Infrastructure Development in Africa (PIDA) is starting to address this investment shortfall in a systematic manner.

6. Conclusions

The root of smallholder vulnerability lies in the political marginalization of farmers, pastoralists, and other rural groups in power and decision-making. This is a fundamental problem for smallholders everywhere because, despite their large numbers, each has very little power and it is costly to get organized. In any negotiation, smallholders are faced with asymmetric information and power, whether they face government or corporations. New forms of communication offer a potential means to establish larger and more powerful organizations and structures able to represent smallholder interests, and to demand greater accountability from local and national government. Several projects aim to open up government decisions and administration to greater transparency, such as the computerization of land registers in several Indian states, participatory budget design in West Africa, and the Twaweza initiative in East Africa (Foresight, 2011). Equally, programmes to empower smallholders by providing support to their organizations and training in basic legal rights could make a difference in helping to level the playing field (Cotula, 2009).

Climate change will generate increased difficulties for many smallholders, particularly those in low-income tropical nations. Increased uncertainty and volatility as regards rainfall and length of growing season will add to existing challenges they face due to political marginalization, such as competition for land, and changing market access. The higher the level of GHG emissions, the more extreme will be the impacts on farming systems. We should all be very seriously concerned about the lack of any progress in reining in GHG emissions at a global level. Neither government, business, nor most citizens have understood the enormous and growing costs of inaction, compared to the needed investment to transform our economic and productive system towards a low, or zero carbon system.

Addressing the impacts of climate change demands a combination of measures. These include building social institutions for learning and accountability, and investment in securing property rights, and in infrastructure, agricultural research, and developing new markets. Support for adaptation is fundamental to achieving a fair balance between mitigation and adaptation within the UNFCCC agreements on addressing climate change. Reliance on market opportunities alone to achieve effective adaptation and mitigation would be both unethical and unwise. It is not a question of making a choice between either market-focused interventions and measures, or publicly funded collective, social infrastructure, and investment. Both will need to be combined in varying mixtures, and public policy is key to setting the frame for many new market opportunities to evolve. This public–private mix is essential in generating greater resilience within broader livelihood systems, which recognize the spread and mix of activities and revenues through which smallholders have been able to survive and prosper to date.

REFERENCES

Agoumi, A. (2003). 'Vulnerability of North African countries to climate changes, adaptation and implementation challenges', IISD Climate Change Knowledge Network. Retrieved from: <http://www.cckn.org>.

Aker, J. (2008). 'Does digital divide or provide? The impact of cell phones on grain markets in Niger', *Working Paper 177.* Durham NC, BREAD (Bureau for Research and Economic Analysis of Development), Duke University, USA.

Anderson, K. and A. Bowes (2011). 'Beyond "dangerous" climate change: emission scenarios for a new world', *Phil.Trans Royal Society,* 369, pp. 20–44.

Anseeuw, W., L. A. Wily, L. Cotula, and M. Taylor (2011). *Land rights and the rush for land.* Rome: International Land Coalition.

CCAFS (2010). 'Agriculture, food security and climate change: Outlook for knowledge, tools and action', *Report No. 3.* Denmark: CCAFS.

Conway, G. (2009). 'The science of climate change in Africa: impacts and adaptation', Discussion paper no. 1. London, Grantham Institute for Climate Change.

Cotula, L. (2009). 'Legal tools for citizen empowerment: Getting a better deal for natural resource investment in Africa'. London: IIED.

Cotula, L. et al. (2009). 'Land grab or development opportunity?' London: IIED.

Dryzek J. (2000). *Deliberative Democracy and Beyond.* Oxford: Clarendon Press.

Dufey, A., S. Vemeulen, and B. Vorley (2007). 'Biofuels: strategic choices for commodity dependent developing countries', *Common Fund for Commodities.* London: IIED.

Edwards, M. (2010). 'Why social transformation is not a job for the market', *openDemocracy,* 25 January. Retrieved from: <http://www.opendemocracy.net>.

Ericksen, P., P. Thornton, A. Notenbaert, L. Cramer, P. Jones, and M. Herrero (2011). 'Mapping hotspots of climate change and food insecurity in the global tropics', CCAFS. [Online] Retrieved from: <http://ccafs.cgiar.org/resources/climate_hotspots>.

Fan, S., B. Omilola, and M. Lambert (2009). 'Public spending for Agriculture in Africa: Trends and composition', *ReSAKSS Paper no. 28,* Regional Strategic Knowledge and Support System, USAID.

FAO (2010). *Global Survey of Agricultural Mitigation Projects.* Rome: FAO.

Ferroni, M. (2010). *Affordable 'Pay as You Plant' Index Insurance.* Basel: Syngenta Foundation for Sustainable Agriculture.

Foresight (2011). *The Future of Food and Farming.* London: The Government Office for Science.

GebreMichael, Y. et al. (2010). 'Emerging responses to climate change in pastoral systems', ccaa Keynote address, Retrieved from: <http://www.prolinnova.net/sites/default/files/documents/resources/publications/2010/ccaa_final_keynote_address_emerging_responses_150410.doc>.

Godfray, C. J., R. J. Beddington, I. R. Crute, L. Haddad, D. Lawrence, J. F. Muir, J. Pretty, S. Robinson, and C. Toulmin (2010). 'Food security: feeding the world in 2050', *Science*, vol. 327:5967, pp. 812–18.

Hazell, P., J. Anderson, N. Balzer, A. Hastrup Clemmensen, U. Hess, and F. Rispoli (2010). *Potential for scale and sustainability in weather index insurance for agriculture and rural livelihoods*. Rome: IFAD and WFP.

IAMP (2010). 'Statement of the health co-benefits of policies to tackle climate change', Inter Academy Medical Panel. Retrieved from: <http://www.iamp-online.org>.

IFAD (2010). *Rural Poverty Report 2011*. Rome: IFAD.

IFPRI (2009). *Climate change: Impact on Agriculture and Costs of Adaptation*. Washington, DC: IFPRI.

IIED (2012). *Investing in Locally Controlled Forestry: Natural Protection for People and Planet*. London: IIED.

ILC (2011). *Land Rights and the Rush for Land*. Rome: International Land Coalition.

IPCC (2007). 'Summary for policy makers. Climate Change 2007. Impacts, adaptation and vulnerability', Contribution of *Working Group II to the Fourth Assessment*. Geneva.

Joireman, S. F. (2007). 'The mystery of capital formation in sub-Saharan Africa: Women, property rights and customary law', *World Development*, 36(7), pp. 1233–46.

Jones, A., M. Pimbert, and J. Jiggins (2010). *Virtuous Circles: Values, Systems and Sustainability*. London: IIED.

Lavigne, P. et al. (2002). *Negotiating Access to land in West Africa*. Paris: GRET and London: IIED.

Lenton, T. (2011). *Earth System Tipping Points*. Norwich: University of East Anglia.

Mahul, O. and C. Stutley (2010). *Government Support to Agricultural Insurance: Challenges and Options for Developing Countries*. Washington, DC: World Bank.

Mayers, J. (2010). Turn REDD on its head. IIED website blog. Retrieved from: <http://www.iied.org>.

Meizen-Dick, R. et al. (2010). 'Engendering agricultural research', Paper prepared for the *Global Conference on Agriculture and Rural Development, Montpellier France, 28–31 March, 2010.*

Murphy, S. (2010). *Changing Perspectives: Small-Scale Farmers, Markets and Globalization*. The Netherlands: Hivos and London: IIED.

OXFAM (2007). 'Biofuelling Poverty—EU plans could be disastrous for poor people'. Retrieved from: <http://www.oxfam.org.uk/applications/blogs/>.

Parry, M., N. Arnell, P. Berry, D. Dodman, S. Fankhauser, C. Hope, S. Kovats, R. Nicholls, D. Satterthwaite, R. Tiffin, and T. Wheeler (2009). 'Assessing the costs of adaptation to climate change. A review of the UNFCCC and other recent estimates'. London: IIED and the Grantham Institute.

Porras, I. et al. (2008). *All that glitters. A review of payments for watershed services in developing countries*. London: IIED.

Pretty, J. C. Toulmin, and S. Williams (2011). 'Sustainable intensification in African agriculture. Analysis of cases and common lessons', Overview paper, *International Journal of Agricultural Sustainability*, Vol. 9(1).

PROLINNOVA (2010). 'Strengthening local adaptive capacities'. Retrieved from: <http://www.prolinnova.net/sites/default/files/documents/thematic_pages/climate_change_pid/2011/prolinnova_policy_brief_climate_change_june_2011_lowres.pdf>.

Reid, H. and S. Huq (2010). 'Community Champions: Adapting to Climate Challenges', *Participatory Learning and Action (PLA)*, no. 60. London: IIED.

Reij, C. (2008). 'Proposal: Regreening the Sahel'. Amsterdam: Free University.

Reij, C., I. Scoones, and C. Toulmin (1996). *Sustaining the Soil*. London: Earthscan.

Robbins, M. (2011). *Crops and Carbon: Paying Farmers to Combat Climate Change*. London: Earthscan.

Rockstrom, J. et al. (2009). 'A safe operating space for humanity', *Nature*, Vol. 461, pp. 472–5.

Saunders, J. and R. Nussbaum (2007). 'Forest Governance and Reduced Emissions from Deforestation and Degradation (REDD)', Chatham House Briefing paper, December, London.

Scoones, I. (1998). 'Sustainable rural livelihoods: A framework for analysis', IDS Working Paper no. 72. Brighton: IDS.

Scoones, I. and J. Thompson (2009). *Farmer First Revisited*. Rugby: Practical Action.

Shames, S., E. Woolenberg, L. E. Buck, B. Kristjanson, M. Masiaga, and B. Biryahaho (2012). 'Institutional innovations in African smallholder carbon projects', *CCAFS Report no. 8*. Copenhagen, Denmark: CGIAR Research programme on Climate Change, Agriculture and Food Security, (CCAFS). Retrieved from: <http://www.ccafs.cgiar.org>.

Skinner, J. et al. (2009). *Sharing the Benefits of Large Dams in West Africa*. London: IIED.

State of Science Review (2010). 'Education, training and extension', prepared for the UK Government Foresight Programme, London.

Steer, A. (2010). Speech to the Hague Conference on Climate Change and Agriculture. November.

Stern, N. (2009). *A Blueprint for a Safer Planet*. Oxford: Bodley Head.

Tennigkeit, T. and A. Wilkes (2008). 'An assessment of the potential for carbon finance in rangelands', *Working Paper No. 68*, Nairobi, Kenya: World Agroforestry Centre.

Thornton, P. et al. (2011). 'Agriculture and food systems in sub-Saharan Africa in a 4°C+ world', *Phil Trans Royal Society*, 369, pp. 117–36.

Toulmin, C. (1992). *Cattle women and wells. Managing household survival in the Sahel*. Oxford: Clarendon Press.

United Nations (2007). 'Developing Index-Based Insurance for Agriculture in Developing Countries', in *Sustainable Development Innovation Briefs* (2007). Retrieved from: <http://sustainabledevelopment.un.org/content/documents/no2.pdf>.

Vermeulen, S. and L. Cotula (2010). 'Making the most of agricultural investment: A survey of business models that provide opportunities for smallholders', London: IIED.

Vorley, B. et al. (2012). *Small Producer Agency in the Globalized Market: Making Choices in a Changing World*. Hague/London: Hivos and IIED.

Vorley, B. et al. (eds) (2007). *Regoverning Markets: A Place for Small-Scale Producers in Modern Agrifood Chains?* Farnham, UK: Gower.

World Bank (2010a). *Ethiopia. Economics of Adaptation*. Washington, DC: World Bank.

World Bank (2010b). *Rising Global Interest in Farmland: Can it Yield Sustainable and Equitable Benefits?* Washington, DC: World Bank.

World Bank (2012). *4°C Turn Down the Heat: Why a 4°C Warmer World Must be Avoided*. Washington, DC: World Bank.

17 Rural non-farm economy: current understandings, policy options, and future possibilities

STEVE WIGGINS[1]

1. Introduction

Very few smallholders dedicate all their time to farming. Most have other jobs, both on and off their farms. These activities are often seasonal, undertaken when there is little to do on the fields. Although they sometimes involve temporary migration, including commuting from the village to urban jobs, most of these activities are rural, part of the rural non-farm economy (RNFE). So to appreciate the livelihoods, options, and possible dynamics of small farm households, it is necessary to understand the RNFE. Such understanding helps answer three of the five questions posed by Hazell (Hazell, 2011), namely: how will the transitions from agrarian to urban economy take place, and how quickly? How large is a viable smallholding? And what policies and programmes are needed to support small farmers?

To anticipate the argument that follows, a vibrant RNFE that generates plenty of decently rewarded jobs and business opportunities will allow marginal farmers to earn their livelihoods from non-agricultural activity without necessarily having to leave the countryside, although some may move to market towns. It may allow them to continue to farm part-time on small plots that could never offer them incomes above the poverty line, but which, combined with non-farm work, does offer a reasonable livelihood. Policies to support small farmers, most of whom have non-farm earnings as well, need to take into account those applying to the RNFE as well.

[1] The author thanks Peter Hazell for helpful comments on a draft of this chapter and for previous joint work on the RNFE that informs much of this piece. He also thanks the anonymous reviewer for useful suggestions to improve the chapter. The views expressed in the chapter are solely of the author and do not necessarily reflect those of anyone else, including the Overseas Development Institute.

Since the early 1990s, awareness that rural economies are something more than agriculture has grown, prompting studies of the RNFE, its links to agriculture, and its likely future development. Signs, very clear in Asia and Latin America, that agriculture may not be generating as many jobs as it once did have injected urgency into such investigations. Study of the RNFE can be rewarding, since the questions it raises are central to those of how economies grow, how benefits are transmitted from enterprises to households and individuals and thereby affect equity and poverty, and how different places experience varying economic fortunes.

Reports often show the non-farm part of the rural economy to be large and growing. There have, however, been temptations to exaggerate: to announce the rapid decline of smallholder farming (Bryceson, 2002) or to see the RNFE as a dynamic new opportunity that will drive rural economic growth, reduce poverty, and improve regional and social equity in ways not seen before. As evidence accumulates, judgments have become more measured. Agriculture, as recent price shocks have reminded us, cannot be ignored. Moreover, it seems that in many rural areas, the RNFE is not an autonomous force, but depends on thriving agriculture to generate demand for its goods and services.

This chapter reviews what is currently known about the RNFE. Four questions are posed. How important is the RNFE for rural incomes and especially for the poor? What affects the growth of the RNFE? What policies are needed to stimulate the RNFE? And, what may be future development of the RNFE?

This chapter builds on the impressive review of the rural non-farm economy, in depth and range, compiled by Haggblade et al. (2007), which stands as the reference work for those wanting further guidance. It reflects the presentations by Dirven (2011) and Mahajan and Gupta (2011), and subsequent discussions of the RNFE session at the conference on 'Future Directions for Smallholder Agriculture' held at IFAD, Rome in early January 2011.

2. How important is the rural non-farm economy?

2.1 CONTRIBUTION TO JOBS AND INCOMES

Non-farm activities are increasingly important in rural areas: a growing share of households participates in them, while they provide increasing proportions of rural household income. Most statistics from comparable surveys show this. That said, in most developing countries measurements of the RNFE are infrequent, imprecise, and not easily compared owing to different definitions

BOX 17.1 MEASURING THE RURAL NON-FARM ECONOMY

Definitions. Non-farm activities can be defined as all those other than agriculture, forestry, and fishing, category A of the International Standard Industrial Classification (ISIC). Remittances derived from migration and wages from commuting-dependent work should be part of non-farm income, so long as the earner can be treated as still belonging to the rural household.

While collecting data on non-farm activities, some surveys include wages from agricultural activities carried out on the farms of others. Strictly speaking, these are on-farm activities and therefore should not be reported as part of the non-farm economy.

More problematic is the definition of 'rural', for which there is no standard international measure so definitions vary between countries. Rural is often defined as a residual, i.e. those areas that are not urban. These areas in turn may be defined by economic and administrative functions, density of settlement, and by size of locality—and often by some combination of these. When size of settlement is the primary measure, thresholds set before the locality becomes classified as urban vary greatly between 2,000 and 20,000 persons. Hence cross-country comparisons of rural statistics are inexact.

A common problem is the treatment of small market centres in rural areas. Some surveys consider these centres as part of the rural economy, others see them as urban. Since some non-farm activities such as trading tend to concentrate in these centres, their classification as urban or rural can make a considerable difference to the estimated size of the RNFE.

Measurement. Surveys to establish rural non-farm incomes are prone to errors, primarily since many non-farm activities are part-time, seasonal, intermittent, micro-scale, and informal, where often few written records are kept, so that those surveyed have only a rough idea of the time they spend on non-farm work or their earnings from it.

Employment is often taken as a more reliable way to establish the scale of non-farm activity, but it is difficult to obtain accurate recall of the occupations of working people who engage in seasonal and part-time work.

Finally, few developing countries have regular, national surveys of employment or economic activity that allow the RNFE to be tracked.

For all these reasons, understanding of the sector is incomplete and the statistics quoted need to be taken as broad estimates, subject to generous margins of error.

and the difficulties of accurate measurement (see Box 17.1). Hence in this section, statistics on the RNFE need to be treated as showing broad features, sufficient to distinguish between times and categories, rather than being precise observations.

Primary employment data, which offer the most widely available indicator of the scale of rural non-farm activity, suggest that the RNFE accounts for about 30 per cent of full-time rural employment in Asia and Latin America, 20 per cent in West Asia and North Africa, and 10 per cent in Africa; see Figure 17.1. Inclusion of rural towns—which frequently depend on the rural hinterlands for both inputs and markets—raises non-farm employment shares by an additional 10–15 per cent (Hazell and Haggblade, 1993).

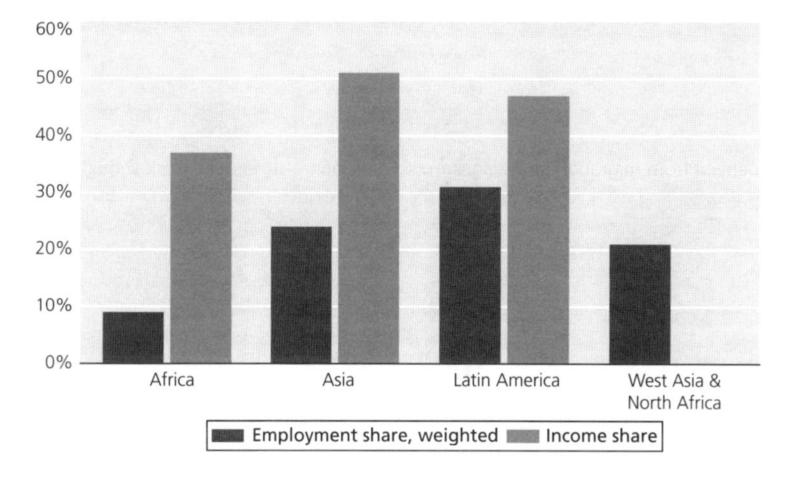

Figure 17.1 Shares of rural employment and incomes from non-farm sources, 1980s to 2001

Source: Haggblade et al. (2007). Tables 1.1 and 1.2, drawing on multiple sources and taking simple averages of reported statistics. Countries included and dates of reported statistics:

- Africa: Cameroon (1987), Ethiopia (1994), Ivory Coast (1986), Malawi (1998), Mozambique (1980), Namibia (1981), South Africa (1996), and Zambia (2000).
- Asia: Bangladesh (2001), India (1991), Indonesia (1995), Iran (1986), Korea (1980), Nepal (1981), Pakistan (1998), Philippines (1981), Sri Lanka (1981), Thailand (1996), and Vietnam (1997).
- Latin America: Argentina (1980), Bolivia (1988), Chile (1984), Dominican Republic (1981), Ecuador (1990), Honduras (1988), Uruguay (1985), and Venezuela (1990).
- West Asia and North Africa: Egypt (1986), Morocco (1994), and Turkey (1990).

Note: No data for non-farm share of income for West Asia and North Africa.

These statistics probably underestimate the share of rural labour spent on non-farm activities, since they report only main occupations, leaving out secondary and seasonal activities. Farming households in particular are often found to spend substantial amounts of time on non-farm occupations.

In addition, rural households are increasingly likely to have migrants who have moved within the country, to neighbouring countries or to a distant international destination. Household surveys show substantial percentages of households have one or more members who migrate. For example, a survey of some 39 publications dealing with migration in selected developing countries concluded that a median of around 25 per cent of households had a member who was a migrant of one kind or another (Wiggins and Proctor, 1999).

Income data, which include earnings from seasonal and part-time activity, offer a more complete picture of the scale of the RNFE. Evidence from a wide array of rural household surveys suggests that non-farm income accounts for about 35 per cent of rural income in Africa and roughly 50 per cent in Asia and Latin America; see Figure 17.2. Standing roughly 20 per cent higher than comparable employment data, probably because most on-farm activity is

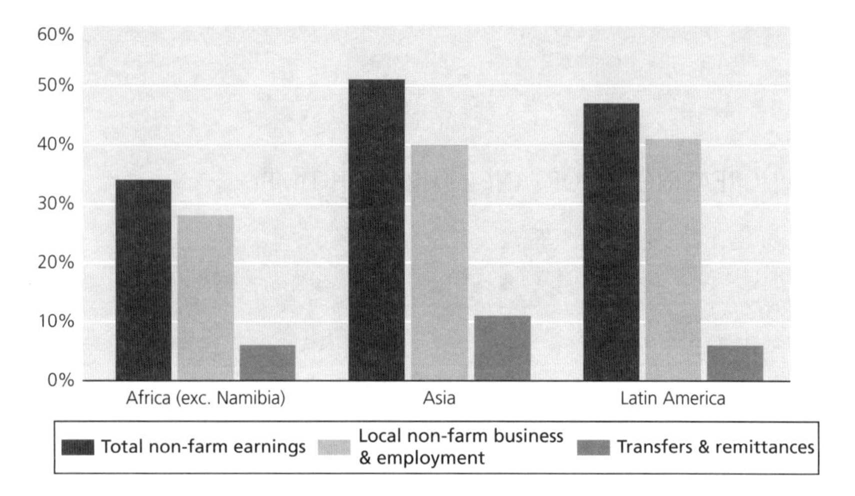

Figure 17.2 Non-farm share of rural income

Source: Reardon et al. (2007), Table 6.1, summarizing 54 rural income surveys from the 1990s and 2000s:

- Africa: Ethiopia 1989–90, 1999; Ghana 1992, 1998; Côte d'Ivoire 1993–5; Kenya 1994–6, western 1993; Malawi 1990–1, 2004; Mali, southern 1994–6; Mozambique 1991; Namibia, unfavourable zones 1992–3 and favourable zones 1992–3; Niger, unfavourable zones 1989–90 and favourable zones 1989–90; Rwanda 1991, 1999–2001; Tanzania 1991, 2000; Uganda 1996, 1999–2000; Zimbabwe 1990–1, poor areas 1990–1.
- Asia: Bangladesh 2000; China 1993, 1997; India 1993–4; Republic of Korea 2003; Nepal 1996; Pakistan 1990–1, 1999; Philippines 1994, 1998; Sri Lanka 1999–2000; Vietnam 1998, northern uplands 2002.
- Latin America: Brazil 1997; Chile 1996; Colombia 1997; Ecuador 1994, 1995; El Salvador 1995, 2001; Guatemala 2000; Haiti 1996; Honduras 1997, 1998; Mexico 1997, 2003; Nicaragua 1998, 2001; Panama 1997; Peru 1997.

Note: Remittance earnings from commuting and temporary migration by rural household members are usually treated as non-farm income. Transfers from government, former household members, or relatives are often classified separately. Practice, however, is not standard: hence Figure 2.2 shows remittances and transfers separately. Latin American estimates appear low, but the average hides their uneven distribution. For countries with many emigrants, remittances can contribute 10 per cent or more to GDP: Guyana (30%), Honduras (25%), El Salvador (18%), and Nicaragua (15%) are good examples (IFAD, 2007).

better remunerated than farm work, these income shares confirm the economic importance of part-time and seasonal non-farm activities.

These statistics come from surveys of rural households, whether farming or not. If the share of household incomes for households that have no land, and those with varying amounts of land, are compared—see Figure 17.3—two things are apparent. First, the share of income from non-farm sources declines in almost all cases with increasing access to land, with a particularly sharp decline from the landless to those with even the smallest amount of land. The median share of incomes from non-farm sources for landless rural households is 46 per cent, while for landed households the median declines from 26 per cent for those in the quintile with least land to 17 per cent to those with the most land. Second, even if farming households have a lesser share of their incomes from non-farm activities, the shares are substantial, rarely less than 10 per cent of incomes, and often double that.

2.2 INCREASING IMPORTANCE THROUGH TIME

Surveys at both national and household level usually show an increasing proportion of rural incomes coming from non-farm activities over time; see, for example, the six Asian countries plotted in Figure 17.4. All show similar patterns of an increasing share of farm household incomes from non-farm activities through time, even if the rate of increase and timing differ across the cases.

The same often applies to rural employment: an increasing fraction in non-farm activities. Figure 17.5 shows something remarkable: in many countries of Latin America from 2000 to 2008, it was only the growth of non-farm jobs that created more jobs in rural areas, since farming was actually shedding

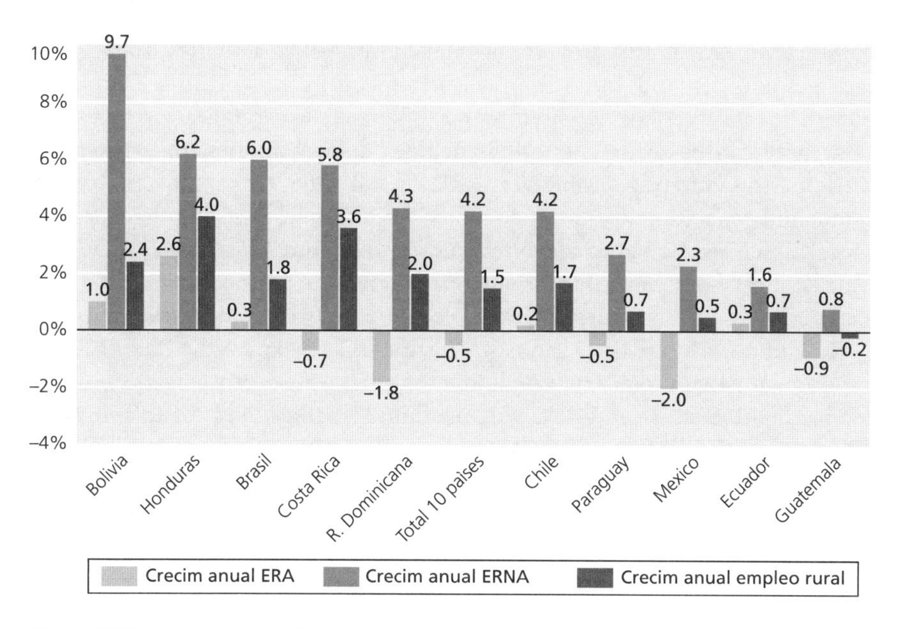

Figure 17.5 Annual growth in total rural employment, farm and non-farm, Latin America 2000–8 (Annual growth, %)

Source: Dirven (2011), own calculations based on data from Javier Meneses and Adrian Rodríguez, Agricultural Development Unit, ECLAC, based on household surveys by ECLAC Statistics division.

Legend: Crecim. Anual ERA = annual growth farm jobs; Crecim. Anual ERNA = growth non-farm rural jobs; Crecim. Anual Empleo Rural = annual growth all rural jobs.

Note: Total growth for all ten countries is approximate, calculated by adding first-year and last-year values for each country, considering 2001 and 2007 as first and last year, respectively, for reported amounts.

labour at the time (Dirven, 2011). Mexico is the exemplar: losing farm jobs at the rate of 2 per cent a year, but gaining non-farm jobs in rural areas at 2.3 per cent, and thereby achieving a small net gain in rural employment. These statistics apparently show that for some countries, the growth rate of non-farm jobs in rural areas was surprisingly high: six countries registered 4 per cent a year or more, rates that should have done much to reduce rural underemployment.

2.3 COMPOSITION OF THE RURAL NON-FARM ECONOMY

The rural non-farm economy comprises a highly heterogeneous collection of agro-processing, manufacturing, trading, and services. No single classification is widely accepted: Table 17.1 sorts activities by primary processing, manufacturing, and services.

In this classification three things stand out. One, only a few non-farm activities are linked directly to agriculture. Two, many of the categories and activities listed respond to demand in the village and surrounding area. That, of course, means that such activities depend on other activities to generate incomes and demand: often that will be agriculture which can sell produce outside the local economy. Three, major differences in capitalization and scale of activity can be seen. The RNFE includes many activities undertaken by single-person enterprises using very little capital, often facing intense competition, and generating low returns—think, for example, of selling snack foods. More capitalized activities are often carried out on a larger scale; they also usually offer higher returns to labour. Some agricultural processing fits in here: sugar mills and tea factories, for example, are usually large industrial units.

How important are the various sub-sectors that make up the RNFE? Most surveys show that most activity—in the range 50–75 per cent—is made up of services, with trading and transport prominent, rather than manufacturing. The share of the rural economy made up by manufacturing may even decline through time as crafts are replaced by mass-produced goods from urban factories (Haggblade et al., 2007, Chapter 12). India is a case in point, where the share of rural manufacturing in rural net domestic product fell in the 20 years between 1980/1 and 1999/2000 (Figure 17.6).

The concentration of jobs in manufacturing in urban areas stands out clearly when national surveys distinguish locations by metropolitan cities, intermediate cities, and rural areas, as shown in Figure 17.7. In the three countries shown in the figure, manufacturing makes up no more than 10 per cent of rural employment, being concentrated in urban areas and above all in the main cities.

Although not always identified separately from other services, public services can provide significant rural employment in some instances. In rural

Table 17.1 Rural non-farm activity, classified

Rural non-farm activity	Typical activities	
Processing, trading, and transport of farm outputs Mostly carried out prior to shipping produce to urban markets, but some processing for local consumption—especially grain milling, butchery, oil extraction, brewing, and soft drinks	Milling grains Sugar refining, jaggery making. Slaughtering, butchery, salting, drying (ham, bacon, sausage) Dairy processing to cream, cheese, yoghurt Coffee, tea processing Fruit and veg packing and canning	Brewing and distilling Soft drink making Rolling cigars and cigarettes Honey cleaning Oil crushing and extraction Fish drying, salting Timber sawing, drying Cotton ginning
Non-farm primary and processing Often small-scale, but quarrying may be industrial	Mining of minerals Quarrying and production of building materials: stone, sand, gravel, bricks, clay tiles, lime, cement	Charcoal production Salt extraction Fuel wood gathering and trading Water collection
Production, trading, and transport of farm inputs May be little of this when market towns and cities are close	Simple tool making and repair Tractor and ox ploughing and other mechanical hire services	Animal feed making Wholesale and retail of fertilizer, agro-chemicals, veterinary medicines
Manufacture and repair of consumer goods for local rural market	Furniture-making Domestic utensils	Mats, baskets Pottery
Almost always artisan work carried out in small workshops	Clothes, blankets Shoes	Repairs—tools, clothes, shoes, electrical, vehicle Ice blocks
Manufacture of consumer goods for domestic and export markets: • ***Utilitarian, artisan*** • ***Artistic, fine crafts*** Artisan, micro-scale	Textiles: blankets, clothes Leatherwork Furniture Mats, baskets	Ceramics Wood carvings Decorations Tourist items
• ***Industrial*** Uncommon: subcontracting from urban businesses, peri-urban areas	Textiles and clothing Glass Metals	Plastics Electronics
Trading and transport of manufactured goods, mainly from cities to local rural market Mainly small-scale, owner-operated, low capital. Often comprises 20% or more of all village economic activity	Transport and haulage Wholesale trading and storage Retailing	
Provision of services for local rural market • ***Private services*** Micro-scale usually. Another major component of the RNFE	Barbers, beauty salons Healing Cooked food sale, café, tea-stall, tea-shop, bars, restaurants, etc. Lodgings and accommodation Transport: taxi, bus, etc. Cleaning, cooking and child-minding	Construction and building repairs Photography Musicians Religious instructors, teachers, priests Pawnbroking, moneylending, deposit-taking Typing, photocopying, fax, phones
• ***Public services: education, health, roads, etc.*** Relatively well-paid, dependable jobs. Many posts occupied by outsiders, often not resident in village	Primary and secondary schools Health posts and centres Road maintenance	Communications (posts, phones, radio) Police Extension services, usually agricultural and veterinary

(continued)

Table 17.1 *Continued*

Rural non-farm activity	Typical activities	
Services for outsiders	Tourism: hotels, restaurants, entertainment, etc. Maintenance of parks and other valued habitats and landscapes	Commuter or weekend homes Watershed protection

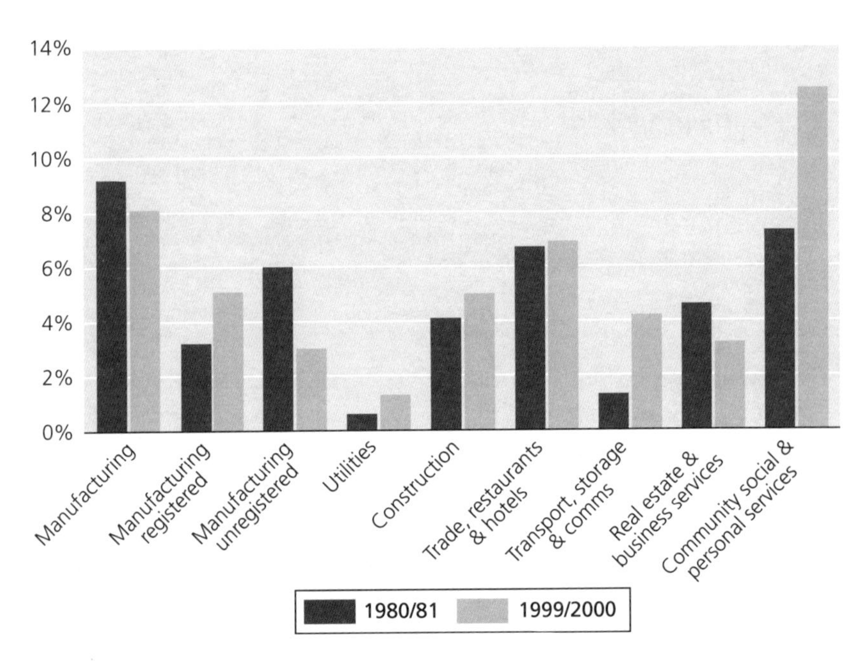

Figure 17.6 India, changing composition of the rural non-farm economy, 1980/1 to 1999/2000 (Share of rural net domestic product, %)

Source: Mukherjee and Zhang (2007), Table 13.2, reporting data from Chadha (2003).

Egypt, government employment generates over 45 per cent of rural non-farm incomes; in rural Pakistan about 25 per cent of rural non-farm earnings (Adams and He, 1995; Adams, 2002); and in rural India nearly 20 per cent of rural non-farm employment (Fisher et al., 1997).

Some rural non-farm activities are seasonal, either because, as with agricultural processing, they fluctuate with the availability of agricultural raw materials or because they are carried out in the slack season for farmers, when household labour has little to do on the farm. The latter feature explains in part the historical popularity of 'putting out' of some manufacturing, whereby factory agents take raw materials to rural workshops and collect finished products to market in towns and cities. When spinning and weaving were not yet mechanized, this system dominated English textiles. It was also used in Japan in the early phases of industrialization, in Taiwan from the 1890s onwards, and contemporary China (Grabowski, 1995). Contemporary examples include electronics workshops in rural Taiwan (Otsuka and Reardon, 1998; Otsuka, 2007) and household silk spinning in Thailand (Haggblade, 2007).

2.4 CONTRIBUTION TO POVERTY AND INEQUALITY

How much of a difference does the rising importance of the RNFE make to the poor? Most of the poor in the developing world live in rural areas and typically lack land or livestock, so that non-farm activities might be the only way that they can earn enough to escape poverty.

There are three ways that the RNFE might allow the poor to escape poverty (Lanjouw, 2007):

- Directly, from earnings from occupations in the RNFE;
- By depending on non-farm activities when shocks occur, using the non-farm activity as a safety net that prevents destitution; and
- Indirectly, as a thriving RNFE creates demand for produce from farmers, or tightens the rural labour market to the benefit of agricultural labourers.

Direct employment in the RNFE for the poor may not lift them out of poverty, since poor people, lacking skills and capital, tend to work in low-return activities such as cottage industries, small-scale trading, and unskilled wage labour used in construction, portering, and many personal services. Wage labour, in both agriculture and non-farm business, also accrues primarily to the poor. In contrast, white-collar jobs such as medicine, teaching, accounting, and administration tend to be taken up by those from households where people have formal education and qualifications.

Rural India illustrates this; see Figure 17.8. All income quintiles have similar shares of incomes from non-farm activities, between 30 and 39 per cent; yet the composition of both agricultural and non-farm income differs markedly across the quintiles. The poor rely disproportionately on earnings from labouring on and off farms, while they are least likely to have earnings from regular jobs off the farm.

The second way that the RNFE can benefit the rural poor is by providing a safety net when times are hard. For example, in 1992 in Chivi District, Zimbabwe, following one of the worst droughts of the twentieth century that led to widespread and severe crop failure, non-farm activities were a principal means—together with selling livestock and working on public schemes—of coping. People traded in vegetables, panned for gold, brewed beer, sewed and crocheted, searched for casual jobs, went building, sold sex, made clay pots, and collected tree pods for sale—in that order of frequency (Scoones et al., 1996).

Even in normal times, some non-farm pursuits fulfil this function. For example, in rural Mexico embroidering leather belts was a common activity in one village surveyed, yet the estimated returns per hour worked were low at US$0.25, compared to US$0.41 an hour for agricultural labour. Members of farm households, however, appreciated the chance to use some of the time not working on the farm to gain additional income, however modest. Embroidery had the advantage that it could be done at almost any time, picked up and set down (Wiggins et al., 1999).

The third pathway, however, may be the strongest for some of the rural poor—those who depend in large part on farm labouring. As the rural non-farm economy grows and offers more jobs, it can drive up rural wages, including those for agricultural labouring. Village studies from Tamil Nadu, India show this. When in the 1970s the green and white (dairy) revolutions allowed modest increases in agricultural output and farm incomes, strong linkages to the non-farm economy meant additional jobs in processing and in providing goods and services for farmers with more to spend. Agricultural wages rose by 20 per cent (Hazell and Ramasamy, 1991). More recently, Harriss et al. (2010) report a doubling of farm wages between 1981 and 2008 in a village of northern Tamil Nadu that has been repeatedly surveyed since 1916, where wages had persisted at the equivalent of three kilos of grain a day for decades. Given that landowners had mechanized many operations to save on the now 'expensive' labour, the pressure on farm wages came from off-farm opportunities in the local economy that drew labourers away from the fields. In central Tamil Nadu, resurveys of villages between 1979 and 2004 (Djurfeldt et al., 2008) confirm this: while mechanization meant there was less farm work than before, wages for farm labouring were rising. Loss of jobs on farm were compensated by jobs in non-farm activities.

The findings from Tamil Nadu are striking: in an agrarian economy, the Green Revolution boosted farm output and incomes significantly for many farmers from the late 1960s to the early 1980s. Subsequently it seems that agricultural growth has been more modest, at least on the very small farms that poor people have. Much of the improvement in rural incomes seen in the last quarter century since the mid-1980s has come from the dynamism of the non-farm economy, in the process starting to reduce long-standing inequalities and deep poverty. There is still some way to go, but in these accounts, the position would have been much worse had the rural non-farm economy not thrived.

These results for one region of India may apply more generally in Asia. Otsuka and Yamano (2006b) report similar patterns of fewer jobs in farming,

Table 17.2 Rural non-farm income as a share of total income: impact on income equality

	Equity-enhancing		Neutral		Inequitable	
Quintile (a)	Kenya 1975	Pakistan 1989	India 1999	Ethiopia 1990	Ecuador 1995	Vietnam 1997
Poorest	82	75	32	32	22	40
Q2	80	63	39	—	37	42
Q3	45	36	38	30	37	50
Q4	40	33	39	—	46	60
Highest	—	21	31	31	64	82

Source: Lanjouw (2007), drawing on FAO (1998), Lanjouw (1999), Lanjouw and Shariff (2004).

Note: (a) Kenya data by quartile, Ethiopian data by tercile.

rising farm wages, and increasing shares of rural incomes coming from off the farm for both high potential and marginal areas of the Philippines, Thailand, Bangladesh, and Tamil Nadu, India.

Clearly the three pathways will operate to greater or lesser effect depending on local variations that determine the extent of development of the rural non-farm economy, the vigour of agriculture, and other processes that affect labour markets, such as migration. Do rural non-farm jobs make the distribution of rural income more or less equitable? Patterns vary across countries (see Table 17.2): in some cases non-farm earnings constitute a larger share of rural incomes amongst the poor than amongst the rich; in other cases it is the reverse, so that non-farm earnings tend to widen inequalities. And in some settings, the relationship between household welfare and the share of income from off the farm proves U-shaped (Reardon et al., 1998). Empirically, no consistent pattern emerges.

3. **How does the rural non-farm economy develop?**

How the rural non-farm economy develops, in different circumstances and spatial patterns, matters for the options that small farmers have to earn additional income locally—and hence for the viability of very small farms. In many cases the relation between farming and the RNFE is synergistic, in that the latter responds in large part to local demand, especially services that are often a large part of the rural non-farm economy. Hence the local economic base is a prime consideration. But it can be that in some cases the RNFE responds to forces beyond the rural economy. These possibilities will be discussed in turn.

3.1 INTERNAL FORCES: TRADABLES AND THE ECONOMIC BASE

Tradables that can be sold outside the local economy, the 'economic base' (Richardson, 1985), set the potential for growth of non-tradable activities through linkages. In many rural areas, agriculture—and in some cases forestry or fishing—produces the tradable goods. There are, however, other possibilities: mining, tourism, and leisure may also be drivers of the economy. Mining, however, often develops in an enclave with few of the revenues spent locally and hence does little for the economy of the surrounding area.[2] Tourism and leisure usually develop significantly only in areas with notable landscapes and attractions, such as coasts and mountains. For the large areas that lack these attractions, there are thus only minor possibilities.

How might these opportunities vary by location? Three variations can be picked out. One are resource-poor regions, where lack of water, infertile soils or other problems impede farming, forestry, or fisheries, such as northern parts of the West African Sahel and the southern Altiplano of Bolivia. Outside the peri-urban areas of such regions, prospects for development may be limited to a few products serving niche markets: in the case of the Altiplano, examples are alpaca meat and wool, and quinoa—specialist produce that can be produced in an otherwise inhospitable environment. Local crafts may be another possibility. But these niche products are unlikely to support a large economy. In the short term, these areas are likely to see out-migration.

Two, some rural regions have unexploited potential. This may happen, for example, where fertile soils, minerals, a strategic location, or great natural beauty exist, but lack of physical infrastructure, technology, or human capital have prevented development, or where conflict has deterred investment. Prominent examples include north-east Thailand and the Cerrado of northern Brazil; in both cases their potential lay dormant until a combination of road building and better management of modest soils allowed investment. Both areas have subsequently thrived from agricultural exports (World Bank, 2009). Understanding distant markets can be another ingredient: Chile has long produced wine, but the scale of exports is a recent development, founded partly on agricultural research and technical improvements, but driven by increased appreciation of demand in distant markets.

A third case occurs when a dynamic economic base already exists, as in the agricultural boom of the central Chilean plain during the 1980s and 1990s, the Punjab of the 1960s and 1970s, and in Uganda's agricultural recovery during the 1990s.

[2] Which is not to say that over a wider area the demand for labour from mining may not have a strong impact on local economies, as has been the case with South African and Zambian mines and the rest of southern Africa, or with Andean mines and the economies of Bolivia and Peru.

3.2 EXTERNAL FORCES: LINKS TO URBAN AREAS AND THE RISE OF PERI-URBAN AREAS

In fast-growing, urbanizing countries like China and India, the rural economy may have more possibilities. This applies especially in peri-urban areas, defined as those where it is possible to travel to a reasonably large town within two hours or so. Indeed, as urbanization increases and better transport links are developed, an increasing fraction of rural areas become peri-urban. In densely settled parts of rural Asia, the majority of the rural population may now be living within a two-hour journey of a city.

With easier interaction to cities, new possibilities present themselves: commuting from villages to cities for day jobs;[3] weekend and day leisure in the countryside by city-dwellers; out-of-town manufacturing, warehousing, offices—where facilities are relocated for lower rents and often more pleasant surroundings, but still sufficiently close to the city to benefit from economies of agglomeration; and provision of environmental services for the city such as water supply. Peri-urban farms can use their proximity to market to produce highly perishable items such as vegetables and milk.

These trends apply strongly in the last two decades in densely settled parts of India, such as Tamil Nadu (Djurfeldt et al., 2008; Harriss et al., 2010) where there has been a great increase in the off-farm opportunities open to villagers thanks to urban developments. Rural wages have been pushed up in response. Even in less densely settled parts of Andhra Pradesh, such as Nalgonda District, the start of regular bus services has allowed commuting and access to towns:

> Finally, the role of transport infrastructure as a double edged sword, creating some opportunities for non-farm work both within, and outside the village, and destroying other opportunities, is confirmed by the field survey evidence. Particularly in a relatively low non-farm income area such as Nalgonda, the bus service has become the lifeline, not only for the procurement of inputs by small scale enterprises, but also for commuters and school children attending, or wishing to attend, schools outside their home village. (Bhalla, 2003)

It is not just ease of travel to urban centres that matters: the vigour of the industrial and urban economy also influences interactions, in both strength of demand for rural outputs and for additional labour, thereby affecting the attractiveness of migration or commuting for rural workers. In Vietnam, between 1993 and 1998 poverty fell by more than 8 per cent, one of the largest falls seen in the 1990s. Since most of the poor were rural, many of those

[3] In rural Chihuahua, Mexico, buses from assembly plants in the city of Chihuahua bring in workers from villages 90 minutes' drive away or further, collecting staff at first light and returning them home in the evening.

leaving poverty were employed in agriculture, yet growth of agriculture was pulled by demand from successful export manufacturing which saw many more informal workshops set up to service the main export firms (Bernabè and Krstic, 2005).

For rural regions located in less dynamic national economies and distant from cities—as in much of Africa until recently—the strength of urban links is much weaker and hence motors of growth will probably lie in the local economy.

3.3 LINKAGES AND MULTIPLIERS: SPREADING THE BENEFITS OF A DYNAMIC ECONOMIC ACTIVITY

Given a thriving tradable sector, much of the RNFE can be expected to grow through multipliers generated by linkages. Since agriculture is, beyond peri-urban areas, the most common motor, the points are illustrated by reference to this sector but with similar modifications to other activities.

Linkages include those in production: farming requires inputs and services such as seeds, fertilizer, credit, pumps, farm machinery, marketing, and processing of produce. While some of these inputs may be provided from distant towns and cities, there are local jobs in distribution of them. Artisans and workshops to repair farm machinery need to be locally based: for example, in Bangladesh in the early 2000s there were around 760,000 tube-wells with pumps, the maintenance of which employed 160,000 mechanics (Mandal, 2002).

Consumption links, as farmers spend additional income, can be stronger, above all when small farmers spend, since they are less likely to spend on luxury goods shipped from cities and further afield. Typically they demand housing improvements, clothing, schooling, health services, prepared foods, visits to town, cinema, and tea shops, all of which dramatically increase demand for rural transport. Hence in regions where agriculture has grown robustly, the RNFE has also typically enjoyed rapid growth. A large literature suggests that each dollar of additional value added in agriculture generates US$0.60 to US$0.80 of additional RNFE income in Asia, and US$0.30 to US$0.50 in Africa and Latin America (Haggblade et al., 2007).

Linkages do not end there: thriving farming can mean increased savings which, when channelled through banks, make capital available for investment—although these would not then contribute to linkages in consumption. These savings rates have reached up to 25–35 per cent in many areas of green revolution in Asia (Bell et al., 1982; Hazell and Ramasamy, 1991).

When farming produces more staples, this may push down local food prices, to the benefit of households buying in food—as applies to many of the rural poor. Not all rural non-farm activities thrive as economies grow. With better links to urban centres, some cottage industries wither in the face of competition from goods brought in from cities. Labour-intensive household manufacturing of baskets, pottery, and roof thatching die out, displaced by the import of cheap plastic pails, iron pots, and corrugated roofing.[4] Household manufacturing typically shrinks over time, pushed also by household members opting for work in better-rewarded activities. The demise of low-productivity household manufacturing explains, in part, why employment in services and commerce frequently grows faster than in manufacturing. Evidence from 22 countries for which 10–20-year series of data are available suggest that rural manufacturing employment typically shrinks slightly, while overall employment in non-farm activities grows at around 1.2 per cent a year (Haggblade et al., 2007, Table 4.4).

Asset distribution affects the strength of such links and their outcomes. Where wealth and power are unequal, growth of the tradables sector may trigger accelerating inequality as differential access to education, technology, capital, commercial, and political power translate into first mover advantages for the elite, both as producers of tradables or as owners of non-farm businesses responding to local demand. Linkages may have less effect, since when wealth is concentrated, additional incomes are less likely to be spent locally.

With a more even distribution of assets, on the other hand, broad-based rural non-farm growth is more likely, as in rural Taiwan from the 1940s to the 1960s (Ranis and Stewart, 1993; Ho, 1986; Johnston and Kilby, 1975).

3.4 STAGNATION AND THE PUSH INTO NON-FARM WORK

In regions without a dynamic economic base, events unfold very differently. Sluggish growth of farm incomes leads to anaemic consumer demand, limited agro-processing and agricultural input requirements, and stagnant wages. Where population growth continues unabated, land availability diminishes. Without technological advance in agriculture, labour productivity and per capita farm production fall: the soil may also be overworked, losing fertility and becoming more susceptible to erosion.

Growing landlessness pushes labour into non-farm activity by default. For lack of a thriving economic base, however, there are few productive opportunities, so diversification sees many take up low-return, labour-intensive activities such as petty trading, snack foods, basket-making,

[4] Hymer and Resnick (1969) dubbed these declining rural manufactures 'Z-goods'. See Anderson (1982) for evidence of this bifurcated transition.

gathering fuel wood, pottery, weaving, embroidery, and mat-making. Intense competition drives down returns to these occupations. In this case, prospects for growth of the RNFE are bleak: migration becomes the only safety valve.

3.5 SUPPLY OF RURAL NON-FARM GOODS AND SERVICES

The discussion so far stresses the importance of demand for the outputs of the RNFE, implicitly assuming that if something is in demand, it will be supplied. Clearly supply response from the RNFE, as with agriculture, may be limited if workers lack skills, if physical infrastructure such as transport and power are poor, and perhaps most critically, if would-be entrepreneurs cannot get capital to invest. It does not help, either, if there are official restrictions on setting up businesses or if local taxation falls heavily on small businesses, as has been reported for Tanzania and Uganda (Balihuta and Sen, 2001; Ellis and Bahiigwa, 2001; Ellis and Mdoe, 2002).

Since deficiencies in physical infrastructure, human development, and market failures are likely to be more serious in low-income countries, it may be imagined that these will be more of a hindrance to developing the RNFE in these cases compared to middle-income countries.

3.6 RURAL NON-FARM DEVELOPMENT: POSSIBLE PATTERNS

With several major influences on the development of the RNFE, it is not surprising to see different patterns. Table 17.3 illustrates how two factors may interact: one, opportunities in agriculture, forestry, and fisheries; the other, the attractiveness of off-farm opportunities, thanks to multipliers from agriculture or good links to cities. These determine how much people feel pushed to look for non-farm work or pulled into non-farm jobs.

This gives four stylized possibilities. One, A, is peri-urbanization, where a combination of stagnant farming with good links to growing cities sees the rural area cease to be agrarian, with rural households undertaking a wide range of activities with much migration and commuting. Farming households increasingly become part-time, weekend, and hobby farmers. A second case, B, has stagnant farming but weak links to cities. Here the RNFE acts as a safety valve, providing some work albeit poorly rewarded. These regions are likely to see strong out-migration as those able to travel seek better opportunities. Case C has a thriving agriculture but weak links to urban areas, so the rural economy is likely to be dominated by farming, but with a range of non-farm activities very much linked to demands from the farm economy and following the seasonal patterns of farming. Finally, case

Table 17.3 Rural non-farm development under differing conditions

		Attractiveness of economic opportunities in non-farm activities	
		Strong pull	Weak pull
		Good links to growing cities	Cities relatively remote, costly to reach
Need to find livelihood beyond farming, forestry, fishing	Strong push Stagnant farming, smaller farm sizes, growing population	**A. Peri-urbanization** Out-migration by educated persons for urban white-collar jobs, and by the less educated to find jobs in construction, factories, and the urban informal sector. May be permanent moves, or daily commuting. Possible jobs locally in rural factories linked to urban industries. Those unable to migrate take up poorly rewarded local jobs such as cutting firewood. Some jobs in construction as remittances from migrants are spent on improved housing.	**B. Rural non-farm activity as a safety net** Some out-migration, but the prospects for those with little education and few skills are limited and they may have to take up badly paid or hazardous work. Widespread resort to poorly rewarded crafts or carrying out local services for low returns—firewood, laundry, cooking snacks.
	Weak push Agriculture (forestry, fishing) offers a reasonable livelihood	**D. Highly diverse, thriving rural economy** Jobs in agro-processing, rural services and recreation, etc., allow people to work full or part time off the farm. Some migration, especially of educated youth. Commuting to nearby towns. Possible jobs in rural factories linked to urban industries.	**C. Traditional rural economy** Most off-farm work responds to seasonality, spreading risk and overcoming failures in markets for finance and land. Widespread engagement in a range of low-paid activities to generate some earnings, but for limited periods.

D has both thriving agriculture and good connections to cities, producing an increasingly diverse rural economy.

Recent studies of Bangladesh shed interesting light on these processes. In the 1990s when a green revolution was taking place, observers tended to see non-farm jobs being created as a result of the increased production of rice (Hossain et al., 2003; Mandal, 2002). By the 2000s, however, a test of the comparative strength of agricultural potential versus proximity to large cities, showed that areas closer to the metropolitan areas had greater incidence of non-farm jobs that paid above the median agricultural earnings (Deichmann et al., 2009)—which the authors attribute to the increasing importance of manufacturing and services to economic growth in Bangladesh.

4. What policies may stimulate the rural non-farm economy?

It is easy to compile a familiar list of desirable conditions for the growth of the RNFE. Given the diversity of circumstances and possible non-farm activities, the list is also likely to be long. As Ahmed and colleagues (2007) reviewing policies ruefully note:

> And so we are forced to conclude, like a long parade of past observers, that growth in the RNFE will require favorable public policies, adequate infrastructure, human skills, and well-functioning market and credit institutions. [p. 253].

A familiar complaint from those making policy for the RNFE is that the sector is in most countries orphaned, lacking either a specific public agency responsible for its development, or an organized interest group to promote it. Typically responsibilities for non-farm rural businesses are divided across ministries of agriculture (for agro-processing), industry, commerce, business development, regional development agencies, etc. Some countries have ministries of rural development, but they have their attention as much taken up by social investments as economic—partly since the economy in rural areas is taken to be agriculture and forestry. When Dirven (2011) reviewed policies that might apply to the RNFE for Colombia, Ecuador, Mexico, and El Salvador, she found that there was almost no mention of its existence; almost all references to rural economic activity were to agriculture.

This lack of focus, however, may also arise from the inability to reduce the long list of desirable actions to a shorter one of critical issues; it is difficult to focus political attention on long lists. Is there, then, a way to focus on the more critical factors that policy might then try to influence? One approach is to see potential interventions in a hierarchy, or more specifically to distinguish between those applying at national level, and those that apply at district, village, enterprise, and household levels, as shown in Table 17.4 (broadly following the proposals of Haggblade et al., 2007a).

Four key points stand out about policy for the RNFE, as follows:

a) Higher-level policy is necessary, if not sufficient

Central government needs to focus on three main things: ensuring a favourable rural investment climate; providing rural public goods; and resolving or mitigating failures in rural markets. Without progress on these three fronts, initiatives at lower levels are likely to be stymied.

Comparing the development of the RNFE in Cambodia and Laos to that in Thailand and Vietnam, Acharya (2003) concluded that while in the latter two countries, the economy and agriculture had been growing quite rapidly, creating opportunities for rural businesses to develop and flourish; in the

Table 17.4 Options for policymakers

Issues	Policies, programmes, projects	Key actors in planning and implementation
National issues		
Favourable rural investment climate	Peace, stability 'Good-enough' governance Macro-economic and general economic policies Fostering formal economic institutions Trade policy	Central government Donors, especially International Financial Institutions
Rural public goods	Physical infrastructure: roads, electricity, etc. Education, health Water, sanitation Research, extension	Central government and donors Decentralized government
Rural market failures: monopoly power, property rights, high transactions costs	Supply chain improvement Rural financial systems Property registration Competition policy	Government Formal private enterprise NGO Informal enterprise Groups of farmers, citizens
District, village, enterprise, household issues		
Enterprise promotion	Information Skills and training	Government Formal private enterprise NGO
Participation by poor households	Social protection	Central government NGO
	Absence of discrimination in labour markets, land rights, credit market, etc.	Central government NGO, civil society Grassroots groups

former two, economic growth had been less and from a lower base, so that their rural economies remained rudimentary and only a small range of non-farm enterprises could be developed. In a similar vein, Bland's (1995) review of USAID programmes for employment creation, some concerned with small-scale business in rural areas, found that too often programmes did not pay enough attention to the overall economic environment affecting the success of the enterprises.

This may be obvious, but it is easy to lose sight of it. From the early days of development efforts in the 1950s to at least the 1990s, policy for the RNFE has tended to focus on promoting supply from individual enterprises (see Haggblade et al., 2007 for a history of RNFE policy). Industrial estates in rural market centres, training in skills, grants, and cheap credit for business start-ups and business advisory services have been the instruments deployed.

By and large, these measures have not been that successful. It is not hard to imagine why. By focusing on supply, it was easy to overlook the importance of demand. These programmes tended to see non-farm enterprises as rural manufacturing, which is not surprising since the implementing agency was often a ministry of industry. This meant that services, the majority of rural businesses, were ignored. Coverage was limited. With efforts focused on individual enterprises, or at best, a cluster of businesses, limited budgets meant the programmes lavished attention on a small fraction of non-farm businesses, while the rest were left unattended.

Times have changed: there is now more awareness of the importance of a decent rural investment climate than before, following on from a general rise of interest in business and investment climates. The great advantage of prioritizing the overall business climate is that this can potentially benefit all enterprises and households, at relatively low cost per business served. The investment climate does not have to be perfect. Most Asian economic successes of the last 20 years have occurred with conditions of governance, at least initially, that fell far short of 'good governance' (Chang, 2003; Khan, 2002). Moreover, in many developing countries contemporary standards of governance are better than those in place at the time the industrialized countries achieved their economic take-off. The point is the need for 'good-enough governance': feasible conditions sufficient to allow initial investment and growth that can later be improved (Grindle, 2007; Moore and Schmitz, 2007).

The climate for rural investment depends in large part on national conditions, but also has its own distinctive features where issues such as cattle rustling, disputes over land ownership, predatory local politicians running what amount to protection rackets, and local taxation that weighs heavily on business and especially on small enterprises become prominent. Recent reviews of rural investment climates have proved revealing. A detailed enquiry into the situation of small and very small businesses in the Indonesian countryside (World Bank, 2006) stresses the difference between the limitations faced by small rural business compared to that faced by large corporations in urban areas:

> Demand constraints, access to credit, poor roads and unreliable electricity top the list of concerns faced by micro and small enterprises at the Kabupaten (i.e. district or municipality) level. This is in marked contrast to macroeconomic instability, policy uncertainty, corruption, the legal system and taxation issues, which are the main concerns of large formal firms. Thus a different emphasis is needed both at the national and the regional level to stimulate the growth of micro and small non-farm enterprises in the regions.

b) Some challenges are straightforward, others are more complex

Many of the key areas for policy to address are relatively straightforward: there is not much disagreement over the objectives or the ways to go about achieving them. For example, building and maintaining rural roads, providing clean rural water, primary health care, are all relatively uncontentious. Some elements of an encouraging investment climate are also straightforward: the perils of gross macro-economic imbalances are well understood, as is how to avoid them. There may be debate over details, but not over the fundamentals.

Other things are more difficult, largely since the way to achieve them is disputed or not fully known. Mitigating rural market failures falls in this category: disputes arise over how much the state should intervene directly as opposed to encouraging private solutions through institutional innovations.

Since not all doubts will be clarified in the short term, the implication is that action should progress on two tracks: move forward on the straightforward, while tackling the more complex issues as learning experiences. A recent interpretation of experiences of growth (Rodrik, 2003) suggests that it is not necessary to get all things exactly right; what matters is to get important things broadly correct, or even just to avoid major pitfalls.[5]

c) Decentralize to deal with local variations

National issues may be all important, but locally there may well be specific things that need attention. Since it would overburden the centre to consider these nationally, and since central policymakers are likely to get it wrong through lack of appreciation of local detail, there is much to be said for decentralizing some authority and funds to local governments to attend to local matters (Bardhan, 1996; Binswanger et al., 2009).

Again, this message is now quite well understood: in many countries specified fractions of central revenues are passed down to local authorities for them to allocate as they see fit. Decentralization is rarely a smooth process, and there is much to discuss about how far to decentralize and how to help local authorities become effective and efficient, but the dangers of over-centralized government are well understood.

d) There is little trade-off between policy for agriculture and for RNFE

This may surprise, yet a comparison of priority actions for both sectors shows heavy overlap, with few things not common to both lists. Above all, the two sectors share a common need for an amenable rural investment climate, for public goods in physical infrastructure and human development, and for mitigating rural market failures—above all in financial services.

[5] Would anyone consider the changes that China made in 1978–9, or the economic reforms that were signalled in the 1980s and put into effect in the 1990s in India as ideal? Most observers saw the changes as steps in the right direction, rather than complete programmes. Yet they have resulted in remarkable accelerations of economic growth in both countries (Rodrik, 2004; Bromley and Yao, 2006; Khan, 2002; Chang, 2003).

Where they differ, it is on items that are not particularly costly: agriculture needs research and extension services specific to farming; some business services may be fairly specific to services and manufacturing and of little value to farmers. But in both cases, the share of public budgets spent in rural areas on these items is likely to be low, less than one-tenth of the whole, if that.[6]

Hence actions for both sectors do not, for the most part, trade off. This might not be apparent from the vigour of debates over which sector is likely to be more dynamic. These arguments may matter for overall understanding, but for immediate policy purposes, they are largely beside the point.

Having made these overall points, discussion now turns to rural public goods, finance, and location.

4.1 RURAL PUBLIC GOODS: PHYSICAL INFRASTRUCTURE AND INVESTING IN PEOPLE

Investments in roads and other means of transport reduce the costs of rural business. While in some cases this can open rural markets to competition from urban manufactures, it allows links to urban areas to function more powerfully. As the previous section argues, as countries become increasingly urbanized, such links become more important to rural economies.

Electricity supply may be more important for some non-farm activities than for agriculture, both to power machinery in manufacturing, desk equipment in offices, and to provide indoor light for longer hours of work in offices and workshops. The key challenge is not so much to establish connections, as to provide reliable power so that businesses do not have to invest in back-up generators.

Education and vocational training may also take on more importance for non-farm activities, especially when these produce goods and services to meet the demand of clients with higher incomes, both locally and in towns and cities. Some skills can be taught through formal education, others may come from experience. In both mainland China and Taiwan the establishment of rural industry has been helped by the return of migrants with experience from work in urban workshops (Murphy, 1999; Otsuka and Reardon, 1998).

[6] Fan and Rao (2003) report estimates that public spending on agriculture in developing countries ran at an average of 8–10 per cent of the value of agricultural product between 1980 and 1998: agricultural research spending averaged less than 1 per cent of agricultural product. Hence, even within the agricultural budget, research is a small element. Compared to total government spending in rural areas, taking in for example the very large budgets for roads, health, education, and water, spending on research rapidly becomes very small indeed.

4.2 FINANCE: DEALING WITH MARKET FAILURES

Rural enterprises, especially small firms, find their access to capital largely limited to whatever savings can be mobilized from family and friends. For rural Bangladesh, for example, a survey in 2000 showed that more than 70 per cent of enterprises reported that their main source of start-up capital, as well as more recent injections, had been their own savings (Hossain, 2010). Only one in ten reported getting loans from banks.

It is not hard to understand why rural businesses find it difficult to get formal credit. Bankers are reluctant to lend to small businesses since they know too little about their abilities and capacity to repay loans, or about their moral character and willingness to repay debts. Wholesalers and traders may similarly be unwilling to advance goods on credit to retailers.

Three approaches may alleviate capital restrictions for small rural businesses: government supply of credit; vertical integration of small with larger firms; and microfinance. There are long-standing experiences of the state directly providing farms and other rural businesses with credit and other financial services through public banks and agencies. By the 1980s (see von Pischke et al., 1983 and Adams et al., 1984), there was mounting evidence of the high cost, inefficiency, and ineffectiveness of such agencies and their programmes. They rarely overcame the underlying high costs of rural banking. When they tried to reduce administrative costs, the quality of lending fell and bad debts mounted. Moreover, they were often politically directed to lend no matter what the risks, and often at subsidized interest rates. As bad debts rose and costs outran earnings, these agencies found themselves dependent on public subsidies to continue to operate. As these costs rose, many were closed in the 1980s and 1990s.

A quite different approach is vertical integration, where a single dominant enterprise organizes the supply chain, usually advancing inputs on credit and guaranteeing to buy up outputs. Examples of this are quite frequent with small-scale and craft manufacturing, in putting-out arrangements whereby rural workshops get inputs from agents who later return to collect produce. This can even be used with some quite sophisticated manufacturing, as applies with the rural subcontractors of electronics component manufacturers in Taiwan (Otsuka, 2007).

It is easy to see the scope for vertical integration when the rural business manufactures for a distant market, where there may be large manufacturers or traders looking for additional supplies from low-cost rural workshops. But the model is not applicable to the very many rural businesses that produce goods and services for local markets.

A third approach has been microfinance, an umbrella term for banking that reduces transaction costs and offers financial services suitable for small-scale and often poor clients. Microfinance tries to reduce problems of moral hazard by working with groups where members screen each other for their character and behaviour, to improve financial management through training of clients, and by looking to offer credit, savings, remittance services, and insurance tailored to the needs of clients. The early experiences of the 1980s and 1990s, with the Grameen Bank of Bangladesh as the exemplar, were pioneered by non-governmental organizations, but increasing numbers of private banks are adopting these innovations to reach unbanked clients.

Microfinance has had its successes, but also its disappointments (Economist, 2005). A persistent tension arises between the ideals of serving the poor, and the temptation for those running the schemes to move up the chain and deal with better-off clients where the risks and transaction costs may diminish and economies of scale apply. An associated issue is that of the cost of lending and the interest rate charged. No matter how good the institutional innovations may be for reducing transaction costs, they cannot be eliminated completely, and when these are added to the cost of capital, microfinanciers often find themselves having to lend at what may seem like high rates of interest—even if they are usually lower than those charged by moneylenders and pawnbrokers.[7]

If there is one overall lesson from experiences of microfinance, it is that there are few if any blueprints—several attempts to replicate the Grameen Bank have failed. Hence effective responses need adaptation to local circumstances. The key point may be learning: promising approaches have to be tried, monitored, and adjusted as lessons become clear.

4.3 LOCATION: THREE DIFFICULT QUESTIONS FOR POLICYMAKERS

Agglomeration can reduce considerably costs of transport as well as recruitment of labour, exchange of information, and carrying out joint research. This is why so much economic activity that is not tied to natural resources, as is agriculture, concentrates in cities (Henderson et al., 2001; Krugman, 1993; World Bank, 2008). This poses three additional questions about the geography of rural businesses: what can be done in remote areas with poor access to cities and ports? Is it possible to influence the hierarchy of urban centres to gain local agglomeration economies? Alternatively, can industrial clusters allow rural businesses to gain compensating advantages?

(a) Remote areas

There may be few things remote areas can produce that will bear the transport costs: most of those are likely to come from primary production and

[7] High interest rates can attract criticism that this is not microfinance so much as loan sharking.

processing. Hence most non-farm businesses in such areas will serve local economies with goods—probably rather limited—and services. Policy may seek to offset high transport costs by subsidizing transport, or offering tax breaks and similar incentives to locate in remote areas. These can be costly and therefore limited to those OECD countries that can afford them. Future developments in information technology may offset the disadvantages of remoteness for some activities.

(b) Urban hierarchies

Hopes have long been held that by encouraging urban centres of certain scale, these would have productive interactions with their rural surroundings to create 'growth poles'. Although popular, attempts to put the idea into operation by, for example, locating significant state industries in market towns, were not that successful. Later in the 1970s and 1980s a similar strategy was proposed. The 'agropolitan' approach sees the district town as a centre for political and administrative functions, with sufficient local economies to encourage non-agricultural activity (Douglass, 1998). Once again, however, attempts to put the concepts into operation had limited success.

(c) Industrial clusters

Notwithstanding the apparently overwhelming agglomeration economies of major cities, industrial clusters have been observed with relatively small factories located in smaller centres: well-known examples include the network of firms in northern Italy producing knitted textiles or shoes in southern Brazil (Schmitz and Nadvi, 1999). Studies of such clusters suggest that it is not necessary to invest in large plants with economies of scale to have rapid and productive industrial growth; progress can be made from sequences of smaller-scale actions.

This has spurred interest in 'local economic development' or 'territorial development' ('desarrollo territorial'). Interactions of factors within regions, it is proposed, allow very different rates and patterns of economic growth even within the same national economies. Territorial development takes a slightly wider view of local interactions than do cluster studies. It is not just economic factors, but regional and local social networks, cultural norms, and the institutions that arise out of such matrices that are the focus in this case (see de Janvry and Sadoulet, 2004; Schejtman and Berdegué, 2004).

The insights so far are intriguing; whether they can be put into operation is another question. It may be that the reasons some regions prosper are both complex and partly fortuitous, offering few lessons for other areas.

A final point: the principles of developing a vibrant and diversified rural economy are known, but the relevant details for particular cases can be elusive. The challenge—and it is no simple matter—is to adapt and tailor principles to the particular context. Only by luck will any plan get it right first time; in most cases, monitoring and evaluating outcomes and adjusting programmes accordingly are critical. Learning must take place. Good environments for learning will engage a diversity of actors—since this will spur innovation as different agencies try different measures, while some institutional rivalry may lead to informal monitoring of the efforts of other agencies and stimulate improved performance.

5. What is the future of the rural non-farm economy in developing countries?

What is the future of the RNFE likely to be, given probable developments within national and the international economies? Over the next ten years or so, the most likely factors to change rural economies comprise, at national level, urbanization, economic growth, migration, and demographic transitions; and internationally, globalization, the increasing economic importance of rapidly growing emerging economies, above all in Asia, and technical advances. Less certain and more likely to apply over a longer time period are climate change, energy costs, and more distant technical advances. This section focuses on the former, with a brief discussion of the latter at the end.

5.1 URBANIZATION, ECONOMIC GROWTH, AND MIGRATION

Rising urbanization and national economic growth, together with improved transport and communication networks, provide important economic linkages between urban and rural areas, opening up new opportunities for rural households (Table 17.5 and Tacoli, 1998; Tacoli and Satterthwaite, 2003).

Evidence from India, for example, suggests that rapid rural non-farm growth is occurring along transport corridors linked to major urban centres, largely independent of their agricultural base (Bhalla, 1997). Similarly, in South-East Asia and in China high population density and low transport costs have led to rapid growth in urban-to-rural subcontracting for labour-intensive manufactures destined for export markets (Otsuka, 2007).

Rural-to-urban commuting has also become significant in some densely settled Asian and Latin American settings. A recent study of six villages in Andhra Pradesh and Madhya Pradesh, India, found that 10 per cent of rural households have at least one member commuting daily to work in nearby towns and cities, while a further 47 per cent deploy members in temporary migration (Deshingkar, 2004). Similar studies from Chile and Peru document

Table 17.5 Expected changes and their implications for the rural non-farm economy

Changes	Implications for the RNFE
National	
Urbanization, economic growth, and migration	Rural economies increasingly respond to urban fortunes and correspondingly rely less on agriculture and other natural resource-based activities.
	More diverse rural economies, more stimulus to RNFE.
Demographic transitions	Window of opportunity as dependency ratios fall: higher average productivity with either increased savings to fuel investment or consumption that stimulates the economy.
	Ageing population with increasing demand for leisure and health, retirement homes.
International developments	
Globalization and concentration in supply chains	Threats to rural manufacturing, some rural retailing and brokering.
	Some potential for exports, providing standards for quality and processes can be met.
	Increased urban/rural interaction in densely settled areas. Remote areas left out.
Rise of China, India, and other large emerging economies	Less chance for manufacturing in smaller countries? Initially yes, but later FDI (Foreign Direct Investment) from Asian giants: transfer of capital and know-how.
	Demand for energy and raw materials push up prices. More opportunities for agriculture, mining, forestry.
Technical advances in near future	Information and communications technologies allow remote areas to provide services, better and more timely advice on opportunities.

sharp increases in daily commuting from rural areas to nearby towns, particularly during the 1990s (Berdegué et al., 2001; Escobal, 2005).

Opportunities for migration are likely to multiply, even if the importance of migration varies considerably by place. Evidence suggests that migrant remittances may serve to increase rural investment, finance schooling, house construction, and agricultural inputs in some locations (Francis and Hoddinott, 1993; Marenya et al., 2003).[8] In some cases, migrants from the same locality form an association and contribute to common funds for local public investments. Remittances may stimulate the rural economy through linkages, such as when funds are invested in improved housing, thus creating jobs for local masons and carpenters—as seen, for example, in rural Egypt in the 1970s and through much of central Mexico in the 1990s.[9]

[8] In Ghana migrants expelled from Nigeria in the early 1980s brought back vehicles, chainsaws, and generators that allowed them to create new businesses (Dei, 1991). In Qwaqwa, South Africa, Slater (2002) recounts life histories that show how those returning or expelled from the cities brought back taxies or funds to set up small stores. For villages in Michoacán, Mexico, Taylor and Wyatt (1996) argue that when capital is scarce, the shadow value of remittances may be high, and demonstrate that migration allows households to overcome capital market limitations.

[9] Models can capture the full impact of remittances on the rural economy. In China, labour loss from migration harms farms and rural businesses, but remittances have strong multipliers in the rural economy—of 2.78 for farming, and 4.50 for self-employment income flows (Taylor, 2002 reporting

On the debit side, migration can expose those moving to additional risks and hazards: illness, accidents, and crime. Migrants may lose their rights and entitlements at destinations, such as access to health care and schools—even when they remain within their own country. They may also be exploited through low pay and poor conditions, since migrants often have little bargaining power with employers, on whom they may also depend for temporary shelter and food. At destinations migration may exacerbate social tensions as newcomers rub shoulders with long-standing residents. Large numbers of migrants may place a strain on services.

The challenge will be to facilitate beneficial movements, and mitigate the less favourable ones.

5.2 DEMOGRAPHIC TRANSITIONS

As fertility rates fall in most developing countries, populations will age. Dependency ratios will probably dip to a historic low before rising somewhat as the fraction of retired persons increases in a generation or two. There is thus a window of opportunity with relatively large numbers of workers that could accelerate growth, raise savings to fuel investment or consumption that stimulates the economy, and allow higher public spending on the education of children with benefits in the future. The challenge will be to make sure that there is productive work for them to do. A flourishing RNFE will be a boon.

When the next stage of an increasingly aged population in the developing world is reached, there will be increasing demand for leisure, health services, and retirement homes. Rural areas may well be preferred to cities, for their restful environment, so long as they can provide rapid access to medical care.

5.3 LIBERALIZATION AND GLOBALIZATION

Globalization has opened up the RNFE both to new opportunities and threats. Relaxed controls on foreign exchange and investment have unleashed foreign direct investment across the developing world. As a result, large exporters, agribusiness firms, and supermarket chains increasingly operate in the rural economies of the developing world, altering the scale and structure of rural supply chains (Reardon et al., 2003; Reardon and Timmer, 2005; Reardon et al., 2007).

Threats from competition are readily apparent. Some rural non-farm activities have thrived in the past because of protection from outside

De Brauw et al., 2000). Similarly for villages in Michoacán, Mexico, models show that migration reduces labour supply and encourages activities less intensive in labour, in this case grazing livestock. But remittances stimulate local incomes and encourage investments, including in livestock and schooling (Adelman et al., 1988).

competition by high transport costs, restrictive regulations such as reserved handicraft industries in India, trade barriers to cheap imported consumer goods, subsidized inputs and credit, and preferential access to key markets. Some small-scale rural manufacturing cannot compete with mass-produced goods which, apart from low cost, can be better quality as well. This can hit some poor households and women workers hard, since they often make up the bulk of those working in rural manufacturing.

In retailing, the advance of supermarkets may put some rural shops out of business. Over sixty thousand small food retailers closed their doors in Argentina from 1984 to 1993 (Gutman, 2002), while over five thousand small food retailers ceased operations in Chile between 1991 and 1995 (Faiguenbaum et al., 2002). Although many of these bankruptcies affected urban traders, emerging evidence suggests that small rural traders and the wholesale markets they serve likewise risk being displaced by larger, specialized wholesalers.

Brokers and traders in agricultural supply chains may also suffer. The leading supermarket chains in Guatemala and Nicaragua, for example, have shifted away from rural small-scale brokers for procuring fresh produce. Instead they now procure through large, specialized wholesalers (Balsevich et al., 2006; Flores and Reardon, 2006; Hernandez et al., 2006). The same trend has emerged in the beef sector in Nicaragua and Costa Rica (Balsevich et al., 2006).

Even in sub-Saharan Africa, South African supermarket chains have expanded aggressively northward following the majority rule in 1994 and the end of economic sanctions that had previously prevented these investments. Two major chains, Shoprite and Pick 'N Pay, have opened outlets in cities and rural towns in Zambia, Malawi, Mozambique, and Uganda and are considering forays into West Africa. In each locality, their entry has altered product selection and market share in favour of imported South African brands at the expense of local farmers, processors, food suppliers, and retailers (Weatherspoon and Reardon, 2003). Evidence from Kenya suggests that local supermarkets there have similarly tended to replace rural traders and wholesale markets with specialized large-scale suppliers. While traditional brokers and wholesale markets supplied 70 per cent of fresh fruits and vegetables in 1997, managers project that share will fall to 10 per cent by 2008 (Neven and Reardon, 2004).

There are, however, opportunities as well. For example, rural Taiwan has seen small workshops grow with their links to electronics companies which export (Otsuka, 2007). In Thailand following the Second World War, former OSS (Office of Strategic Services) operative Jim Thompson began working with silk weavers in north-east Thailand, improving production technology and quality in rearing, reeling, and weaving, creating an international reputation for the quality and design of Thai silk. The resulting growth in silk exports created jobs for over sixty thousand village women, roughly quadrupling returns to labour (Haggblade and Ritchie, 1992).

If rural business is to take up such opportunities to export to international markets, and even to some of the national markets, they will increasingly find that they need to meet standards for quality and process—such as core labour standards. Certification, such as that applying to fair trade or organic markets—may be needed.

Liberalization can also allow technical innovations into rural areas, with widespread ramifications. In Bangladesh during the second half of the 1980s, liberalized import of small diesel engines to run irrigation pumps quite unexpectedly revolutionized two major non-farm activities. After the cropping season, millers harnessed the new diesel engines to power 30,000 seasonal rice hullers, dramatically increasing competition in rice markets. Later, during the rainy season, metal smiths, and boat makers adapted the engines to power river boats (Jansen et al., 1989; Haggblade et al., 2007).

5.4 RISE OF CHINA AND INDIA AND OTHER LARGE EMERGING ECONOMIES

The rapid growth of the economies of some large developing countries, notably China and India, holds both a threat and a promise to rural economies elsewhere. The threat lies in the large countries' economies of scale in manufacturing, whereas smaller countries cannot develop manufacturing in the face of cheap imports. Without a thriving urban economy, the prospects for rural economies worsen, with less demand for rural tradables and fewer productive links. That said, at some point China and India will run out of cheap labour, encouraging some industries to move to places with unemployed labour. A combination of Asian investment with industrial know-how could transform manufacturing in other parts of Asia, and perhaps also Africa.

The promise lies in the rising demand of Asian economies for raw materials, which already appears to be driving up the cost of energy and raw materials on world markets. This represents a market for developing countries with natural resources, most of them rural. Agriculture, mining, and forestry will be stimulated with the potential for strong multipliers into local rural economies.

Just how powerful are the new opportunities provided by urbanization and globalization, and to what extent can they substitute for agricultural growth as the main driver of the RNFE? These forces seem to be working powerfully in densely populated, rapidly growing countries. The rise of the Chinese township and village enterprises (TVE) has been remarkable (Fu and Balasubramanyam, 2003). The correlation between growth of agriculture and non-farm employment has weakened in many parts of rural India (Harriss-White and Janakarajan, 1997; Foster and Rosenzweig, 2004). Where urban congestion, soaring rents, and higher wages raise the cost of doing business in cities, rural-to-urban commuting, temporary migration, and urban-to-rural subcontracting become attractive (Otsuka, 2007).

Opportunities for rural growth led largely by urban development appear more limited in poorer, less densely populated and slow-growing economies, such as much of Africa. Rural areas with better infrastructure and market access seem more likely to benefit from urbanization and globalization, often the better-off regions to begin with. Remote, lagging regions in poor, slow-growing countries will probably gain the least (Reardon et al., 2007).

5.5 TECHNOLOGY ADVANCES

Currently, rural areas in developing countries are seeing an impressive diffusion of information and communications technologies in cell-phone coverage and the arrival of the Internet. These technologies are already having an impact on trading, with faster diffusion of accurate information creating opportunities in distant markets. In the last few years, schemes to use cell-phones to make money transfers have been started, which could reduce the often high cost and uncertainty of sending remittances home, either from abroad or from a distant part of the same country.

Some urban services may relocate to rural areas once reliable Internet access can be provided. For example, entrepreneurs in the US send digital photographs to China, where rural painters produce oil portraits and then ship them back by courier service. Linked by a former NGO employee turned businessman, Bangladeshi draftsmen produce architectural drawings to order and then digitally transmit them to the US. Valued at US$600 million per year, these offshore, electronically transmitted services require good external contacts and communications facilities (Chowdhury, 2000).

5.6 CHANGES POSSIBLE IN THE MEDIUM TERM

Demand is likely to mount for rural areas of the developing world to provide environmental services—in particular biodiversity conservation and water supplies as well as carbon capture to mitigate climate change. While much of the response will take place within agriculture and forestry, some additional jobs may be created in conservation activities.

The rise in the oil price since the mid-2000s may continue if it be that easily accessible reserves have largely been tapped. This could have contrasting impacts on rural economies. An increase in transport costs would leave rural areas at an increasing disadvantage to urban areas, and especially so for the more distant and remote areas. On the other hand, the likely search for alternative energy may create new rural activities, some within agriculture such as biofuels, others such as hydroelectricity, geo-thermal, solar, wind, wave, and tidal where feasible. Technical advances that lower the cost of some of these renewable energy sources might allow some rural manufacturing to take advantage of cheaper energy supplies.

5.7 FUTURE RURAL ECONOMIES

Taking these considerations into account, it is possible to imagine three evolving rural economies, where the RNFE will be quite distinct, as follows.

***Agrarian rural economy*:** In most areas beyond the peri-urban surround, the motor of the rural economy will be farming, forestry, or fisheries, although occasionally mining or tourism will serve a similar function. The RNFE will develop largely as a response to activities producing tradable goods or services. It will be dominated by services, probably small businesses for the most part. Manufacturing will be restricted to any primary processing carried out locally on grounds of perishability or where major savings can be made on bulk to be transported out.

In some regions this may become a lonely economy. Where agriculture becomes more mechanized and sheds jobs, where reduced transport costs mean that many goods can be shipped in from cities, then the RNFE may shrink to provide only those day-to-day services for a declining farm population that cannot be satisfied by the weekly drive to the nearest regional centre.[10]

***Peri-urban economy*:** Increasing parts of the countryside in the more densely settled developing world will become peri-urban, where intense interactions with the city dominate the economy. Here the rural economy will probably become highly diversified, with close and dynamic links to the urban economy, responding more to urban influences than rural. Increasingly, young people whose parents were low-status agricultural labourers locked into long-standing client relations with their village patrons, now board a bus early in the morning to go to work in the nearest town. It can be a world of new possibilities unimagined a generation before.

On the other hand, it may be seen more gloomily through Bernstein's 'fragmentation of labour' lens in which the rural poor have to look for

[10] In OECD countries, concern surrounds the future of small market towns in rural areas. They can be found every 15 km or so across the landscape, where a weekly market for the surrounding farms would be held, with people walking, riding, or catching a bus to make the trip. With better highways and shorter journey times, the weekly outing by bus or minibus from the village may bypass these, heading for a larger regional centre with more services on offer. Hence some of the market towns seem to be in decline.

increasingly informal and precarious work, spread across a wider landscape, bridging the agrarian and the urban, with fewer of the old certainties and patronage to reassure them that destitution will not be their fate (Harriss et al., 2010 reporting Bernstein, 2004). This is already the reality of quite large areas of rural Asia, as well as the more densely settled areas of Latin America and a few parts of Africa.

Remote regions rebooted?: Remote regions often have few possibilities at present, with out-migration taking places at disconcertingly high rates. Might their fortunes be revived in the future? As advances take place in communications, some of these areas will be attractive to relocating service workers on account of their landscapes, low congestion, and the promise of less hurried living. This scenario is already emerging in some remote areas of OECD countries.

As economic growth and urbanization take place, it is likely that some of these remote areas may develop tourism as an activity, their very isolation and low populations making them attractive for some holiday-makers.

It is just possible that a combination of high fossil-fuel prices and advances in renewable energy generation may create new opportunities. Some of these areas are inhospitable precisely on account of their endowments for renewable power: deserts with long hours of sunshine, uplands with strong winds, coastal areas with strong waves and tides. Moreover, their low population means there may be few limitations on construction of windmills, tidal barrages, wave machines, solar capture towers, etc.

6. **Conclusions: small farmers and the rural non-farm economy**

What, then, does this review say in response to three of Peter Hazell's five questions about small farmers, set out in the introduction?

How will the transition from agrarian to urban economy take place, and how quickly? A thriving RNFE, whether driven by dynamic agriculture or the proximity of a city, will both stimulate the transition—allowing farmers to rely increasingly on other means of income, while stemming mass emigration from the countryside. Policymakers concerned about young cities being overwhelmed by indigent newcomers thus need to study how to promote the RNFE. It is equally clear that differing geography will see different transitions, depending on access to cities and ports, and quality of natural resources. A single model is not to be expected.

How large is a viable smallholding? In the presence of a thriving RNFE, this question loses significance. If rural households have multiple sources of income, then the size of the farm only affects just how much of their income they get from agriculture. It does not necessarily determine their livelihood, so viability will no longer depend on land. Parts of Europe still have many small family farms that generate low incomes, yet they are viable since the household has other means of support. There may be questions here about the efficiency of farming when much of the land is in small plots, but that is a different question. In practice, the very small farms of Europe are often not occupying prime agricultural land. In areas of high agricultural potential, most of the land has been consolidated into larger, but still usually family operated, farms.

What policies and programmes are needed to support small farmers? This review addresses this point only indirectly but reaches an important conclusion, and perhaps, for some readers, an unexpected one. That is that most policy for supporting small farms, including almost all rural public investment, serves the RNFE as well as it does farming or any other economic activity. For the most part, a government that focuses on the triple challenge for agriculture of investment-climate-plus-rural-public-goods-plus-solving-market-failures will help those starting rural businesses as well as farmers.

Two qualifications may be added, however. One is that at the margin, there will be differences: agricultural research and extension for farming, for example; or some business services for non-farm enterprises. But this is at the margin; the fundamentals are common.

The other is about future developments. Additional policies will be needed to adapt to and mitigate climate change, to make economies in general more environmentally sustainable, and in the medium term, to facilitate a transition to renewable sources of energy. For farmers, these additions will become part of the weft and weave of policies for agriculture. For the RNFE, these policies may well see new enterprises and jobs in providing environmental services and in producing renewable energy.

As a final word, it will increasingly be non-farm activities that will define the rural economy, and that will allow a relatively untroubled transition from an agrarian to an urban and industrialized economy and society. The best way that policymakers can assist in this, is to ensure that sufficient attention is paid to the three things that make a difference to all rural enterprise: a decent rural investment climate; rural public goods; and the solution, or mitigation, of rural market failures.

REFERENCES

Acharya, S. (2003). 'Off-farm and Non-farm Jobs in SEATEs and Thailand: Rationale and Synthesis of Country Studies', in Development Analysis Network, *Off-farm and Non-farm Employment in Southeast Asian Transitional Economies and Thailand*, DAN—Cambodia

Development Resource Institute, Cambodian Institute for Cooperation and Peace, Central Institute for Economic Management (Vietnam), Institute of Economics (Vietnam), National Economic Research Institute (Lao PDR), National Statistical Research Center (Lao PDR) and Thailand Development Research Institute. Supported by the International Development Research Centre of Canada, and the Rockefeller Foundation, Phnom Penh.

Adams, D. W., D. H. Graham, and J. D. von Pischke (1984). *Undermining Rural Development with Cheap Credit.* Boulder, Colorado and London: Westview Press.

Adams, Richard H. Jr (2002). 'Nonfarm income, inequality and land in rural Egypt', *Economic Development and Cultural Change*, pp. 339–63.

Adams, Richard H. Jr and J. J. He (1995). 'Sources of income inequality and poverty in rural Pakistan', Research Report No. 102. Washington, DC: International Food Policy Research Institute.

Adelman, I., J. E. Taylor, and S. Vogel (1988). 'Life in a Mexican village: a SAM perspective', *Journal of Development Studies*, 25(1), pp. 5–24.

Ahmed, R., S. Haggblade, P. Hazell, R. L. Meyer, and T. Reardon (2007). 'The policy and institutional environment affecting the rural non-farm economy', Chapter 11 in Haggblade, S., P. B. R. Hazell, and T. Reardon, *Transforming the Rural Non-farm Economy: Opportunities and Threats in the Developing World.* Baltimore: Johns Hopkins University Press.

Anderson, D. (1982). 'Small Industry in developing countries: A discussion of issues', *World Development*, 10(11), pp. 913–48.

Balihuta, A. and K. Sen (2001). 'Macroeconomic policies and rural livelihood diversification: an Ugandan case study', LADDER Working Paper No. 3, October 2001, Overseas Development Group, University of East Anglia, Norwich, UK.

Balsevich, F., J. Berdegue, and T. Reardon (2006). 'Supermarkets, new-generation wholesalers, tomato farmers and NGOs in Nicaragua', Department of Agricultural Economics Staff Paper. March 2006-03. East Lansing: Michigan State University.

Balsevich, F., P. Schuetz, and E. Perez (2006). 'Cattle producers' participation in market channels in Central America: Supermarkets, processors, and auctions', Department of Agricultural Economics Staff Paper, January 2006. East Lansing: Michigan State University.

Banco Central de Chile (1986). *Indicadores económicos y sociales, 1960–1985.* Santiago: Banco Central de Chile, Dirección de Estudios.

Banco Central de Chile (2002). *Chile social and economic indicators 1960–2000.* Santiago: Banco Central de Chile.

Bangladesh Bureau of Statistics (2003). *Bangladesh Population Census, 2001: National Report.* Dhaka: Bangladesh Bureau of Statistics.

Bardhan, P. (1996). 'Decentralised development', *Indian Economic Review*, 31(2), pp. 139–56.

Bell, C., P. Hazell, and R. Slade (1982). *Project Evaluation in Regional Perspective: A Study of an Irrigation Project in Northwest Malaysia.* Baltimore: The Johns Hopkins University Press.

Berdegué, J., E. Ramírez, T. Reardon, and G. Escobar (2001). 'Rural non-farm employment and incomes in Chile', *World Development*, 29(3), pp. 411–25.

Bernabè, S. and G. Krstic (2005). 'Labor productivity and access to markets matter for pro-poor growth; The 1990s in Burkina Faso and Vietnam', Sector Paper, *Operationalising Pro-Poor Growth*, World Bank. Retrieved from: <http://siteresources.worldbank.org/INTPGI/Resources/342674-1115051862644/Labormarketsjune2005.pdf>.

Bernstein, H. (2004). '"Changing before our very eyes": Agrarian questions and the politics of land in capitalism today', *Journal of Agrarian Change*, 4, 1 and 2, pp. 190–225.

Bhalla, S. (1997). 'The rise and fall of workforce diversification processes in rural India: A regional and sectoral analysis', Centre for Economic Studies and Planning, DSA Working Paper. New Delhi: Jawaharlal Nehru University.

Bhalla, S. (2003). 'Linked livelihoods: Trajectories in farm and non-farm activities in eight villages of Andhra Pradesh and Haryana, 1992–2002', Draft Working Paper.

Binswanger-Mkhize, H. P., A. F. McCalla, and P. Patel (2009). 'Structural transformation and African agriculture', Africa Emerging Markets Forum, Cape Town, South Africa, 13–15 September 2009.

Bland, J. (1995). 'Employment generation in developing countries: A synthesis of findings from selected USAID projects', PN-ABX-010, October 1995, Research & Reference Services, United States Agency for International Development, Washington, DC, 20523–1820.

Bromley, D. W. and Y. Yao (2006). 'Understanding China's economic transformation. Are there lessons here for the developing world?', *World Economics*, 7(2), pp. 73–95.

Bryceson, D. F. (2002). 'The scramble in Africa: reorienting rural livelihoods', *World Development*, 30(5), pp. 725–39.

Carletto, G., K. Covarrubias, B. Davis, M. Krausova, K. Stamoulis, P. Winters, and A. Zezza (2007). 'Rural income generating activities in developing countries: re-assessing the evidence', electronic, *Journal of Agricultural and Development Economics*, 4(1), pp. 146–93.

Chadha, G. K. (2003). 'Rural nonfarm sector in the Indian economy: Growth, challenges and future direction', Mimeo, International Food Policy Research Institute, Washington, DC.

Chang, Ha-Joon (2003). 'Institutional development in historical perspective', in H.-J. Chang (ed.), *Rethinking Development Economics*. London: Anthem Press.

Chowdhury, N. (2000). 'Information and communications technologies and IFPRI's mandate: A conceptual framework', Mimeo, International Food Policy Research Institute, Washington, DC.

De Brauw, A., J. E. Taylor, and S. Rozelle (2000). 'Migration and incomes', in *Source communities: a new economics of migration perspective from China*. University of California, Davis, CA: Department of Agricultural and Resource Economics.

de Janvry, A. and E. Sadoulet (2004). 'Toward a territorial approach to rural development', Paper to the 85th Seminar of the European Association of Agricultural Economists, *Agricultural development and rural poverty under globalisation: asymmetric processes and differentiated outcomes*, Florence, 8–11 September 2004.

Dei, G. J. S. (1991). 'The re-integration and rehabilitation of migrant workers into a local domestic economy: lessons for "endogenous" development', *Human Organization*, 50(4), pp. 327–36.

Deichmann, U., F. Shilpi, and R. Vakis (2009). 'Urban Proximity, Agricultural Potential and Rural Non-farm Employment: Evidence from Bangladesh', *World Development*, 37(3), pp. 645–60.

Deshingkar, P. (2004). 'Rural–Urban Links in India: New Policy Challenges for Increasingly Mobile Populations', Paper presented at the World Bank Rural Week, 2 March 2004. Retrieved from: <http://www.livelihoods.org/hot_topics/docs/UR_IndiaCaseStudy.doc>.

Dirven, M. (2011). 'Rural non-farm employment and rural poverty reduction: What we know in Latin America in 2010', Paper presented at the IFAD Conference on *New Directions for Smallholder Agriculture*, 24–25 January 2011, Rome.

Djurfeldt, G. V. A., N. Jayakumar, S. Lindberg, A. Rajagopal, and R. Vidyasagar (2008). 'Agrarian Change and Social Mobility in Tamil Nadu', 8 November 2008, *Economic & Political Weekly (EPW)*, 45, pp. 50–61.

Douglass, M. (1998). 'A regional network strategy for reciprocal rural-urban linkages', *Third World Planning Review*, 20(1), pp. 1–33.

The Economist (2005). 'The hidden wealth of the poor. A survey of microfinance', 5 November 2005.

Ellis, F. and G. Bahiigwa (2001). 'Livelihoods and rural poverty reduction in Uganda', LADDER Working Paper No. 5, November 2001, Overseas Development Group, University of East Anglia, Norwich, UK.

Ellis, F. and N. Mdoe (2002). 'Livelihoods and rural poverty reduction in Tanzania', LADDER Working Paper No. 11, February 2001, Overseas Development Group, University of East Anglia, Norwich, UK.

Escobal, J. (2005). 'The Role of Public Infrastructure in Market Development in Rural Peru', PhD dissertation, Development Economics Group, University of Wageningen, NL.

Faiguenbaum, S., J. A. Berdegué, and T. Reardon (2002). 'The rapid rise of supermarkets in Chile and its effects on the dairy, vegetable and beef chains', *Development Policy Review* (4), 20 (4), pp. 459–71, Blackwell for Overseas Development Institute.

Fan, S. and N. Rao (2003). 'Public spending in developing countries: trends, determination, and impact', EPTD Discussion Paper No. 99, Environment and Production Technology Division. International Food Policy Research Institute, Washington, DC.

FAO (Food and Agriculture Organization) (1998). *The State of Food and Agriculture, Part 3, Rural Nonfarm Income in Developing Countries.* Rome: FAO.

Fisher, T., V. Mahajan, and A. Singha (1997). *The Forgotten Sector: Non-farm Employment and Enterprises in Rural India.* London: Intermediate Technology Publications.

Flores, L. and T. Reardon (2006). 'Supermarkets, new-generation wholesalers, farmers' organizations, contract farming and lettuce in Guatemala: Participation by and effects on small farmers', Department of Agricultural Economics Staff Paper 2006–07. East Lansing: Michigan State University.

Foster A. D. and M. R. Rosenzweig (2004). 'Agricultural productivity growth, rural economic diversity and economic reforms: India, 1970–2000', *Economic Development and Cultural Change*, 52(3), pp. 509–42.

Francis, E. and J. Hoddinott (1993). 'Migration and differentiation in Western Kenya: A tale of two sub-locations', *Journal of Development Studies*, 30(1), pp. 115–45.

Fu, X. and V. N. Balasubramanyam (2003). 'Township and village enterprises in China', *Journal of Development Studies*, 39(4), pp. 27–46.

Grabowski, R. (1995). 'Commercialization, non-agricultural production, agricultural innovation, and economic development', *Journal of Developing Areas*, 30, pp. 41–62.

Grindle, M. (2007). 'Good enough governance revisited', *Development Policy Review*, 25(5), pp. 553–74.

Gutman, G. E. (2002). 'Impacts of the rapid rise of supermarkets on the dairy products supply chains in Argentina', *Development Policy Review*, 20, pp. 409–27, Blackwell for Overseas Development Institute.

Haggblade, S. (2007). 'Sub-sector supply chains: operational diagnostics for a complex rural economy', chapter 15 in Haggblade, S., P. B. R. Hazell, and T. Reardon, *Transforming the Rural Nonfarm Economy: Opportunities and Threats in the Developing World.* Baltimore: Johns Hopkins University Press.

Haggblade, S., P. Hazell, and T. Reardon (eds) (2007). *Transforming the Rural Nonfarm Economy.* Baltimore: Johns Hopkins University Press.

Haggblade, S., P. Hazell and T. Reardon (2007a). 'Strategies for stimulating equitable growth in the rural non-farm economy', chapter 17 in *Transforming the Rural Nonfarm Economy:*

Opportunities and Threats in the Developing World. Baltimore: Johns Hopkins University Press.

Haggblade, S., D. C. Mead, and R. Meyer (2007). 'An overview of programs for promoting the rural non-farm economy', chapter 17 in Haggblade, S., P. B. R. Hazell, and T. Reardon (2007). *Transforming the Rural Non-farm Economy: Opportunities and Threats in the Developing World.* Baltimore: Johns Hopkins University Press.

Haggblade, S., T. Reardon, and E. Hyman (2007). 'Technology as a motor of change in the rural nonfarm economy', in Haggblade, S., P. B. R. Hazell, and T. Reardon (eds), *Transforming the Rural Nonfarm Economy.* Baltimore: Johns Hopkins University Press.

Haggblade, S. and N. Ritchie (1992). 'Opportunities for intervention in Thailand's silk subsector', GEMINI Working Paper 27, Bethesda, Md.: Development Alternatives.

Harriss, J., J. Jeyaranjan, and K. Nagaraj (2010). 'Land, labour and caste politics in rural Tamil Nadu in the 20th century: Iruvelpattu (1916–2008)', *Economic & Political Weekly,* 45(31).

Harriss-White, B. and S. Janakarajan (1997). 'From Green Revolution to rural industrial revolution in south India', *Economic and Political Weekly,* 32(25), pp. 1469–77.

Hazell, P. (2011). 'Five big questions about five hundred million small farms', paper presented at the *IFAD Conference on New Directions for Smallholder Agriculture,* 24–25 January 2011, Rome.

Hazell, P. and C. Ramasamy (1991). *Green Revolution Reconsidered.* Baltimore: Johns Hopkins University Press.

Hazell, P. and S. Haggblade (1993). 'Farm-nonfarm growth linkages and the welfare of the poor', in Lipton, M. and J. van der Gaag (eds), *Including the Poor.* Washington, DC: World Bank.

Henderson, J. V., Z. Shalizi, and A. J. Venables (2001). 'Geography and development', *Journal of Economic Geography,* 1, pp. 81–105.

Hernandez, R. and J. Berdegué (2006). 'Tomato farmer participation in supermarket market channels in Guatemala: Determinants and technology and income effects', Department of Agricultural Economics Staff Paper, April 2006. East Lansing: Michigan State University.

Ho, S. P. S. (1986). 'Off-farm employment and farm households in Taiwan', in Shand R. T. (ed.), *Off-Farm Employment in the Development of Rural Asia,* Vol. 1. Canberra: Australian National University.

Hossain, M. (2010). 'Non-farm economy in Bangladesh: nature, determinants and impact on poverty reduction', Report of the Regional Workshop *on Micro Credit Delivery System and Good Governance in Rural Development,* 7–18 February 2010, Comilla, Bangladesh, New Delhi: Afro-Asian Rural Development Organization (AARDO).

Hossain, M., D. Lewis, M. L. Bose, and A. Chowdhury (2003). 'Rice research, technological progress, and impacts on the poor: the Bangladesh case (Summary Report)', Discussion Paper No. 110, Environment and Production Technology Division, International Food Policy Research Institute, Washington, DC.

Hymer, S. and S. Resnick (1969). 'A Model of an Agrarian Economy with Nonagricultural Activities', *American Economic Review,* 59(4), pp. 493–506.

IFAD (2007). 'Sending money home. Worldwide remittance flows to developing and transition countries', IFAD, Rome, December 2007.

Jansen, E. G., A. J. Dolman, A. M. Jerve, and N. Rahman (1989). *The Country Boats of Bangladesh: Social and Economic Development and Decision-making in Inland Water Transport.* Dhaka: University Press Limited.

Johnston, B. F. and P. Kilby (1975). *Agriculture and Structural Transformation.* New York: Oxford University Press.

Khan, M. H. (2002). 'State failure in developing countries and strategies of institutional reform', Draft Paper for World Bank ABCDE Conference, Oslo, June 2002.

Krugman, P. (1993). 'First nature, second nature, and metropolitan location', *Journal of Regional Science*, 33(2), pp. 129–44.

Lanjouw, P. (1999). 'Rural non-agricultural employment and poor in Ecuador', *Economic Development and Cultural Change*, 48(1), October.

Lanjouw, P. (2007). 'Does the rural nonfarm economy contribute to poverty reduction?', chapter 3 in Haggblade, S., P. Hazell, and T. Reardon (eds), *Transforming the Rural Nonfarm Economy*. Baltimore: Johns Hopkins University Press.

Lanjouw, P. and A. Shariff (2004). 'Rural non-farm employment in India: Access, incomes and poverty impact', *Economic and Political Weekly*, 39(40), October.

Mahajan, V. and R. K. Gupta (2011). 'Non farm opportunities for smallholder agriculture', Paper presented at the *IFAD Conference on New Directions for Smallholder Agriculture*, 24–25 January 2011, Rome.

Mandal, M. A. S. (2002). 'Agricultural machinery manufacturing and farm mechanization: a case of rural non-farm economic development in Bangladesh', Paper presented to the *Workshop on Fostering Rural Development Agriculture-based Enterprises and Services*, GTZ/DFID/IFAD/CTA/IBRD, Berlin, November 2002.

Marenya, P. P., W. Oluoch-Kosura, F. Place, and C. B. Barrett (2003). 'Education, nonfarm income and farm investment in land-scarce western Kenya', Basis Brief No. 14.

Moore, M. and H. Schmitz (2007). 'Can we capture the spirit of capitalism? The investment climate debate', Draft 25 May 2007, Brighton, UK: Institute of Development Studies.

Mukherjee, A. and X. Zhang (2007). 'Contrasting rural nonfarm policies and performance in China and India: Lessons for the future', in chapter 13, Haggblade, S., P. B. R. Hazell, and T. Reardon (2007), *Transforming the Rural Nonfarm Economy: Opportunities and Threats in the Developing World*. Baltimore: Johns Hopkins University Press.

Murphy, R. (1999). 'Return migrant entrepreneurs and economic diversification in two countries in South Jiangxi, China', *Journal of International Development*, 11, pp. 661–72.

Neven D. and T. Reardon (2004). 'The rise of Kenyan supermarkets and evolution of their horticulture product procurement systems: Implications for agricultural diversification and smallholder market access programs', *Development Policy Review*, 22(6), pp. 669–99.

Otsuka, K. (2007). 'The Rural Industrial Transition in East Asia: Influences and Implications', chapter 10 in Haggblade, S., P. Hazell, and T. Reardon (eds), *Transforming the Rural Nonfarm Economy*. Baltimore: Johns Hopkins University Press.

Otsuka, K. and T. Reardon (1998). 'Lessons from rural industrialization in East Asia: Are they applicable to Africa?', Paper No. 13, *IFPRI Conference, Strategies for Stimulating Growth of the Rural Nonfarm Economy in Developing Countries*, May 1998, Airlie House, Virginia.

Otsuka, K. and T. Yamano (2006b). 'The role of rural labor markets in poverty reduction: evidence from Asia and East Africa', First Draft, Background Paper for World Development Report 2008.

Ranis, G. and F. Stewart (1993). 'Rural non-agricultural activities in development. Theory and application', *Journal of Development Economics*, 40, pp. 75–101.

Reardon, T., J. Berdegué, C. B. Barrett and K. Stamoulis (2007). 'Household income diversification into rural non-farm activities', in Haggblade, S., P. Hazell, and T. Reardon (eds), *Transforming the Rural Non-farm Economy. Opportunities and Threats in the Developing World*, Johns Hopkins University Press, Baltimore, for the International Food Policy Research Institute.

Reardon, T., K. Stamolous, A. Balisacan, M. E. Cruz, J. Berdegué, and B. Banks (1998). 'Rural nonfarm income in developing countries', special chapter in *The State of Food and Agriculture 1998*. Rome: Food and Agricultural Organization of the United Nations, pp. 283–356.

Reardon, T., K. Stamoulis, and P. Pingali (2007). 'Rural nonfarm employment in developing countries in an era of globalization', in Otsuka K. and K. Kalirajan (eds), *Contributions of Agricultural Economics to Critical Policy Issues*. Malden, MA: Blackwell.

Reardon, T. and P. Timmer (2005). 'Transformation of markets for agricultural output in developing countries since 1950: How has thinking changed?' in Evenson, R. and P. Pingail (eds), *Handbook of Agricultural Economics*, Vol. 3A. Amsterdam: North Holland Press.

Reardon, T. C., P Timmer, C. Barrett, and J. Berdegué (2003). 'The rise of supermarkets in Africa, Asia and Latin America', *American Journal of Agricultural Economics*, 85(5).

Richardson, H. W. (1985). 'Input–output and economic base multipliers: Looking backward and looking forward', *Journal of Regional Science*, 25, pp. 607–61.

Rodrik, D. (2003). *In Search of Prosperity: Analytical Narratives on Economic Growth*. Princeton, New Jersey: Princeton University Press.

Rodrik, D. (2004). 'Rethinking growth policies in the developing world', Mimeo of October 2004, Draft of the Luca d'Agliano Lecture in Development Economics to be delivered on 8 October 2004, in Torino, Italy.

Schejtman, A. and J. Berdegué (2004). 'Desarrollo territorial rural', Temas y Debates No 1, RIMISP, Santiago de Chile.

Schmitz, H. and K. Nadvi (1999). 'Clustering and industrialization: introduction', *World Development*, 27(9), pp. 1503–14.

Scoones, I., with C. Chibudu, S. Chikura, P. Jeranyama, D. Machaka, W. Machanja, B. Mavedzenge, B. Mombeshora, M. Mudhara, C. Mudziwo, F. Murimbarimba, and B. Zirereza (1996). *Hazards and opportunities. Farming livelihoods in dryland Africa: lessons from Zimbabwe*. London and New Jersey: Zed Books.

Shand, R. T. (1986). *Off-farm employment in the development in rural Asia*, Canberra: Australian National University.

Slater, R. (2002). 'Differentiation and diversification: changing livelihoods in Qwaqwa, South Africa, 1970–2000', *Journal of Southern Africa Studies*, 28(3), pp. 599–614.

Tacoli, C. and D. Satterthwaite (2003). 'The urban part of rural development: The role of small and intermediate urban centres in rural and regional development and poverty reduction', Rural–Urban Interactions and Livelihood Strategies Working Paper 9. London: International Institute for Environment and Development.

Tacoli, C. (1998). 'Rural–Urban Interactions: A Guide to the Literature', *Environment and Urbanization*, 10(1), pp. 147–66.

Taylor, J. E. (2002). 'The microeconomics of globalisation: evidence from China and Mexico', Paper for discussion at the Global Forum on Agriculture: Agricultural trade reform, adjustment and poverty, May 2002, OECD.

Taylor, J. E. and T. J. Wyatt (1996). 'The shadow value of migrant remittances, income and inequality in a household-farm economy', *Journal of Development Studies*, 32(6), pp. 899–912.

von Pischke, J. D., D. W. Adams, and G. Donald (1983). 'Changing perceptions of rural financial markets', in von Pischke, J. D., W. Adams, and G. Donald, *Rural financial markets in developing countries: their use and abuse*. Baltimore and London: Johns Hopkins University Press for the Economic Development Institute of the World Bank.

Weatherspoon, D. D. and T. Reardon (2003). 'The Rise of Supermarkets in Africa: Implications for Agrifood Systems and the Rural Poor', *Development Policy Review*, 21(3), pp. 333–55.

Wiggins, S., K. Preibisch, and S. Proctor (1999). 'The impact of agricultural policy liberalization on rural communities in Mexico', *Journal of International Development*, 11(7), pp. 1029–42.

Wiggins, S. and S. Proctor (1999). Literature Review, 'Migration and the rural non-farm sector', Mimeo, University of Reading.

World Bank (2006). 'Revitalizing the rural economy: An assessment of the investment climate faced by non-farm enterprises at the district level', Consultative Draft, World Bank Office, Jakarta.

World Bank (2008). Overview, *Reshaping Economic Geography, World Development Report 2009.* Washington, DC: World Bank.

World Bank (2009). *Awakening Africa's Sleeping Giant. Prospects for Commercial Agriculture in the Guinea Savannah Zone and Beyond.* Washington, DC: World Bank.

Zambia, Republic of (2003). Zambia 2000, *Census of Population and Housing: Main Census Report.* Lusaka: Central Statistics Office.

Part IV
The Policy Agenda

18 Concluding chapter: the policy agenda

PETER HAZELL AND ATIQUR RAHMAN

Chapter 1 reviewed the arguments for and against small farms. This chapter revisits some of the same issues, drawing on the evidence presented in the book to see what conclusions can be drawn, as well as the outlook for small farms in the years ahead. This leads to a discussion of appropriate policy options for supporting the small farm sector.

Defining small farms

Any discussion has to begin with an agreed definition of a small farm. A common practice, and one followed in much of this book, is to define small farms as holdings less than 2 ha in size. A simple land size definition enables the number of small farms to be enumerated using widely available agricultural census and household survey data, and facilitates comparisons across countries and over time. However, it can also be a misleading definition because the economics of size depends on the quality of the available land, the prevailing agro-ecological and market conditions, and the off-farm income-earning opportunities available to farmers. A 'viable' small farm might vary from just a couple of hectares in Asia to several hundred hectares in parts of Latin America. But it can be even smaller than 2 ha if farming is combined with non-farm sources of income, as with many very small, part-time farmers in Asia.

Berdegue and Fuentealba (Chapter 5) argue for a broader definition of smallholders based on the concept of the family farm, which they define as farms that are operated by families using largely their own labour. This definition is consistent with a long tradition of peasant studies, starting with Chayanov's work in the USSR in the 1920s (Chayanov, 1986) and extending into the recent work of Brookfield (2008) and Lipton (2009).

Since just about every small farm is a family farm, this definition leads to higher estimates of the total number of 'small' farms. For example, in Latin America there are about 5 million small farms less than 2 ha, but about 16 million family farms (Chapter 5).

A problem with using family farms as a definition of smallness is that they are not all small. In fact, many of the largest commercial farms found around the world today are also family farms as defined earlier; they are just highly capitalized and mechanized. Attempts to draw a line between 'small' and 'non-small' family farms are sometimes made on the basis of assets (e.g. capital stock or machines), farm income, or gross turnover. Berdegue and Fuentealba (Chapter 5) show how useful a flexible approach to defining small farms can be in the Latin American context. But this does require access to data that is not widely available, and definitions that are not easily comparable across countries and over time. The broader family farm approach is also less compelling for Asia and the more populous countries of sub-Saharan Africa, where the average and modal farm sizes are well under 2 ha.

Competitiveness of small farms

Most of the evidence presented in this book (and especially Chapters 3 to 5) supports the hypothesis that small farms are still the more efficient producers, despite the fact that on average they are getting smaller. However, it has to be recognized that large commercial farms are rarely included in farm size studies in Asia or Africa, so there is little hard evidence about the relative cost structures and yields of some of today's new types of large commercial farms. Some of these use the latest precision farming technologies, machinery, best seeds, and agrochemicals to increase yields and lower production costs (Byerlee, 2011; Deininger and Byerlee, 2010). For example, Brazilian investors are bringing their large-scale farm models from the Cerrado region, where they have been hugely successful, to Angola and Mozambique. We do not know how the production efficiency of these farms compares with that of competing small farms. It also has to be recognized that in contexts where land is being degraded through increasing population pressure and unsustainable farming practices, small farms may be losing efficiency.

However, even where small farms are the most efficient producers, that is not necessarily enough. With modern value chains and liberalized markets, there is growing evidence that small farms are facing major disadvantages in accessing modern inputs and credit and in selling products, while large farms are capturing significant economies of scale in their links to value chains. The most threatening examples are the large corporate farms that not only capitalize on the latest production technologies, but are also able to back this with investments and political connections that give them privileged access to markets, modern inputs, insurance, and credit, resulting in yields and cost structures that small farms simply may not be able to beat (Byerlee et al., 2012; Deininger and Byerlee, 2010).

Given potential economies of scale in production and marketing or both, many small farms are in danger of becoming less competitive and hence less viable as businesses. But there are important differences across commodities. For example, small farms are more likely to be competitive in producing labour-intensive, high-value tree, crop, and livestock products than in producing cereals or industrial crops. Moreover, some chapters provide many examples of small farmers successfully linking with and competing in modern value chains. In our policy section we summarize key lessons about the conditions under which these kinds of successes occur and how they might be scaled up.

Farm size—a reverse transition

Despite all the challenges they face, small farms are proving to be surprisingly resilient. They not only persist, but continue to increase in number. As Chapters 3 to 5 show, farm sizes are diminishing across much of Asia and sub-Saharan Africa, and have stagnated in South America (see also Figure 1 in Eastwood et al., 2010). Small farms in Asia today are less than half the size of the small farms that drove the Green Revolution in the 1960s and 1970s.

Small farms are also becoming more diversified, with off-farm sources of income, often because they are now too small to provide an adequate living from farming alone. In China, non-farm income shares for farm households increased from 33.7 per cent in 1985 to 63 per cent in 2000 and 70.9 per cent in 2010 (Huang et al., 2012). This is a more extreme example, but non-farm income shares have reached 40 per cent or more in many other Asian, African, and Latin American countries, and are often much higher for the smallest farms (see Chapter 17 and Haggblade et al., 2007). In much of Asia, absolute levels of employment are increasing in both farming and the rural non-farm economy, whereas in parts of Latin America, farm employment is falling but is being more than offset by growth in non-farm jobs (Chapter 17).

The overwhelming story is one of more small farms, shrinking farm sizes, and increased income diversification. Despite growth, sometimes quite rapid growth, in national per capita incomes, there is little sign yet of a shift to the patterns of farm consolidation and matching levels of rural–urban migration that occurred during the economic transformation of today's industrialized countries. Rather, relatively few workers are leaving their farms for the cities and instead are diversifying into non-farm activity from a small farm base. This is leading to farm size distributions that look more and more like Figure 18.1. There is a general drift in the farm size distribution towards the origin on the horizontal axis, while off-farm diversification is leading to a simultaneous movement along the depth axis. Even in land-abundant countries where the average farm size is increasing, many small farms persist in lagging regions. In some countries (e.g. Bangladesh, India, and the Philippines), even the total agricultural land area is becoming more concentrated among small farms, and it is the large farms that are being squeezed out.

There are many factors driving this reverse farm size transition:

- Rapid rural population growth, especially in already populous countries.
- Insufficient growth in urban jobs to enable faster rural–urban migration. Even relatively fast-growing countries like India have not generated sufficient growth in non-farm jobs. Bangladesh and China may be two recent exceptions.
- Other constraints on rural–urban migration, such as language, racial, and cultural barriers; legal restrictions on resettlement (e.g. China).
- Inheritance systems that lead to sub-division of farms amongst multiple heirs.
- Dense rural settlement patterns that provide enough income-earning opportunities in the local non-farm economy so that farm-based workers do not need to migrate to urban areas.
- Growing high-value opportunities in farming that create significant new employment opportunities in agriculture.
- Restrictions on land market transactions, such as caps on farm size (India), or indigenous land rights systems that limit opportunities for land consolidation (Africa).
- An ageing and immobile population of farmers. Farm exits tend to be an intergenerational phenomena; land is consolidated when farmers retire or die.
- Constraints on women's employment opportunities that keep them on the farm.
- Inadequate social security systems so that farms are kept as a retirement hedge.
- Agricultural support policies in some emerging middle-income countries that make small-scale farming more attractive than its real economic worth.

Many of these drivers are very powerful and seem unlikely to diminish in the near future. In poor, heavily populated countries experiencing rapid rural population growth (much of South Asia, sub-Saharan Africa, and the Andes), the pressure on land seems likely to keep growing. How many small farms will remain trapped in low-productivity farming and poverty, and how many will successfully escape poverty by diversifying into high-value agriculture or productive non-farm activities, will depend critically on national and regional rates of economic growth and urban/rural linkages (Chapter 17). In slow-growing countries and in lagging regions more generally, large numbers of small and marginal farmers seem likely to remain trapped in subsistence farming and poverty.

The earlier experiences of Japan, Taiwan, and South Korea suggest that the reverse farm size transition could continue until well into middle-income status (Otsuka, 2012). In Japan, for example, the average farm size only bottomed out around 1960 at 1 ha, and then increased to 1.2 ha in 1980 and 1.8 ha in 2005, while the percentage of farms less than 3 ha in size fell from 97.6 per cent to 90.5 per cent over the same period. China may finally have reached a tipping point in that the average farm size, which had fallen from 0.7 ha in 1985 to 0.55 ha in 2000, increased to 0.6 in 2010 (Huang et al., 2012). However, it is difficult to obtain data to determine whether the actual number of small farms less than 2 ha is now falling in China.

The outlook for land-abundant countries in Africa and Latin America seems likely to follow a different and dualistic path, much like Brazil, Argentina, and South Africa. If present trends continue, land will become increasingly concentrated amongst a declining number of large farms, while there will be a growing number of small farms concentrated on a diminishing land base, particularly in lagging regions (e.g. north-east Brazil).

Does reverse transition matter?

From the perspective of economic efficiency or growth it does not really matter that farms are getting smaller, unless there are economies of scale in farming. As indicated, there is not much evidence for this on the production side, but small farms are facing growing challenges in linking to value chains, especially high-value chains. Large farms seem able to capture economies of scale and scope in linking to value chains, so unless small farms are organized into marketing groups, it is possible that they are becoming less efficient than large farms. If so, then the reverse transition does matter from an efficiency perspective.

Another economic growth concern is that as small farms get smaller, they may not have the kinds of cash income and expenditure patterns that can help drive growth in the rural non-farm economy. During Asia's Green Revolution, for example, small farms generated significant marketed surpluses and cash incomes, much of which was spent locally on a range of agricultural inputs, consumer goods and services, and investment goods for their farm and household. These expenditure and investment patterns generated significant secondary rounds of employment-intensive growth in the rural non-farm economy—or large growth multipliers (see Haggblade et al., 2007 for a review of the literature). Small farms today are less than half the size of the small farms of the Green Revolution era, and many are subsistence- rather than market-oriented. Much may depend on how off-farm sources of income are spent, but the possibility arises that it is now the commercially oriented and medium-sized farms that are able to generate significant growth multipliers.

From food security perspectives the reverse transition poses a difficult dilemma. Small farms provide for the food security of huge numbers of rural poor. But many small farms are net buyers of food and they generate relatively little of the food required to feed large urban populations. Urban population shares are projected to grow strongly across the developing world (United Nations, 2011), and feeding these populations will require rapid growth in marketed food supplies. For most foods, these supplies will need to come from larger farms and commercially oriented small farms that can generate net surpluses. It follows that a food security agenda needs to walk on two legs. One leg is to provide support to the many smallholders who farm largely to meet their own subsistence needs. The other leg is to invest in large- and medium-sized farms and commercially oriented smallholdings that can produce marketed surpluses for the cities. Today, about half the malnourished people in the developing world live on small farms (IFPRI, 2005), so support for subsistence-oriented farms is crucial for meeting the current global food security challenge. But as urbanization proceeds, an increasing share of the poor will become urban based and detached from the land, so support for commercial farms will become increasingly important for meeting the food security needs of the poor.

From poverty and income equality perspectives the reverse transition also poses difficult challenges. Although diversification into non-farm activities is a useful way of supplementing farm income, it may not be enough to maintain an adequate income, to escape poverty, or prevent widening rural–urban income gaps. Local diversification opportunities into high-value farming and non-farm activity are higher in fast-growing countries, and in dynamic and more densely populated rural areas (Chapter 17). Small farms in such areas may be achieving adequate livelihoods despite having little land. Elsewhere, opportunities for diversifying into high-value farming or local non-farm opportunities are more limited, leaving many small farms trapped in subsistence-oriented farming and poverty. This is especially common in lagging regions where most of Asia's rural poor now live (Ghani, 2010).

In India and some other Asian countries there seems to have been sufficient growth in remittances and rural non-farm income in recent years to enable farm households successfully to avoid any widening gap between rural and urban per capita incomes. Rural poverty rates have also declined in tandem with urban poverty rates (Otsuka, 2012; Binswanger-Mkhize, 2012). But this is not true in many slow-growing countries, particularly in Africa, where rural–urban income gaps are widening and rural poverty rates remain stubbornly high. The relatively slow growth of the agricultural sector and the generally sparser rural population densities in Africa also constrain growth in rural non-farm opportunities.

Evidence from Japan, South Korea, and Taiwan suggests that income diversification by small farms is not a long-term solution to the rural–urban income gap problem. In these countries, governments eventually had to introduce income-support measures to narrow the income gap, and China and some other Asian countries are now beginning to follow suit (Otsuka, 2012).

From an environmental perspective, more small and marginal farms can lead to mixed outcomes. Many small farms retain complex farming systems that are ecologically well balanced and serve to conserve in situ many underutilized and neglected foods and indigenous crop varieties and animal species. On the other hand, many highly intensified small farms are an important source of environmental pollution and zoonotic diseases. Many other small farms struggle to make a basic living, and can become trapped in downward spirals of resource degradation and poverty (Cleaver and Schrieber, 1994). Yet other small farms encroach into forests and are an important cause of deforestation. A larger number of small farms in a landscape also increases the difficulties of introducing knowledge-intensive natural resource management (NRM) practices, and can make it more difficult to undertake the kinds of collective action needed sustainably to manage and improve watersheds and common properties. On the other hand, it needs to be noted that many large farms also cause significant environmental damage.

In sum, the reverse transition is not a uniformly good thing, and is creating new tensions and potential trade-offs between important economic, social, and environmental goals. Since the Green Revolution era, small farm growth has been seen as a winning proposition for growth, poverty alleviation, and food security outcomes, and concern has focused largely on adverse environmental outcomes. This is now changing and the future outlook is for less complementary outcomes at national scales between growth, poverty alleviation, and food security goals, posing more difficult choices for policymakers.

The widening fault line between these goals is most evident in the recent emergence of two very different agricultural agendas. On the one hand, recent increases in world food and energy prices have made agricultural growth an imperative for food security. Since most of the food-insecure households live in rural areas and mostly on farms, improving the productivity of subsistence-oriented farms has become a high priority. This has led to several international initiatives to promote food security, such as the Comprehensive Framework for Action developed by the UN Secretary General's High-Level

Task Force (HLTF) on the Global Food Security Crisis[2], and the New Alliance for Food Security and Nutrition ('New Alliance'), which was born out of the Camp David Summit of the G-8 in May 2012[3]. On the other hand, higher agricultural and energy prices have turned agricultural growth into a 'business' opportunity for producing food, raw materials, and biofuels, with significant growth in agricultural investment by sovereign wealth funds and foreign and national corporate-sector investors.

Unfortunately, these two drivers of change are not necessarily complementary. Many donors and NGOs are pushing for a broad social, environmental, and climate-change agenda based on subsistence-oriented farmers for food security and poverty alleviation reasons, but with little thought about increasing agricultural growth (Badiane, 2008). On the other hand, the private sector is pushing a new business agenda, often with an emphasis on large commercial farms, integrated value chains, and exports. Many governments seem uncertain which way to go; should it be a food security or a business-oriented strategy?

The business-oriented strategy does not have to be inconsistent with a pro-poor, food security approach, as long as it engages with large numbers of smallholders who are, or can, become commercially viable. Already, private-sector investments along value chains are opening up new market opportunities for some smallholder farms, particularly for high-value products. However, it is also becoming apparent that many more smallholders are being left behind while larger farms are gaining market shares. Many smallholders are not only missing out on new high-value chains, but in many countries have also lost access to modern inputs, credit, and market outlets even for their traditional food staples (Djurfeldt et al., 2011). There has also been growth in land grabbing and the development of corporate-sized farms which threaten to displace smallholders from their land as well as their markets (Deininger and Byerlee, 2010).

If more smallholder farms are to become commercially successful, governments will need to do more to support them by investing in the kinds of R&D and rural infrastructure that small farmers need, helping to organize small farmers for the market, and incentivizing the private sector to link with more small farmers.

What of the smallholders who cannot become commercially viable? Some are successfully diversifying their livelihoods out of farming, but there are many instances where this is not yet possible on the scale required or where the returns on non-farm activities remain too low for them to escape poverty. Many others are sinking into deeper poverty and subsistence modes of production because of higher food prices and reduced access to land, markets,

[2] <http://www.un.org/en/issues/food/taskforce/index.shtml>.

[3] <http://iif.un.org/content/new-alliance-food-security-and-nutrition>.

and modern inputs. Yet investing in this type of farming is often little more than a productive safety-net approach, particularly in remote and more marginal agricultural regions. It may be more cost effective to invest in improving subsistence farming rather than to spend on income-transfer programmes or facilitating farm exits, but that is something that needs to be determined on a case-by-case basis. There seems to have been very little work comparing the two approaches.

The context of assisting small farms

Policies towards small farms need to be guided by country economic context and the enormous diversity of small farms in terms of their assets and aspirations.

COUNTRY CONTEXT

Country context matters in determining the kinds of opportunities available to small farms. An important difference arises between countries where agriculture can serve as a major engine of national economic growth versus countries where agriculture is a secondary sector. This difference defines both the types of agricultural-sector investments that are worthwhile, as well as the relevant roles that small farms might play.

Table 18.1 highlights different types of country situations. Historical evidence from around the world shows that agriculture plays its largest role in the early stages of a country's development, and diminishes significantly in relative (though rarely in absolute) importance as economies diversify and workers migrate to the non-agricultural sector. But even in the early stages of development when its importance is potentially high, agriculture's contributions are affected by a country's resource endowments and its access to international markets. Table 18.1 captures these characteristics by differentiating between countries at early versus later stages of development, and among the former, between countries that have significant minerals or urban-based manufacturing sectors versus those that must rely more on agriculture as their lead sector.

Two key roles for small farms are identified. One is a growth or development role. This role arises when agriculture itself has a growth role to play and when commercially oriented small farms are efficient and can compete in the market. This role will be greatest in countries at early stages of development with good agricultural potential and lacking large mineral or urban-based manufacturing alternatives. In these cases, the best opportunities for small

Table 18.1 Priorities for small farms by country economic characteristics

Early stages of development		Later stages of development
Leading sector		
Minerals or urban-based manufacturing	Agriculture	All types of countries
Commercial opportunities for small farms to sell high-value products in domestic markets. Social value in retaining small farms as a reserve employer, and to spread mineral wealth and provide subsistence for the rural poor.	Commercial opportunities for small farms in export crops, food staples, and some high-value products for the domestic market.	Remaining small farms gradually squeezed out, and those that survive focus on high-value products and part-time farming. Social value in retaining small farms as a reserve employer until sufficient exit opportunities have been created.

Source: Adapted from Hazell et al. (2007).

farms are likely to be in food staples for the domestic market and high-value production for export. Countries starting with large mineral or urban-based manufacturing sectors will have strong currencies and ready access to low-cost food imports, so their best small-farm growth opportunities are likely to be in high-value production for the domestic markets. An important but challenging growth role for small farms lies in countries with limited agricultural potential and which also lack significant minerals or urban-based manufacturing.

As countries industrialize, small farms typically play a shrinking role in all kinds of countries. Rising real wages within the wider economy tend to drive farm consolidation, and the small farms that survive find niches in high-value markets or become part-time farms.

In Asia, the Green Revolution of the 1960s to 1980s was predominantly led by small farms at a time when agriculture, and cereals in particular, served as a leading growth sector. But now that many Asian countries are successfully industrializing, small farm-led agricultural growth based on cereals is becoming less relevant and less able to avoid widening rural–urban income gaps (Otsuka, 2012). The reverse transformation that is occurring in Asia is potentially leading to a growing backlog of workers who will eventually need to exit from agriculture (Headey et al., 2010). In this case, policies for small farms need to be designed to help improve the productivity of small farms and help them find high-value opportunities, but remain cognizant of the possibility that many of them will need to diversify out of agriculture, or leave farming altogether.

In Africa, by contrast, government and donor neglect of agriculture in recent decades has created a different situation. Except in relatively few countries with large mineral sectors or urban-based manufacturing, agriculture needs to become a leading growth sector. By neglecting to invest, a situation has arisen in which too many workers have either left farming for the cities, or diversified out of farming from a small farm base. This has happened in the context of stagnating or even declining per capita incomes, suggesting a premature exit of agricultural workers into low-productivity non-farm jobs (Headey et al., 2010). In this situation there is still a good case for productive investments in agriculture and small farms.

A second role for small farms arises from their potential social contributions. Small farms can provide a way for governments to spread the benefits from a large mineral or urban-based manufacturing sector during the early stages of development when most people are still engaged in agriculture. As economies grow, small farms can also serve as a useful reserve employer until sufficient exit opportunities exist. Finally, small farms may provide a social safety net, or subsistence living, for many of the rural poor, even when they are too small to be commercially viable. These social roles are most important in countries with a poor agricultural productivity potential, or a large mineral or urban-based manufacturing sector. These social roles do not necessarily require that all small farms be commercially viable. There are also important environmental considerations in countries where poverty and environmental degradation is associated with small farms, and these problems may also warrant targeted interventions of their own. But as we shall see, policies and investments to support these kinds of roles may often need to be structured differently from those for commercially oriented small farm growth.

DIVERSITY OF SMALL FARMS

As this book has made clear, small farms form a very diverse group, and they face varying prospects that depend on their own assets and aspirations as well as on their country and regional context. Policies need to take this diversity and context into account. A variety of farm typologies has been offered in the literature to help manage this diversity. Vorley (2002) distinguishes between farmers operating in three rural worlds. In rural world 1, commercial farmers are globally competitive, linked to export markets, and use modern technologies; in rural world 2, farmers sell primarily in local, regional, and national markets and use intermediate technologies; in rural world 3, farmers are subsistence-oriented and use traditional technologies. This is very similar to the typology used by FAO and IFAD (2008) and presented in Chapter 1. The World Development Report 2008 (World Bank, 2007) identifies five smallholder groups—market oriented, subsistence oriented, off-farm labour oriented, migration oriented, and diversified households that combine multiple income sources. Berdegué and Escobar (2002) (see also Chapter 5) identify three groups of family farms based on regional context and household assets. The first category is family farms with good assets (land, labour, and/or access to capital) located in places with good agricultural potential and access to markets. These farmers are usually fully integrated in a market economy and make a substantial contribution to the production of food for domestic and international markets. The second category comprises family farms that have reasonable assets and agricultural potential but are constrained by being located in slow-moving regional economies with limited market access. The third category comprises resource-poor farmers located in places where conditions are adverse not only for agriculture, but often for non-farm activities. The majority of smallholders in this group are poor, subsistence oriented and may be diversified into low-productivity non-farm sources of income. Torero (Chapter 6) also classifies farms on the basis of the agricultural potential and market connectedness of the region in which they live, but also in terms of the management efficiency of the farmers themselves.

Key elements in these typologies are the characteristics of the region in which a farmer lives—especially its agricultural potential and access to markets, household assets, business orientation and acumen, and the degree of diversification into off-farm sources of income. Drawing on this work, it is proposed to classify smallholders into three groups for the purposes of targeting small farm assistance:

- *Commercial small farmers* who are already successfully linked to value chains, or who could link if given a little help. Commercially oriented small farms may be full- or part-time farmers.
- *Small farmers in transition* who have or will soon have favourable off-farm opportunities and would do better if they were to either exit farming completely or obtain most of their income from off-farm sources. Most transition farmers are likely to leave farming, and it is just a question of when and how. Those that remain will farm part time and may not be very market driven.
- *Subsistence-oriented small farms* are marginalized for a variety of reasons that are hard to change, such as ethnic discrimination, affliction with HIV/AIDS, or being located in remote areas with limited agricultural potential. Many of the same factors also prevent them from becoming transition farmers. Subsistence-oriented farms frequently sell small amounts of produce at harvest to obtain cash income, but they are invariably net buyers of food over the entire year.

The relative size of these three groups will vary by country context. With economic growth and urbanization, significant numbers of commercially oriented small farms are likely to prosper through diversification into high-value agriculture. The most successful small farmers will tend to be located in areas with good agricultural potential and market access. Over time, some commercially oriented small farmers will expand their farms while others will eventually become transition farmers or successfully exit farming to the

Table 18.2 Transitions from small farm groups

Type small farm		Period t + 1				
		Subsistence	*Transition*	*Commercial*	*Large Farm*	*Non-farm*
Period t	*Subsistence*	O	X	X		
	Transition	O	O			X
	Commercial	O	O	X	X	X

Note: X = desired transition; O = undesired transition

non-farm economy. Transition farmers will either have, or will be able to develop, suitable skills and assets for undertaking non-farm activity, and they are likely to live in well-connected areas with access to off-farm opportunities. Their farming activities are likely to be oriented towards their own consumption rather than the market. Subsistence-oriented farmers are more likely to persist in less-favoured and tribal areas, and to grow traditional food staples (both crop and livestock) for their own consumption.

Table 18.2 summarizes the kinds of transitions that are possible for each of the three small farm groups. Over finite periods of time, shown as a move from period t to period t + 1 in the table, it is desired that subsistence farms should become transition or commercial farms; that transition farms should successfully move to the non-farm economy; and that commercial small farms should either prosper as such, transform into larger farms, or find successful exit strategies to the non-farm economy. To be avoided are situations where many small farms revert to or remain trapped in subsistence farming, or where transition farms fail to find successful exits to the non-farm economy.

Approaches to assisting small farms

SOME GUIDING PRINCIPLES

Table 18.3 highlights the kinds of interventions that may be relevant for each of the three groups of small farms. Commercially oriented small farms need support as farm businesses. They need access to improved technologies and natural resource management (NRM) practices, modern inputs, financial services, markets, and secure access to land and water. Much of this assistance will need to be geared towards high-value production and provided on a commercial basis. Many smallholders will also require help acquiring the necessary knowledge and skills to become successful business entrepreneurs in today's value chains, especially women and other disempowered groups.

Managing market and climate risk is a challenge for many small farms, and in addition to insurance and access to safety nets, they need to develop resilient farming systems.

Transition farmers need help developing appropriate skills and assets to succeed in the non-farm economy, including in many cases assistance in developing small businesses. This can be especially important for women and other disempowered groups who have little experience working off farm. The transition to the non-farm economy may also be facilitated by securing land rights and developing efficient land markets so that they can more easily dispose of their farms. Since many transition farmers seem likely to remain part-time farmers, they can also benefit from improved technologies and NRM practices that improve their on-farm productivity.

Subsistence farmers are predominantly poor and will mostly need some form of social protection, often in the form of safety nets, food subsidies, or cash transfers. Interventions that help improve the productivity of their farms (e.g. better technologies and NRM practices) can make important contributions to their own food security and perhaps provide some cash income, and may in many cases prove more cost effective than some forms of social protection. But subsistence farmers have limited ability to pay for modern inputs or credit, so intermediate technologies that require few purchased inputs may be needed, or inputs will need to be heavily subsidized. Subsistence farmers are typically the most exposed and vulnerable to climate risks, and in addition to safety nets, they need help developing resilient farming systems.

Although the choice of assistance policies will need to be different for the three groups of small farms, not all interventions need to be as carefully targeted as others. Figure 18.2 shows how possible interventions to assist small farmers might impact on the three groups of small farms. Some interventions will benefit all three groups, and these are the interventions that fall in area A. Other types of interventions will benefit two groups (areas B, C, and D) and others will benefit only one group (areas E, F, and G). Interventions that benefit only one group (areas E, F, and G) may be relatively easy to target, but interventions that benefit two or more groups can be more problematic. If an intervention can benefit other groups at little or no additional cost beyond the cost of reaching the primary target group (e.g. some types of agricultural R&D), then the benefits captured by other groups can be viewed favourably as 'spillover' benefits, and careful targeting may not be required. But if the benefits captured by other groups represent a diversion of benefits from the primary target group, then this must be viewed as a 'leakage' that needs to be minimized through careful targeting. Cash transfers, food subsidies, and fertilizer vouchers intended for the poor typically fall into this category.

Table 18.3 Types of assistance relevant for different small farm groups

Type small farm	Types of assistance
Commercial	Farming as a business
	Better technologies and NRM practices
	Organizing small farmers for marketing purposes
	Incentivizing large agribusiness to link with small farms
	Accessing seeds, fertilizer, finance, and insurance
	Securing land rights and development of efficient land markets
	Encouraging entrepreneurship
	Empowering women and other vulnerable groups
	Building resilient farming systems
	Safety nets
Transition	Stepping out of farming
	Training and support for non-farm activity, including development of small businesses
	Encouraging entrepreneurship
	Empowering women and other vulnerable groups
	Securing land rights and development of efficient land markets
	Better technologies and NRM practices
	Safety nets
Subsistence	Social protection
	Safety nets and transfers
	Better technologies and NRM practices
	Subsidized inputs for own food crops
	Securing land rights
	Building resilient farming systems
	Empowering women and other vulnerable groups
	Support for non-farm diversification

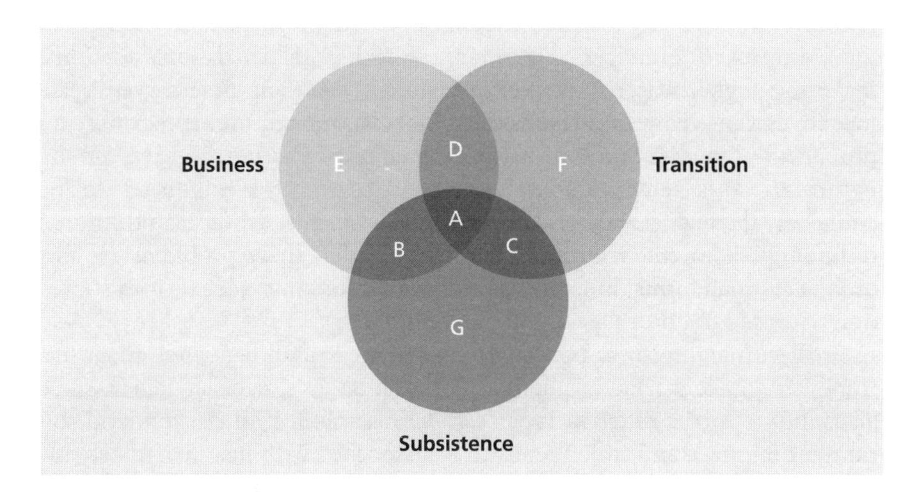

Figure 18.2 Potential benefits accruing to different types of small farms from assistance interventions

With these principles in mind, we turn now to a detailed discussion of the main forms of assistance appropriate for each of the three groups of small farms.

ASSISTING BUSINESS-ORIENTED SMALL FARMS

Left to market forces alone, far fewer small farms are likely to succeed as businesses than is desirable from an economic and social perspective, and too many will be pressured to find non-farm livelihoods. Small farms face a tilted playing field that favours large farms, and correcting this imbalance requires concerted action in a number of important areas by governments, donors, NGOs, corporate businesses, and farmer organizations, among others.

Accessing improved technology

Small farms need access to game-changing technologies that can significantly raise their farm productivity and incomes.

In irrigated and high-potential rain-fed areas, small farmers have benefited enormously from R&D systems that have provided them with improved inputs such as crop varieties, fertilizers, pesticides, animal breeds and feeds, veterinary medicines, and farm machines. However, the resulting intensification has often led to environmental and health problems, such as soil degradation, pollution of waterways, and pesticide poisoning. To reduce these problems, improved management practices have been developed, such as precision use of fertilizers, low or zero tillage, improved water management, and integrated pest management (Hazell, 2008). A continuing challenge is the generally poor adoption rates by farmers of these improved practices. There are several possible reasons for this, including high levels of knowledge required for their practice, perverse incentives caused by input subsidies, high labour requirements, insecure property rights, difficulties of organizing collective action, and the off-site nature of some environmental damage. These problems need to be addressed through policy and institutional reforms as well as additional technology research. It is important to note that these problems are not unique to small farms; in fact large farms are often the more serious source of environmental damage.

Small farmers in lower potential rain-fed areas have benefited much less from R&D systems. Environmental degradation is also occurring in these areas, but it is of a different type (e.g. deforestation, land erosion, and soil nutrient mining), and which tends to be associated with insufficient agricultural intensification in the face of growing population pressure (Hazell and Wood, 2008). A lot of research has been directed at ways of improving farm productivity in rain-fed areas that do not require high levels of modern input use. These include crop improvement research, agroforestry, water capture, soil erosion control, conservation farming, improved rotations, crop-livestock integration, and soil fertility management. But again, levels of adoption have been disappointing. One problem is that many improved NRM practices are labour intensive, and yield returns to labour that are too low to be attractive to many farmers. Other constraints include insecure property rights, risk, and difficulties in organizing collective action.

An important reason for the poorer performance of agricultural R&D systems in rain-fed areas is the much more complex nature of the research problem, given both the high crop diversity and spatial heterogeneity that characterize smallholder farming systems in these areas. This complexity also leads to results that tend to be site specific, making it difficult to justify the cost of the research. The challenge is compounded by the increasing number of often conflicting objectives expected of agricultural research today. To overcome these problems, Lynam and Twomlow (Chapter 11) argue that new institutional modalities are needed, based on innovative partnerships between different agents, and better integration of research design, adaptation, diffusion, and scaling-up activities. They give examples of the new CGIAR research programmes (CRPs) and of new scientific platforms for innovation (see also Chapter 2).

In the future, new technologies for smallholders will need to be more resilient to changing climate risks. Much adaptive research to date has focused on breeding new varieties for drought or flood tolerance as the best means to address climate change. As Conway (Chapter 2) and Toulmin (Chapter 16) argue, additional research is needed on making farming systems more resilient, such as ways of managing soil and water in the wider landscapes, and building more effective nutrient recycling systems.

Linking small farmers to markets

For many smallholders, the best market opportunities still lie with their local markets for food staples and some perishable high-value products such as livestock and horticultural products. Linking small farmers to urban and export markets is more challenging, in part because many governments have adopted liberalization reforms that have removed direct marketing assistance to smallholders, and in part because many value chains have become more integrated and controlled by large agribusiness interests.

Clustering smallholders into groups is a useful way of aggregating output for the market and reducing the costs of supplying inputs and credit, making them more competitive with large farms. Suitably organized and managed, such clusters can also provide a useful channel and enforcement mechanism for entering into collective contract arrangements with corporate agribusinesses and financial institutions. Singh (Chapter 7) reviews a variety of successful examples illustrating how different actors—state and parastatal organizations, agribusinesses, NGOs, cooperatives, and farmer companies—can play important roles in organizing small farmers and linking them to markets for inputs, outputs, and credit.

The public sector can facilitate smallholder links to markets by investing in the kinds of R&D and rural infrastructure that can make them more competitive in the market place. Given also the difficulties and costs of working with small farmers, agribusinesses may need to be encouraged through moral persuasion and financial incentives, at least during the initial stages of setting up their value chains (Chapter 10). One option is for government to enter into specific public–private partnership arrangements whereby it subsidizes the incremental costs to corporate agribusinesses of setting up linkage arrangements with small farms. Publicly provided credit guarantees, insurance, or co-financing also seem to be emerging as instruments of choice for reducing risk along value chains, and can help leverage additional commercial bank lending (Chapters 8 and 9).

Accessing seeds and fertilizer

Interlinked transactions like contract farming provide one mechanism for smallholders to access the improved seeds, fertilizers, and other modern farm inputs they need. However, more general solutions are still required to serve a broader range of farmers. Public-sector distribution systems for seeds and fertilizer were discredited in the past because of problems with poor quality, late delivery, and high cost. More recently, governments and donors have emphasized market-assisted programmes that seek to build up private input supply and distribution systems. Agro-dealers can offer a range of support services for small farmers, aside from selling agricultural inputs, such as the provision of expert advice and credit. The challenges facing agro-dealers in serving small farms include inadequate access to credit, limited knowledge about use of the inputs they sell, high costs of certification, and lack of business management skills. Chapter 9 discusses how these constraints can be overcome through setting up investment funds and training programmes to support networks of agro-dealers, and illustrates this with some of AGRA's (Alliance for a Green Revolution for Africa's) recent work in Africa.

Additionally, a number of institutional and policy interventions are needed. Among these are policies to strengthen national seed laws, facilitate access to foundation seed, and implementation of harmonized seed laws between countries with similar agro-ecologies. In Africa, making fertilizers more affordable and accessible to small farmers will require development of regional trade corridors and regional collaboration between countries on bulk-buying and transportation of fertilizers (Morris et al., 2007). Packaging inputs in small packs can also make them more affordable to small farms.

In principle, subsidies can help encourage the development of private-sector distribution systems. In practice, input subsidies have typically been used indiscriminately and without exit strategies, driven more by political agendas than economic purposes. This inevitably turns out to be very expensive, and there is now considerable interest in so-called 'smart' subsidies. So far, vouchers look most promising (Minot and Benson, 2009). They can be targeted to specific types of farmers, rationed on a per farmer basis, and the subsidy component can be phased out over time. Moreover, if they are redeemed through private agro-dealers, they can also help build up fully privatized procurement and distribution systems.

Another use of seed and fertilizer subsidies is to help poor subsistence-oriented farmers become more food self-sufficient. This typically involves allocating a small amount of heavily subsidized fertilizer and seed (starter packs), as in Malawi during 1998–2004. This type of subsidy is essentially an alternative to a cash or food transfer, and may sometimes be more cost effective.

Accessing finance

Since the demise of the agricultural development banks, there has been considerable progress in developing market-facilitated approaches to rural finance with support for informal and semi-formal credit institutions (including village banking schemes, savings and credit cooperatives, and microfinance) and increased lending by commercial banks. In many respects this new approach has been successful, increasing the range of rural financial services available in rural areas and for poor people, but it has encountered problems in reaching emerging smallholders—the missing middle who are too large for microfinance but at the same time remain outside the formal financial intermediation—and in making sufficient financing available along value chains (Chapter 8).

In Chapter 8, Rahman and Smolak argue that a promising approach to solving these problems is value-chain financing (VCF), made possible by new opportunities for interlinking markets for inputs, outputs, and credit in today's more integrated value chains. In practice, VCF can be as simple as a trader providing a cash advance and accepting payment in kind at harvest time. Or it can be a highly sophisticated configuration of farmers, traders, and agribusinesses that leverages formal financial flows. VCF requires an enabling business environment for domestic and international investors. Experience to date suggests that VCF works better for high-value chains than food staples.

Credit guarantees that underwrite part of the risk along value chains are also a promising way of leveraging more commercial bank lending to small farmers and SMEs, and Chapter 9 reviews some of AGRA's promising experiences with credit guarantees in Africa. Guarantee schemes need to be carefully designed so as not to encourage moral hazard in commercial bank lending. Weather index insurance is also developing as a promising way of managing some of these risks.

Warehouse receipt systems have also emerged as another way to solve the missing middle problem. This approach requires the public sector to devise an appropriate institutional framework, legislation to recognize a receipt as legal tender, licensing and inspection of warehouses, performance oversight, and collaboration with the private sector to establish commodity quality standards.

Despite many innovative efforts to develop market-oriented approaches to finance, there are still insufficient levels of credit for many smallholders and SMEs. This has led to renewed interest in reforming government credit programmes and agricultural development banks to fill the gap. A common thread across successful reforms of development banks is changes in ownership and governance structures, sometimes to public–private partnerships. As part of their reform, agricultural development banks have sought funds from capital markets in order to limit the political influence that contributed to past failures (Chapter 8).

Accessing land

Many smallholder farmers do not have secure access to land and water, making it difficult for them to pursue new business opportunities, borrow credit, or make investments in irrigation infrastructure and other long-term land improvements. Farmers also need secure ownership rights in order to lease or trade land, which is important if more efficient farm sizes and spatial configurations are to emerge. In some cases, weak land rights make retention of land a challenge in the face of potential land grabs by the rich and powerful, something that seems to be happening on a growing scale. Conditions vary widely across cultural, economic, and social contexts, and can be particularly challenging where farmers hold and use their land under customary tenure arrangements. Under customary systems, the rights and obligations pertaining to land are not recorded and do not require any documentation since they are guaranteed by their local communities. Often, multiple users have rights over the same parcels of land; for example, grazing, cropping, and tree rights on the same land may belong to different individuals or groups. Deininger (2003) estimates that over 90 per cent of the farm land in Africa is held under customary arrangements.

Land titling programmes that privatize land to individuals can undermine customary systems and strip multiple users of their rights. They may further disempower women and can also legitimize or even encourage encroachment on communal or public land such as forests. Recent years have seen new innovations and experiences in reformulating national land laws to reconcile overlapping and competing rights between the formal and informal systems, while strengthening the rights of communities and individual farmers. A number of African governments have initiated land policy and legislative reforms along these principles, but so far implementation of new laws has been rather slow. However, given the huge variability in agro-ecological conditions, demographic dynamics, kinship structures, and inheritance systems, even within the same country, compounded by financial and human capacity limitations, effective decentralization of land administration to local communities may require protracted and participatory approaches. In Chapter 14, Deininger reviews practical and cost-effective approaches to securing land rights and emphasizes three points. First, the importance of a legal and institutional framework that recognizes existing rights, enforces them at low cost, and allows users to exercise them in line with their aspirations and in a way that is transparent, equitable, and benefits society as a whole. Second, reliable and complete information on land and property rights needs to be broadly accessible, comprehensive, reliable, current, and cost-effective in the long run. Finally, transparent management of the public-good aspect of land requires both regulations to avoid negative externalities from uncoordinated action by private parties and clear limits to, and rules for, the way the state exercises its right of pre-emption and manages or disposes of public land.

Encouraging entrepreneurship

To succeed as commercial farms in today's more competitive, diversified, and volatile economy, small farmers must be vested with the skills and education needed to experiment, adapt, and compete in markets where both knowledge and effort are rewarded. Entrepreneurship is also needed along agricultural value chains. The SMEs that service many of the needs of small farmers also need to be adept within their own spheres of influence if they are to prosper and provide the innovative services and market opportunities that small farmers need. It is also crucial to attract many more young people into farming and allied activity if innovation and entrepreneurship are to be secured for the future.

Juma and Spielman (Chapter 12) identify four key areas where policymakers can help develop and support entrepreneurship. One is through strategic investments in better infrastructure to generate new and more diversified opportunities beyond the farm. Another is enhancing the technical capability of individuals. Technical capacity begins with basic education in literacy and numeracy skills, but rural entrepreneurs also need quality education relevant to their needs. One way of encouraging greater numbers of skilled entrepreneurs to enter the agriculture and allied sectors is to teach agriculture as a formal subject in schooling at all levels, vesting them with a range of practical farming and business management skills. Formal education also needs to be supplemented with continuing technical training throughout adulthood.

Policymakers can also foster business development. Innovative capabilities need to be supported by access to capital and expertise. Although entrepreneurs are often the individuals willing to take risks and launch new commercial ventures, they are rarely the actual financiers of these ventures. For the entrepreneur's financing needs, individual investors, banks, venture capital firms, philanthropies, and governments have a major role to play. Business development also requires access to viable technologies and supportive institutions.

Finally, policy-makers need to create an enabling business environment to foster entrepreneurship and innovation. This covers the legal and regulatory environment and public policies on science, industry, business, labour and agriculture, as well as the informal institutions that govern the behaviours, practices and attitudes that condition the ways in which individuals and organisations act and interact.

Empowering women and vulnerable groups to become successful farmers

Poor people and especially poor women farmers are frequently disempowered and have limited options for developing new business opportunities. They are often excluded from access to land, water, credit, and other financial services, extension advice, and markets (Chapter 15; Quisumbing and Pandonfelli, 2009). Some NGOs have developed successful and innovative programmes for organizing disempowered groups and helping them to develop market opportunities. Hosain and Jaim (Chapter 15) review the experience of one such programme in Bangladesh, the Northwest Crop Diversification Project (NCDP). This programme provides credit, training, and marketing assistance to some of the poorest and smallest farms, with an emphasis on women farmers, with the aim of helping them to diversify out of rice into high-value crops for the market.

Land issues can be especially important for women and marginalized groups. In many societies there is little tradition of co-ownership of property by husband and wife. In this context, formalizing customary arrangements without first changing women's inheritance and ownership rights can easily lead to catastrophic outcomes for women (Joireman, 2007). Although many countries now give attention to gender equality in national land legislation, implementation remains constrained by deeply rooted cultural norms, compounded by women's lack of access to legal institutions, especially in rural areas. Also needed is the education of legal and traditional authorities and men and women in areas where customary law might conflict with new statutes. Yet changing attitudes on gender rights and related culturally sensitive issues can never be a short-term enterprise.

Special efforts are needed to encourage agribusinesses to link to women and poorer farmers. NGOs and social enterprises can play useful intermediary roles in making these market links, and there may be good arguments for government or donors to subsidize the development of these market linkages (Chapter 10).

Building resilience

Production and market risks permeate most smallholder farming systems, holding back investment and business development. Climate change seems destined to make things worse, particularly in arid and semi-arid areas. Yet the institutions for managing risk in agriculture are not well developed, and farmers and rural communities often have to rely largely on their own devices. While there is evidence that they are quite effective in managing many types of risk, they are much less able to manage severe and covariate risks that impact on most farmers within a region at the same time (e.g. regional droughts or floods). Unfortunately, one of the major impacts of climate change is likely to be an increase in the frequency of these types of severe weather events.

Several book chapters (2, 8, 9, and 16) review options for helping small farmers adapt and manage risk, and in making their farming systems more resilient. Adaptation is likely to involve a range of strategies, pursued by actors at different scales, some market-based and others relying more on social and institutional networks. Local governments need to work with communities in strengthening local capacities to adapt and cope with risk, and provide complementary investments in infrastructure, knowledge, markets, and policy. Regional and national governments need to ensure there are adequate safety-net and relief programmes in place to help people survive catastrophic events, replace key assets, and generally speed up recovery. Governments can also facilitate the development and spread of risk-management aids like insurance. Weather index insurance may not yet be at a stage of development where it can be scaled up for small farms, but it may be immediately relevant for underwriting relief and safety-net programmes (see also Hazell et al., 2010).

On the positive side, climate change may offer some farmers new income-earning opportunities. Market demand for biofuel feedstock is growing and there are emerging opportunities to tap into carbon payments by reducing deforestation and forest degradation, and perhaps eventually for sequestered carbon in farming landscapes (Chapter 16).

ASSISTING SMALL FARMS IN TRANSITION

Many small farms are already diversified into wage employment and local non-farm activity, often to the point where farming has become a part-time activity. Diversification is to be encouraged as countries develop and small-scale farming becomes less competitive and less able to provide incomes that are commensurate with other sectors. Fortunately, diversification opportunities also grow as countries develop, so there is a great deal of spontaneous diversification that needs little policy support. The policy challenges in these cases are to help ensure that sufficient workers find exit strategies from agriculture over time, and that poorer households and women are not left out of the process. Another policy challenge arises in slow-growing countries where growth in rural non-farm opportunities is more limited, and rural population growth is putting additional pressure on households to find supplementary sources of income. Growing the non-farm economy under these conditions is much more challenging for policymakers. Wiggins (Chapter 17) explores ways of growing the rural non-farm economy (RNFE) to create more income-earning opportunities for small farmers.

A first priority in promoting the RNFE is good rural infrastructure and an enabling policy environment. An enabling macroeconomic environment, local law and order, contract enforcement, labour laws, sector-specific taxes, tariffs, quotas and regulations, etc. all influence opportunities for rural non-farm firms. Past attempts to support directly the growth of the RNFE have focused on incentives to attract new industries to rural regions, support programmes for individual enterprises, especially small manufacturing firms, and technology development.

But supply side interventions alone are rarely enough to promote non-farm activity. This is because much non-farm activity produces goods and services that are consumed almost entirely within the region in which they are produced (e.g. many retailing and personal services, highly perishable agricultural products and the processing of local agricultural outputs). The profitable expansion of these activities depends on growth in local demand, which in turn depends on growth in regional income and in the volume of goods produced that need to be processed and traded. Without local agricultural growth or access to new markets (e.g. from tourism, mining, government jobs), incomes and the demand for non-farm goods and services remain low, and rapid expansion of non-farm activity can quickly depress local prices and wages, making them more a refuge occupation than a productive alternative to agriculture.

As successfully growing countries transform, other important sources of demand have emerged for the RNFE. Rising urbanization and national economic growth, together with improved transport and communication networks, provide important economic linkages between urban and rural areas, opening up new opportunities for rural households. These changes have been reinforced since the 1990s by widespread economic liberalization, which has opened up the rural non-farm economy as never before to new opportunities, but also to increased competition.

Opportunities for small farmers and women to reduce their dependence on agriculture are constrained by the paucity of their human, financial, and physical assets. Lack of human and financial assets restricts many of the poor to low-productivity, low-growth market segments from which there are few pathways out of poverty, simply a means of bare survival. In this environment, the policy challenge becomes one of equipping poor households to move from these 'refuge' non-farm jobs to more remunerative ones. To do so, they require access to education and start-up funds, public assets such as roads and electricity, and information about how to access dynamic market segments. Gender, caste, and social status can restrict access by the poor to the most lucrative non-farm activities in some settings. Discrimination, weak asset base, and restrictions on geographic and occupational mobility all conspire to limit access by key disadvantaged social groups to more remunerative rural non-farm activities.

ASSISTING SUBSISTENCE-ORIENTED SMALL FARMS

With the right kinds of support, many smallholders should be able to succeed in growing their farm incomes and diversifying into productive sources of non-farm income. Yet there are many who, for various reasons, are not likely to be so successful. Many of these will be poorer households with limited human capital and assets; many will be women, from ethnic minorities and other disempowered groups; and many will be located in lagging regions where they have limited access to markets and non-farm opportunities. Many of these disadvantaged small farms are already sinking into poverty and subsistence modes of production because of higher food prices and reduced access to land. If neglected, they can become a long-term poverty trap for many and a cause of considerable environmental damage.

Investing in these types of disadvantaged small farms is often little more than a productive safety-net approach. It may be more cost-effective to invest in improving subsistence farming rather than to spend on income transfer programmes or facilitating farm exits, but that is something that needs to be determined on a case-by-case basis. Where agricultural investments are warranted, that support may be aimed primarily at enabling them to produce enough food for their own household consumption needs and to reduce the losses they suffer post-harvest.

One of the sources of failure in many past support programmes for small farms is that they have not differentiated between the needs of business-oriented and subsistence-oriented small farms. While some kinds of interventions are beneficial to both groups of small farms (e.g. rural roads and some kinds of R&D), many others need to be tailored differently to the two situations (e.g. market mediated approaches to financial and insurance services, and fertilizer subsidies). For the poorest households, access to modern inputs, credit, and insurance may have to be heavily subsidized (e.g. the heavily subsidized fertilizer and seed starter packs given to poor farmers in Malawi), whereas subsidies for business-oriented farms can play much smaller and market development roles. Subsistence-oriented small farms also need supporting safety nets designed to help them manage shocks and cope with chronic poverty and food insecurity.

Targeting and delivering small farm assistance

Households pursue diverse livelihood strategies relevant to their circumstances. Given the enormous diversity of small farm circumstances, as determined by variations in agroclimatic conditions, market access, and social and economic conditions, there are few if any 'one size fits all' strategies for assisting small farms. Different types of assistance need to be blended into strategies that are locally relevant. This requires the ability to target different types of small farms, and an ability to blend different interventions into a holistic strategy for each group.

TARGETING ASSISTANCE

As mentioned earlier, some kinds of assistance need to be better targeted than others. But where targeting is required, this calls for the construction of a relevant typology or classification of small farms and the ability to identify the different groups on the ground. There has been a lot of recent work using GIS and spatial analysis methods to identify target areas for rural development purposes. Most of this work focuses on mapping different regions in terms of their agro-ecology, market access, and rural population density (see, for example, Omamo et al., 2006), but does not disaggregate according to differences in farmer endowments and orientation. Torero (Chapter 6) provides an important contribution by disaggregating regions further, according to differences in the management efficiencies of the farmers themselves, which are conditioned by such factors as levels of education and local institutional development. Case studies illustrate how this typology can be used spatially to target different types of business, and non-farm and humanitarian interventions.

DELIVERING ASSISTANCE

A profound challenge facing those who would intervene to support agriculture and small farms is how to integrate various needs and approaches into holistic packages of intervention. For example, if small farms are to exploit growth opportunities, then they not only need access to markets but also access to key inputs and technologies to increase their productivity and to meet required market standards. Interventions that seek to help farm households as farmers also need to be integrated with interventions that seek to enhance their non-farm employment opportunities or to protect them in emergency situations. Different interventions can have positive cross-impacts on each other. For example, relief programmes that help protect a farmer's assets against losses due to catastrophic weather events like drought or flood may enhance their opportunities as farmers. On the negative side, safety-net programmes might crowd out more market-based alternatives (e.g. drought relief vs insurance or investments in risk-reducing technologies).

Many past government-led attempts to assemble integrated packages to assist small farms (e.g. the integrated rural development projects [IRDPs] of the 1970s and 1980s) did not fare well. Key lessons are that they were top-down approaches that overreached in terms of coordinating many different government agencies and over-simplified in the face of considerable diversity in local agroclimatic and socio-economic conditions. They also gave too little attention to the problems of the poor and the inherent weaknesses of many public institutions.

There have since been important changes in the kinds of agents contributing to the development of agriculture and small farms, with the restructuring and decentralizing of government agencies and the emergence of civil society (including non-governmental organizations [NGOs] and community- and voluntary producer-based organizations [CBOs]) and large private firms (e.g. agro-processing firms, supermarkets, and tourism promoters) as important players. This has opened up new opportunities for more participatory, multi-agency, decentralized, and market-oriented approaches that build on local knowledge of needs, opportunities, and constraints that are far more relevant for coping with diversity and changing economic conditions. The challenge for rural development experts is how to build on this new institutional landscape and create new kinds of approaches towards the agricultural and rural sector that meet the different needs of commercial, transition, and subsistence-oriented small farms.

Two other key issues need to be mentioned. One is the problem of timely delivery of support services for small farms. Chapter 3 identifies the critical importance of 'right place' and 'right time' as key determinants of the success of small farm support programmes. Some of the spatial mapping techniques described above can help solve the 'right place' challenge, but 'right time' is something that has to be built into the operation of programme interventions.

Finally, there is the issue of scaling up from proven successes. Given the levels of complexity needed to develop effective support programmes for small farms, there is much to be said for encouraging innovation by a wide range of public, private, community, and NGO players operating at different spatial levels. Once a successful approach has been identified, the question then is how it can be replicated on a larger scale. Although there is considerable experience of going up in scale, there is no simple recipe. Conway (Chapter 2) identifies some emerging principles for scaling up from local successes, and highlights the important role the private sector can play in scaling up technologies or processes that significantly increase income and which are 'marketable' and hence saleable. However, he recognizes that the private sector on its own may be insufficient if the scaling up is to be equitable and environmentally benign, and hence various forms of public–private partnerships may be required. The more general problem of scaling up from successful small farm interventions calls for a systematic process, beginning with the design and testing of promising new approaches, rigorous monitoring, and evaluation (M&E) frameworks and impact assessments, formation, and management of the right partnerships for implementation, adequate and sustained funding over realistic timeframes, and a willingness to take risks and learn by doing (Linn, 2012).

Conclusions

The case for smallholder development as one of the main ways to achieve agricultural growth and reduce poverty and food insecurity remains compelling. The use of public funds for this purpose can yield high returns, both in terms of economic growth and poverty alleviation. The best evidence for this comes from Asia (Fan, 2008), but more recent evidence shows that it can also hold true in Africa (Mogues and Benin, 2012).

However, the gathering forces of rapid urbanization, a reverse farm size transition towards ever smaller and more diversified farms, and an emerging corporate-driven business agenda in response to higher agricultural and energy prices, are creating a situation where policymakers need to differentiate more sharply between the needs of different types of small farms, and between growth, poverty, and food security goals.

Many smallholdings today are too small to provide adequate livelihoods, and their farm families have either begun a transition out of farming into the non-farm economy, or they are trapped in subsistence modes of farming, often in lagging regions. Both kinds of smallholders may need assistance in developing new off-farm opportunities, and in overcoming poverty and food insecurity. These smallholders account for large shares of the total rural poor and food-insecure people in the developing world, and they are an important target group for international efforts to achieve the MDGs and promote food security (e.g. the New Alliance for Food Security and Nutrition). However, transition and subsistence-oriented farms play a relatively minor role in producing marketed surpluses to drive economic growth and feed growing urban populations, and are unlikely to link successfully to modern value chains. Interventions to improve on-farm productivity can be helpful to the food security of both groups, but will need to be complemented by other interventions that more directly alleviate poverty and facilitate off-farm transitions.

In contrast, there are also many small farmers who, because of their resource endowments, good location, or sheer entrepreneurial skill, are succeeding as commercial farm businesses, even if only on a part-time basis. These kinds of small farms are much more aligned with the new corporate-driven business agenda. As with small farms in Green Revolution days, they can play important roles in driving economic growth and feeding urban populations. The greatest challenge facing these types of smallholders is accessing modern value chains. Private-sector investments along value chains are opening up new market opportunities for some smallholder farms, particularly for high-value products, but it is also becoming apparent that many more commercially oriented smallholders are being left behind while larger farms are gaining market shares. In Africa, many smallholders are not only missing out on new high-value chains, but have lost access to modern inputs, credit, and market outlets even for their traditional food staples (Djurfeldt et al., 2011).

If more smallholder farms are to become commercially successful, policy-makers will need to do more to support them. Key areas for support include improving the workings of markets for outputs, inputs, land, and financial services to overcome market failures that discriminate against small farms, investing in the kinds of R&D and rural infrastructure that small farmers need, helping to organize small farmers for the market, and incentivizing the private sector to link with more small farmers. The best way to achieve these is for government to work through private-sector and civil society partners, creating an enabling policy and business environment, and scaling up proven successes.

REFERENCES

Badiane, O. (2008). 'Sustaining and accelerating Africa's agricultural growth recovery in the context of changing global food prices', IFPRI Policy Brief 9, Washington, DC: International Food Policy Research Institute.

Berdegué, J. A. and G. Escobar (2002). 'Rural diversity, agricultural innovation policies, and poverty reduction', AgREN Network Paper No. 122, London: Overseas Development Institute.

Binswanger-Mkhize, H. (2012). 'India 1960–2010: Structural change, the rural non-farm sector, and the prospects for agriculture', Center on Food Security and the Environment, Stanford Symposium Series on *Global Food Policy and Food Security in the 21st Century*. Stanford University.

Brookfield, H. (2008). 'Family farms are still around: Time to invert the old agrarian question', *Geography Compass*, 2/1 (2008), pp. 108–26.

Byerlee, D. (2011). 'Private investment and corporate farming. PowerPoint presentation', *FAO Technical Workshop on Policies for Promoting Investment in Agriculture*, 12–13 December 2011, Rome.

Byerlee, D., A. Lissitsa, and P. Savanti (2012). 'Corporate models of broadacre crop farming: International experience from Argentina and Ukraine', *Farm Policy J*, 9(2), pp. 13–15.

Chayanov, A. V. (1986). *The Theory of Peasant Economy*. Madison: The University of Wisconsin Press.

Cleaver, K. and G. Schrieber (1994). *Reversing the spiral: The population agriculture, and environment nexus in sub-Saharan Africa*. Washington, DC: World Bank.

Deininger, K. (2003). 'Land policies for growth and poverty reduction', a World Bank Policy Research report, World Bank, Washington, DC.

Deininger, K. and D. Byerlee (2010). *The Rise of Large Farms in Land Abundant Countries: Do they have a future?* Washington, DC: World Bank, unpublished.

Djurfeldt, G., E. Aryeetey, and A. Isinika (eds) (2011). *African Smallholders: Food crops, markets and policy*. Wallingford, Oxford: Centre for Agricultural Bioscience International (CABI).

Eastwood, R., M. Lipton, and A. Newell (2010). 'Farm size', in Pingali and Evenson (eds), *Handbook of Agricultural Economics, Volume 4*. Amsterdam: Elsevier.

Fan, S. (ed.) (2008). *Public expenditure, growth and equity: Lessons from developing countries*. Baltimore: Johns Hopkins University Press.

FAO and IFAD (2008). *Water and the Rural Poor*. Rome.

Ghani, E. (ed.) (2010). *The poor half billion in South Asia: What is holding back lagging regions?* New Delhi: Oxford University Press.

Haggblade, S., P. Hazell, and T. Reardon (eds) (2007). *Transforming the Rural Nonfarm Economy*. Baltimore: Johns Hopkins University Press.

Haggblade, S., P. Hazell, and P. Dorosh (2007). 'Sectoral growth linkages between agriculture and the rural nonfarm economy', in Haggblade, S., P. Hazell, and T. Reardon (eds), *Transforming the Rural Nonfarm Economy*. Baltimore: Johns Hopkins University Press.

Hazell, P. B. R. (2008). *An Assessment of the Impact of Agricultural Research in South Asia Since the Green Revolution*. Rome, Italy: CGIAR Science Council Secretariat.

Hazell, P., J. Anderson, N. Balzer, A. Hastrup Clemmensen, U. Hess, and F. Rispoli (2010). *Potential for scale and sustainability in weather index insurance for agriculture and rural livelihoods*. Rome: International Fund for Agricultural Development and World Food Programme. Retrieved from: <http://www.ifad.org/ruralfinance/pub/weather.pdf>.

Hazell, P., C. Poulton, S. Wiggins, and A. Dorward (2007). 'The future of small farms for poverty reduction and growth', 2020 Discussion Paper 42, Washington, DC: International Food Policy Research Institute.

Hazell, P. and S. Wood (2008). 'Drivers of change in global agriculture', *Philosophical Transactions of the Royal Society* B, 12 February, 363 (1491), pp. 495–515.

Headey, D., D. Bezemer, and P. Hazell (2010). 'Agricultural employment trends in Asia and Africa: Too fast or too slow?', *The World Bank Research Observer*, 25, pp. 57–89.

Huang, J., X. Wang, and H. Qui (2012). *Small-scale Farmers in China in the Face of Modernisation and Globalisation.* London/The Hague: IIED/HIVOS.

IFPRI (2005). *The Future of Small Farms: Proceedings of a Research Workshop.* Washington, DC.

Joireman, S. F. (2007). 'The mystery of capital formation in Sub-Saharan Africa: Women, property rights and customary law', *World Development*, 36(7), pp. 1233–46.

Linn, J. (ed.) (2012). 'Scaling up in agriculture, rural development, and nutrition', 2020 Focus 19, Washington, DC: International Food Policy Research Institute.

Lipton, M. (2009). *Land Reform in Developing Countries: Property Rights and Property Wrongs.* London and New York: Routledge.

Minot, N. and T. Benson (2009). 'Fertilizer subsidies in Africa: Are vouchers the answer?', IFPRI Issue Brief 60. Washington, DC: International Food Policy Research Institute.

Mogues, T. and S. Benin (eds) (2012). *Public Expenditures for Agricultural and Rural Development in Africa.* London and New York: Routledge.

Morris, M., V. Kelly, R. Kopicki, and D. Byerlee (2007). *Promoting Increased Fertilizer Use in Africa.* Washington, DC: World Bank.

Omamo, S. W., X. Diao, S. Wood, J. Chamberlin, L. You, S. Benin, U. Wood-Sichra, and A. Tatwangire (2006). *Strategic Priorities for Agricultural Development in Eastern and Central Africa.* Washington, DC: International Food Policy Research Institute.

Otsuka, K. (2012). 'Food insecurity, income inequality, and the changing comparative advantage in world agriculture', Presidential Address at 27th International Conference of Agricultural Economists, Foz do Iguaçu, Brazil, August.

Quisumbing, A. and L. Pandonfelli (2009). 'Promising approaches to address the needs of poor female farmers: resources, constraints, and interventions', Discussion Paper 882, International Food Policy Research Institute, Washington, DC.

United Nations (2011). *World Urbanization Prospects; The 2011 Revision.* New York: Economic and Social Affairs, United Nations.

Vorley, Bill (2002). 'Sustaining Agriculture: Policy, Governance and the Future of Family Farming', A synthesis report of the collaborative research project 'Policies that work for sustainable agriculture and regenerating rural livelihoods', London: International Institute for Environment and Development (IIED).

World Bank (2007). *World Development Report 2008: Agriculture for Development.* Washington, DC: The World Bank.

INDEX